# CHARACTERIZATIONS OF C*-ALGEBRAS

# PURE AND APPLIED MATHEMATICS

*A Program of Monographs, Textbooks, and Lecture Notes*

## MONOGRAPHS AND TEXTBOOKS IN PURE AND APPLIED MATHEMATICS

1. *K. Yano,* Integral Formulas in Riemannian Geometry (1970)*(out of print)*
2. *S. Kobayashi,* Hyperbolic Manifolds and Holomorphic Mappings (1970) *(out of print)*
3. *V. S. Vladimirov,* Equations of Mathematical Physics (A. Jeffrey, editor; A. Littlewood, translator) (1970) *(out of print)*
4. *B. N. Pshenichnyi,* Necessary Conditions for an Extremum (L. Neustadt, translation editor; K. Makowski, translator) (1971)
5. *L. Narici, E. Beckenstein, and G. Bachman,* Functional Analysis and Valuation Theory (1971)
6. *D. S. Passman,* Infinite Group Rings (1971)
7. *L. Dornhoff,* Group Representation Theory (in two parts). Part A: Ordinary Representation Theory. Part B: Modular Representation Theory (1971, 1972)
8. *W. Boothby and G. L. Weiss (eds.),* Symmetric Spaces: Short Courses Presented at Washington University (1972)
9. *Y. Matsushima,* Differentiable Manifolds (E. T. Kobayashi, translator) (1972)
10. *L. E. Ward, Jr.,* Topology: An Outline for a First Course (1972) *(out of print)*
11. *A. Babakhanian,* Cohomological Methods in Group Theory (1972)
12. *R. Gilmer,* Multiplicative Ideal Theory (1972)
13. *J. Yeh,* Stochastic Processes and the Wiener Integral (1973) *(out of print)*
14. *J. Barros-Neto,* Introduction to the Theory of Distributions (1973) *(out of print)*
15. *R. Larsen,* Functional Analysis: An Introduction (1973) *(out of print)*
16. *K. Yano and S. Ishihara,* Tangent and Cotangent Bundles: Differential Geometry (1973) *(out of print)*
17. *C. Procesi,* Rings with Polynomial Identities (1973)
18. *R. Hermann,* Geometry, Physics, and Systems (1973)
19. *N. R. Wallach,* Harmonic Analysis on Homogeneous Spaces (1973) *(out of print)*
20. *J. Dieudonné,* Introduction to the Theory of Formal Groups (1973)
21. *I. Vaisman,* Cohomology and Differential Forms (1973)
22. *B.-Y. Chen,* Geometry of Submanifolds (1973)
23. *M. Marcus,* Finite Dimensional Multilinear Algebra (in two parts) (1973, 1975)
24. *R. Larsen,* Banach Algebras: An Introduction (1973)
25. *R. O. Kujala and A. L. Vitter (eds.),* Value Distribution Theory: Part A; Part B: Deficit and Bezout Estimates by Wilhelm Stoll (1973)
26. *K. B. Stolarsky,* Algebraic Numbers and Diophantine Approximation (1974)
27. *A. R. Magid,* The Separable Galois Theory of Commutative Rings (1974)
28. *B. R. McDonald,* Finite Rings with Identity (1974)
29. *J. Satake,* Linear Algebra (S. Koh, T. A. Akiba, and S. Ihara, translators) (1975)

30. *J. S. Golan,* Localization of Noncommutative Rings (1975)
31. *G. Klambauer,* Mathematical Analysis (1975)
32. *M. K. Agoston,* Algebraic Topology: A First Course (1976)
33. *K. R. Goodearl,* Ring Theory: Nonsingular Rings and Modules (1976)
34. *L. E. Mansfield,* Linear Algebra with Geometric Applications: Selected Topics (1976)
35. *N. J. Pullman,* Matrix Theory and Its Applications (1976)
36. *B. R. McDonald,* Geometric Algebra Over Local Rings (1976)
37. *C. W. Groetsch,* Generalized Inverses of Linear Operators: Representation and Approximation (1977)
38. *J. E. Kuczkowski and J. L. Gersting,* Abstract Algebra: A First Look (1977)
39. *C. O. Christenson and W. L. Voxman,* Aspects of Topology (1977)
40. *M. Nagata,* Field Theory (1977)
41. *R. L. Long,* Algebraic Number Theory (1977)
42. *W. F. Pfeffer,* Integrals and Measures (1977)
43. *R. L. Wheeden and A. Zygmund,* Measure and Integral: An Introduction to Real Analysis (1977)
44. *J. H. Curtiss,* Introduction to Functions of a Complex Variable (1978)
45. *K. Hrbacek and T. Jech,* Introduction to Set Theory (1978)
46. *W. S. Massey,* Homology and Cohomology Theory (1978)
47. *M. Marcus,* Introduction to Modern Algebra (1978)
48. *E. C. Young,* Vector and Tensor Analysis (1978)
49. *S. B. Nadler, Jr.,* Hyperspaces of Sets (1978)
50. *S. K. Segal,* Topics in Group Rings (1978)
51. *A. C. M. van Rooij,* Non-Archimedean Functional Analysis (1978)
54. *L. Corwin and R. Szczarba,* Calculus in Vector Spaces (1979)
53. *C. Sadosky,* Interpolation of Operators and Singular Integrals: An Introduction to Harmonic Analysis (1979)
54. *J. Cronin,* Differential Equations: Introduction and Quantitative Theory (1980)
55. *C. W. Groetsch,* Elements of Applicable Functional Analysis (1980)
56. *I. Vaisman,* Foundations of Three-Dimensional Euclidean Geometry (1980)
57. *H. I. Freedman,* Deterministic Mathematical Models in Population Ecology (1980)
58. *S. B. Chae,* Lebesgue Integration (1980)
59. *C. S. Rees, S. M. Shah, and C. V. Stanojević,* Theory and Applications of Fourier Analysis (1981)
60. *L. Nachbin,* Introduction to Functional Analysis: Banach Spaces and Differential Calculus (R. M. Aron, translator) (1981)
61. *G. Orzech and M. Orzech,* Plane Algebraic Curves: An Introduction Via Valuations (1981)
62. *R. Johnsonbaugh and W. E. Pfaffenberger,* Foundations of Mathematical Analysis (1981)
63. *W. L. Voxman and R. H. Goetschel,* Advanced Calculus: An Introduction to Modern Analysis (1981)
64. *L. J. Corwin and R. H. Szcarba,* Multivariable Calculus (1982)
65. *V. I. Istrătescu,* Introduction to Linear Operator Theory (1981)
66. *R. D. Järvinen,* Finite and Infinite Dimensional Linear Spaces: A Comparative Study in Algebraic and Analytic Settings (1981)

67. *J. K. Beem and P. E. Ehrlich,* Global Lorentzian Geometry (1981)
68. *D. L. Armacost,* The Structure of Locally Compact Abelian Groups (1981)
69. *J. W. Brewer and M. K. Smith, eds.,* Emmy Noether: A Tribute to Her Life and Work (1981)
70. *K. H. Kim,* Boolean Matrix Theory and Applications (1982)
71. *T. W. Wieting,* The Mathematical Theory of Chromatic Plane Ornaments (1982)
72. *D. B. Gauld,* Differential Topology: An Introduction (1982)
73. *R. L. Faber,* Foundations of Euclidean and Non-Euclidean Geometry (1983)
74. *M. Carmeli,* Statistical Theory and Random Matrices (1983)
75. *J. H. Carruth, J. A. Hildebrant, and R. J. Koch,* The Theory of Topological Semigroups (1983)
76. *R. L. Faber,* Differential Geometry and Relativity Theory: An Introduction (1983)
77. *S. Barnett,* Polynomials and Linear Control Systems (1983)
78. *G. Karpilovsky,* Commutative Group Algebras (1983)
79. *F. Van Oystaeyen and A. Verschoren,* Relative Invariants of Rings: The Commutative Theory (1983)
80. *I. Vaisman,* A First Course in Differential Geometry (1984)
81. *G. W. Swan,* Applications of Optimal Control Theory in Biomedicine (1984)
82. *T. Petrie and J. D. Randall,* Transformation Groups on Manifolds (1984)
83. *K. Goebel and S. Reich,* Uniform Convexity, Hyperbolic Geometry, and Nonexpansive Mappings (1984)
84. *T. Albu and C. Năstăsescu,* Relative Finiteness in Module Theory (1984)
85. *K. Hrbacek and T. Jech,* Introduction to Set Theory, Second Edition, Revised and Expanded (1984)
86. *F. Van Oystaeyen and A. Verschoren,* Relative Invariants of Rings: The Noncommutative Theory (1984)
87. *B. R. McDonald,* Linear Algebra Over Commutative Rings (1984)
88. *M. Namba,* Geometry of Projective Algebraic Curves (1984)
89. *G. F. Webb,* Theory of Nonlinear Age-Dependent Population Dynamics (1985)
90. *M. R. Bremner, R. V. Moody, and J. Patera,* Tables of Dominant Weight Multiplicities for Representations of Simple Lie Algebras (1985)
91. *A. E. Fekete,* Real Linear Algebra (1985)
92. *S. B. Chae,* Holomorphy and Calculus in Normed Spaces (1985)
93. *A. J. Jerri,* Introduction to Integral Equations with Applications (1985)
94. *G. Karpilovsky,* Projective Representations of Finite Groups (1985)
95. *L. Narici and E. Beckenstein,* Topological Vector Spaces (1985)
96. *J. Weeks,* The Shape of Space: How to Visualize Surfaces and Three-Dimensional Manifolds (1985)
97. *P. R. Gribik and K. O. Kortanek,* Extremal Methods of Operations Research (1985)
98. *J.-A. Chao and W. A. Woyczynski, eds.,* Probability Theory and Harmonic Analysis (1986)
99. *G. D. Crown, M. H. Fenrick, and R. J. Valenza,* Abstract Algebra (1986)
100. *J. H. Carruth, J. A. Hildebrant, and R. J. Koch,* The Theory of Topological Semigroups, Volume 2 (1986)

101. *R. S. Doran and V. A. Belfi,* Characterizations of C*-Algebras: The Gelfand-Naimark Theorems (1986)
102. *M. W. Jeter,* Mathematical Programming: An Introduction to Optimization (1986)

*Other Volumes in Preparation*

# CHARACTERIZATIONS OF C*-ALGEBRAS

## The Gelfand-Naimark Theorems

ROBERT S. DORAN
VICTOR A. BELFI
Texas Christian University
Fort Worth, Texas

CRC Press
Taylor & Francis Group
Boca Raton London New York

CRC Press is an imprint of the
Taylor & Francis Group, an **informa** business

First published 1986 by Marcel Dekker, Inc.

Published 2019 by CRC Press
Taylor & Francis Group
6000 Broken Sound Parkway NW, Suite 300
Boca Raton, FL 33487-2742



First issued in paperback 2019



ISBN 13: 978-0-367-45163-9 (pbk)
ISBN 13: 978-0-8247-7569-8 (hbk)



**Visit the Taylor & Francis Web site at**
**http://www.taylorandfrancis.com**

**and the CRC Press Web site at**
**http://www.crcpress.com**

Library of Congress Cataloging in Publication Data

Doran, Robert S., [date]
Characterizations of C* -algebras--the Gelfand-Naimark theorems.

(Monographs and textbooks in pure and applied mathematics ; 101)
Bibliography: p.
Includes index.
1. C*-algebras. I. Belfi, Victor A., [date]
II. Title. III. Title: Gelfand-Naimark theorems.
IV. Series: Monographs and textbooks in pure and applied mathematics ; v. 101.
QA326.D67 1986 512'.55 85-29234
ISBN 0-8247-7569-4

To

ROBERT B. BURCKEL

IZRAIL' M. GELFAND

MARK A. NAIMARK

# Preface

This book is devoted to giving an account of two characterization theorems which have had a substantial impact upon our thinking in modern analysis. These theorems, due to I. M. Gelfand and M. A. Naimark in their original form, were published in 1943, and served notice to the world that C*-algebras had arrived and were a voice (among many, to be sure) to be heard regarding future mathematical developments.

This volume is admittedly specialized. Our goal is to discuss the Gelfand-Naimark theorems and related results, old and new, which have been stimulated by them. It is our belief that readers with an interest in Banach algebras and C*-algebras will appreciate and enjoy having these results collected in a single volume and treated in a unified way. The book is a natural outgrowth of a paper by Doran and Wichmann [1] which appeared in 1977.

In order to achieve maximum generality we have treated algebras without identity as well as algebras with identity. Further, much of the theory requires that one deal with algebras with arbitrary (possibly discontinuous) involutions and we have done this. However, as the experts know, at these levels of generality there is a real danger in obscuring the main ideas and principal lines of thought. We have tried to counteract this by first giving complete proofs of the (now) classical Gelfand-Naimark theorems (Chapters 2 and 3), and then proceeding to the more refined results in later chapters. We continually try to be attentive to the needs of the beginning reader who simply wants to know what the Gelfand-Naimark theorems are all about. Therefore we do not hesitate to repeat definitions if necessary or remind the reader of what is going on in particular situations.

An introduction to those parts of the general theory of Banach algebras needed to understand the remainder of the book is provided in Appendix B. The theory of Banach algebras with involution is treated in the text. A reader with no previous knowledge of Banach algebras could read Chapter 1 on the history of the Gelfand-Naimark theorems and then should go to Appendix B before proceeding to Chapter 2. It is expected that the reader is familiar with basic real and complex analysis and has been exposed to a first course in functional analysis. For the reader's convenience we have summarized in Appendix A the main results from functional analysis which will be used.

It should be emphasized that this book is not intended, in any way, to replace the existing works and monographs on C*-algebras. Rather, it is meant to complement and supplement them in a particular area. Even so, in view of the fairly complete appendix on Banach algebras, the book could be used as an introductory text on Banach algebras and elementary C*-algebra theory.

The authors are deeply grateful for the help of many friends in the writing of this book. Among these we would like to cite Professor Robert B. Burckel for his constant support and interest in the project. He has read the manuscript carefully and has offered many suggestions which have clarified proofs and generally improved the exposition. We wish to thank him also for bringing relevant papers to our attention which we might otherwise have overlooked. His unselfish giving of his time, even when busy with large projects of his own, is the sign of a true friend. Professor Leo F. Boron was very helpful in securing photographs and biographical information on Gelfand and Naimark. We thank him for his efforts.

The first draft of this book was written while the first author was a member of the Institute for Advanced Study at Princeton. Portions of the book were also written while he was a visitor in the Department of Mathematics at the Massachusetts Institute of Technology. The hospitality and support extended by these institutions is deeply appreciated.

Finally we wish to thank Texas Christian University for partial financial support during the writing and Shirley Doran for an exceptionally nice job of typing.

Robert S. Doran

Victor A. Belfi

# Contents

# CHARACTERIZATIONS OF C*-ALGEBRAS

# 1
# The Gelfand-Naimark Theorems: Historical Perspective

## §1. *Introduction.*

$C^*$-algebras made their first appearance in 1943 in the now famous paper of Gelfand and Naimark [1]. Since then hundreds of mathematicians have contributed more than 2500 publications to the subject. It continues to grow at a phenomenal rate and has permeated many branches of mathematics through its connections with group representations, abstract harmonic analysis, operator theory, algebraic topology, and quantum physics.

Our primary purpose is to give an account of two celebrated theorems of Gelfand and Naimark, their tangled history, generalizations and applications, in a form accessible not only to specialists but also to mathematicians working in various applied fields, and also to students of pure and applied mathematics.

There are several reasons why one might wish to study the Gelfand-Naimark theorems. They are, for example, beautiful in statement, mathematically elegant, and rich in applications. Furthermore, the theorems are central to the general theory of $C^*$-algebras. Mathematicians were quick to recognize the power implicit in the theorems. Spectral theory of linear operators in Hilbert space was particularly affected by them. Indeed, the Gelfand-Naimark theorem for commutative $C^*$-algebras is essentially the spectral theorem for normal operators in slight disguise. Another aspect of the theorems concerns the canonical and instructive nature of their proofs. The basic constructions in these proofs have been imitated in many different and widely divergent contexts.

What do the Gelfand-Naimark theorems say and how did they achieve their present form? We shall have a great deal to say about these questions soon. In this introduction we content ourselves with the following general discussion. Definitions of all terms will be given in Section 2.

At the turn of the century the abstract tendency in analysis, which developed into what is now known as functional analysis, began with the work of Volterra, Fredholm, Hilbert and F. Riesz, to mention some of the principal figures. They studied eigenvalue problems, integral equations, orthogonal expansions, and linear operators in general. In 1918 the axioms for a normed linear space appeared for the first time in F. Riesz's work on compact operators. The first abstract treatment of normed linear spaces was given in Banach's 1920 thesis, and later, in 1932, he published his celebrated book *Théorie des opérations linéaires* which was to be tremendously influential.

Many of the Banach spaces studied by Banach and others were at the same time algebras under some multiplication---a fact which they neither mentioned nor used. It is of some interest to record that in 1932 N. Wiener [1, p. 10], in his work on the tauberian theorem, observed the fundamental inequality $||xy|| \leq ||x||\cdot||y||$ for the algebra of absolutely convergent Fourier series; however, no systematic use of the ring theoretic structure was made in his study---certainly a missed opportunity!

The notion of an abstract Banach algebra was introduced by M. Nagumo [1] in 1936 under the name "linear metric ring" in connection with Hilbert's fifth problem. During the late 1930's the term "normed ring" was introduced by the Soviet mathematicians. The present term "Banach algebra" was used for the first time in 1945 by W. Ambrose [1] in his work on generalizing the $L^2$-algebra of a compact group.

One of the early fundamental results in the general theory of Banach algebras was a generalization of a classical theorem of Frobenius that any finite-dimensional division algebra over the complex numbers is isomorphic to the field of complex numbers. S. Mazur [1] announced in 1938 that every complex normed division algebra is isomorphic to the field of complex numbers. (He also dealt with real normed division algebras and showed that they were isomorphic to either the reals, the complexes, or the quaternions.) As an immediate consequence one obtains the following beautiful characterization of the complex field among normed algebras: any complex normed algebra satisfying the norm condition $||xy|| = ||x||\cdot||y||$ for all elements $x$ and $y$ is isometrically isomorphic to the field of complex numbers. It was Gelfand [1], in his 1939 thesis, who nearly single-handedly developed the general theory of commutative Banach algebras much as it is presented in Appendix B.

Many important Banach algebras carry a natural involution. In the case of an algebra of functions the involution is the operation of taking the complex-conjugate and in the case of an algebra of operators on a Hilbert space it is the operation of taking the adjoint operator. Motivated by these observations and by the earlier work of Gelfand on the representation of commutative Banach algebras, Gelfand and Naimark [1], working together, proved, under some additional assumptions, the following two theorems:

*GELFAND-NAIMARK THEOREM I. Let* $A$ *be a commutative Banach algebra with involution satisfying* $||x^*x|| = ||x^*||\cdot||x||$ *for all* $x$ *in* $A$. *Then* $A$ *is isometrically *-isomorphic to* $C_0(X)$, *the algebra of all continuous complex-valued functions which vanish at infinity on some locally compact Hausdorff space* $X$.

*GELFAND-NAIMARK THEOREM II. Let* $A$ *be a Banach algebra with involution satisfying* $||x^*x|| = ||x^*||\cdot||x||$ *for all* $x$ *in* $A$. *Then* $A$ *is isometrically *-isomorphic to a norm-closed *-subalgebra of bounded linear operators on some Hilbert space.*

As mentioned earlier, in this book we shall present a thorough discussion of these two representation theorems. We shall trace, as carefully as we can, the interesting and rather tangled history which led to their present form. Full proofs of the theorems will be given, as well as a survey, with proofs, of recent generalizations and developments which have been inspired by the theorems. Finally a few applications of the theorems will be given.

## §2. *Definitions.*

We set down in this brief section the basic definitions that the reader needs to know to read this chapter with understanding. A discussion of the elementary theory of Banach algebras is given in Appendix B.

A linear space A over a field F is said to be an *associative linear algebra* over F (or simply an *algebra* if no confusion can arise) if for each pair x, y of elements from A a product xy is defined on A, i.e., a map from $A \times A$ into A, such that for all $x, y, z \in A$ and all $\lambda \in F$:

(i) $x(yz) = (xy)z$;

(ii) $x(y + z) = xy + xz$; $(y + z)x = yx + zx$;

(iii) $\lambda(xy) = (\lambda x)y = x(\lambda y)$.

*Real* and *complex algebras* are algebras over R and C respectively. Whenever the field is unspecified we will be working with complex algebras. An algebra A is *commutative* if $xy = yx$ for all $x, y \in A$. If there exists an element e in an algebra A such that $ex = x = xe$ for all $x \in A$, then A is said to be an *algebra with identity*. Throughout this book we shall assume that our algebras are nontrivial, i.e., that they do not consist of the zero element alone.

A linear subspace I of an algebra A is a *left ideal* if $x \in I$, $z \in A$ imply $zx \in I$. A linear subspace I is a *right ideal* if $x \in I$, $z \in A$ imply $xz \in I$. A *two-sided ideal* is a left ideal that is also a right ideal. An ideal I of A such $I \neq A$ is a *proper ideal*. An algebra possessing no proper two-sided ideals except {0} is said to be *simple*. Of course in a commutative algebra the definitions of left, right, and two-sided ideals are equivalent. In this case, we simply write "ideal" for these.

A **-algebra* is an algebra over C with a mapping $x \to x^*$ of A into itself such that for all $x, y \in A$ and complex $\lambda$:

(a) $(x + y)^* = x^* + y^*$;

(b) $(\lambda x)^* = \bar{\lambda} x^*$;

(c) $(xy)^* = y^* x^*$;

(d) $x^{**} = x$.

The map $x \to x^*$ is called an *involution*; because of (d) it is clearly bijective. A subalgebra B of A is called a **-subalgebra* if $x \in B$ implies $x^* \in B$. Similarly, an ideal I is called a **-ideal* if $x \in I$ implies $x^* \in I$. A **-homomorphism* of a *-algebra A into a *-algebra B is a linear mapping $\phi: A \to B$ such that $\phi(xy) = \phi(x)\phi(y)$ and $\phi(x^*) = \phi(x)^*$ for all x, y in A. If $\phi$ is bijective, $\phi$ is a **-isomorphism* of A onto B, and A and B are said to be **-isomorphic*.

An algebra A which is also a normed (resp. Banach) space satisfying

$$||xy|| \leq ||x||\cdot||y|| \quad (x, y \in A)$$

is called a *normed algebra* (resp. *Banach algebra*). A normed algebra which is also a $^*$-algebra is called a *normed $^*$-algebra*. Of course, if the algebra is complete it is called a *Banach $^*$-algebra*.

The involution in a normed $^*$-algebra is *continuous* if there exists a constant $M > 0$ such that $||x^*|| \leq M\cdot||x||$ for all $x$; the involution is *isometric* if $||x^*|| = ||x||$ for all $x$. Two normed $^*$-algebras $A$ and $B$ are *isometrically $^*$-isomorphic*, denoted $A \simeq B$, if there exists a $^*$-isomorphism $f\colon A \to B$ such that $||f(x)|| = ||x||$ for all $x \in A$.

A norm on a $^*$-algebra $A$ is said to satisfy the *$C^*$-condition* if

$$||x^*x|| = ||x^*||\cdot||x|| \quad (x \in A). \tag{1}$$

A *$C^*$-algebra* is a Banach $^*$-algebra whose norm satisfies the $C^*$-condition. The norm in a $C^*$-algebra with isometric involution clearly satisfies the condition

$$||x^*x|| = ||x||^2 \quad (x \in A). \tag{2}$$

It turns out that in a Banach $^*$-algebra (1) also implies (2) without the assumption that the involution is isometric; however this is highly non-trivial and will be shown in Chapter III. It is easily seen that condition (2) implies that the involution is isometric and hence (2) implies (1). Therefore conditions (1) and (2) turn out to be equivalent.

The Banach space $C(X)$ of continuous complex-valued functions on a compact Hausdorff space $X$ is a commutative $C^*$-algebra under point-wise multiplication $(fg)(t) = f(t)g(t)$, involution $f^*(t) = \overline{f(t)}$, and sup-norm. A function $f$ on a locally compact Hausdorff space $X$ is said to "vanish at infinity" if for each $\varepsilon > 0$ there is a compact set $K \subseteq X$ such that $|f(x)| < \varepsilon$ for all $x \in X \setminus K$, the complement of $K$ in $X$. As with $C(X)$, the algebra $C_o(X)$ of continuous complex-valued functions which vanish at infinity on a locally compact Hausdorff space is a commutative $C^*$-algebra. While $C(X)$ possesses an identity, $C_o(X)$ does not unless $X$ is compact, and in this case $C_o(X) = C(X)$.

Examples of noncommutative C*-algebras are provided by the algebra B(H) of bounded linear operators on a Hilbert space H. Multiplication in B(H) is operator composition, the involution $T \to T^*$ is the usual adjoint operation, and the norm is the operator norm

$$||T|| = \sup\{||T\xi||: ||\xi|| \leq 1, \xi \in H\}.$$

A norm-closed *-subalgebra of B(H) is called a *concrete* C*-*algebra*; clearly, every concrete C*-algebra is a C*-algebra in the abstract sense. If X is a locally compact Hausdorff space, then the algebra $C_o(X)$ can be viewed as a subalgebra of B(H); indeed, if $\mu$ is a regular Borel measure on X, with support X, and $H = L^2(X,\mu)$, the Hilbert space of $\mu$-square-integrable complex functions on X, then $C_o(X)$ may be interpreted as an algebra of multiplication operators on H. Hence $C_o(X)$ is a concrete C*-subalgebra of B(H) which is commutative. The remarkable message of Theorems I and II of Section 1 is that these examples exhaust the class of C*-algebras.

We end this section with a few historical notes regarding the above terminology. The term "C*-algebra" was introduced in 1947 by I. E. Segal [2] and was reserved for what we have called a "concrete C*-algebra". The "C" stood for "closed" in the norm topology of B(H). It has been speculated by some authors that the "C" was meant to indicate that a C*-algebra is a noncommutative analogue of C(X); however, Professor Segal has assured the first named author that he didn't have this in mind--although he agreed that it was certainly a reasonable supposition. The term "B*-algebra" was introduced in 1946 by C. E. Rickart [1] for Banach algebras satisfying condition (2). This terminology is still in wide usage today. Finally, we mention that the term "B'*-algebra" was used by T. Ono [1] in 1959 to describe Banach *-algebras whose norm satisfies (1). This somewhat cumbersome notation was only to be temporary as his goal was to show that every B'*-algebra was, in fact a B*-algebra. We shall not use the terminology "B*-algebra" in the sequel except possibly in the historical notes.

## §3. *Historical development: a tangled trail.*

In 1943 Gelfand and Naimark [1] published (in English!) a groundbreaking paper in which they proved that a Banach *-algebra with an identity element $e$ is isometrically $^*$-isomorphic to a concrete $C^*$-algebra if it satisfies the following three conditions:

1° $||x^*x|| = ||x^*||\cdot||x||$ (the $C^*$-condition);

2° $||x^*|| = ||x||$ (isometric condition);

3° $e + x^*x$ is invertible (symmetry)

for all $x$. They immediately asked in a footnote if conditions 2° and 3° could be deleted--apparently recognizing that they were of a different character from condition 1° and were needed primarily because of their method of proof. This indeed turned out to be true after considerable work. To trace the resulting history in detail it is convenient to look at the commutative and noncommutative cases separately.

*Commutative algebras:* In their paper Gelfand and Naimark first proved that every commutative $C^*$-algebra with identity is a $C(X)$ for some compact Hausdorff space $X$. They were able to show quite simply that in the presence of commutativity the $C^*$-condition implies that the involution is isometric (hence continuous). Utilizing a delicate argument depending on the notion of "Shilov boundary" (cf. Naimark [1, p. 231]) they proved that every commutative $C^*$-algebra is symmetric. Thus in the commutative case they were able to show that conditions 2° and 3° follow from condition 1°.

A much simpler proof for symmetry of a commutative $C^*$-algebra was published in 1946 by Richard Arens [1]. It may be of some historical interest that Professor Arens--as he mentioned in a conversation with the first named author--had not seen the Gelfand-Naimark proof when he found his. In 1952, utilizing the exponential function for elements in a Banach algebra, the Japanese mathematician Masanori Fukamiya [2] published yet another beautiful proof of symmetry. The arguments of Arens and Fukamiya will be given in full in Chapter II.

*Noncommutative algebras:* The 1952 paper of Fukamiya [2] implicitly contained the key lemma needed to eliminate condition 3° for noncommutative algebras. In essence this lemma states that if $x$ and $y$ are "positive elements" in a $C^*$-algebra with identity and isometric involution, then the sum $x + y$ is also positive. Independently and nearly simultaneously this lemma was discovered by John L. Kelley and

Robert L. Vaught [1]. The Kelley-Vaught argument is extremely brief and elegant, and is the one that we shall give in Section 12.

The nontrivial observation that this lemma was the key to eliminating condition 3° was due to Irving Kaplansky. His ingenious argument was recorded in Joseph A. Schatz's [2] review of Fukamiya's paper, making it an amusing instance where a theorem was first "proved" in the *Mathematical Reviews*.

In marked contrast to the commutative case, the redundancy of condition 2° for noncommutative algebras did not follow easily; in fact, the question remained open until 1960 when a solution for $C^*$-algebras with identity was published by James G. Glimm and Richard V. Kadison [1]. Their proof was based on a deep "n-fold transitivity" theorem for unitary operators in an irreducible $C^*$-algebra of operators on a Hilbert space. A beautiful theorem, proved in 1966, by Bernard Russo and Henry A. Dye [1] made it possible to by-pass the Glimm-Kadison transitivity theorem; an elementary proof of their result was given in 1972 by Lawrence A. Harris [1], and an extremely short elegant proof was given in 1984 by L. Terrell Gardner [ 1 ]. The paper of Harris contained powerful new techniques that simplified and unified several other parts of the theory of Banach algebras with involution. Another paper concerning the elimination of 2° (and also 3°) was published by the Japanese mathematician Tamio Ono [1] in 1959. This paper contained useful techniques but was flawed by errors in the arguments of both of the main theorems (see the Mathematical Review of Ono [1]). Ten years later Ono [2] acknowledged these mistakes and corrected them from the viewpoint of 1959.

The original 1943 conjecture of Gelfand and Naimark was, at this time, completely solved for algebras with identity. What about algebras without identity? This question is of considerable importance since most $C^*$-algebras which occur in applications do not possess an identity. An answer was provided in 1967 by B. J. Vowden [1]. He was able to utilize the notion of "approximate identity" and several arguments from Ono [1] to embed a $C^*$-algebra without identity in a $C^*$-algebra with identity. He then applied the known case for algebras with an identity to complete the proof. Hence after nearly twenty-five years of work by outstanding mathematicians, the mathematical community had the theorems as we have stated them in the introduction.

## §4. *Gelfand and Naimark: the mathematicians.*

The preceding section contained an account of how the Gelfand-Naimark theorems originated. In this section we will discuss Gelfand and Naimark themselves. Our purpose is not to give a complete biographical account, but rather to give the reader a brief glimpse into their mathematical background and work. Far more complete accounts of their lives and mathematical accomplishments can be found in the references given at the end of this section.

### IZRAIL' MOISEEVICH GELFAND

Izrail' Moiseevich Gelfand was born at Krasnye Okny in the province of Odessa on August 20, 1913. After an incomplete secondary education, he went to Moscow in 1930 and, at first, took casual work as a doorkeeper at the Lenin Library. During this time he began to teach elementary mathematics at evening institutes. Soon he was teaching higher mathematics as well, and he began to attend lectures and seminars at the University of Moscow. His first serious encounter with research-level mathematics occurred in M. A. Lavrent'ev's seminar on the theory of functions of a complex variable.

In 1932, Gelfand, at the age of 18, was formally admitted as a research student to the university. His supervisor was A. N. Kolmogorov, who directed him toward the field of functional analysis. At that time, functional analysis was just emerging, and enjoyed the attention of only a very small group of mathematicians in Moscow. Two of these were L. A. Lyusternik and A. I. Plesner, who were influential in Gelfand's choice of topics of his first research papers.

In his thesis for the candidate's degree, written in 1935, Gelfand developed a theory of integration of functions. The thesis contained several theorems on the general form of linear operators in normed spaces. His method of proof of these theorems was, perhaps, more important than the theorems themselves. Indeed, Gelfand was the first to apply linear functionals to vector-valued functions, and thus reduce their theory to ordinary function theory.

The subject of Gelfand's 1939 doctoral thesis was commutative Banach algebras. He was the first to recognize the central role played by maximal ideals. Utilizing their properties, he created the modern structure theory of commutative Banach algebras. This theory, now called Gelfand theory, unified what previously appeared to be unrelated facts in several areas of mathematics and revealed close connections between functional analysis and classical analysis. As a striking example of this, Gelfand showed that Wiener's classical theorem, which states that if $f$ is never zero and has an absolutely convergent Fourier expansion, then its reciprocal $1/f$ also has these properties, could be proved in a few lines with his theory. This demonstrated the power of the theory and simultaneously brought it to the attention of the mathematical world.

Next to come was the joint work with M. A. Naimark in 1943 which resulted in the Gelfand-Naimark theorems as described in the preceding section. This work initiated the theory of $C^*$-algebras and, as we have already pointed out, has had an influence on many areas of modern mathematics.

After this, Gelfand turned his attention to the theory of group representations. For compact groups the theory was already quite well understood by 1940, and involved only finite-dimensional representations. However, for noncompact groups, the situation was far more complicated. On the one hand, it had been shown that such groups cannot, in general, have non-trivial finite-dimensional unitary representations, and on the other hand, upon examining the infinite-dimensional representations of such groups, substantial complications of a set-theoretic nature were revealed. Thus, even the formulation of the basic problems were not clear. It was Gelfand who succeeded in finding the correct approach. He noticed that unitary representations were of fundamental importance, and he developed a deep and important theory for locally compact groups. He and D. A. Raikov showed in 1943 that every locally compact group $G$ has enough irreducible unitary representations to separate the points of $G$. The next problem was to describe and classify these representations for the most important groups. It must be emphasized that it was not at all clear whether a sufficiently explicit description could be given, even for well-known groups such as the group of complex second order matrices. However, from 1944 to 1948, working together, Gelfand and Naimark constructed a theory of infinite-dimensional representations of the classical complex Lie groups. They established that irreducible unitary representations of these groups can be given by simple, explicit formulae. Many additional papers by Gelfand and his collaborators followed, dealing with both real and complex Lie groups and the classification of their unitary representations. It is not possible here even to indicate the importance and influence of the work in these papers. Suffice it to say that out of them came deep and significant studies on zonal harmonics, spherical functions, homogeneous spaces, automorphic functions, general noncommutative harmonic analysis on Lie groups, and the general theory of group representations.

Gelfand has, over the years, proposed and solved many problems in the theory of ordinary and partial differential equations. Work that he and his collaborators began has been continued by many of the world's best known mathematicians (e.g., L. D. Fadeev, L. Hörmander, N. Levinson, M. Atiyah, I. Singer, and others). Gelfand was also one of the first Soviet mathematicians to appreciate the future prospects and importance of the work of S. L. Sobolev and L. Schwartz on the theory of generalized functions (distributions). Once again his papers and those of his students and collaborators played a leading part in the development of the subject. These ideas, which found applications in the theories of partial differential equations, representations, stochastic processes, and integral geometry, have been recorded in a series of five books, entitiled "Generalized Functions", co-authored by Gelfand, Shilov, Graev, and Vilenkin. This series, which began around 1954, has achieved international recognition.

In another direction, Gelfand has made substantial contributions to the development of computational mathematics. He found general methods for the numerical solution of equations of mathematical physics, and also solved particular applied problems. For more than twenty years he was Head of a section of the Institute of Applied Mathematics of the USSR Academy of Sciences. The contributions of Gelfand and his collaborators formed an essential stage in the development of the whole of computational mathematics.

Around 1960, Gelfand began research in biology and on complicated physiological systems. On the basis of actual biological results, he developed important general principles of the organization of control in complex multi-cell systems. Gelfand's biological work is characterized by the same clarity in posing problems, the ability to find non-trivial new approaches, and the combination of concreteness and breadth of general concepts that distinguish his mathematical research.

When speaking of Gelfand's creative work as a scholar, one cannot help but mention also his teaching activities. One of the characteristic features of these activities is the extremely close bond between his research and his teaching. A distinctive feature of Gelfand's creative work has been his skill in organizing purposeful, concerted work in a team. A large number of Gelfand's papers have been written in collaboration with his colleagues and students, often quite young ones, for whom such combined work has been an exceedingly valuable experience. Thus, it is practically impossible to separate his own research work proper from his teaching and supervising activities. The first of Gelfand's students was G. E. Shilov, who came to him as a research student almost 45 years ago. Since that time he has supervised dozens of students, many of whom have already become prominent scholars in their own right, and who have also had distinguished students.

As founder of an extensive scientific school, Gelfand showed great powers of organizational, public, and pedagogical work. For many years he was a member of the editorial board of the "Uspekhi Matematicheskikh Nauk", chief editor of the "Journal of Functional Analysis", and director of the Inter-Faculty Laboratory of Mathematical Methods in Biology at the Moscow State University. From 1968 to 1970, he was President of the Moscow Mathematical Society and is now an Honorary Member of it.

Gelfand's scientific achievements have received wide international recognition: he is an Honorary Member of the American National Academy of Sciences, the American Academy of Sciences and Arts, the Royal Irish Academy, and a member of several mathematical societies. On the eve of his sixtieth birthday, he was awarded an honorary doctorate from the University of Oxford. He has been awarded the order of Lenin three times, the Order of the Badge of Honor, the Lenin Prize, and other prizes. He had published nearly three hundred papers and several well-known books on linear algebra, the calculus of variations, distribution theory, and the theory of group representations.

## MARK ARONOVICH NAIMARK

Mark Aronovich Naimark was born into an artist's family in Odessa on December 5, 1909. While a child in school he already displayed a great aptitude for mathematics. At the age of fifteen he enrolled in a technical college, while simultaneously working in a foundry, and studied mathematical analysis from 1924 until 1928. Completing this work, he enrolled in 1929 in the Physico-Mathematical Faculty of the Odessa-Institute of

National Education, which soon after became known as the Physico-Chemico-Mathematical Institute of Odessa.

Upon graduation in 1933, Naimark enrolled as a post-graduate student in the Department of Mathematics at the University of Odessa, where his supervisor was the well-known Soviet mathematician M. G. Krein. His first scientific papers on "the theory of separation of the roots of algebraic equations" were written with Krein during the next two years. While an undergraduate and post-graduate student, Naimark also studied pedagogical methods in various institutes of higher education. His interest in pedagogy endured throughout his lifetime, and was reflected in his teaching and writing. In 1936 he defended his Ph.D. thesis on the theory of normal operators in Hilbert space.

Two years later, in 1938, a new period began in Naimark's life when he moved from Odessa to Moscow to study for his doctor's degree at the Steklov Mathematical Institute of the USSR Academy of Sciences. During the next three years his main scientific interests, focusing on spectral theory of operators and the representation theory of locally compact groups would be formed. After completing the preliminary study for a doctor's degree in April, 1941, Naimark brilliantly defended his D.Sc. thesis, had the title of Professor conferred upon him and, at the direction of the Academy of Sciences, began work in the Theoretical Department of the Seismological Institute of the USSR Academy of Sciences.

With the start of World War II, Naimark signed up for special duty (called the home-guard), and worked on the labor front. From the end of 1941 until April of 1943, he worked in Tashkent, where the Seismological Institute had been relocated after evacuation. He then returned to Moscow where he worked in a number of institutes, including the Institute of Chemical Physics and the USSR Academy for the Arms Industry. Once more he was involved with pedagogical work at various provincial institutes of higher education. On the other hand, during this time Naimark also worked closely with Gelfand on Banach algebras with involution, their representations, and the representations of the classical matrix groups in Hilbert spaces. As we pointed out earlier in the article on Gelfand, these papers laid the foundations of the modern theory of infinite-dimensional representations of algebras and groups, and studied basic properties of $C^*$-algebras. They also described a remarkable non-commutative analogue of the Fourier transform on the complex classical Lie groups, and obtained analogues of the Plancherel formula for this transform. Among other things, they investigated special problems of harmonic analysis. Specifically, they described the characters of infinite-dimensional representations, and contributed to the understanding of spherical functions on groups. This research on harmonic analysis was systematized in Gelfand and Naimark's well known survey article "Unitary representations of the classical groups" which appeared in 1950.

Naimark's papers on seismology, which concern the oscillations of a fine elastic layer on an elastic half-space, led him to new studies on the spectral theory of differential operators. In this area he studied the second-order singular differential operators having a non-empty continuous part of the spectrum, for which he described the spectral decompositions by eigenfunctions, studied the perturbation of these operators, and discovered singular points of a new type (the so-called spectral singularities). The results of this work are reflected in Naimark's famous monograph "Linear Differential Operators" which was published in 1954.

From 1954 on, Naimark was a professor in the Department of Mathematics at the Physico-Technical Institute of Moscow. He regularly gave courses in mathematical analysis, partial differential equations, and functional analysis. He also supervised a group of post-graduate students and organized research seminars in these subjects. In 1956 Naimark's fundamental monograph "Normed Rings" appeared. This book contained the first comprehensive treatment of Banach algebras, and it played an enormous role in the development of the new theory. It was striking in its depth, the beauty of its structure, and the breadth of its scope.

Among the results on the representation theory of the classical groups, an essential role for the subsequent development of the theory was played by the introduction and analysis of the so-called representations of the fundamental series of the complex classical groups. This construction, later generalized by Harish-Chandra to all reductive Lie groups with finite center, is the basis for harmonic analysis on these groups and, at the same time, is a basic model for the study of completely reducible representations of these groups. The representations of the fundamental series, which depend on a definite collection of continuous and discrete parameters, are irreducible for parameters in general position and, in general, have a finite Jordan-Hölder series. Naimark undertook the study of these representations, extended to complex-valued parameters, in the first instance for the group SL(2,C), which is locally isomorphic to the Lorentz group. The results of this study are contained in Naimark's monograph "Linear representations of the Lorentz group" which appeared in 1958. Later, jointly with D. P. Zhelobenko, he generalized these results to all complex semisimple Lie groups.

During 1962 Naimark began working at the Steklov Institute of Mathematics of the USSR Academy of Sciences, in the Department of the theory of functions and functional analysis, which was headed by S. M. Nikol'skii. Here, until near the end of his life, Naimark was actively engaged in scientific work, gave special courses in functional analysis, and led seminars on the theory of group representations and operator algebras. A number of his research papers were devoted to general questions on the decomposition of unitary representations of a locally compact group into a direct integral with respect to factor representations (which in the infinite-dimensional case are an analogue of representations). During the sixties Naimark's interests were directed to a considerable extent toward the representation theory of groups and algebras in spaces with an indefinite metric. His last monograph "The theory of group representations" was published in 1976. It is both a textbook and a detailed reference on the classical theory of Lie groups and their finite-dimensional representations.

As with Gelfand, Naimark's research interests were formed during the mid-1930's, at a time when functional analysis was just beginning to develop rapidly. He became an important specialist in the field, and one of the initiators of a number of branches. His name is linked with fundamental results in each of these branches. He was the author of 130 papers and five books, each of which has been translated into many languages.

Naimark's scientific activity was inseparably linked with his pedagogical work and his education of young people. He was a widely educated man, a connoisseur of literature, painting, and music.

On December 30, 1978, Naimark died in his sixtieth year after a prolonged illness. Those who knew him best remember him as a man of spiritual qualities, unusual honesty, sympathy, high morals, and kindness. He was a model person and scientist.

We invite the reader to consult the following references for more information concerning the lives and mathematical accomplishments of Gelfand and Naimark.

M. I. Vishik, A. N. Kolmogorov, S. V. Fomin, and G. E. Shilov, *Izrail' Moiseevich Gelfand (on his fiftieth birthday).* Russian Math. Surveys 19: 3(1964), 163-180.

S. G. Gindikin, A. A. Kirillov, and D. B. Fuks, *The work of I. M. Gelfand on functional analysis, algebra, and topology (on his sixtieth birthday).* Russian Math. Surveys 29: 1(1974), 5-35.

O. V. Lokutsievskii and N. N. Chentsov, *The work of I. M. Gelfand in applied and computational mathematics (on his sixtieth birthday).* Russian Math. Surveys 29: 1(1974), 36-61.

I. M. Gelfand, M. I. Graev, D. P. Zhelobenko, R. S. Ismagilov, M. G. Krein, L. D. Kudryatsev, S. M. Nikol'skii, Ya. Khelemskii, and A. V. Shtraus, *Mark Aronovich Naimark.* Russian Math. Surveys 35: 4(1980), 157-164.

N. N. Bogolyubov, S. G. Ginkikin, A. A. Kirillov, A. N. Kolmogorov, S. P. Novikov and L. D. Faddeev, *Izrail' Moiseevich Gelfand (on his seventieth birthday).* Russian Math. Surveys 38: 6(1983), 145-153.

## EXERCISES

(I.1) Let $A$ be a normed $^*$-algebra such that $||x^*x|| = ||x||^2$ for all $x$ in $A$. Prove that the involution is isometric, and hence that $||x^*x|| = ||x^*||\cdot||x||$ for all $x$ in $A$.

(I.2) Let $A$ be a normed $^*$-algebra such that $||x^*x|| \geq ||x||^2$ for all $x$ in $A$. Prove that:

(a) the involution is isometric;

(b) $||x^*x|| = ||x||^2$ for all $x$ in $A$.

(I.3) Let $A$ be nonzero normed $^*$-algebra such that $||x^*x|| = ||x||^2$ for all $x$ in $A$. Prove that:

(a) $||x|| = \sup\{||xy||: y \in A, ||y|| \leq 1\}$ for all $x$ in $A$.

(b) if $e$ is an identity in $A$, then $e^* = e$ and $||e|| = 1$.

(c) if $xx^* = x^*x$, then $||x^2|| = ||x||^2$.

(I.4) Let $A$ be a normed $^*$-algebra such that $||x^*x|| = ||x||^2$ for all $x$ in $A$.

(a) If $x \in A$ and $(x^*x)^n = 0$ for some positive integer $n$, prove that $x = 0$.

(b) If $x^*x = xx^*$ and $x^n = 0$ for some positive integer $n$, prove $x = 0$.

(I.5) Let $A$ be a normed $^*$-algebra such that $||x^*x|| = ||x||^2$ for all $x$ in $A$. An element $x$ in $A$ is *idempotent* if $x^2 = x$; an idempotent element $x$ such that $x = x^*$ is a *projection*. Prove that:

(a) any nonzero projection $x$ in $A$ has norm 1;

(b) if $x$ is an idempotent and $x^*x = xx^*$, then $x$ is a projection.

(I.6) Let $A$ be a $^*$-algebra. Prove the following "parallelogram law" for elements $x$ and $y$ in $A$:

$$(x + y)^*(x + y) + (x - y)^*(x - y) = 2(x^*x + y^*y).$$

# 2
# The Gelfand-Naimark Theorem for Commutative C*- Algebras

## §5. *Gelfand structure theory: a brief review.*

Given a commutative $C^*$-algebra $A$ we wish to construct a locally compact Hausdorff space $X$ from $A$ and then show that $A$ is isometrically $*$-isomorphic to the function algebra $C_o(X)$.

Construction of the locally compact space $X$ from $A$ is canonical and a thing of considerable beauty. This construction together with other results needed from Gelfand theory are reviewed briefly here. For the full story the reader should consult Appendix B. The main task before us, then, is to establish that $A$ and $C_o(X)$ are isometric and $*$-isomorphic.

Let $A$ be a commutative Banach algebra and let $\hat{A}$ be the set of all nonzero multiplicative linear functionals, i.e., all nonzero linear maps $\phi: A \to C$ such that $\phi(xy) = \phi(x)\phi(y)$ for all $x, y \in A$. If $\phi \in \hat{A}$, then $\phi$ is necessarily continuous; in fact, $||\phi|| \leq 1$ (B.6.3).

For each $x$ in $A$ define $\hat{x}: \hat{A} \to C$ by $\hat{x}(\phi) = \phi(x)$; $\hat{x}$ is called the *Gelfand transform* of $x$ and is the abstract analogue of the usual Fourier transform (B.6.6).

The *Gelfand topology* on $\hat{A}$ is defined to be the weakest topology on $\hat{A}$ under which all of the functions $\hat{x}$ are continuous; it is the relative topology which $\hat{A}$ inherits as a subset of the dual space $A^*$ with the weak $*$-topology. The set $\hat{A}$ endowed with the Gelfand topology is called the *structure space* of $A$. Since the maximal ideals in $A$ are in one-to-one correspondence with elements in $\hat{A}$ the structure space is often called the *maximal ideal space* of $A$.

If the algebra $A$ has no identity element, it is frequently convenient (and necessary!) to adjoin one. This can be done by considering the algebra $A_e = A \oplus C$ with product

$$(x,\lambda)(y,\mu) = (xy + \lambda y + \mu x, \lambda\mu)$$

and involution

$$(x,\lambda)^* = (x^*,\overline{\lambda})$$

if $A$ is a $^*$-algebra. Identifying $x$ in $A$ with $(x,0)$ in $A_e$, we see that $A$ is a maximal two-sided ideal in $A_e$ with $e = (0,1)$ as identity; further, $(x,\lambda)$ can be written as $x + \lambda e$ or simply as $x + \lambda$ when no confusion is possible. If $A$ is actually a Banach algebra (as is the case here), then $A_e$ can also be made into a Banach algebra by extending the norm on $A$ to $A_e$; for example, by setting

$$||(x,\lambda)|| = ||x|| + |\lambda|.$$

The algebra $A_e$ is called the *unitization* of $A$ and can be constructed whether $A$ has an identity or not.

Clearly, every multiplicative linear functional $\phi$ on a commutative Banach algebra $A$ can be extended uniquely to a multiplicative linear functional $\phi_e$ on $A_e$ by setting

$$\phi_e((x,\lambda)) = \phi(x) + \lambda, \quad (x,\lambda) \in A_e.$$

It follows from the Banach-Alaoglu theorem (A.3) that the structure space $\hat{A}$ of a commutative Banach algebra $A$ is a locally compact Hausdorff space which is compact if $A$ has an identity. Furthermore, the functions $\hat{x}$ on $\hat{A}$ vanish at infinity (B.6.7).

The mapping $x \to \hat{x}$, called the *Gelfand representation*, is an algebra homomorphism of $A$ into $C_0(\hat{A})$. Moreover, if $||\cdot||_\infty$ denotes the sup-norm on $C_0(\hat{A})$, then $||\hat{x}||_\infty \leq ||x||$, and so $x \to \hat{x}$ is continuous (B.6.10). In general, the Gelfand representation is neither injective, surjective, nor norm-preserving. However, in the case of a commutative $C^*$-algebra it will be seen to be an isometric $^*$-isomorphism of $A$ onto $C_0(\hat{A})$.

For this purpose let us introduce the *spectrum of an element* $x$ in an algebra $A$ with identity as the set

$$\sigma_A(x) = \{\lambda \in C\colon\ x - \lambda e \text{ is not invertible in } A\};$$

if $A$ has no identity define $\sigma_A(x) = \sigma_{A_e}(x)$ (B.4.2).

The spectrum of an element x in a Banach algebra A is a compact subset of the complex plane and furthermore the following basic Beurling-Gelfand formula holds:

$$|x|_\sigma = \lim_{n\to\infty}||x^n||^{1/n} \leq ||x||$$

where

$$|x|_\sigma = \sup\{|\lambda| : \lambda \in \sigma_A(x)\}$$

is called the *spectral radius of* x (B.4.12).

The multiplicative linear functionals on a commutative Banach algebra A are related to the points in the spectra of elements in A. If $\lambda \neq 0$, then $\lambda \in \sigma_A(x)$ if and only if there exists $\phi \in \hat{A}$ such that $\phi(x) = \lambda$. Hence

$$\hat{x}(\hat{A}) \cup \{0\} = \sigma_A(x) \cup \{0\}$$

and so

$$||\hat{x}||_\infty = |x|_\sigma \leq ||x||.$$

## §6. *Unitization of a C*-algebra.*

Many C*-algebras with which we will be working may not possess an identity element. Thanks to the efforts of B. Yood we know how to extend the norm to a C*-norm on the unitization of the algebra, at least if the given C*-algebra has isometric involution. The purpose of this section is to show how this is done. The result is essential for much that follows. In the proof we use the simple fact that a Banach *-algebra such that $||x||^2 \leq ||x^*x||$ is a C*-algebra with isometric involution.

(6.1) PROPOSITION. *(Yood). Let* A *be a C*-algebra without identity with isometric involution, and consider the unitization* $A_e$ *(without norm) of* A. *Then there exists a C*-norm on* $A_e$ *under which the involution is isometric and which extends the norm on* A.

*Proof.* Identifying x with (x,0) and e with (0,1) we may write $x + \lambda e$ for the element $(x,\lambda)$ in $A_e$. Define

$$||x + \lambda e||_1 = \sup\{||xy + \lambda y|| : ||y|| \leq 1, y \in A\}. \qquad (1)$$

We assert that $||\cdot||_1$ is a norm on $A_e$ which extends the norm on A and makes $A_e$ into a C*-algebra with isometric involution. Since $||x|| = \sup\{||xy||: ||y|| \leq 1, y \in A\}$ for x in A, it is clear that $||\cdot||_1$ extends the norm on A.

Assume $||x + \lambda e||_1 = 0$, where $x \in A$ and $\lambda \in C$. If $\lambda \neq 0$, then from (1) we have $xy + \lambda y = 0$ or $(-x/\lambda)y = y$ for all $y \in A$; hence the element $u = -x/\lambda$ is a left identity for A. But then $u^*$ is a right identity and $u = uu^* = u^*$ shows that u is an identity for A, a contradiction of the hypothesis. Thus $\lambda = 0$; it follows that $||x|| = 0$ or $x = 0$. Hence $||x + \lambda e||_1 = 0$ implies $x + \lambda e = 0$.

From the definition of $||\cdot||_1$ as the norm of the left multiplication operator, it is clear that $||\cdot||_1$ is a normed algebra norm on $A_e$.

Next we shall show that the norm defined in (1) is a C*-norm with isometric involution. It suffices to show that if $x + \lambda e \in A_e$, then

$$||x + \lambda e||_1^2 \leq ||(x + \lambda e)^*(x + \lambda e)||_1. \tag{2}$$

Let c be a real number, $0 < c < 1$. By (1) there is an element $y \in A$ such that $||y|| = 1$ and $c\cdot||x + \lambda e||_1 < ||xy + \lambda y||_1$. It follows that since $||y||_1 = ||y||$ we have

$$\begin{aligned} c^2\cdot||x + \lambda e||_1^2 &\leq ||xy + \lambda y||_1^2 = ||(xy + \lambda y)^*(xy + \lambda y)||_1 \\ &= ||y^*(x + \lambda e)^*(x + \lambda e)y||_1 \\ &\leq ||y^*||_1||(x + \lambda e)^*(x + \lambda e)||_1||y||_1 \\ &= ||(x + \lambda e)^*(x + \lambda e)||_1. \end{aligned}$$

Letting $c \uparrow 1$ gives (2).

To show that $A_e$ is complete, let $\{x_n + \lambda_n e\}$ be Cauchy in $A_e$. Then $\{\lambda_n\}$ is bounded, for if not, by passing to a subsequence, we may assume $|\lambda_n| \to \infty$. Then, since $\{x_n + \lambda_n e\}$ is bounded, we have

$$(1/\lambda_n)x_n + e = (1/\lambda_n)(x_n + \lambda_n e) \to 0$$

or $-x_n/\lambda_n \to e$ as $n \to \infty$. Since A is complete, it is closed in $A_e$

and thus $e \in A$, an impossibility. Thus $\{\lambda_n\}$ is bounded. Passing to a subsequence, we may assume $\{\lambda_n\}$ converges to $\lambda$ in C. But then $x_n = (x_n + \lambda_n e) - \lambda_n e$ shows that $\{x_n\}$ is also a Cauchy sequence; since A is complete, $x_n \to x$ for some x in A. Therefore $x_n + \lambda_n e \to x + \lambda e$ in $A_e$; hence $A_e$ is complete. □

§7. *The Gelfand-Naimark theorem.*

We are now prepared to prove the first Gelfand-Naimark theorem. It turns out to be the key which unlocks the whole subject of C*-algebras. Indeed, it could legitimately be called "the fundamental theorem of C*-algebras," for nearly everything that follows depends on it in one way or another.

(7.1) THEOREM. *(Gelfand-Naimark). Let* A *be a commutative C*-algebra. Then the Gelfand representation* $x \to \hat{x}$ *is an isometric *-isomorphism of* A *onto* $C_0(\hat{A})$. *In particular,* $(x^*)^\wedge = \overline{\hat{x}}$ *for every* x *in* A.

*Proof.* We have seen in Section 5 that $x \to \hat{x}$ is a homomorphism of A into $C_0(\hat{A})$. That the involution in A is isometric is proved quite simply by the following argument of Gelfand and Naimark [1]. For every $h \in A$ with $h^* = h$ the C*-condition gives $||h^2|| = ||h||^2$; by iteration $||h^{2^n}|| = ||h||^{2^n}$ or $||h|| = ||h^{2^n}||^{1/2^n}$ and so $||h|| = |h|_\sigma$. In particular, for $h = x^*x$ we have $||x^*x|| = |x^*x|_\sigma$. Since $\sigma(x^*) = \overline{\sigma(x)}$, we see that $|x^*|_\sigma = |x|_\sigma$. Hence using the sub-multiplicativity of the spectral radius on commuting elements

$$||x^*||\cdot||x|| = ||x^*x|| = |x^*x|_\sigma \leq |x^*|_\sigma |x|_\sigma$$
$$= |x|_\sigma^2 \leq ||x||^2$$

and so $||x^*|| \leq ||x||$. Replacing x by $x^*$, we also have $||x|| \leq ||x^*||$; thus $||x^*|| = ||x||$.

If A has an identity element, we now show that $x \to \hat{x}$ is a *-map. We first show by two different arguments that $\phi(h)$ is real for $h \in A$ with $h^* = h$ and $\phi \in \hat{A}$.

*Arens' argument* [1]: Set $z = h + ite$ for real $t$. If $\phi(h) = \alpha + i\beta$ with $\alpha$ and $\beta$ real, then $\phi(z) = \alpha + i(\beta + t)$ and $z^*z = (h - ite)(h + ite) = h^2 + t^2e$ so that

$$\alpha^2 + (\beta + t)^2 = |\phi(z)|^2 \leq ||z||^2 = ||z^*z|| \leq ||h^2|| + t^2$$

or $\alpha^2 + \beta^2 + 2\beta t \leq ||h^2||$ for all real $t$. Thus $\beta = 0$ and $\phi(h)$ is real.

*Fukamiya's argument* [2]: Recall that in a Banach algebra $\exp(x) = \sum_{n=0}^{\infty} x^n/n!$. Set $u = \exp(ih)$. Then $u^* = \exp(-ih)$ and so $u^*u = e = uu^*$. Since $1 = ||u^*u|| = ||u^*||\cdot||u|| = ||u||^2$, we see that $||u|| = 1 = ||u^{-1}||$. Hence $|\hat{u}(\phi)| \leq 1$ and $|\hat{u}^{-1}(\phi)| \leq 1$, which implies $|\hat{u}(\phi)| = 1$. Since $1 = |\hat{u}(\phi)| = |\phi(u)| = |\exp(i\phi(h))|$, it follows that $\phi(h)$ is real.

Now, if $x \in A$, then $x = h + ik$, with $h = (x + x^*)/2$ and $k = (x - x^*)/2i$. Since $h^* = h$, $k^* = k$, and $x^* = h - ik$, we have for every $\phi \in \hat{A}$,

$$(x^*)\hat{}(\phi) = \phi(x^*) = \phi(h - ik) = \overline{\phi(h + ik)} = \overline{\phi(x)} = \overline{\hat{x}(\phi)}.$$

Thus $(x^*)\hat{} = \overline{\hat{x}}$; i.e., the Gelfand representation is a $*$-map.

Next, if $A$ has no identity element, we may extend the norm on $A$ to a C*-norm on $A_e$ by (6.1) since the involution on $A$ is isometric. Since every $\phi$ in $A$ can be extended to a multiplivative linear functional on $A_e$, this shows that $x \to \hat{x}$ is a $*$-map even if $A$ has no identity.

It is now easily seen that $x \to \hat{x}$ is an isometry. Indeed:

$$||x||^2 = ||x^*x|| = |x^*x|_\sigma = ||(x^*x)\hat{}||_\infty = ||(x^*)\hat{}\hat{x}||_\infty$$

$$= ||\overline{\hat{x}}\hat{x}||_\infty = ||\hat{x}||_\infty^2$$

or $||x|| = ||\hat{x}||_\infty$.

Summarizing, we have shown that the Gelfand representation is an isometric $*$-isomorphism of $A$ into $C_o(\hat{A})$. Let $B$ denote the range of $x \to \hat{x}$. Then $B$ is clearly a norm-closed subalgebra of $C_o(\hat{A})$ which separates the points of $\hat{A}$, vanishes identically at no point of

$\hat{A}$, and is closed under complex conjugation. By the Stone-Weierstrass theorem [Appendix A.7] we conclude that $B = C_o(\hat{A})$ and hence that $x \to \hat{x}$ is onto. Thus the proof of the representation theorem for commutative algebras is complete. □

The reader who is interested in an unconventional proof of the preceding theorem for $C^*$-algebras with identity and isometric involution may consult Edward Nelson [1]. A "constructive" proof of the theorem for algebras with identity is given in the book by D. S. Bridges [1, p. 157]. Quite simple proofs of the Gelfand-Naimark theorem in the special case of function algebras have been given by Nelson Dunford and Jacob T. Schwartz [1, p. 274] and Karl E. Aubert [1].

## §8. *Functional calculus in $C^*$-algebras.*

Now that Theorem (7.1) has been established, what is our next step? The answer is: use it to develop a "functional calculus" for certain elements in a $C^*$-algebra. This powerful technique will enable us to reduce much of the theory of $C^*$-algebras to the familiar setting of function algebras.

An element $x$ in a $C^*$-algebra $A$ is said to be *normal* if $x^*x = xx^*$. Let $B$ be any closed commutative $*$-subalgebra $A$ which contains a normal element $x$ in $A$. For example, one can take $B$ to be the closed $*$-subalgebra of $A$ generated by $x$. Then $B$ is isometrically $*$-isomorphic to $C_o(\hat{B})$, and the function $\hat{x}$ in $C_o(\hat{B})$ which corresponds to the element $x$ is called a *functional representation* of $x$. It clearly depends on the choice of $B$. In case $x$ is *hermitian*, i.e., $x^* = x$, it follows from the last statement of (7.1) that $\hat{x}$ is a real-valued function on $\hat{B}$, and that $||x|| = ||\hat{x}||_\infty = \sup\{|\hat{x}(\phi)|: \phi \in \hat{B}\}$. Recall that if $A$ is unital, then an element $u$ is *unitary* if $u^*u = e = uu^*$. As a first application we prove:

(8.1) PROPOSITION. *Let* $A$ *be a* $C^*$*-algebra.*

*(a) If* $h \in A$ *is hermitian, then* $\sigma_A(h)$ *is real.*

*(b) If* $A$ *is unital, and* $u \in A$ *is unitary, then* $\sigma_A(u) \subseteq \{\lambda \in C: |\lambda| = 1\}$.

*Proof.* (a) Let $B$ be a closed commutative $*$-subalgebra of $A$

containing h, and let $\hat{h}$ be the corresponding functional representation of h. Then, as was noted above, $\hat{h}$ is a real-valued function. Since $\sigma_B(h) = \hat{h}(\hat{B}) \cup \{0\}$ (B.6.6), then $\sigma_B(h)$ is real. Since $\sigma_A(h) \subseteq \sigma_B(h) \cup \{0\}$ (B.4.3), part (a) is proved.

(b) Again, let B be a closed commutative *-subalgebra of A containing u. Then, as in the first part of the proof of (7.1), $||u|| = ||u^*||$ in B. Hence $||u||^2 = ||u^*u|| = ||e|| = 1$ and so $||u|| = 1$. Similarly, $||u^{-1}|| = ||u^*|| = 1$. By the Beurling-Gelfand theorem (B.4.12), $|u|_\sigma \leq 1$ and $|u^{-1}|_\sigma \leq 1$. It follows that $\sigma_A(u)$ and $\sigma_A(u^{-1})$ are both subsets of $\{\lambda \in C: |\lambda| \leq 1\}$. Since $\sigma_A(u^{-1}) = \sigma_A(u)^{-1}$ (B.4.10), we obtain (b). □

In general the spectrum of an element in a Banach algebra may become larger upon passing to a subalgebra (B.4.3). For C*-algebras this is not the case.

(8.2) PROPOSITION. *Let* A *be a C*-algebra,* B *a closed *-subalgebra of* A *and* $x \in B$. *Then:*

*(a)* $\sigma_A(x) \cup \{0\} = \sigma_B(x) \cup \{0\}$

*(b) If* A *is unital with* $e \in B$, $\sigma_A(x) = \sigma_B(x)$.

*Proof.* (a) It is clear that $\sigma_A(x) \subseteq \sigma_B(x) \cup \{0\}$, so it suffices to show that $\sigma_B(x) \subseteq \sigma_A(x) \cup \{0\}$. This will follow if we show that any $z \in B$ which is quasi-regular in A is quasi-regular in B. In the special case that z is hermitian with quasi-inverse z', we see by (8.1) that for any $\varepsilon > 0$ $1 + i\varepsilon \notin \sigma_B(z)$ so that $z/(1 + i\varepsilon)$ is quasi-regular in B. By the continuity of quasi-inversion (B.3.2),(c), $(z/(1 + i\varepsilon))' \to z'$ as $\varepsilon \to 0$; so $z' \in B$. If z is any element of B which is quasi-regular in A, $z \circ z^*$ is an hermitian element of B which is quasi-regular in A. Thus $z \circ z^*$ is quasi-regular in B and so z is right quasi-regular in B. Applying the same argument to $z^* \circ z$, we find that z is also left quasi-regular in B. Now if $\lambda \notin \sigma_A(x) \cup \{0\}$, $x/\lambda$ is quasi-regular in A, hence quasi-regular in B. Therefore $\lambda \notin \sigma_B(x)$.

(b) By part (a) we need only show x is invertible in B if and only if x is invertible in A. If x is invertible in B, it is clearly invertible in A. If x is invertible in A, we argue analogously to part (a): For hermitian x, $i\varepsilon$ does not belong to $\sigma_B(x)$, so $(x - i\varepsilon)^{-1}$ exists in B for $\varepsilon > 0$. Since $(x - i\varepsilon)^{-1} \to x^{-1}$, we have $x^{-1} \in B$. Now for any $x \in B$ which is invertible in A, the hermitian $x^*x \in B$ is invertible in A, hence in B. Thus x is left invertible in B. Applying the same reasoning to $xx^*$ we find that x is also right invertible in B. Hence $x^{-1} \in B$. □

Let A be a C*-algebra with identity e, x a normal element in A, and B any closed commutative *-subalgebra of A which contains x and e. By (7.1) $B \simeq C(\hat{B})$. By (8.2) we have $\sigma_A(x) = \sigma_B(x)$ and hence

$$\sigma_A(x) = \{\hat{x}(\phi): \phi \in \hat{B}\}.$$

If $f \in C(\sigma_A(x))$, then $f\circ\hat{x}$ is a continuous complex-valued function on $\hat{B}$, i.e., $f\circ\hat{x} \in C(\hat{B})$. Hence there exists a unique element y in B such that $\hat{y} = f\circ\hat{x}$. It is customary to denote this element y in A by f(x).

(8.3) PROPOSITION. *Let* A *be a C*-algebra with identity* e *and* x *a normal element of* A. *Then the mapping* $f \to f(x)$ *is an isometric *-isomorphism of the commutative C*-algebra* $C(\sigma_A(x))$ *into* A *with the following properties:*

*(a) If* $f(\lambda) = 1$ *for all* $\lambda \in \sigma_A(x)$, *then* $f(x) = e$.

*(b) If* $f(\lambda) = \lambda$ *for all* $\lambda \in \sigma_A(x)$, *then* $f(x) = x$.

*(c)* $\sigma_A(f(x)) = f(\sigma_A(x))$.

*(d)* f(x) *is contained in every closed commutative *-subalgebra of* A *which contains* x *and* e; *thus* f(x) *is independent of the C*-algebra* B *used in its definition.*

*Proof.* It is easy to check that $f \to f(x)$ is a *-homomorphism of $C(\sigma_A(x))$ into B. For example if $f, g \in C(\sigma_A(x))$, then

$$[(f + g)(x)]\hat{} = (f + g)\circ\hat{x} = f\circ\hat{x} + g\circ\hat{x}$$

$$= f(x)\hat{} + g(x)\hat{} = [f(x) + g(x)]\hat{}$$

and by uniqueness, $(f + g)(x) = f(x) + g(x)$. Similarly, $(\lambda f)(x) = \lambda f(x)$, $\lambda \in C$, $(fg)(x) = f(x)g(x)$, and $\overline{f}(x) = f(x)^*$.

Statements (a) and (b) are immediate consequences of the definition.

(c) We have

$$\sigma_A(f(x)) = \sigma_B(f(x))$$

$$= f(x)\hat{}(\hat{B}) = (f\circ\hat{x})(\hat{B})$$

$$= f(\sigma_B(x)) = f(\sigma_A(x)).$$

The mapping $f \to f(x)$ is an isometry (hence injective) since

$$||f(x)|| = \sup\{|f(x)(\phi)|: \phi \in \hat{B}\}$$

$$= \sup\{|f(\lambda)|: \lambda \in \sigma_B(x)\}$$

$$= \sup\{|f(\lambda)|: \lambda \in \sigma_A(x)\} = ||f||_\infty.$$

(d) Let $f \in C(\sigma_A(x))$, and suppose $B_1$, $B_2$ are arbitrary closed commutative *-subalgebras of A which contain x and e, and let $f_1(x)$, $f_2(x)$ be the corresponding elements in A. The problem is to show that $f_1(x) = f_2(x)$. Since $B_1 \cap B_2$ is also a closed commutative *-subalgebra of A, we may assume without loss of generality that $B_2 \subseteq B_1$. If $\phi \in \hat{B}_1$, let $\phi' = \phi|B_2$. Then, for each $\phi \in \hat{B}_1$,

$$\phi(f_2(x)) = f_2(x)\hat{}(\phi') = f(\hat{x}(\phi')), \text{ all } \hat{} \text{ being in } B_2,$$

$$= f(\phi(x)) = f(\hat{x}(\phi)) = f_1(x)\hat{}(\phi) = \phi(f_1(x)), \text{ all } \hat{} \text{ being in } B_1.$$

Hence, $f_2(x) = f_1(x)$ and the proof is complete. □

We next extend (8.3) to C*-algebras without identity. If S is a compact subset of the complex plane which contains zero, then $C(S)_o$ will denote the C*-algebra of functions f in C(S) such that $f(0) = 0$.

(8.4) PROPOSITION. *Let A be a C*-algebra with isometric involution but without identity. Let x be a normal element of A. Then there exists an isometric *-isomorphism* $f \to f(x)$ *of* $C(\sigma_A(x))_o$ *into A with the following properties:*

*(a) If* $f(\lambda) = \lambda$ *for all* $\lambda \in \sigma_A(x)$, *then* $f(x) = x$.

*(b)* $\sigma_A(f(x)) = f(\sigma_A(x))$.

*(c)* $f(x)$ *is contained in every closed commutative *-subalgebra of A which contains* x.

*Proof.* Let B be a closed commutative *-subalgebra of A containing x, and consider the C*-algebras $B_e$ and $A_e$ obtained from B and A by adjoining the identity to each. Note that both $A_e$ and $B_e$

have the same identity element. Regarding $x$ as an element of $B_e$ it follows from (8.3) that the element $y = f(x)$ exists in the algebra $B_e$. We assert that $y \in B$; indeed, since $x \in B$, then $\hat{x}(\phi_\infty) = \phi_\infty((x,0))$, where $\phi_\infty$ is as in (B.6.2), and hence $\hat{y}(\phi_\infty) = (f \circ \hat{x})(\phi_\infty) = f(\hat{x}(\phi_\infty)) = f(0) = 0$, i.e., $y \in B$. The verification that $f \to f(x)$ is an isometric $^*$-isomorphism and of parts (a), (b), and (c) proceeds as in the proof of (8.3). □

We remark that the meaning of $f(x)$ in (8.3) and (8.4) is consistent with that given for general Banach algebras by the holomorphic functional calculus (see Appendix B.8).

## EXERCISES

(II.1) Prove that the commutative $C^*$-algebra $C_0(X)$ of continuous complex-valued functions on a locally compact Hausdorff space $X$ has an identity element iff $X$ is compact.

(II.2) A compact Hausdorff space $X$ is called a *Stonean space* if the closure of every open set is open. Prove that if $X$ is a Stonean space, then every element of the $C^*$-algebra $C(X)$ can be uniformly approximated by finite linear combinations of projections, i.e., self-adjoint idempotents.

(II.3) Let $A$ be a commutative $C^*$-algebra which is generated by its projections. Let $x \in A$ and $\varepsilon > 0$. Prove that $A$ contains a projection $e$ which is a multiple of $x$ and satisfies $||x - ex|| < \varepsilon$.

(II.4) Let $X$ be a compact Hausdorff space. Show that if $M$ is a maximal ideal in the $C^*$-algebra $C(X)$, then there exists $t \in X$ such that $M = \{f \in C(X): f(t) = 0\}$.

(II.5) Let $X$ be a compact Hausdorff space. Show that there is a bijective correspondence between closed ideals in the $C^*$-algebra $C(X)$ and closed subsets $F$ of $X$ given by

$$F \leftrightarrow I_F = \{f \in C(X): f(F) = \{0\}\}.$$

(II.6) If $A$ is a commutative $C^*$-algebra which is generated by one element $a$, prove that $A$ is isometrically $^*$-isomorphic to the $C^*$-algebra of continuous complex-valued functions on the spectrum $\sigma_A(a)$ of $a$, which vanish at 0.

(II.7) Consider the C*-algebra $A = C([0,1])$. For each $f \in A$, define

$$(Tf)(t) = \int_0^t f(s)ds, \quad t \in [0,1].$$

Show that $T \in B(A)$ and that $\lim_{n\to\infty} ||T^n||^{1/n} = 0$.

(II.8) Let $X$ be a compact Hausdorff space and $T$ a linear operator on $C(X)$. Prove that if $T(1) = 1$ and $||T|| = 1$, then $T$ is positive, i.e., $Tf \geq 0$ whenever $f \geq 0$.

(II.9) Let $A$ be a C*-algebra with identity $e$ and $x$ an element in $A$ which is not left invertible. Prove

(a) $x^*x$ is not invertible in the closed commutative *-subalgebra $B$ of $A$ generated by $e$ and $x^*x$.

(b) there is a sequence $\{y_n\}$ of elements of $B$ such that $||y_n|| = 1$ and $y_n(x^*x) \to 0$ as $n \to \infty$.

(II.10) If $A$ is a C*-algebra with identity $e$, prove that $e$ is an extreme point of the closed unit ball of $A$.

(II.11) Let $X$ be a metric space and $C_b(X)$ the C*-algebra of bounded continuous complex-valued functions on $X$. Prove that if $C_b(X)$ is separable, then $X$ is compact.

# 3
# The Gelfand-Naimark Theorem: Arbitrary C*-Algebras

## §9. *Introduction.*

Having established the representation theorem for commutative C*-algebras and a few of its consequences, we turn now to the case of a general C*-algebra A. We must construct, from A, a Hilbert space and then faithfully represent A as a norm-closed *-subalgebra of bounded linear operators on this space. The proof is substantially more involved than the commutative case and it will be divided into several steps.

It is far from obvious that the involution in a general C*-algebra is continuous; accordingly we handle this problem first. Then a new equivalent C*-norm with isometric involution is introduced. An investigation of the unitary elements will show that the original norm on the algebra coincides with the new norm. The representation theorem will then be effected by the well-known Gelfand-Naimark-Segal construction.

## §10. *Continuity of the involution.*

The strong C*-norm condition $||x^*x|| = ||x||^2$ which is usually used in the definition of a C*-algebra, implies immediately that $||x^*|| = ||x||$ and hence $x \to x^*$ is certainly continuous. On the other hand, the norm condition $||x^*x|| = ||x^*||\cdot||x||$ is not so generous in handing out its secrets. We show in Proposition (10.1) that it does indeed imply that the involution in a C*-algebra is continuous. Ultimately we wish to show that it even implies $x \to x^*$ is isometric.

(10.1) PROPOSITION. *The involution in a C*-algebra is continuous.*

*Proof*. First we show that the set $H(A) = \{h \in A: h^* = h\}$ of hermitian elements in $A$ is closed. Let $\{h_n\}$ be a convergent sequence in $H(A)$ whose limit is $h + ik$, with $h, k \in H(A)$. Since $h_n - h \to ik$, we may assume (by putting $h_n$ for $h_n - h$) that $h_n$ converges to $ik$, and also that $||h_n|| \le 1$. The spectral mapping theorem for polynomials (B.4.9) gives

$$\sigma_A(h_n^2 - h_n^4) = \{\lambda^2 - \lambda^4 : \lambda \in \sigma_A(h_n)\};$$

since $||x|| = |x|_\sigma$ and $\sigma_A(x)$ is real, for all $x \in H(A)$, ((8.1), (a)), we have

$$||h_n^2 - h_n^4|| = \sup\{\lambda^2 - \lambda^4 : \lambda \in \sigma_A(h_n)\}$$

$$\le \sup\{\lambda^2 : \lambda \in \sigma_A(h_n)\} = ||h_n^2||.$$

Letting $n \to \infty$ we obtain $||-k^2 - k^4|| \le ||k^2||$. Hence

$$\sup\{\lambda^2 + \lambda^4 : \lambda \in \sigma_A(k)\} \le \sup\{\lambda^2 : \lambda \in \sigma_A(k)\}.$$

Choose $\mu \in \sigma_A(k)$ such that $\mu^2 = \sup\{\lambda^2 : \lambda \in \sigma_A(k)\}$. Then $\mu^2 + \mu^4 \le \mu^2$, so $\mu = 0$. It follows that $||k|| = |k|_\sigma = 0$ and so $k = 0$. This shows $H(A)$ is closed.

Now it is easy to prove that the graph of the map $x \to x^*$ of $A$ onto $A$ is closed. For suppose $x_n \to x$ and $x_n^* \to y$. Then $x_n + x_n^* \to x + y$ and $(x_n - x_n^*)/i \to (x - y)/i$. Since $H(A)$ is closed, $x + y$ and $(x - y)/i$ are hermitian and so $x + y = x^* + y^*$ and $x - y = y^* - x^*$, whence $y = x^*$. Thus by the closed graph theorem, for real Banach spaces, the involution in $A$ is continuous. □

## §11. *An equivalent C*-norm.*

Given a C*-algebra we show that there is an isometric C*-norm on the algebra equivalent to the original norm which coincides with the original norm on hermitian elements.

(11.1) PROPOSITION. *Let* A *be a* $C^*$*-algebra. Then*

$$||x||_o = ||x^*x||^{1/2}$$

*is an equivalent* $C^*$*-norm on* A *such that* $||x^*||_o = ||x||_o$ *for all* $x \in A$, *and* $||h||_o = ||h||$ *for all hermitian* h *in* A.

*Proof.* By (10.1) there exists a constant $M \geq 1$ such that $||x^*|| \leq M\cdot||x||$ for all $x \in A$. Then

$$M^{-1/2}||x|| \leq ||x^*||^{1/2}||x||^{1/2} = ||x||_o \leq M^{1/2}||x||$$

so that $||\cdot||_o$ and $||\cdot||$ are equivalent. Clearly $||\cdot||_o$ is homogeneous and submultiplicative. To prove the triangle inequality, let $x, y \in A$. Then

$$||x + y||_o^2 = ||(x + y)^*(x + y)|| \leq ||x^*x|| + ||y^*y|| + ||x^*y + y^*x||,$$

so it is enough to prove that $||x^*y + y^*x|| \leq 2||x||_o||y||_o$. For any positive integer n

$$||(x^*y)^{2^{n-1}} + (y^*x)^{2^{n-1}}||^2$$

$$= ||(x^*y)^{2^n} + (y^*x)^{2^n} + (x^*y)^{2^{n-1}}(y^*x)^{2^{n-1}} + (y^*x)^{2^{n-1}}(x^*y)^{2^{n-1}}||$$

$$\leq ||(x^*y)^{2^n} + (y^*x)^{2^n}|| + 2(||x^*x||\cdot||y^*y||)^{2^{n-1}}. \quad (1)$$

For every $\varepsilon > 0$ there is an integer n such that

$$||(x^*y)^{2^n}|| \leq (|x^*y|_\sigma^2 + \varepsilon)^{2^{n-1}} \quad \text{and} \quad ||(y^*x)^{2^n}|| \leq (|y^*x|_\sigma^2 + \varepsilon)^{2^{n-1}}.$$

Then, by (B.4.8),

$$||(x^*y)^{2^n}|| \leq (|x^*y|_\sigma|y^*x|_\sigma + \varepsilon)^{2^{n-1}} \leq (||x^*y||\cdot||y^*x|| + \varepsilon)^{2^{n-1}}$$

$$\leq (||x^*x||\cdot||y^*y|| + \varepsilon)^{2^{n-1}}$$

and similarly

$$||(y^*x)^{2^n}|| \leq (||x^*x||\cdot||y^*y|| + \varepsilon)^{2^{n-1}}$$

so that

$$||(x^*y)^{2^n} + (y^*x)^{2^n}|| \leq 2(||x^*x||\cdot||y^*y|| + \varepsilon)^{2^{n-1}}. \qquad (2)$$

Beginning with (2) and applying (1) recursively we obtain

$$||(x^*y)^{2^{k-1}} + (y^*x)^{2^{k-1}}||^2 \leq 4(||x^*x||\cdot||y^*y|| + \varepsilon)^{2^{k-1}}$$

for any $k$, $1 \leq k \leq n+1$. Thus, in particular

$$||x^*y + y^*x||^2 \leq 4(||x^*x||\cdot||y^*y|| + \varepsilon)$$

for arbitrary $\varepsilon > 0$. Hence $||x^*y + y^*x|| \leq 2||x||_o||y||_o$. So we have seen that $||\cdot||_o$ is an equivalent algebra norm on A. Furthermore, $||h||_o = ||h^*h||^{1/2} = ||h||$ for all hermitian $h \in A$ and so $||x||_o^2 = ||x^*x|| = ||x^*x||_o$; i.e., $||\cdot||_o$ is a $C^*$-norm on A with $||x^*||_o = ||x||_o$ for all $x \in A$. □

§12. *Positive elements and symmetry.*

In the late nineteen forties and early fifties several mathematicians were actively working to eliminate the "symmetry axiom" (see §3, 3°) from the definition of a $C^*$-algebra. These investigations led to a careful study of the "positive" elements of the algebra. It was recognized that if the positive elements could be characterized as precisely those elements of the form $x^*x$, then symmetry would follow. This was shown to be the case in 1953 by I. Kaplansky whose argument was communicated in a Mathematical Review of J. Schatz [2]. Kaplansky's proof was based on a lemma obtained independently by Fukamiya [1] and Kelley-Vaught [1].

A systematic presentation of positivity in $C^*$-algebras, including the results of Kelley-Vaught and I. Kaplansky, will be given in this section. Certain additional results which will be needed later involving the natural order induced on the algebra by the positive cone will also be proved.

(12.1) DEFINITION. *Let* A *be a $C^*$-algebra. An element* x *in* A *is said to be* positive, *denoted* $x \geq 0$, *if* $x$ *is hermitian and* $\sigma_A(x) \geq 0$. *The set of positive elements in* $A$ *will be denoted by* $A^+$. *We write* $y \geq x$ *or* $x \leq y$ *if* $y - x \in A^+$.

If $B$ is a C*-subalgebra of a C*-algebra $A$ and if $x \in B$, then (8.2) implies that $x \geq 0$ in $B$ if and only if $x \geq 0$ in $A$. Hence, the notion of positivity is independent of the particular subalgebra the element lies in.

(12.2) PROPOSITION. *Let* $A$ *be a C*-algebra and let* $x \in A$.

*(a)* $x$ *is positive if and only if* $x = h^2$, *where* $h$ *is an element of* $A$ *and* $h = h^*$.

*(b) If* $A$ *has an identity* $e$, *then* $x$ *is positive if and only if* $x$ *is hermitian and* $||(||x||e) - x|| \leq ||x||$.

*Proof.* (a) If $x$ is positive, any functional representation of $x$ satisfies $\hat{x} \geq 0$. Take $h$ to be the element of $A$ corresponding to $\sqrt{\hat{x}}$; then $h^* = h$ and $x = h^2$. Conversely, let $x = h^2$, where $h^* = h$. If $\hat{h}$ is a functional representation of $h$, then $\hat{h}$ is real-valued and so $\hat{x} = (\hat{h})^2 \geq 0$. Hence, $\sigma_A(x) \geq 0$ by (B.6.6).

(b) Assume that $A$ has an identity $e$ and let $x$ be a nonzero hermitian element of $A$. In terms of a functional representation of $x$ we have: $x \in A^+$ iff $x \geq 0$ iff $||1 - (\hat{x}/||x||)||_\infty \leq 1$ iff $||e - (x/||x||)|| \leq 1$ iff $||(||x||e) - x|| \leq ||x||$. □

Recall that a nonempty subset of a vector space is a *cone* if it is closed under multiplication by nonnegative scalars. The cone is *convex* if it is also closed under addition.

The proof of the following theorem is due to Kelley and Vaught [1].

(12.3) THEOREM. *If* $A$ *is a C*-algebra, then* $A^+$ *is a closed convex cone in* $A$ *such that* $A^+ \cap (-A^+) = \{0\}$.

*Proof.* We consider $A$ with the equivalent C*-norm $||\cdot||_o$ of (11.1) and adjoin an identity $e$ if $A$ does not have one, using (6.1). Note that $A^+ = A \cap A_e^+$ and that the norms $||\cdot||$ and $||\cdot||_o$ applied to hermitian elements of $A$ need not be distinguished. That $A^+$ is a closed subset of $A$ follows immediately from (10.1) and (12.2), (b). It is obvious that if

$x \in A^+$ and $\lambda \geq 0$, then $\lambda x \in A^+$. Let $x, y \in A^+$. We will show that $x + y \in A^+$. Now $x + y$ is certainly hermitian. By normalizing $x$ and $y$ we may assume that $||x|| \leq 1$ and $||y|| \leq 1$. Taking a functional representation of $x$ we have $||\hat{x}||_\infty \leq 1$ and $\hat{x} \geq 0$ so that $||1 - \hat{x}||_\infty \leq 1$, i.e., $||e - x|| \leq 1$. Likewise $||e - y|| \leq 1$. Therefore $||e - \frac{1}{2}(x + y)|| = \frac{1}{2}||(e - x) + (e - y)|| \leq \frac{1}{2}||e - x|| + \frac{1}{2}||e - y|| \leq 1$. Considering a functional representation of $z = \frac{1}{2}(x + y)$, $||1 - \hat{z}||_\infty \leq 1$ implies that $\hat{z} \geq 0$, i.e., $z \in A^+$. Since $A^+$ is a cone, $2z = x + y \in A^+$. Finally if $x \in A^+ \cap (-A^+)$, then $\sigma_A(x) \subseteq [0,\infty)$ and $-\sigma_A(x) \subseteq [0,\infty)$, hence $\sigma_A(x) = \{0\}$. Since $x$ is hermitian, $||x|| = |x|_\sigma = 0$ and so $x = 0$. □

(12.4) PROPOSITION. *Let* A *be a* $C^*$*-algebra.*

*(a) If* A *has identity* e *and* $x \in A$ *is hermitian with* $||e - x|| \leq 1$, *then* $x \in A^+$.

*(b) If* $x, y \in A^+$ *and* $||x|| \leq 1$, $||y|| \leq 1$, *then* $||x - y|| \leq 1$.

*Proof.* (a) If $x = x^* \in A$ and $||e - x|| \leq 1$, then the functional representation of $x$ shows that $x \in A^+$ (as in the proof of (12.3)).

(b) If A is not unital, we form $A_e$ using the norm $||\cdot||_0$. Since $x \geq 0$ and $y \geq 0$, we have $-y \leq x - y$ and $-x \leq y - x$. Also $0 \leq x \leq ||x||e \leq e$ and $0 \leq y \leq ||y||e \leq e$ (both by functional calculus). Now the convexity of $A^+$ shows that " $\leq$ " is transitive so that

$$-e \leq -||y||e \leq -y \leq x - y \leq x \leq ||x||e \leq e$$

implies that $-e \leq x - y \leq e$. We conclude from the functional calculus on $x - y$ that $||x - y|| \leq 1$. □

(12.5) PROPOSITION. *Let* A *be a* $C^*$*-algebra and let* x *be an hermitian element in* A. *Then there exist elements* $x^+, x^- \in A^+$ *such that* $x = x^+ - x^-$ *and* $x^+x^- = x^-x^+ = 0$. *Further,* $||x^+|| \leq ||x||$, $||x^-|| \leq ||x||$, *and the norm of the element* $|x| = x^+ + x^-$ *in* $A^+$ *is equal to the norm of* x.

*Proof.* Since $x$ is hermitian, any functional representation $\hat{x}$ of $x$ is real-valued. Writing the function $\hat{x}$ as the difference of its positive and negative parts

$$\hat{x} = \hat{x}^+ - \hat{x}^-,$$

where $\hat{x}^+ = \max\{\hat{x},0\}$, $\hat{x}^- = -\min\{\hat{x},0\}$, and then passing back to the algebra A, it follows that we can write $x = x^+ - x^-$ with $x^+, x^- \in A^+$. Since $\hat{x}^+\hat{x}^- = \hat{x}^-\hat{x}^+ = 0$, this property is also reflected in A. The remaining statements of the theorem follow similarly. □

The elements $x^+$ and $x^-$ in (12.5) are often called the *positive* and *negative parts*, respectively, of the hermitian element x.

The proof of the next theorem is due to Kaplansky (see Schatz [2]).

(12.6) THEOREM. *If* A *is a C*-algebra, then* $A^+ = \{y^*y: y \in A\}$. *In particular,* A *is symmetric.*

*Proof.* If $x \in A^+$ then $x = h^2$, where h is hermitian by (12.2), (a). Hence, $A^+ \subseteq \{y^*y: y \in A\}$.

Conversely, suppose that $x = y^*y$ for some y in A. Since x is hermitian, (12.5) implies that there are $u, v \in A^+$ such that $x = y^*y = u - v$ with $uv = vu = 0$. Now $(yv)^*(yv) = v^*y^*yv = vy^*yv = v(u - v)v = -v^3$. Since $v \geq 0$, then $(yv)^*(yv) = -v^3 \leq 0$. The elements $(yv)^*(yv)$ and $(yv)(yv)^*$ have the same nonzero spectrum (B.4.8); hence $(yv)(yv)^* \leq 0$ also. Write $yv = h + ik$ with h and k hermitian. Then $h^2, k^2 \in A^+$ and by (12.4) we have

$$0 \geq (yv)^*(yv) + (yv)(yv)^* = 2(h^2 + k^2) \geq 0.$$

Thus $h = 0 = k$ or $yv = 0$. But then $0 = (yv)^*(yv) = -v^3$ and so $v = 0$. Hence $x = y^*y = u \in A^+$, i.e., $\{y^*y: y \in A\} \subseteq A^+$. In particular, $e + y^*y$ is invertible for all $y \in A$. □

The positive elements in a C*-algebra A induce an order relation $\leq$ on the algebra in a natural way. (See (12.1).) Since $A^+$ is a convex cone in A such that $A^+ \cap (-A^+) = \{0\}$, an easy verification shows that $\leq$ is a partial order on A, a fact already used in the proof of (12.4).

If X is a locally compact Hausdorff space and A is the C*-algebra $C_0(X)$, then an element $f \in A$ is positive in A if and only if $f(t) \geq 0$ for all $t \in X$. The proof of this statement is an immediate consequence of the relation $\sigma_A(f) = f(X) \cup \{0\}$. We show next that this situation carries over to the algebra B(H) of operators on a Hilbert space.

(12.7) PROPOSITION. *Let* $H$ *be a Hilbert space,* $A$ *the* $C^*$*-algebra* $B(H)$*, and* $T$ *an element of* $A$*. Then* $T$ *is a positive element in* $A$ *if and only if* $(T\xi|\xi) \geq 0$ *for all* $\xi \in H$*. Hence,* $T \in A^+$ *if and only if* $T$ *is a positive operator on* $H$.

*Proof.* Assume that $T \in A^+$. Then $T = S^*S$ for some $S \in A$ and we have $(T\xi|\xi) = (S^*S\xi|\xi) = ||S\xi||^2 \geq 0$ for all $\xi \in H$.

Conversely, suppose that $(T\xi|\xi) \geq 0$ for all $\xi \in H$. Then $(T\xi|\xi) = \overline{(T\xi|\xi)} = \overline{(\xi|T^*\xi)} = (T^*\xi|\xi)$ and by polarization $T = T^*$, i.e., $T$ is self-adjoint. By (12.5) we may write $T = T^+ - T^-$ where $T^+, T^- \in A^+$. For $\eta \in H$, let $\xi = T^-\eta$. Then

$$0 \leq (T\xi|\xi) = ((T^+ - T^-)\xi|\xi) = -(T^-(T^-\eta)|T^-\eta)$$

$$= -((T^-)^3\eta|\eta).$$

Since $(T^-)^3 \geq 0$, we also have $((T^-)^3\eta|\eta) \geq 0$; hence $((T^-)^3\eta|\eta) = 0$ or $(T^-)^3 = 0$. Thus $T^- = 0$ and we have $T = T^+ \geq 0$. □

(12.8) PROPOSITION. *Let* $A$ *be a* $C^*$*-algebra.*

*(a) If* $x, y, z \in A$ *and* $x \leq y$*, then* $z^*xz \leq z^*yz$.

*(b) If* $x, y \in A^+$ *and* $x \leq y$*, then* $||x|| \leq ||y||$.

*Proof.* (a) Since $x \leq y$, there exists $w \in A$ such that $y - x = w^*w$. Hence

$$z^*yz - z^*xz = z^*(y - x)z = z^*(w^*w)z = (wz)^*(wz) \geq 0.$$

(b) We may assume that $A$ has an identity $e$ by using $||\cdot||_0$ as in (12.3). By considering a functional representation of $y$ we clearly have $y \leq ||y||e$. By assumption $0 \leq x \leq ||y||e$, and again by considering a functional representation of $x$ it follows that $||x|| \leq ||y||$. □

(12.9) COROLLARY. *Let* $A$ *be a* $C^*$*-algebra with identity* $e$.

*(a) If* $x \in A$ *and* $e \leq x$*, then* $x$ *is invertible and* $x^{-1} \leq e$.

*(b) If* $x, y \in A$ *are invertible and* $0 \leq x \leq y$*, then* $y^{-1} \leq x^{-1}$.

*Proof.* (a) Let $z = x^{-1/2}$ and apply (12.8), (a).

(b) By (12.8), (a) and part (a) of this corollary we have $e \leq x^{-1/2}yx^{-1/2}$, $e \geq x^{1/2}y^{-1}x^{1/2}$, and finally $y^{-1} \leq x^{-1}$. □

§13. *Approximate identities in $C^*$-algebras.*

Recall that an *approximate identity* in a normed algebra $A$ is a net $\{e_\alpha\}$ in $A$ such that $\lim_\alpha e_\alpha x = x = \lim_\alpha xe_\alpha$ for every $x$ in $A$. If there is a finite constant $M$ such that $||e_\alpha|| \leq M$ for all $\alpha$, then the approximate identity is said to be *bounded*. If $A$ is a $C^*$-algebra, an approximate identity $\{e_\alpha\}$ in $A$ is said to be *increasing* if each $e_\alpha$ is positive, and if $\alpha \leq \beta$ implies $e_\alpha \leq e_\beta$. It is an important and useful fact that every $C^*$-algebra admits an increasing approximate identity bounded by 1. The construction in the proof of the following proposition is due to I. E. Segal [2] with refinements due to J. Dixmier [2].

(13.1) PROPOSITION. *Let* $A$ *be a* $C^*$*-algebra and* $I$ *a dense two-sided ideal in* $A$*. Then there exists an increasing approximate identity* $\{e_\alpha\}$ *in* $A$ *bounded by* 1*, consisting of hermitian elements in* $I$*. Further, if* $A$ *is separable, then* $\{e_\alpha\}$ *can be indexed by the positive integers.*

*Proof.* Let $A_e$ be formed by adjoining an identity $e$ to $A$ equipped with the equivalent $C^*$-norm $||\cdot||_o$. Let $F$ denote the family of all finite subsets of $I$ ordered by inclusion. If $\alpha = \{x_1,\ldots,x_n\}$ is an arbitrary element of $F$, let

$$h_\alpha = \sum_{i=1}^{n} x_i x_i^* \in I.$$

Since $h_\alpha \geq 0$ then $-1/n \notin \sigma_A(h_\alpha)$, and hence $e/n + h_\alpha$ is invertible in $A_e$. Define

$$e_\alpha = h_\alpha(e/n + h_\alpha)^{-1}.$$

Then $e_\alpha \in I$ for each $\alpha$ since $I$ is an ideal of $A_e$ also. Further, each $e_\alpha$ is hermitian; since the values of the real function $t \to t(1/n + t)^{-1}$ lie between $0$ and $1$ for $t \geq 0$, a functional representation of $e_\alpha$ gives $0 \leq e_\alpha \leq e$. By (12.8), (b), and (11.1), $||e_\alpha||_o = ||e_\alpha|| \leq 1$. From the definition of $h_\alpha$ and $e_\alpha$ we have

$$\sum_{i=1}^{n}[(e_\alpha - e)x_i][(e_\alpha - e)x_i]^* = (e_\alpha - e)h_\alpha(e_\alpha - e)^*$$
$$= \frac{1}{n^2} h_\alpha(e/n + h_\alpha)^{-2}, \qquad (1)$$

where the last equality is obtained by considering a functional representation of $(e_\alpha - e)h_\alpha(e_\alpha - e)^*$. Now we consider a functional representation of the element $h_\alpha(e/n + h_\alpha)^{-2}$. The real function $t \to t(1/n + t)^{-2}$ has a maximum value $n/4$ at $t = 1/n$; hence it follows from (1) that

$$\sum_{i=1}^{n} [(e_\alpha - e)x_i][e_\alpha - e)x_i]^* \leq e/4n.$$

For each $i$, $1 \leq i \leq n$, we then have

$$[(e_\alpha - e)x_i][(e_\alpha - e)x_i]^* \leq e/4n,$$

and applying (12.8), (b) gives $||(e_\alpha - e)x_i||_o^2 \leq 1/4n$, or

$$||(e_\alpha - e)x_i||_o \leq 1/2n^{1/2}.$$

Now, for an arbitrary $x \in I$ and $\varepsilon > 0$, let $\alpha_\varepsilon$ be any finite set of $n$ elements of $I$ such that $x \in \alpha_\varepsilon$ and $n > \varepsilon^{-2}$. Then, for any $\alpha \geq \alpha_\varepsilon$, $||e_\alpha x - x||_o < \varepsilon$; that is,

$$\lim_\alpha e_\alpha x = x, \quad (x \in I). \tag{2}$$

Since $I$ is dense in $A$, and $\{e_\alpha\}$ is bounded, then (2) holds for all $x$ in $A$, i.e., $\lim_\alpha e_\alpha x = x$ for all $x$ in $A$. Applying the involution to both sides of this result and using its continuity, we obtain $\lim_\alpha xe_\alpha = x$ for all $x$ in $A$. Hence the net $\{e_\alpha\}$ in $I$ is an approximate identity for $A$ which is bounded by one.

Now let $\alpha, \beta \in F$ with $\alpha \leq \beta$. Then $\alpha = \{x_1,\ldots,x_n\}$ and $\beta = \{x_1,\ldots,x_p\}$, where $n \leq p$. Clearly $h_\alpha \leq h_\beta$, and by (12.9), (b)

$$(e/n + h_\alpha)^{-1} \geq (e/n + h_\beta)^{-1}. \tag{3}$$

For all real numbers $t \geq 0$ we have

$$n^{-1}(n^{-1} + t)^{-1} \geq p^{-1}(p^{-1} + t)^{-1}$$

and so

$$en^{-1}(en^{-1} + h_\beta)^{-1} \geq ep^{-1}(ep^{-1} + h_\beta)^{-1}. \tag{4}$$

Combining (3) and (4), we have

$$e_\alpha = e - \frac{e}{n}(\frac{e}{n} + h_\alpha)^{-1} \leq e - \frac{e}{n}(\frac{e}{n} + h_\beta)^{-1} \leq e - \frac{e}{p}(\frac{e}{p} + h_\beta)^{-1} = e_\beta;$$

hence $\alpha \leq \beta$ implies $e_\alpha \leq e_\beta$. Therefore $\{e_\alpha\}$ is an increasing approximate identity.

Finally, assume that $A$ is separable and let $D$ be a countable dense subset of $A$. Since $I$ is dense in $A$ each element of $D$ can be approximated by sequences in $I$. Taking the (countable) union of all of these sequences we obtain a countable dense subset $M = \{y_1, y_2, \ldots\}$ of $I$. Set $e_n = e_{\{y_1,\ldots,y_n\}}$. The above argument shows that, for each $i$, $||e_n y_i - y_i||_o \to 0$ as $n \to \infty$. Since $||e_n|| \leq 1$, it follows that $e_n x \to x$ for each $x$ in $A$. The remainder of the proof is the same as before. □

Repeating the proof of (13.1) verbatim gives the following:

(13.2) PROPOSITION. *Let* $A$ *be a* $C^*$*-algebra and* $I$ *a right ideal of* $A$. *Then there is a net* $\{e_\alpha\}$ *in* $I \cap A^+$ *such that the following hold:*

*(a)* $||e_\alpha|| \leq 1$ *for all* $\alpha$;

*(b)* $\alpha \leq \beta$ *implies* $e_\alpha \leq e_\beta$;

*(c) for all* $x \in \overline{I}$, $||e_\alpha x - x|| \to 0$.

## §14. *An embedding theorem for C*-algebras.*

It was shown in (6.1) that every $C^*$-algebra with *isometric involution* can be embedded in a $C^*$-algebra with identity. It is essential to have this result for general $C^*$-algebras. The result in this section establishes precisely this fact. The argument is due to B. J. Vowden [1].

(14.1) PROPOSITION. *Every* $C^*$*-algebra without identity can be isometrically embedded in a* $C^*$*-algebra with identity.*

*Proof.* Let A be a C*-algebra without identity. By (13.1), A has an approximate identity $\{e_\alpha\}$ consisting of hermitian elements such that $||e_\alpha|| \leq 1$. Utilizing the approximate identity, observe that, for every $x \in A$,

$$||x|| = \sup\{||xy||: y \in A, ||y|| \leq 1\} = \sup\{||yx||: y \in A, ||y|| \leq 1\}$$

and extend the norm on A to $A_e$ by

$$||x + \lambda e|| = \sup\{||(x + \lambda e)y||: y \in A, ||y|| \leq 1\}$$

$$= \sup\{||y(x + \lambda e)||: y \in A, ||y|| \leq 1\}.$$

The last equality can be seen as follows. For each $y \in A$ with $||y|| \leq 1$ we have

$$||(x + \lambda e)y|| = \sup\{||z(x + \lambda e)y||: z \in A, ||z|| \leq 1\}$$

$$\leq \sup\{||z(x + \lambda e)||: z \in A, ||z|| \leq 1\},$$

thus

$$\sup\{||(x + \lambda e)y||: y \in A, ||y|| \leq 1\} \leq \sup\{||z(x + \lambda e)||: z \in A, ||z|| \leq 1\}$$

and the reverse inequality is established similarly.

Thus $A_e$ is a Banach *-algebra with identity in which A is isometrically embedded as a closed ideal of codimension one. To see that the C*-condition holds in $A_e$, we first prove that

$$||x + \lambda e|| = \lim_\alpha ||(x + \lambda e)e_\alpha|| = \lim_\alpha ||e_\alpha(x + \lambda e)||.$$

Given any $\varepsilon > 0$ there exists $y \in A$ with $||y|| \leq 1$ such that

$$||(x + \lambda e)y|| > ||x + \lambda e|| - \varepsilon.$$

Since $\lim_\alpha (x + \lambda e)e_\alpha y = (x + \lambda e)y$, there exists $\alpha_0$ such that for all $\alpha \geq \alpha_0$,

$$||(x + \lambda)e_\alpha|| \geq ||(x + \lambda e)e_\alpha y|| > ||x + \lambda e|| - \varepsilon.$$

Since

$$||(x + \lambda e)e_\alpha y|| \leq ||(x + \lambda e)e_\alpha|| \leq ||x + \lambda e||,$$

it follows that $\lim_\alpha ||(x + \lambda e)e_\alpha||$ exists and is equal to $||x + \lambda e||$. Similarly, $\lim_\alpha ||e_\alpha(x + \lambda e)|| = ||x + \lambda e||$. Thus

$$\begin{aligned} ||(x + \lambda e)^*||\cdot||(x + \lambda e)|| &= \lim_\alpha ||e_\alpha(x + \lambda e)^*||\cdot\lim_\alpha ||(x + \lambda e)e_\alpha|| \\ &= \lim_\alpha ||e_\alpha(x + \lambda e)^*(x + \lambda e)e_\alpha|| \\ &= ||(x + \lambda e)^*(x + \lambda e)||. \end{aligned}$$

Therefore $||(x + \lambda e)^*(x + \lambda e)|| = ||(x + \lambda e)^*||\cdot||x + \lambda e||$, and so $A_e$ is a $C^*$-algebra. □

§15. *The unitary seminorm.*

A remarkable seminorm is constructed from the group of unitary elements in a $C^*$-algebra which turns out to be the key, together with (11.1), to establishing that the involution is isometric in a $C^*$-algebra.

(15.1) PROPOSITION. *Let* A *be a* $C^*$*-algebra with identity* e. *Let* U(A) *denote the group of unitary elements in* A. *Then every element in* A *is a linear combination of unitary elements and* $||x|| = ||x||_u$, *where*

$$||x||_u = \inf\{\sum_{n=1}^N |\lambda_n| : x = \sum_{n=1}^N \lambda_n u_n, \lambda_n \in C, u_n \in U(A)\}.$$

*Proof.* To prove that every x in A is a linear combination of unitary elements it clearly suffices to show that every hermitian h in A with $||h|| < 1$ can be written as a linear combination of unitary elements. If $||h|| < 1$, then $||h^2|| \leq ||h||^2 < 1$ and so

$$k = \sum_{n=0}^{\infty} \binom{1/2}{n} (-h^2)^n$$

is a well-defined element in A. Clearly, k is an hermitian element commuting with h such that $k^2 = e - h^2$. Thus $u = h + ik$ is unitary and $h = \frac{1}{2}u + \frac{1}{2}u^*$.

It now follows that $||x||_u$ is well-defined for each x in A; further, it is clear from the definition that $||\cdot||_u$ is a seminorm on A. We shall call it the *unitary seminorm*. Since the unitary elements U(A) form a group under multiplication, $||\cdot||_u$ is submultiplicative.

The proposition will be proved if we can show that $||x|| = ||x||_u$ for each x in A. To this end let us compare the unitary seminorm with the $C^*$-norm on A. Observe that $||h||_u \leq ||h||$ for every hermitian h in A. Indeed if $||h|| < 1$, then $h = \frac{1}{2}u + \frac{1}{2}u^*$ for some unitary u, so $||h||_u \leq 1$. Thus $||h||_u \leq ||h||$ for every hermitian h in A. Further there is a positive constant K for which $||x||_u \leq K||x||$ for every x in A. For if $x = h + ik$ with hermitian h and k, then $h = \frac{1}{2}(x + x^*)$ and $k = \frac{1}{2i}(x - x^*)$ and we have

$$||x||_u \leq ||h||_u + ||k||_u \leq ||h|| + ||k|| \leq ||x|| + ||x^*|| \leq ||x|| + M||x||$$

for a positive constant M because the involution is continuous (10.1). Letting $K = 1 + M$, we obtain $||x||_u \leq K||x||$. On the other hand $||x|| \leq ||x||_u$ for all x in A. First observe that $||u|| = 1$ for every $u \in U(A)$. Since $uu^* = u^*u$ and $||e|| = |e|_\sigma = 1$, we have

$$||u^*||\cdot||u|| = ||u^*u|| = |u^*u|_\sigma \leq |u^*|_\sigma |u|_\sigma = |u|_\sigma^{\,2} \leq ||u||^2,$$

so that $||u^*|| \leq ||u||$. Replacing u by $u^*$, we also have $||u|| \leq ||u^*||$. Then $||u||^2 = ||u^*||\cdot||u|| = ||u^*u|| = 1$ so that $||u|| = 1$ as desired. Now if $x = \sum_{n=1}^{N} \lambda_n u_n$, $\lambda_n \in C$, $u_n \in U(A)$, then

$$||x|| = ||\sum_{n=1}^{N} \lambda_n u_n|| \leq \sum_{n=1}^{N} |\lambda_n|\cdot||u_n|| = \sum_{n=1}^{N} |\lambda_n|$$

so $||x|| \leq ||x||_u$. Hence the unitary seminorm and the $C^*$-norm on A are equivalent norms with

$$||x|| \leq ||x||_u \leq K||x||$$

for all $x$ in $A$. To see that these two norms are actually equal, we need the following result of Russo and Dye [1] about the closure of the convex hull of the group $U$ of unitary elements in $A$.

(15.2) THEOREM. *(Russo-Dye). Let* $A$ *be a $C^*$-algebra with identity* $e$. *Then the closed unit ball of* $A$ *is the closed convex hull of the group* $U$ *of unitary elements in* $A$.

*Proof.* Let $A_1$ denote the open unit ball of $A$ and $\overline{\text{conv}}(U)$ the closed convex hull of $U$. It suffices to show that $A_1 \subset \overline{\text{conv}}(U)$, i.e., if $x \in A_1$, then $x \in \overline{\text{conv}}(U)$. To show this it is enough to show that for every $u \in U$, the element $y = \frac{x + u}{2}$ is in $\overline{\text{conv}}(U)$. For then $U \subset 2\overline{\text{conv}}(U) - x$, which is closed and convex, so $\overline{\text{conv}}(U) \subset 2\overline{\text{conv}}(U) - x$ or $\frac{x + \overline{\text{conv}}(U)}{2} \subset \overline{\text{conv}}(U)$; if $x_o = u$ and a sequence $\{x_n\}$ is defined iteratively by $x_{n+1} = \frac{x + x_n}{2}$ for $n = 1,2,\ldots$, then $x_n \in \overline{\text{conv}}(U)$ and $x_n \to x$.

However the element $y$ can be rewritten as $y = \frac{xu^{-1} + e}{2}\, u$, and since $||xu^{-1}|| \leq ||x|| < 1$, then $||y|| < 1$ and $||yu^{-1} - e|| < 1$, so $y$ is invertible. Hence $y = v|y|$, with $v$ unitary and $|y| = (y^*y)^{1/2} = \frac{w + w^*}{2}$ where $w = |y| + i(e - |y|^2)^{1/2}$ is also unitary. □

This clever proof of the Russo-Dye Theorem substantially simplifies and shortens the original one, and is due to L. T. Gardner [1]. C. K. Fong has pointed out the following: if $x$ and $u$ are as in the proof of (15.2) and $\varepsilon > 0$ is small, then the element $x' = tx + (1 - t)u$ lies in the open unit ball $A_1$, where $1 < t < 1 + \varepsilon$. Applying the given argument inductively to $x'$ instead of $x$ it can be shown that the elements $x'_n$ lie in $\text{conv}(U)$. By the inductive definition of $x'_n$ the element $x$ will lie between $x'_n$ and $u$ for sufficiently large $n$, showing that $x$ is also in $\text{conv}(U)$. It follows that $A_1$ is contained in the convex hull of $U$. Therefore, we have:

(15.3) THEOREM. *If* $A$ *is a $C^*$-algebra with identity, then* $A_1 \subseteq \text{conv}(U)$.

REMARK. The Russo-Dye Theorem will be generalized later (see §22) to Banach *-algebras with identity and isometric involution, and a weaker version of the theorem will be given for arbitrary Banach *-algebras with identity. Theorem (15.3) was proved in 1970 in a slightly different form involving the exponential function by T. W. Palmer [2]. Further results of this type will be given in §35 in the context of symmetric Banach *-algebras.

The equality of the unitary seminorm and the C*-norm on A is an immediate consequence of the Russo-Dye Theorem. Recall that $||x|| \le ||x||_u \le K||x||$. Let $x \in A$ with $||x|| < 1$. Then, for every $\varepsilon > 0$, there is a positive integer m and unitary elements $u_k$ such that $||x - \sum_{k=1}^{m} \lambda_k u_k|| < \varepsilon$ and so

$$||x||_u \le ||\sum_{k=1}^{m} \lambda_k u_k||_u + ||x - \sum_{k=1}^{m} \lambda_k u_k||_u$$

$$\le \sum_{k=1}^{m} \lambda_k ||u_k||_u + K||x - \sum_{k=1}^{m} \lambda_k u_k||$$

$$\le 1 + K\varepsilon \qquad (\sum_{k=1}^{m} \lambda_k = 1 \text{ and } \lambda_k \ge 0);$$

since $\varepsilon$ was arbitrary, $||x||_u \le 1$. This proves that $||x||_u \le ||x||$ and so $||x|| = ||x||_u$ for all x in A. □

§16. *The involution in a C*-algebra is isometric.*

We are now prepared to reap some of the fruit from our work. Indeed, from (15.1) and the fact that the involution is isometric relative to the unitary seminorm we obtain the following theorem.

(16.1) THEOREM. *The involution in a C*-algebra is isometric.*

Therefore, no distinction exists between the two conditions $||x^*x|| = ||x^*||\cdot||x||$ for all x and $||x^*x|| = ||x||^2$ for all x.

§17. *The Gelfand-Naimark-Segal construction.*

We have seen that the involution in a $C^*$-algebra A is isometric. Further, if A has no identity we can embed A isometrically as a closed ideal of codimension one in the $C^*$-algebra $A_e$ with identity e. Thus we can (and will) assume that $||x^*x|| = ||x||^2$ and that A has an identity e.

The representation of such an algebra A as a norm-closed *-subalgebra of bounded linear operators on a Hilbert space is effected by means of positive functionals on A and a construction due to Gelfand-Naimark [1] and Segal [2]. Our purpose here is to present only what is needed to prove the representation theorem. We shall return to a study of positive functionals and the GNS-construction later (see (§27)) for a more detailed analysis.

A *positive functional* on A is a linear functional p such that $p(x^*x) \geq 0$ for all x in A. For x, y in A set

$$(x|y) = p(y^*x).$$

This inner product on A is linear in x, conjugate-linear in y and $(x|x) \geq 0$ for all x. Thus in particular

$$p(y^*x) = \overline{p(x^*y)}$$

and

$$|p(y^*x)|^2 \leq p(x^*x)p(y^*y) \qquad \text{(Cauchy-Schwarz)}$$

for all x, y in A (for explicit proofs of these see (21.5)). Setting $y = e$, we get

$$p(x^*) = \overline{p(x)}$$

and

$$|p(x)|^2 \leq p(e)p(x^*x).$$

In general this inner product on A is degenerate, i.e., $(x|x) = 0$ for some $x \neq 0$, so a reduction is necessary to obtain nondegeneracy. To this end we define the associated null ideal

$$I = \{x \in A: p(x^*x) = 0\}.$$

Since by the above properties of positive functionals

$$I = \{x \in A\colon p(y^*x) = 0 \quad \text{for all} \quad y \in A\},$$

the null ideal is clearly a left ideal in A. Then the quotient space $X = A/I$ is a pre-Hilbert space with respect to the induced inner product

$$(x + I|y + I) = p(y^*x)$$

and, further, for each $a \in A$ we can define a linear operator $\pi(a)$ on X by $\pi(a)(x + I) = ax + I$. The fact that these are well-defined is easily checked. The map $a \to \pi(a)$ has the following properties which follow directly from the definition:

$$\pi(a + b) = \pi(a) + \pi(b),\ \pi(\lambda a) = \lambda\pi(a),\ \pi(ab) = \pi(a)\pi(b),$$

and $\pi(e)$ is the identity operator; also

$$(\pi(a)(x + I)|y + I) = (x + I|\pi(a^*)(y + I))$$

so that $a \to \pi(a)$ is a $^*$-representation of A on the pre-Hilbert space X.

Let H be the Hilbert space completion of X. We want to show that every operator $\pi(a)$ on X can be extended to a bounded operator on H. We claim that $||\pi(a)|| \leq ||a||$. Note that

$$||\pi(a)(x + I)||^2 = (ax + I|ax + I) = p(x^*a^*ax).$$

For any $\alpha > ||a^*a|| = ||a||^2$ there exists an hermitian h in A such that $h^2 = \alpha e - a^*a$ (h is the square root of the positive element $\alpha e - a^*a$, which exists by functional calculus). Hence

$$\alpha p(x^*x) - p(x^*a^*ax) = p(x^*(\alpha e - a^*a)x) = p((hx)^*(hx)) \geq 0$$

and so $p(x^*a^*ax) \leq ||a||^2p(x^*x)$. Thus $||\pi(a)|| \leq ||a||$. Denote the extended operator on H also by $\pi(a)$.

The preceding discussion has shown that for every positive functional on A there is associated a $^*$-representation of A as a $^*$-subalgebra of bounded linear operators on a Hilbert space H such that $||\pi(a)|| \leq ||a||$. In general this representation is neither injective nor norm-preserving. By constructing appropriate positive functionals and showing every $C^*$-algebra has enough positive functionals to separate the points of the algebra, we will be able to build a representation with these properties.

## §18. *Construction of positive functionals.*

Let $A$ be a $C^*$-algebra with identity $e$. We will construct for every fixed $z$ in $A$ a positive functional $p$ on $A$ such that $p(e) = 1$ and $p(z^*z) = ||z||^2$. The associated GNS $*$-representation $\pi$ has the property $||\pi(z)|| = ||z||$. Indeed,

$$||z||^2 = p(z^*z) = (\pi(z)(e + I)|\pi(z)(e + I)) = ||\pi(z)(e + I)||^2$$

$$\leq ||\pi(z)||^2||e + I||^2 = ||\pi(z)||^2 p(e) = ||\pi(z)||^2,$$

which together with $||\pi(z)|| \leq ||z||$ gives $||\pi(z)|| = ||z||$.

The construction given in the proof of the following theorem is a special case of a construction for an extension theorem for positive functionals due to M. Krein [1].

(18.1) THEOREM. *Let* $A$ *be a* $C^*$*-algebra with identity* $e$. *For each fixed* $z$ *in* $A$ *there exists a positive functional* $p$ *on* $A$ *such that* $p(e) = 1$ *and* $p(z^*z) = ||z||^2$.

*Proof*. Let $H(A)$ denote the real vector space of hermitian elements in $A$ and $A^+$ the positive cone of all positive elements in $A$. On the subspace $\mathbb{R}e + \mathbb{R}z^*z$ of $H(A)$ generated by $e$ and $z^*z$ define $p$ by

$$p(\alpha e + \beta z^*z) = \alpha + \beta||z^*z||.$$

Note that $p$ is well-defined on $\mathbb{R}e + \mathbb{R}z^*z$ even if $e$ and $z^*z$ are linearly dependent. Since

$$||z^*z|| = |z^*z|_\sigma \in \sigma_A(z^*z),$$

we have that $\alpha + \beta||z^*z||$ lies in $\sigma_A(\alpha e + \beta z^*z)$. In other words, $p(x) \in \sigma_A(x)$ if $x \in \mathbb{R}e + \mathbb{R}z^*z$, so that $p(x) \geq 0$ for all $x \in A^+ \cap (\mathbb{R}e + \mathbb{R}z^*z)$.

Assume $p$ has been extended to a real-linear functional on a subspace $W$ of $H(A)$ such that $p(x) \geq 0$ for all $x \in A^+ \cap W$ and assume that there is a $y \in H(A)$ with $y \notin W$. Set

$$a = \inf\{p(v): y \leq v \in W\} \quad \text{and} \quad b = \sup\{p(u): y \geq u \in W\}.$$

Since $y \leq ||y||\cdot e$ and $y \geq -||y||\cdot e$, the infimum and supremum are taken over nonempty sets, and are finite numbers, clearly satisfying $a \geq b$. Define $p$ on the subspace of $H(A)$ generated by $W$ and $y$ by

$$p(x + \alpha y) = p(x) + \alpha c \qquad (x \in W, \alpha \in R),$$

where $c$ is any fixed number such that $a \geq c \geq b$.

Suppose that $x + \alpha y \geq 0$ $(x \in W, \alpha \in R)$. We shall show that $p(x + \alpha y) \geq 0$. If $\alpha = 0$, then $p(x + \alpha y) = p(x) \geq 0$ by assumption.

If $\alpha > 0$, then $x + \alpha y \geq 0$ implies $y \geq -\frac{x}{\alpha} \in W$, so that $p(-\frac{x}{\alpha}) \leq c$, or $p(x + \alpha y) \geq 0$.

If $\alpha < 0$, then $x + \alpha y \geq 0$ implies $y \leq -\frac{x}{\alpha} \in W$, so that $p(-\frac{x}{\alpha}) \geq c$, or $p(x + \alpha y) \geq 0$.

By Zorn's Lemma we conclude that $p$ can be extended to a real-linear functional $p$ on $H(A)$ such that $p(x) \geq 0$ for all $x \in P$.

Finally set $p(x) = p(h) + ip(k)$ if $x = h + ik$ with $h, k \in H(A)$. Then $p$ is a positive functional on $A$ such that $p(e) = 1$ and $p(z^*z) = ||z^*z|| = ||z||^2$. This completes the construction. □

## §19. *The isometric *-representation.*

Let $A$ be a $C^*$-algebra. In the preceding two sections we constructed for every $z$ in $A$ a positive functional on $A$ such that the associated *-representation $\pi^z$ of $A$ on a Hilbert space $H^z$ is norm-decreasing and $||\pi^z(z)|| = ||z||$ (the superscripts simply denoting dependence of $\pi$ and $H$ on $z$). All of the tools have now been prepared in order to prove the Gelfand-Naimark theorem for noncommutative algebras.

(19.1) THEOREM. *(Gelfand-Naimark). Let* $A$ *be a* $C^*$*-algebra. Then* $A$ *is isometrically *-isomorphic to a norm-closed *-subalgebra of* $B(H)$ *for some Hilbert space* $H$.

*Proof.* Let $H$ be the direct sum of the Hilbert spaces $H^z$. The direct sum of the family $H^z$, $z \in A$, is defined as the set of all mappings $f: A \to \bigcup_{z \in A} H^z$ with $f(z) \in H^z$, $z \in A$, such that

$$\sum_{z \in A} (f(z)|f(z)) < \infty.$$

The algebraic operations in $H$ are pointwise and the inner product is given by

$$(f|g) = \sum_{z \in A} (f(z)|g(z)).$$

One verifies easily that all Hilbert space axioms are satisfied by $H$.

Define the *-representation $\pi$ of $A$ on $H$ by

$$(\pi(a)f)(z) = \pi^z(a)(f(z)).$$

Note that the inequality

$$\sum_{z \in A} ((\pi(a)f)(z)|(\pi(a)f)(z)) \leq ||a||^2 \sum_{z \in A} (f(z)|f(z))$$

shows that with $f$ also $\pi(a)f$ belongs to $H$. Then $\pi(a)$ is a bounded linear operator on $H$ such that

$$||\pi(a)|| = \sup_{z \in A} ||\pi^z(a)|| = ||\pi^a(a)|| = ||a||.$$

Hence the map $a \to \pi(a)$ is a norm-preserving *-representation of $A$ on $H$. This completes the proof. □

## EXERCISES

(III.1) Let $A$ be a unital $C^*$-algebra. Prove:

(a) If $a \in A$ is hermitian, then $\sigma_A(a^2) \subseteq [0, ||a||^2]$

(b) If $a \in A^+$, then $a||a|| \geq a^2$.

(III.2) Let $A$ be a $C^*$-algebra with identity $e$. If $0 \leq b \leq a$ and $\lambda > 0$, prove that $(a + \lambda e)^{-1} \leq (b + \lambda e)^{-1}$.

(III.3) Let $A$ be a $C^*$-algebra. If $a, b, x \in A$ and $0 \leq a \leq b$, prove that $||a^{1/2}x|| \leq ||b^{1/2}x||$.

(III.4) Let $A$ be a $C^*$-algebra. Prove that each element of $A$ is a linear combination of at most four elements of $A^+$.

(III.5) Let $A$ be a $C^*$-algebra and $a \in A$ hermitian. If $f \in C(\sigma_A(a))$, prove that $f(a)$ is a positive element in $A$ iff $f(\lambda) \geq 0$ for all $\lambda \in \sigma_A(a)$.

(III.6) Let $A$ be a $C^*$-algebra and $I$ a closed two-sided ideal of $A$. Show that any positive element of $A$ majorized by a positive element of $I$ belongs to $I$.

(III.7) Give an example of a $C^*$-algebra $A$ and two elements $x$ and $y$ in $A$ such that $x \geq y$ but $x^2 \not\geq y^2$.

(III.8) Prove that any separable $C^*$-algebra $A$ is isometrically $*$-isomorphic to a norm-closed $*$-subalgebra of $B(H)$ for some separable Hilbert space $H$.

(III.9) Let $X$ be a locally compact, noncompact, Hausdorff space. Construct an approximate identity for the $C^*$-algebra $C_o(X)$.

(III.10) Let $A$ be a commutative $C^*$-algebra and $B$ a $C^*$-subalgebra of $A$ which contains a positive increasing approximate identity for $A$ bounded by 1. Then, given $a_o \geq 0$ in $A$, prove there is an element $b \geq 0$ in $B$ with $b \geq a_o$ and $||b|| = ||a_o||$.

(III.11) Suppose that $x$ is an invertible element in a unital $C^*$-algebra $A$.

(a) Show that $x = up$ for some unitary element $u$ in $A$ and some positive element $p$ in $A$.

(b) Show that the elements $u$ and $p$ in $A$ occuring in the "polar decomposition" of $x$ in part (a) are unique.

(c) Show that $x$ is normal iff $up = pu$.

(d) Show that the mappings $x \to u$ and $x \to p$ in the above decomposition are continuous.

(III.12) Let $A$ be a $C^*$-algebra and let $x, y \in A^+$. Prove that $||x + y|| \geq ||x||$.

(III.13) Give a second proof of (18.1) based on Gelfand theory and the Hahn-Banach theorem.

(III.14) Let $A$ be a $C^*$-algebra and consider the GNS-construction. Verify the following details:

(a) $I = \{x \in A\colon p(x^*x) = 0\}$ is a left ideal of $A$.

(b) the inner product $(x + I|y + I) = p(y^*x)$ on $A/I$ is well defined.

(c) the linear operator $\pi(a)$ on $A/I$ given by $\pi(a)(x + I) = ax + I$ is well defined.

(d) the mapping $a \to \pi(a)$ is a $^*$-representation of $A$ on $A/I$.

(III.15) A cone $K$ in the positive part of a $C^*$-algebra $A$ is *hereditary* if $0 \leq x \leq y$ with $y \in K$ implies $x \in K$ for each $x \in A$. Show that if $K$ is an hereditary cone in $A^+$, then the set $J = \{x \in A: x^*x \in K\}$ is a left ideal of $A$.

(III.16) Let $A$ be a unital $C^*$-algebra. If $x$ and $y$ are invertible elements in $A$ and $x^*x = y^*y$, prove that $xy^{-1}$ is unitary.

# 4

# Banach *-Algebras: Generalities

## §20. *Introduction.*

Can we weaken the axioms of $C^*$-algebras further? If so, in what way? In order to answer these and similar questions, and also to provide the necessary tools for a later study of symmetric and hermitian $^*$-algebras, we systematically study general Banach $^*$-algebras in this chapter. We begin with a look at properties of general $^*$-algebras and proceed to normed and Banach $^*$-algebras. We shall not assume that the involution is continuous. In this generality the theory must be treated with some delicacy. Our results will also be developed so as to apply to algebras with and without identity elements.

## §21. $^*$-*algebras.*

We present basic properties of $^*$-algebras in this section. In order to give an organized account we shall repeat several definitions.

Recall that a $^*$-*algebra* is an algebra $A$ with a mapping $x \to x^*$ of $A$ into itself such that for all $x, y \in A$ and complex $\lambda$:

(a) $(x + y)^* = x^* + y^*$;

(b) $(\lambda x)^* = \overline{\lambda} x^*$;

(c) $(xy)^* = y^* x^*$;

(d) $x^{**} = x$.

The mapping $x \to x^*$ is called an *involution* and $x^*$ is called the *adjoint* of $x$.

It follows from (d) that $x \to x^*$ is a bijective mapping of $A$ onto itself.

*Examples of *-algebras*

(1) The field of complex numbers C with involution $\lambda \to \overline{\lambda}$ is a unital commutative *-algebra.

(2) Let X be a topological space. The algebra C(X) of bounded continuous complex functions with involution $f \to f^*$ defined by $f^*(t) = \overline{f(t)}$ for all $t \in X$ is a unital commutative *-algebra. We shall write $f^* = \overline{f}$.

(3) Let H be a Hilbert space. The algebra B(H) of bounded linear operators on H with involution $T \to T^*$, where * denotes the adjoint operation $(T\xi|\eta) = (\xi|T^*\eta)$ for all $\xi, \eta \in H$, is a noncommutative unital *-algebra.

(4) Let G be a locally compact group. The group algebra $L^1(G)$ of G with involution $f \to f^*$ defined by $f^*(t) = \Delta(t^{-1})\overline{f(t^{-1})}$ for all $t \in G$, where $\Delta$ is the modular function of G, is, in general, a noncommutative *-algebra without identity. If G is abelian or compact, then $\Delta(t) \equiv 1$ and the involution reduces to $f^*(t) = \overline{f(t^{-1})}$.

(5) The disk algebra A(D) with involution $f \to f^*$ defined by $f^*(\lambda) = \overline{f(\overline{\lambda})}$ for all $\lambda \in D = \{\lambda \in C: |\lambda| \leq 1\}$ is a unital commutative *-algebra.

(6) For particular topological spaces X, variations of the involution in Example 2 can be given. For example, if $X = [0,1]$ with the usual relative topology, define an involution in C(X) by setting $f^*(t) = \overline{f(1-t)}$. As a second illustration, let $X = [0,1] \cup \{2,3\}$ with the usual relative topology of the reals and, for $f \in C(X)$, define $f^*(t) = \overline{f(t)}$ if $t \in [0,1]$, $f^*(2) = \overline{f(3)}$, and $f^*(3) = \overline{f(2)}$.

(7) Let A be any *-algebra and consider the set $B = A \times A$ with the natural coordinate algebraic operations induced by A. Then B is a *-algebra if we define $(x,y)^* = (y^*,x^*)$.

(8) If A is a *-algebra, then the set $A_n$ of all $n \times n$ matrices with entries from A can be made into an algebra by defining the operations exactly as for matrices of scalars. If we define $(x_{ij})^* = (x^*_{ji})$, then $A_n$ is a *-algebra.

(9) Let $G$ be a locally compact group. The measure algebra $M(G)$ becomes a *-algebra if we define an involution $\mu \to \mu^*$ by $\mu^*(E) = \overline{\mu(E^{-1})}$ for each Borel set $E$ in $G$.

We remark that there exist algebras which admit no involutions at all, and others which admit uncountably many distinct involutions (see Civin and Yood [1]).

Let $A$ be a *-algebra. If $S$ is a nonempty subset of $A$, let $S^* = \{x^*: x \in S\}$. The set $S$ is *self-adjoint* if $S = S^*$, i.e., if $x \in S$, then $x^* \in S$. A self-adjoint subalgebra of $A$ is called a **-subalgebra* of $A$; a self-adjoint left [right, two-sided] ideal of $A$ is called a **-ideal* of $A$. An element $x$ in $A$ is called *hermitian* (or *self-adjoint*) if $x = x^*$; $x$ is *normal* if $xx^* = x^*x$. An element $x$ in $A$ is a *projection* if $x = x^*$ and $x^2 = x$. Finally, if $A$ has an identity $e$, an element $x$ in $A$ is *unitary* if $x^*x = e = xx^*$.

The sets of all hermitian, normal and unitary elements of $A$ will be denoted respectively by $H(A)$, $N(A)$ and $U(A)$. Clearly $U(A) \subseteq N(A)$ and $H(A) \subseteq N(A)$.

For each $x \in A$, $x + x^*$, $x^*x$, and $xx^*$ are hermitian; further, the zero element $0$ is hermitian. If $x, y \in H(A)$ then $xy \in H(A)$ iff $xy = yx$. The set $H(A)$ of hermitian elements in $A$ is a *real*-linear subspace of $A$.

Every left [right] *-ideal of $A$ is necessarily a two-sided ideal. Indeed, the involution sends each left ideal into a right ideal and each right ideal into a left ideal. Since the Jacobson radical of $A$ is the intersection of all maximal modular left (as well as right) ideals of $A$, it is a *-ideal of $A$.

(21.1) PROPOSITION. *Let* $A$ *be a *-algebra with identity* $e$. *Then:*

*(a)* $e = e^*$;

*(b)* $(x^*)^{-1} = (x^{-1})^*$ *for each invertible* $x \in A$;

*(c)* $\sigma_A(x^*) = \overline{\sigma_A(x)}$ *for each* $x \in A$;

*(d) The set* $U(A)$ *of unitary elements is a subgroup of the group* $A^{-1}$ *of invertible elements of* $A$.

*Proof.* (a) We have $e = (e^*)^* = (ee^*)^* = e^{**}e^* = ee^* = e^*$.

(b) Applying $*$ to $x^{-1}x = e = xx^{-1}$ yields $x^*(x^{-1})^* = e = (x^{-1})^*x^*$; that is, $(x^{-1})^* = (x^*)^{-1}$.

(c) This follows from (b) and $(x - \lambda e)^* = x^* - \bar{\lambda}e$.

(d) Let $x, y \in U(A)$. Then $(xy)^* = y^*x^* = y^{-1}x^{-1} = (xy)^{-1}$ and $(x^*)^{-1} = (x^{-1})^{-1} = x$. Hence, $xy \in U(A)$ and $x^{-1} \in U(A)$. □

If $A$ is a *-algebra, the algebra $A_e$ is a *-algebra if we define $(x + \lambda e)^* = x^* + \bar{\lambda}e$ for each $x + \lambda e \in A_e$.

By (21.1), (c) we have $\sigma_{A_e}(x^*) = \overline{\sigma_{A_e}(x)}$ for each $x \in A$. Also, $(x')^* = (x^*)'$ for each quasi-regular $x$ in $A$.

Every element $x$ in a *-algebra can be written in the form $x = a + ib$, where $a, b \in H(A)$. Indeed, set $a = (x + x^*)/2$, $b = (x - x^*)/2i$. The element $x$ is normal iff $ab = ba$.

Let $A$ be a *-algebra. The intersection of a family of *-subalgebras of $A$ is again a *-subalgebra of $A$. In particular, if $S$ is a nonempty subset of $A$, then the intersection of all *-subalgebras of $A$ containing $S$ is a *-subalgebra $A_S$ containing $S$, and is clearly contained in every such *-subalgebra. The algebra $A_S$ is called the **-subalgebra generated* by $S$, and consists of all linear combinations of elements of the form $x_1x_2\ldots x_n$, where $x_1x_2,\ldots,x_n \in S \cup S^*$. Hence, if all of the elements in $S \cup S^*$ are pairwise commutative then the *-subalgebra $A_S$ is commutative. In particular, if $S$ consists of one point $x$ then $A_S$ is commutative iff $x$ is normal.

A commutative subalgebra of an algebra $A$ is said to be *maximal commutative* if it is not contained in any other commutative subalgebra of $A$.

A straightforward Zorn's lemma argument shows that every commutative *-subalgebra of a *-algebra $A$ is contained in some maximal commutative *-subalgebra of $A$.

(21.2) PROPOSITION. *Every maximal commutative *-subalgebra of a *-algebra* $A$ *is a maximal commutative subalgebra of* $A$.

*Proof*. Let $B$ be a maximal commutative *-subalgebra of $A$, and suppose that $x$ is any element of $A$ which commutes with every element of $B$. It must be shown that $x \in B$. Since $B$ is a *-subalgebra, $x^*y = (y^*x)^* = (xy^*)^* = yx^*$ for every $y \in B$. Hence, both $x$ and $x^*$

commute with every element of B, and therefore so does the hermitian element $a = (x + x^*)/2$. Since $a$ is hermitian and commutes with B, the subalgebra C of A generated by B and $a$ is commutative and self-adjoint. Since B is a maximal commutative *-subalgebra, then $a \in B$. Similarly, $b = (x - x^*)/2i \in B$; we then have $x = a + ib \in B$. □

(21.3) COROLLARY. *If* B *is a maximal commutative *-subalgebra of a *-algebra* A, *then* $\sigma_A(x) = \sigma_B(x)$ *for all* $x \in B$.

*Proof*. Apply (B.4.3) and (21.2). □

Let A and B be *-algebras. A **-homomorphism* of A into B is a homomorphism $\phi: A \to B$ such that $\phi(x^*) = \phi(x)^*$ for all $x \in A$. A bijective *-homomorphism is called a **-isomorphism*.

If $\phi: A \to B$ is a *-homomorphism, then the image $\phi(A)$ is a *-subalgebra of B, and the kernel of $\phi$ is a *-ideal of A.

If I is a *-ideal of A, an involution can be introduced into the quotient algebra A/I by defining $(x + I)^* = x^* + I$ for each $x + I \in A/I$. The resulting *-algebra A/I is called the *quotient *-algebra* of A modulo I. The canonical homomorphism $\tau: A \to A/I$ defined by $\tau(x) = x + I$ is clearly a *-homomorphism of A onto A/I Hence, the *-homomorphisms of A onto *-algebras are in one-to-one correspondence with *-ideals of A.

Let A be a *-algebra and f a linear functional on A. The *adjoint* of f is defined by $f^*(x) = \overline{f(x^*)}$ for $x \in A$. If $f^* = f$, i.e., $f(x^*) = \overline{f(x)}$ for all $x \in A$, then f is said to be *hermitian*.

(21.4) PROPOSITION. *Let* A *be a *-algebra and* f *a linear functional on* A. *Then* f *can be written uniquely in the form* $f = f_1 + if_2$, *where* $f_1$ and $f_2$ *are hermitian functionals on* A.

*Proof*. Set $f_1(x) = [f(x) + \overline{f(x^*)}]/2$ and $f_2(x) = [f(x) - \overline{f(x^*)}]/2i$. Then $f_1^* = f_1$ and $f_2^* = f_2$, and $f = f_1 + if_2$. To see that $f_1$ and $f_2$ are unique, suppose that $f = g_1 + ig_2$, where $g_1$ and $g_2$ are hermitian functionals. Then $f(x^*) = g_1(x^*) + ig_2(x^*) = \overline{g_1(x)} + i\overline{g_2(x)}$ and so $\overline{f(x^*)} = g_1(x) - ig_2(x)$. Hence, $f(x) + \overline{f(x^*)} = 2g_1(x)$ and $f(x) - \overline{f(x^*)} = 2ig_2(x)$; it follows that $f_1 = g_1$ and $f_2 = g_2$. □

It is easy to verify that a linear functional $f$ on a *-algebra $A$ is hermitian iff $f(x)$ is real for each hermitian $x$ in $A$.

A linear functional $p$ on a *-algebra $A$ is said to be *positive* if $p(x^*x) \geq 0$ for all $x \in A$.

The identity map on the *-algebra of complex numbers is a positive functional. An example of a positive functional on the *-algebra $C([0,1])$ is obtained by setting $p(f) = \int_0^1 f(t)dt$. If $H$ is a Hilbert space and $\xi$ is a fixed vector in $H$, then the mapping $p$ defined on the *-algebra $B(H)$ by $p(T) = (T\xi|\xi)$ is a positive functional.

(21.5) PROPOSITION. *(Cauchy-Schwarz inequality). Let* $p$ *be a positive functional on a *-algebra* $A$. *Then for all* $x, y \in A$ *we have:*

*(a)* $p(y^*x) = \overline{p(x^*y)}$;

*(b)* $|p(y^*x)|^2 \leq p(x^*x)p(y^*y)$.

*Proof*. (a) Let $x, y \in A$. Then for each complex $\lambda$ we have

$$0 \leq p((\lambda x + y)^*(\lambda x + y)) = |\lambda|^2 p(x^*x) + [\lambda p(y^*x) + \overline{\lambda} p(x^*y)] + p(y^*y). \quad (1)$$

Since $|\lambda|^2 p(x^*x) + p(y^*y)$ is real, then $\lambda p(y^*x) + \overline{\lambda} p(x^*y)$ is real for all complex $\lambda$. Setting $\lambda = 1$ in (1) it follows that $p(y^*x) + p(x^*y)$ is real, and hence that $\text{Im } p(y^*x) + \text{Im } p(x^*y) = 0$; that is, $\text{Im } p(y^*x) = -\text{Im } p(x^*y)$. Setting $\lambda = i$ in (1), we also see that $\text{Re } p(y^*x) = \text{Re } p(x^*y)$. Hence, $p(y^*x) = \overline{p(x^*y)}$.

(b) If $p(y^*x) = 0$, (b) is obvious. Suppose that $p(y^*x) \neq 0$. Let $\alpha$ be any real number, and set $\lambda = \alpha\overline{p(y^*x)}/|p(y^*x)|$ in (1) to obtain, using part (a), the following quadratic inequality in $\alpha$:

$$0 \leq \alpha^2 p(x^*x) + 2\alpha|p(y^*x)| + p(y^*y).$$

The discriminant must then satisfy $4|p(y^*x)|^2 - 4p(x^*x)p(y^*y) \leq 0$; that is, $|p(y^*x)|^2 \leq p(x^*x)p(y^*y)$. □

(21.6) COROLLARY. *Let* $A$ *be a *-algebra with identity* $e$ *and* $p$ *a positive functional on* $A$. *Then:*

*(a)* $p$ *is hermitian;*

*(b)* $|p(x)|^2 \leq p(e)p(x^*x)$ *for all* $x \in A$.

*Proof*. Set $y = e$ in (21.5). □

In general, positive functionals need not be hermitian. Indeed, if A is the *-algebra $\mathbb{C}$ with trivial multiplication $xy = 0$ for all $x, y \in A$, then the functional p on A defined by $p(x) = ix$ is positive but not hermitian.

If p is a positive functional on a *-algebra A, then there does not always exist a positive extension of p to the *-algebra $A_e$. The next proposition gives necessary and sufficient conditions for such an extension to exist.

(21.7) PROPOSITION. *Let* A *be a *-algebra and* $A_e$ *the unitization of* A. *Let* p *be a positive functional on* A. *Then* p *can be extended to a positive functional on* $A_e$ *iff:*

*(a)* p *is hermitian; and*

*(b) there is a finite* $k \geq 0$ *such that* $|p(x)|^2 \leq kp(x^*x)$ *for all* $x \in A$.

*Proof.* Assume that p admits an extension p' to a positive functional on $A_e$. Then p is hermitian and $|p(x)|^2 \leq p'(e)p(x^*x)$ by (21.6); hence (b) holds with $k = p'(e)$.

Conversely, assume that p satisfies (a) and (b). Then define $p'(x + \lambda e) = p(x) + k\lambda$. It is clear that p' is linear and coincides with p on A. The functional p' is also positive since

$$\begin{aligned} p'((x + \lambda e)^*(x + \lambda e)) &= p(x^*x) + \lambda p(x^*) + \overline{\lambda} p(x) + |\lambda|^2 k \\ &= p(x^*x) + 2\mathrm{Re}[\lambda p(x^*)] + |\lambda|^2 k \\ &\geq p(x^*x) - 2|\lambda| \cdot |p(x)| + |\lambda|^2 k \\ &\geq p(x^*x) - 2|\lambda| k^{1/2} p(x^*x)^{1/2} + |\lambda|^2 k \\ &= (p(x^*x)^{1/2} - |\lambda| k^{1/2})^2 \geq 0. \quad \square \end{aligned}$$

Let A be a *-algebra. A positive functional p on A is said to be *extendable* if p can be extended to a positive functional on $A_e$.

If A has an identity, (21.6) and (21.7) show that every positive functional on A is extendable.

Consider the set $A = C([0,1])$ with the usual linear operations and trivial multiplication $fg = 0$ for all $f, g \in A$. Introduce an involution into A by setting $f^* = \overline{f}$. Then A is a *-algebra without identity. Fix $t_o \in [0,1]$ and define $p(f) = f(t_o)$ for $f \in A$. Then $p$ is an hermitian positive functional on $A$; however, since $|p(f)|^2 \leq kp(f^*f)$ fails for all $k \geq 0$, $p$ is *not* extendable to $A_e$.

(21.8) PROPOSITION. *Let* $p$ *be a positive functional on a *-algebra* $A$ *and let* $y \in A$. *Then the linear functional* $g$ *on* $A$ *defined by* $g(x) = p(y^*xy)$ *is an extendable positive functional on* $A$.

*Proof.* If $x \in A$, then $g(x^*x) = p(y^*x^*xy) = p((xy)^*(xy)) \geq 0$; hence $g$ is a positive functional. To prove that $g$ is extendable, we will show that conditions (a) and (b) of (21.7) hold. If $x \in A$, then by (21.5), (a)

$$g(x^*) = p(y^*x^*y) = p((xy)^*y) = \overline{p(y^*xy)} = \overline{g(x)}$$

so that $g$ is hermitian. Applying (21.5), (b) we have

$$|g(x)|^2 = |p(y^*xy)|^2 \leq p(y^*y)p(y^*x^*xy) = p(y^*y)g(x^*x)$$

for all $x \in A$. Letting $k = p(y^*y)$, we have $|g(x)|^2 \leq kg(x^*x)$, as required. □

## §22. *Normed *-algebras.*

We turn our attention in this section to *-algebras which are also normed algebras.

A *normed *-algebra* is a normed algebra $A$ which is also a *-algebra. If $A$ also satisfies the condition $||x^*|| = ||x||$ for each $x$, it is called a **-normed algebra*. If $A$ is a complete normed *-algebra (resp. *-normed algebra) it will be called a *Banach *-algebra* (resp. **-Banach algebra*).

The first five examples of *-algebras considered in Section 21 with norms defined as in Appendix B are *-Banach algebras. Example 2 of Appendix B.2, with involution $p^*(t) = \overline{p(t)}$ is an (incomplete) *-normed algebra.

The norm condition $||x^*|| = ||x||$ in a *-normed algebra $A$ clearly implies that the involution is continuous. On the other hand, if $x \to x^*$ is a continuous involution on a normed algebra $A$, an equivalent norm can be introduced on $A$ with respect to which the involution is isometric (for example, let $||x||_o = \max\{||x||, ||x^*||\}$). In view of this there is little loss of generality in assuming the isometric condition when working with continuous involutions. Of course, involutions need not always be continuous as we see below.

If $A$ is a *-normed algebra, then a sequence $\{x_n\}$ of elements in $A$ will converge to an element $x$ iff the hermitian components of $x_n$ converge respectively to the corresponding hermitian components of $x$. In particular, the real-linear space $H(A)$ of hermitian elements will be a Banach space iff $A$ is a Banach algebra. Further, if $A$ is any Banach *-algebra it follows easily from the closed graph theorem that the involution $x \to x^*$ in $A$ is continuous iff the space $H(A)$ is a closed subset of $A$.

If $A$ is a normed *-algebra (resp. *-normed algebra) and $A_e$ is the unitization of $A$, then $A_e$ is a normed *-algebra (resp. *-normed algebra) under the norm $||x + \lambda e|| = ||x|| + |\lambda|$.

We next describe a simple example of a Banach *-algebra with discontinuous involution. Such examples appear to be few and far between.

Let $A$ be an infinite-dimensional Banach space, and let $E$ be a Hamel basis for $A$, chosen so that $||x|| = 1$ for each $x \in E$. Let $\{x_n: n = 1,2,\ldots\}$ be a sequence of distinct elements of $E$ and define $x_n^*$ by

$$x_{2n-1}^* = nx_{2n}, \quad x_{2n}^* = \frac{1}{n}x_{2n-1} \quad (n = 1,2,\ldots).$$

For all other elements of $E$, let $x^* = x$, and then extend the mapping $x \to x^*$ to all of $A$ by conjugate linearity; that is,

$$(\lambda_1 y_1 + \ldots + \lambda_k y_k)^* = \overline{\lambda}_1 y_1^* + \ldots + \overline{\lambda}_k y_k^* \quad (\lambda_i \in C,\ y_i \in E).$$

Finally, we make $A$ into a Banach algebra by introducing the trivial multiplication $ab = 0$ for $a, b \in A$. Then $A$ is an example of a Banach *-algebra with discontinuous involution. By forming its unitization $A_e$, we obtain an example of a Banach *-algebra with discontinuous involution in which the multiplication is not trivial.

In the following proposition we observe that *every* Banach algebra can be embedded in a *-Banach algebra.

(22.1) PROPOSITION. *Every Banach algebra* $B$ *can be isometrically embedded as a closed two-sided ideal of a *-Banach algebra* $A$. *Further, if* $B$ *is unital, so is* $A$.

*Proof.* Let $A = B \times B$ and introduce operations and norm on $A$ by $(x,y) + (w,z) = (x + w, y + z)$, $\lambda(x,y) = (\lambda x, \bar{\lambda} y)$, $(x,y)(w,z) = (xw, zy)$, $(x,y)^* = (y,x)$, and $||(x,y)|| = \max\{||x||, ||y||\}$. Then $A$ is a *-Banach algebra and the map $x \to (x,0)$ is an isometric embedding of $B$ in $A$. □

If $A$ is a *-normed algebra, then the completion $\tilde{A}$ (see B.1) of $A$ is a *-Banach algebra and the closure of a *-subalgebra of $A$ is again a *-subalgebra. If $S$ is a nonempty subset of $A$, the smallest closed *-subalgebra $B$ of $A$ containing $S$ is called the *closed *-subalgebra of* $A$ *generated by* $S$. If $S = \{x^*, x\}$ for some $x \in A$, then $B$ is commutative iff $x$ is normal.

Every maximal commutative *-subalgebra of a normed *-algebra is clearly closed, as can be seen by using (21.2).

If $I$ is a closed [two-sided] *-ideal of a normed *-algebra (resp., *-normed algebra) $A$, then the quotient *-algebra $A/I$ is a normed *-algebra (resp., *-normed algebra) under the usual quotient norm.

Recall that the norm of a continuous linear functional $f$ on a normed space is given by $||f|| = \sup\{|f(x)|: ||x|| \leq 1\}$. Since the closed unit ball $\{x \in A: ||x|| \leq 1\}$ in a *-normed algebra $A$ is self-adjoint, it is clear that the adjoint $f^*$ of $f$ is also continuous and $||f|| = ||f^*||$.

(22.2) PROPOSITION. *Let* $f$ *be a continuous hermitian linear functional on a *-normed algebra* $A$. *If* $g$ *denotes* $f$ *restricted to the hermitian elements of* $A$, *then* $||f|| = ||g||$.

*Proof.* We may clearly assume that $||f|| \neq 0$. Let $\varepsilon > 0$ with $\varepsilon < ||f||$, and select a nonzero $x \in A$ such that $||x|| \leq 1$ and $|f(x)| \geq ||f|| - \varepsilon > 0$. Then $f(x) \neq 0$ and, multiplying $x$ by $\lambda = f(x)/|f(x)|$, we may assume that $f(x) > 0$. Then $\frac{1}{2}(x + x^*)$ is hermitian, $||\frac{1}{2}(x + x^*)|| \leq 1$, and

$$|g(\tfrac{1}{2}(x + x^*))| = \tfrac{1}{2}|f(x) + f(x^*)| = \tfrac{1}{2}|f(x) + \overline{f(x)}|$$

$$= \tfrac{1}{2}|2\mathrm{Re}\, f(x)| = f(x) > ||f|| - \varepsilon.$$

Therefore $||g|| \geq ||f|| - \varepsilon$; since $\varepsilon$ was arbitrary, $||g|| \geq ||f||$. The inequality $||f|| \geq ||g||$ is clear. □

(22.3) PROPOSITION. *If* A *is a Banach* *-*algebra, then the spectral radius satisfies* $|x^*|_\sigma = |x|_\sigma$ *for all* $x \in A$.

*Proof.* Since $(x/\lambda)^* = x^*/\overline{\lambda}$ is quasi-regular iff $x/\lambda$ is quasi-regular, $\sigma_A(x^*) = \overline{\sigma_A(x)}$. The proposition is then immediate from the definition of the spectral radius. □

(22.4) PROPOSITION. *Let* A *be a Banach* *-*algebra. If* A *is commutative and semisimple, then the involution is continuous.*

*Proof.* First note that since A is commutative, the spectral radius is subadditive. Indeed, it suffices to show that if $x, y \in A$, then $\sigma_A(x + y) \subseteq \sigma_A(x) + \sigma_A(y)$. Let $\lambda \in \sigma_A(x + y)$. By (B.6.5) there is an MLF $\phi$ in $\hat{A}$ such that $\lambda = \phi(x + y)$. Hence $\lambda = \phi(x + y) = \phi(x) + \phi(y) \in \sigma_A(x) + \sigma_A(y)$.

Suppose, now, that $x_n \to 0$ and $x_n^* \to y$. Then

$$|y|_\sigma \leq |y - x_n^*|_\sigma + |x_n^*|_\sigma = |y - x_n^*|_\sigma + |x_n|_\sigma$$

$$\leq ||y - x_n^*|| + ||x_n|| \to 0$$

as $n \to \infty$, so that $|y|_\sigma = 0$. By (B.6.11) and the fact that A is semisimple, we have $y = 0$. The result now follows from the closed graph theorem. □

The key to the study of Banach algebras with arbitrary involutions is the following "square-root" lemma due to J. W. M. Ford [1]. This lemma generalizes a classical square-root lemma which required that the involution be continuous. The lemma is stated in terms of the circle operation and quasi-regularity (see (B.3)) as well as for invertible elements. Recall that $\nu(x) = \lim_{n\to\infty} ||x^n||^{1/n}$ and that $\nu(x) = |x|_\sigma$ for Banach algebras (B.4.12).

(22.5) LEMMA. *(Ford). Let* A *be a Banach* *-*algebra. If* $x \in H(A)$ *with* $\nu(x) < 1$, *then there exists a quasi-regular element* $y \in H(A)$ *such that* $xy = yx$ *and* $y \circ y = x$. *If* A *has identity* e *and* $x \in H(A)$ *satisfies* $\nu(e - x) < 1$, *then there exists an invertible* $u \in H(A)$

*such that* $ux = xu$ *and* $u^2 = x$. *Further, if* $\sigma_A(x)$ *is positive, then* $\sigma_A(u)$ *is positive.*

*Proof*. Let B be a (closed) maximal commutative *-subalgebra of A containing x. Consider the function f defined on the complex numbers $\lambda$ of modulus less than 1 by

$$f(\lambda) = -\Sigma_{k=1}^{\infty}\binom{1/2}{k}(-\lambda)^k. \tag{1}$$

For $|\lambda| < 1$ we have $2f(\lambda) - f(\lambda)^2 = \lambda$. Since the binomial series defining f converges absolutely for $|\lambda| < 1$ and since $\nu(x) < 1$, it follows that the sequence $\{x_n\}$ defined by $x_n = -\Sigma_{k=1}^{n}\binom{1/2}{k}(-x)^k$ is Cauchy in B; hence it converges in norm to an element $y + iz$ in B such that

$$(y + iz)\circ(y + iz) = 2(y + iz) - (y + iz)^2 = x,$$

where $y, z \in H(B)$. Computing we have

$$\begin{aligned} x &= 2(y + iz) - (y + iz)^2 \\ &= 2(y + iz) + z^2 - y^2 - 2izy \\ &= z^2 - y^2 + 2y + 2i(z - zy). \end{aligned}$$

Since x and $z - zy$ are hermitian, we obtain $z - zy = 0$ or $z = zy$.

Now, let R denote the radical of B (see (B.6.11)) and consider the commutative semisimple Banach *-algebra B/R. By (22.4) the involution in B/R is continuous and hence the set of hermitian elements in B/R is closed. Since the quotient map $\tau\colon B \to B/R$ is continuous and $x_n \to y + iz$, then $\tau(x_n - y) \to \tau(iz)$. Also, since $\tau(x_n - y)$ is hermitian for every n, its limit $\tau(iz)$ is hermitian. Therefore, $i\tau(z) = \tau(iz) = \tau(iz)^* = \tau((iz)^*) = \tau(-iz) = -i\tau(z)$ which implies that $\tau(z) = 0$ or that $z \in \ker(\tau) = R$.

We assert that y is quasi-regular. Indeed, if y were quasi-singular, then by (B.6.6), (g), there would exist $\phi \in \hat{B}$ such that $\phi(y) = 1$. Since $z \in R$, then $\phi(z) = 0$ by (B.6.11) and therefore $\phi(x) = \phi(z^2 - y^2 + 2y) = \phi(z)^2 - \phi(y)^2 + 2\phi(y) = 1$. This would imply that

$1 \in \sigma_B(x) = \sigma_A(x)$ (21.3) and hence, by (B.4.12), that $\nu(x) \geq 1$, a contradiction. Hence, $y \in Q_A$.

If $w$ denotes the quasi-inverse of $y$, then $z^2 = (w \circ y) \circ z^2 = w \circ (y \circ z^2) = w \circ (y + z^2 - yz^2) = w \circ (y + z^2 - z^2) = w \circ y = 0$ (recall that $z = yz$ so that $z^2 = yz^2$). Therefore, $x = 2y - y^2 = y \circ y$.

If $A$ has an identity $e$ and $x \in H(A)$ satisfies $\nu(e - x) < 1$, then there is a quasi-regular $y \in H(A)$ such that $xy = yx$ and $y \circ y = e - x$. Let $u = e - y$. Then $u = u^*$, $ux = xu$, and $u^2 = e - (y \circ y) = e - (e - x) = x$.

Finally, suppose $\sigma_A(x)$ is positive and $\nu(e - x) < 1$. When an identity is present, series (1) can be replaced by $f(\lambda) = \Sigma_{k=0}^{\infty}\binom{1/2}{k}(\lambda - 1)^k$ and the $x_n$'s by $x_n = \Sigma_{k=0}^{n}\binom{1/2}{k}(x - e)^k$. For any MLF $\phi$ on $B$ we have $0 < \phi(x) < 2$ and so $\phi(u) = \lim \phi(x_n) = \Sigma_{k=0}^{\infty}\binom{1/2}{k}(\phi(x) - 1)^k > 0$. Hence $\sigma_A(u) = \sigma_B(u)$ is positive. □

An element $v$ in a *-algebra is called *quasi-unitary* if $v \circ v^* = 0 = v^* \circ v$; that is, $vv^* = v^*v = v + v^*$. We show next that the quasi-unitary elements in a Banach *-algebra generate the algebra.

(22.6) PROPOSITION. *If* $A$ *is a Banach *-algebra, then every element of* $A$ *is a linear combination of quasi-unitary elements. If* $A$ *has an identity, then every element of* $A$ *is a linear combination of unitary elements of* $A$.

*Proof*. Let $x \in A$. Since elements of $A$ are linear combinations of hermitian elements and since $\nu(\lambda y) = |\lambda|\nu(y)$ for all $y$ and complex $\lambda$, we may assume that $x \in H(A)$ and $\nu(x) < 1$. Then $\nu(x^2) = \nu(x)^2 < 1$ and by (22.5) there is $y \in H(A)$ which commutes with $x$ and $y \circ y = x^2$. Let $v = y + ix$. Then $v \circ v^* = (y \circ y) - x^2 = 0$, and similarly $v^* \circ v = 0$. Hence $v$ is quasi-unitary and since $x = (v - v^*)/2i$, the first statement is proved.

If $A$ has identity $e$, set $u = v - e$. Then $u^*u = (v^*v - v - v^*) + e = e = uu^*$ and $x = (v - v^*)/2i = (u - u^*)/2i$. □

(22.7) PROPOSITION. *Let* $A$ *be a Banach *-algebra and* $p$ *any positive functional on* $A$. *Then*

$$|p(x^*hx)| \le \nu(h)p(x^*x) \quad (x \in A,\ h \in H(A)).$$

*Proof.* Let $x \in A$ and assume first that $h \in H(A)$ with $\nu(h) < 1$. By (22.5) there are elements $y, z \in H(A)$ such that $y \circ y = h$ and $z \circ z = -h$. Adjoin an identity $e$ to $A$ and set $v = ex - yx$, $w = ex - zx$. Then $v^*v = x^*(e - h)x$ and $w^*w = x^*(e + h)x$. Since $p$ is positive, $p(x^*(e - h)x) \ge 0$, $p(x^*(e + h)x) \ge 0$. These inequalities imply that $|p(x^*hx)| \le p(x^*x)$.

Next, for arbitrary $h \in H(A)$ and $\varepsilon > 0$, set $h_\varepsilon = (\nu(h) + \varepsilon)^{-1}h$. Then $\nu(h_\varepsilon) < 1$, so that $|p(x^*h_\varepsilon x)| \le p(x^*x)$ or $|p(x^*hx)| \le (\nu(h) + \varepsilon)p(x^*x)$. Since $\varepsilon$ was arbitrary, we have $|p(x^*hx)| \le \nu(h)p(x^*x)$. □

(22.8) COROLLARY. *Let* $A$ *be a unital Banach *-algebra. If* $p$ *is a positive functional on* $A$*, then for all* $x \in A$:

*(a)* $p(x^*x) \le p(e)\nu(x^*x)$;

*(b)* $|p(x)| \le p(e)\nu(x^*x)^{1/2}$;

*(c)* $|p(x)| \le p(e)\nu(x)$ *for all* $x \in N(A)$;

*(d)* $p$ *vanishes on the radical of* $A$.

*Proof.* (a) Set $h = x^*x$ and $x = e$ in (22.7).

(b) By (21.6), (b) and part (a) we have $|p(x)| \le p(e)^{1/2}p(x^*x)^{1/2} \le p(e)\nu(x^*x)^{1/2}$.

(c) If $x \in N(A)$, i.e., $xx^* = x^*x$, then $\sigma_A(xx^*) \subseteq \sigma_A(x)\sigma_A(x^*)$ (see the proof of (22.4)), and hence $\nu(xx^*) \le \nu(x)\nu(x^*) = \nu(x)^2$. Now apply (b).

(d) If $x \in Rad(A)$, then $x^*x \in Rad(A)$ and hence $\nu(x^*x) = 0$ by (B.5.17), (c). It then follows from (b) that $p(x) = 0$. □

(22.9) COROLLARY. *Let* $p$ *be an extendable positive functional on a Banach *-algebra* $A$. *Then there exists* $k \ge 0$ *such that* $p(x^*x) \le k\nu(x^*x)$ *and* $|p(x)| \le k\nu(x^*x)^{1/2}$ *for all* $x \in A$. *In particular, if* $x \in Rad(A)$, *then* $p(x) = 0$.

(22.10) LEMMA. *Let* $E$ *be a Banach space and suppose* $E_1$ *and* $E_2$ *are closed subspaces of* $E$ *such that* $E = E_1 + E_2$. *Then there is a constant* $\beta$ *such that every* $x \in X$ *has a representation* $x = x_1 + x_2$, *where* $x_1 \in E_1$, $x_2 \in E_2$, *and* $||x_1|| + ||x_2|| \le \beta||x||$.

*Proof*. Let X be the vector space $E_1 \times E_2$ with componentwise linear operations and norm

$$||(x_1,x_2)|| = ||x_1|| + ||x_2|| \qquad (x_1,x_2) \in E_1 \times E_2.$$

Since $E_1$ and $E_2$ are complete, X is complete. The linear mapping $T: X \to E$ defined by $T(x_1,x_2) = x_1 + x_2$ is continuous, since $||x_1 + x_2|| \leq ||(x_1,x_2)||$, and maps X onto E. By the open mapping theorem, there exists $\beta$ such that each $x \in E$ is $T(x_1,x_2)$ for some $(x_1,x_2) \in X$ with $||(x_1,x_2)|| \leq \beta||x||$. □

The following result is the first of several concerning continuity of positive functionals.

(22.11) THEOREM. *Let* A *be a Banach *-algebra with identity* e *and* p *a positive functional on* A. *Then* p *is continuous. In particular, every extendable positive functional on a Banach *-algebra* A *is continuous. Further:*

*(a) If* A *is unital and commutative, then* $||p|| = p(e)$.

*(b) If* A *is unital and* $||x^*|| \leq \alpha||x||$ *for all* $x \in A$, *then* $||p|| \leq \alpha^{1/2}p(e)$.

*(c) If* A *is unital and* $||x^*|| = ||x||$ *for all* $x \in A$, *then* $||p|| = p(e)$.

*Proof*. We establish the statements in parts (a), (b), and (c) before proving that every positive functional on A is continuous.

(a) If A is commutative, then (22.8), (c) implies that $|p(x)| \leq p(e)\nu(x) \leq p(e)||x||$ for *all* x in A. Hence $||p|| \leq p(e)$. Since $||e|| = 1$, then $||p|| \geq p(e)$.

(b) and (c) If $||x^*|| \leq \alpha||x||$, (22.8), (b) implies $|p(x)| \leq p(e)\alpha^{1/2}||x||$ since $\nu(xx^*) \leq ||x||\cdot||x^*||$. Hence $||p|| \leq \alpha^{1/2}p(e)$. When $||x^*|| = ||x||$, it is clear that $||p|| = p(e)$.

Turning to the general case, we note that if p is identically zero, the theorem is obviously true. Suppose p is nonzero. We may assume without loss of generality that $p(e) = 1$; indeed, $p(e) \geq 0$ and if $p(e) = 0$ it follows by (22.8), (c) that $p(x) = 0$ for all x in A.

Let $H$ denote the norm closure of the set $H(A)$ of hermitian elements of $A$. Now $H(A)$ and $iH(A)$ are *real*-linear subspaces of $A$ and $A = H(A) + iH(A)$. By (22.8), (c) the restriction of $p$ to $H(A)$ is a real-linear functional of norm one, which therefore extends to a real-linear functional $q$ on $H$ also of norm one.

We assert that if $K = H \cap iH$, then

$$q(k) = 0 \quad \text{for all} \quad k \in K. \tag{1}$$

Indeed, if $k = \lim h_n = \lim ik_n$, where $h_n, k_n \in H(A)$, then $h_n^2 \to k^2$ and $k_n^2 \to -k^2$; by (b) of (21.5) and (c) of (22.8) we have

$$|p(h_n)|^2 \le p(h_n^2) \le p(h_n^2 + k_n^2) \le ||h_n^2 + k_n^2|| \to 0.$$

Because $q(k) = \lim p(h_n)$ relation (1) follows.

Now let $x \in A$ be arbitrary. By (22.10) there exists a real number $\beta$ such that

$$x = x_1 + ix_2 \quad \text{with} \quad x_1, x_2 \in H, \quad ||x_1|| + ||x_2|| \le \beta||x||.$$

If $x = h + ik$, with $h, k \in H(A)$, then $x_1 - h$, $x_2 - k \in K$ and hence by (1) we have

$$p(x) = p(h) + ip(k) = q(x_1) + iq(x_2).$$

Therefore

$$|p(x)| \le |q(x_1)| + |q(x_2)| \le ||x_1|| + ||x_2|| \le \beta||x||$$

and so $p$ is continuous. □

It is not always true that $||p|| = p(e)$. Indeed, let $E$ be the Banach space $C^2$ with the norm $||\xi|| = |\xi_1| + t|\xi_2|$, where $\xi = (\xi_1, \xi_2)$ and $t > 2$. Let $A = B(E)$. Representing operators $T$ in $A$ as $2 \times 2$ complex matrices $(\lambda_{ij})$, define $T^* = (\overline{\lambda}_{ji})$. Then $A$ is a Banach *-algebra with continuous (but not isometric) involution. Define $p\colon A \to C$ by $p(T) = \Sigma_{i,j}\lambda_{ij}$. Then $p$ is a positive functional on $A$ such that $p(I) = 2$, where $I$ is the identity of $A$. If

$$T = \begin{pmatrix} 0 & t^2 \\ 1 & 0 \end{pmatrix},$$

then $T\xi = (t^2\xi_2, \xi_1)$ so that $||T\xi|| = t^2|\xi_2| + t|\xi_1| = t||\xi||$ and hence $||T|| = t$. Therefore, $||p|| \geq p(T)/||T|| = (t^2 + 1)/t > t > p(I)$.

In the next result we extend (22.11) to Banach *-algebras with bounded approximate identity. A different proof, depending on representation theory of Banach *-algebras, will be given in Section 27 (see (27.4)).

(22.12) PROPOSITION. *(Varopoulos). Let* A *be a Banach *-algebra with bounded approximate identity. If* p *is a positive functional on* A, *then* p *is continuous.*

*Proof.* For a fixed element $a \in A$ the linear functional ${}_{a^*}p_a(x) = p(a^*xa)$ is extendable and positive by (21.8); hence it is continuous by (22.11). Since

$$4axb = \sum_{k=0}^{3} i^k(a^* - i^k b)^* x (a^* - i^k b)$$

for each $a, b, x \in A$, the functional ${}_ap_b$ on A defined by ${}_ap_b(x) = p(axb)$ is also continuous. Now, let $\{x_n\}$ be a sequence in A such that $x_n \to 0$. Applying the left and right versions of (B.7.3), there exist elements $a, b \in A$ and sequences $\{y_n\}$ and $\{z_n\}$ in A with limit 0 such that $x_n = ay_n = az_nb$ for all n. The continuity of ${}_ap_b$ gives $p(x_n) = {}_ap_b(z_n) \to 0$, i.e., p is continuous. □

Although the preceding proposition assures continuity of positive functionals when they exist, there are Banach *-algebras which admit no (nonzero) positive functionals. Indeed, if B is any Banach algebra with identity e, let A be the *-Banach algebra constructed in (22.1). Since $(e,-e)^*(e,-e) = -(e,e)$ and $(e,e)$ is the identity for A, every positive functional on A must vanish at $(e,e)$ and therefore by (21.6), (b) must be identically zero. This example can be modified to give a *-Banach algebra without identity which also admits no (nonzero) positive functionals.

It is essential that A have an approximate identity and be complete in (22.12) as the following examples show.

(22.13) EXAMPLE. Let $A$ denote the Banach space $C([0,1])$ under pointwise operations and supremum norm. If $f, g \in A$, define multiplication by $fg = 0$ and involution by $f^* = \overline{f}$. Then $A$ is a *-Banach algebra without an approximate identity, and on $A$ every linear functional is positive. Since $A$ is infinite-dimensional, it admits a discontinuous linear functional, hence a discontinuous positive functional.

(22.14) EXAMPLE. Let $A$ denote the set of all complex polynomials defined on the complex numbers with pointwise linear operations and multiplication. For $f(z) = \Sigma_{k=0}^n \lambda_k z^k$ in $A$, let $f^*(z) = \Sigma_{k=0}^n \overline{\lambda}_k z^k$ and define $||f|| = \sup\{|f(t)|: t \in [0,1]\}$. Then $A$ is an incomplete unital *-normed algebra. For $f \in A$ define $p(f) = f(3)$. Then $p$ is a discontinuous positive functional on $A$. In fact, let $n$ be a positive integer and set $f(z) = z^n$. Then $f \in A$, $||f|| = 1$, and $p(f) = f(3) = 3^n$. Hence, $||p|| = \sup\{|p(f)|: f \in A, ||f|| = 1\} \geq 3^n$ for all $n$, i.e., $p$ is unbounded.

(22.15) THEOREM. *Let* $A$ *be a *-Banach algebra with approximate identity* $\{e_\alpha\}$ *bounded by* 1. *If* $p$ *is a positive functional on* $A$, *then:*

*(a)* $p$ *is continuous, hermitian, and extendable;*

*(b)* $|p(x)|^2 \leq ||p||p(x^*x)$ *for all* $x \in A$.

*(c) If* $p'$ *is the minimal positive extension of* $p$ *to* $A_e$ *(see (21.7)), then* $||p'|| = ||p||$.

*Proof*. Let $x \in A$. By (22.12) $p$ is continuous; hence we have by (21.5), (a), and the fact that $\{e_\alpha^*\}$ is an approximate identity

$$p(x^*) = \lim_\alpha p(e_\alpha^* x^*) = \lim_\alpha \overline{p(xe_\alpha)} = \overline{p(x)}.$$

Therefore $p$ is hermitian. By (21.5), (b) we have

$$|p(x)|^2 = \lim_\alpha |p(e_\alpha^* x)|^2 \leq \sup_\alpha p(e_\alpha^* e_\alpha) p(x^*x).$$

Since $||e_\alpha^* e_\alpha|| \leq 1$ for all $\alpha$, $\sup p(e_\alpha^* e_\alpha) \leq ||p||$. Then $|p(x)|^2 \leq ||p||p(x^*x)$ and $p$ is extendable by (21.7). This proves (a) and (b).

Let $k = \sup_\alpha p(e_\alpha^* e_\alpha)$. Then $|p(x)|^2 \leq kp(x^*x)$ for all $x \in A$. Let $p'(x + \lambda e) = p(x) + k\lambda$, where $x + \lambda e \in A_e$, and note that $p'(e) = k$.

By (22.11), (c)

$$||p|| \leq ||p'|| = p'(e) = k \leq ||p||,$$

which proves that $||p|| = \lim_\alpha p(e_\alpha^* e_\alpha)$ and also that $||p'|| = ||p||$. Hence (c) is proved. □

(22.16) PROPOSITION. *Let* A *be a* *-*Banach algebra with approximate identity* $\{e_\alpha\}$ *bounded by* 1.

*(a) If* p *is a positive functional on* A, *then* $||p|| = \lim_\alpha p(e_\alpha)$.

*(b) If* $p_1, \ldots, p_n$ *are positive functionals on* A, *then*

$$||\sum_{i=1}^{n} p_i|| = \sum_{i=1}^{n} ||p_i||.$$

*Proof*. (a) By (22.15) p is continuous and extendable. Hence there is $k \geq 0$ such that $|p(x)|^2 \leq kp(x^*x)$ for all $x \in A$. Assume that k is minimal. Let $p'(x + \lambda e) = p(x) + \lambda k$, and define an inner product on $A_e$ by

$$(x + \lambda e | y + \lambda e) = p'((y + \lambda e)^*(x + \lambda e)). \tag{1}$$

Let $||x + \lambda e||_1 = (x + \lambda e | x + \lambda e)^{1/2}$ for $x + \lambda e \in A_e$. Then $||\cdot||_1$ is a seminorm on $A_e$. Given $\varepsilon > 0$ we assert that there exists $x \in A$ such that $||e - x||_1 < \varepsilon$. Indeed, since k is minimal, there exists $y \in A$ such that $p(y^*y) = 1$ and $|p(y)|^2 > k - \varepsilon^2$. Set $x = \overline{p(y)}y$. Then

$$\begin{aligned} ||e - x||_1^2 &= p'((e - x)^*(e - x)) = p'(e - x^* - x + x^*x) \\ &= k - |p(y)|^2 < \varepsilon^2. \end{aligned} \tag{2}$$

Now, since $\{e_\alpha\}$ is an approximate identity, there is $\alpha_o$ such that if $\alpha \geq \alpha_o$ then $||e_\alpha x - x|| < \varepsilon$. Hence

$$\begin{aligned} ||e_\alpha x - x||_1 &= p'((e_\alpha x - x)^*(e_\alpha x - x))^{1/2} \\ &\leq ||p'||^{1/2} ||e_\alpha x - x|| \leq ||p'||^{1/2} \varepsilon. \end{aligned} \tag{3}$$

If $y \in A$, the positive functional $g(z) = p(y^*zy)$, $z \in A$, is extendable to $A_e$ by (21.8), and $g(e) = ||g||$ by (22.11). Since $|g(z)| \leq ||g||\cdot||z||$ and $g(e) = p(y^*y)$ we have $|p(y^*zy)| \leq ||z||p(y^*y)$ for all $y, z \in A$. Applying this inequality to $p'$, we have

$$||e_\alpha x - e_\alpha||_1 = p'((e_\alpha x - e_\alpha)^*(e_\alpha x - e_\alpha))^{1/2}$$

$$\leq ||e_\alpha^* e_\alpha||^{1/2} p'((x - e)^*(x - e))^{1/2} \tag{4}$$

$$= ||e_\alpha^* e_\alpha||^{1/2} ||x - e||_1 < \varepsilon.$$

Thus, if $\alpha \geq \alpha_o$, (2), (3), and (4) imply that

$$||e_\alpha - e||_1$$

$$\leq ||e_\alpha - e_\alpha x||_1 + ||e_\alpha x - x||_1 + ||x - e||_1 < \varepsilon\ (2 + ||p'||^{1/2}).$$

Hence, $e_\alpha \to e$ relative to $||\cdot||_1$. It follows that $p(e_\alpha) = p'(e^*e_\alpha) = (e_\alpha|e\,) \to (e|e) = p'(e) = ||p'|| = ||p||$ (22.15), (c), i.e., $||p|| = \lim_\alpha p(e_\alpha)$ and (a) is proved. Part (b) is an immediate consequence of (a). □

For C*-algebras we have the following characterization of positive functionals.

(22.17) PROPOSITION. *Let* A *be a C*-algebra and* $\{e_\alpha\}$ *an increasing approximate identity bounded by* 1 *in* A. *Then a continuous linear functional* p *on* A *is positive iff* $\lim_\alpha p(e_\alpha) = ||p||$.

*Proof*. If p is positive then $\lim_\alpha p(e_\alpha) = ||p||$ by (22.16), (a). However, the following direct proof is interesting. Since $\{e_\alpha\}$ is increasing, $\{p(e_\alpha)\}$ is an increasing net of nonnegative reals with limit $\beta \leq ||p||$. For $x \in A$ with $||x|| \leq 1$ we have by (21.5), (b) and the fact that p preserves the order of positive elements, noting $e_\alpha e_\alpha^* \leq e_\alpha$, that

$$|p(e_\alpha x)|^2 \leq p(e_\alpha e_\alpha^*)p(x^*x) \leq p(e_\alpha)||p|| \leq \beta||p||.$$

Since $\{e_\alpha\}$ is an approximate identity for A and p is continuous, this implies that $|p(x)|^2 \leq \beta||p||$, hence $||p||^2 \leq \beta||p||$, i.e., $||p|| \leq \beta$.

The proof of the converse is based on Arens' argument in the proof of (7.1). Assume that $\lim_\alpha p(e_\alpha) = ||p||$. Let $h \in H(A)$ with $||h|| \leq 1$ and write $p(h) = \lambda + i\mu$ with $\lambda, \mu \in R$. We may assume that $\mu \geq 0$ by adjusting the sign of h. Choose an index $\alpha$ such that $||he_\alpha - e_\alpha h|| < n^{-1}$. Then, since A is a $C^*$-algebra

$$||ne_\alpha - ih||^2 = ||n^2e_\alpha^2 + h^2 - in(he_\alpha - e_\alpha h)|| \leq n^2 + 2.$$

But

$$\lim_\alpha |p(ne_\alpha - ih)|^2 = (n||p|| + \mu)^2 + \lambda^2,$$

which implies that

$$(n||p|| + \mu)^2 + \lambda^2 \leq (n^2 + 2)||p||^2$$

for all n. Hence $\mu = 0$, i.e., p is real on H(A) and hence is an hermitian functional.

Now, if $x \in A^+$ with $||x|| \leq 1$, then $e_\alpha - x \in H(A)$ and, by (12.4), (b), $||e_\alpha - x|| \leq 1$. Therefore $p(e_\alpha - x) \leq ||p||$. Passing to the limit and using the fact that $p(e_\alpha) \to ||p||$ gives $p(x) \geq 0$. □

(22.18) COROLLARY. *Let* A *be a* $C^*$*-algebra with identity* e *and* p *a continuous linear functional on* A. *Then* p *is positive iff* $||p|| = p(e)$.

Our next objective is to prove a generalization of the Russo-Dye theorem which was stated and proved in Section 15 for $C^*$-algebras. Later, in Section 35, we shall look at the theorem in another context.

(22.19) THEOREM. *(Russo-Dye). Let* A *be a unital *-Banach algebra. Then the closed unit ball of* A *is contained in the closed convex hull of the unitary elements of* A.

*Proof.* It clearly suffices to show that the open unit ball of A is contained in the closed convex hull of U(A). Let $x \in A$ with

$||x|| < 1$. Then $||xx^*|| \leq ||x||\cdot||x^*|| = ||x||^2 < 1$. Hence, by (22.5), the hermitian element $e - xx^*$ is invertible and has the invertible hermitian square root $(e - xx^*)^{1/2} = \Sigma_{n=0}^{\infty}\binom{1/2}{n}(- xx^*)^n$. Similarly $e - x^*x$ has invertible hermitian square root $(e - x^*x)^{1/2} = \Sigma_{n=0}^{\infty}\binom{1/2}{n}(- x^*x)^n$. For complex $\lambda$ with $|\lambda| = 1$ define

$$u_\lambda = (e - xx^*)^{-1/2}(x - \lambda e)(e - \lambda x^*)^{-1}(e - x^*x)^{1/2}.$$

We intend to show that $u_\lambda$ is unitary. Since $\lambda\bar{\lambda} = 1$,

$$u_\lambda^* = (e - x^*x)^{1/2}(e - \bar{\lambda}x)^{-1}(x^* - \bar{\lambda}e)(e - xx^*)^{-1/2}$$

$$= (e - x^*x)^{1/2}(\lambda e - x)^{-1}(\lambda x^* - e)(e - xx^*)^{-1/2}.$$

Observe that

$$(\lambda e - x)^{-1}(\lambda x^* - e) = (\lambda e - x)^{-1}[(\lambda e - x)x^* - (e - xx^*)]$$

$$= x^* - (\lambda e - x)^{-1}(e - xx^*),$$

$$(e - \lambda x^*)(x - \lambda e)^{-1} = [x^*(\lambda e - x) - (e - x^*x)](\lambda e - x)^{-1}$$

$$= x^* - (e - x^*x)(\lambda e - x)^{-1},$$

and

$$x(e - x^*x)^{1/2} = \sum_{n=0}^{\infty}\binom{1/2}{n}x(- x^*x)^n = \sum_{n=0}^{\infty}\binom{1/2}{n}(- xx^*)^n x = (e - xx^*)^{1/2}x$$

which may be conjugated to give the related equality

$$(e - x^*x)^{1/2}x^* = x^*(e - xx^*)^{1/2}.$$

Utilizing these relations it follows easily that $u_\lambda^* = u_\lambda^{-1}$ so $u_\lambda$ is unitary.

Let $u_{k/m}$ denote the unitary element $u_\lambda$ with $\lambda = \exp(2\pi i\frac{k}{m})$, where $k$, $m$ are positive integers. We will show that $x = \lim \Sigma_{k=1}^{m}(1/m)u_{k/m}$.

With $\lambda$ as above, let $x_{k/m}$ denote the element

$$x_\lambda = (x - \lambda e)(e - \lambda x^*)^{-1}.$$

Then

$$x - \sum_{k=1}^{m} \frac{1}{m} u_{k/m} = x - \frac{1}{m}\sum_{k=1}^{m} (e - xx^*)^{-1/2} x_{k/m} (e - x^*x)^{1/2}$$

$$= (e - xx^*)^{-1/2}\left(x - \frac{1}{m}\sum_{k=1}^{m} x_{k/m}\right)(e - x^*x)^{1/2}$$

and so

$$\begin{aligned} &\left\|x - \sum_{k=1}^{m} \frac{1}{m} u_{k/m}\right\| \\ &\leq \|(e - xx^*)^{-1/2}\| \cdot \left\|x - \frac{1}{m}\sum_{k=1}^{m} x_{k/m}\right\| \cdot \|(e - x^*x)^{1/2}\|. \end{aligned} \tag{1}$$

Observe that

$$x_\lambda = \sum_{n=0}^{\infty} (x - \lambda e)(\lambda x^*)^n = \sum_{n=0}^{\infty} \lambda^n x (x^*)^n - \sum_{n=0}^{\infty} \lambda^{n+1} (x^*)^n$$

and so

$$x - x_\lambda = \sum_{n=0}^{\infty} \lambda^{n+1} (x^*)^n - \sum_{n=1}^{\infty} \lambda^n x (x^*)^n$$

$$= \sum_{n=1}^{\infty} \lambda^n [(x^*)^{n-1} - x(x^*)^n]$$

$$= (e - xx^*) \sum_{n=1}^{\infty} \lambda^n (x^*)^{n-1}.$$

Summing over $k$, $1 \leq k \leq m$, and dividing by $m$ we have

$$x - \frac{1}{m}\sum_{k=1}^{m} x_{k/m} = \frac{1}{m}\sum_{k=1}^{m} (x - x_{k/m})$$

$$= (e - xx^*) \sum_{n=1}^{\infty} \frac{1}{m} \sum_{k=1}^{m} [\exp(2\pi i \tfrac{n}{m})]^k (x^*)^{n-1}.$$

Now, if $1 \leq n < m$, then $\exp(2\pi i \frac{n}{m}) \neq 1$ and so by the formula for a finite geometric sum

$$\sum_{k=1}^{m} [\exp(2\pi i\tfrac{n}{m})]^k = \frac{\exp(2\pi i\frac{n}{m}) - \exp(2\pi i\frac{n(m+1)}{m})}{1 - \exp(2\pi i\frac{n}{m})} = 0;$$

hence we have

$$x - \frac{1}{m}\sum_{k=1}^{m} x_{k/m} = (e - xx^*)\sum_{n=m}^{\infty}\frac{1}{m}\sum_{k=1}^{m}[\exp(2\pi i\tfrac{n}{m})]^k (x^*)^{n-1}.$$

Then

$$||x - \frac{1}{m}\sum_{k=1}^{m} x_{k/m}|| \le ||e - xx^*|| \sum_{n=m-1}^{\infty} ||(x^*)^n|| \tag{2}$$

$$\le ||e - xx^*|| \sum_{n=m-1}^{\infty} ||x||^n$$

$$\le ||e - xx^*|| \frac{||x||^{m-1}}{1 - ||x||}.$$

Since the right-hand side converges to 0 as $m \to \infty$, the theorem now follows from relation (1) above. □

We next prove a generalization of (22.19) to arbitrary unital Banach *-algebras.

(22.20) THEOREM. *Let* A *be a unital Banach *-algebra. Then the closed convex hull of the unitary elements in* A *contains all elements* x *in* A *such that* $\nu(x) < 1$ *and* $\nu(xx^*) < 1$.

*Proof*. The proof is nearly identical with the proof of (22.19); we need only point out the changes that are necessary. The assumption $\nu(xx^*) < 1$ guarantees the existence of an invertible hermitian square root $(e - xx^*)^{1/2}$ by (22.5). Similarly, since $\nu(x^*x) = \nu(xx^*) < 1$, the square root $(e - x^*x)^{1/2}$ also exists and is invertible. Since $\nu(x) < 1$ and $\nu(x) = \nu(x^*)$, we have

$$\lim_{n\to\infty} ||(x^*)^n||^{1/n} < 1$$

and so for some $r < 1$ and all sufficiently large $m$

$$\sup_{n\ge m-1} ||(x^*)^n||^{1/n} \le r.$$

For such $m$ we get from relations (1) and (2) in the proof of (22.19) that

$$||x - \frac{1}{m}\sum_{k=1}^{m} x_{k/m}||$$

$$\leq ||(e - xx^*)^{-1/2}||\cdot||e - xx^*||\cdot||(e - x^*x)^{1/2}|| \sum_{n=m-1}^{\infty} r^n$$

$$= ||(e - xx^*)^{-1/2}||\cdot||e - xx^*||\cdot||(e - x^*x)^{1/2}||\cdot r^{m-1}/(1-r)$$

which converges to 0 as $m \to \infty$. □

## §23. $A^*$-*algebras*.

The purpose of this section is to discuss a class of Banach *-algebras which have important applications to the study of positive functionals and representation theory, and which also are of considerable interest in themselves.

A Banach *-algebra $(A, ||.||)$ is said to be an $A^*$-*algebra* provided there exists on $A$ a second norm $|\cdot|$, not necessarily complete, which satisfies, in addition to the condition $|xy| \leq |x|\cdot|y|$ for all $x, y \in A$, the condition $|x|^2 = |x^*x|$ for each $x \in A$. The second norm will be called an *auxiliary norm*.

Examples (1), (2), and (3) of §21, with their usual norms, are obviously $A^*$-algebras. Any $C^*$-algebra is an $A^*$-algebra. An important example of a Banach *-algebra whose norm does not satisfy the condition $||x||^2 = ||x^*x||$, but is an $A^*$-algebra, is the group algebra $L^1(G)$ of a locally compact group $G$ (see (56.5)). Another example of an $A^*$-algebra is provided by any semisimple *real* commutative Banach algebra. In this case the involution is the identity map and the auxiliary norm is the spectral radius $|\cdot|_\sigma$.

If an $A^*$-algebra does not possess an identity element, we may always adjoin one to obtain an $A^*$-algebra with identity. The proof is similar to (6.1) for $C^*$-algebras, but we shall give it a little different treatment.

(23.1) PROPOSITION. *(Yood). Let* $(A,|\cdot|)$ *be a normed *-algebra without identity which satisfies the condition* $|x^*x| = |x|^2$ *for all* $x \in A$. *Then there exists a normed *-algebra* $(B,|\cdot|_1)$ *with identity and an* *isometric *-isomorphism* $\Phi$ *of* A *into* B *such that:*

(a) $|b^*b|_1 = |b|_1^2$ *for all* $b \in B$;

(b) B *is generated by* $\Phi(A)$ *and the identity in* B.

*Further, if* A *is an A*-algebra, then* B *is also an A*-algebra and* $\Phi$ *preserves both norms.*

*Proof.* Consider the left regular representation $a \to L_a$ of A on itself ($L_a x = ax$). If $|L_a|_1$ denotes the norm of $L_a$ as an operator in B(A), then $|L_a|_1 \leq |a|$. On the other hand, since $|a^*a| = |a|^2$, we have

$$|a|^2 = |aa^*| = |L_a a^*| \leq |L_a|_1 |a^*| = |L_a|_1 |a|.$$

Hence $|a| \leq |L_a|_1$ and so $|L_a|_1 = |a|$. This shows that the mapping $\Phi: A \to B(A)$ defined by $\Phi(a) = L_a$ is an isometric isomorphism of A into B(A). Now let B denote the subalgebra of B(A) generated by the identity operator I and the image $\Phi(A)$. Then B consists of all operators of the form $L_a + \lambda I$, $\lambda \in C$ and $a \in A$.
Define

$$(L_a + \lambda I)^* = L_{a^*} + \bar{\lambda} I.$$

Then $L_a + \lambda I \to (L_a + \lambda I)^*$ is an involution in B and $a \to L_a$ is a *-isomorphism of A into B. Property (b) clearly holds. To see that (a) holds, let $b = L_a + \lambda I$, $a \in A$, $\lambda \in C$. Then for all $x \in A$

$$|bx|^2 = |ax + \lambda x|^2 = |(ax + \lambda x)^*(ax + \lambda x)|$$

$$= |x^*(b^*bx)| \leq |b^*b|_1 |x|^2.$$

Therefore $|b|_1^2 \leq |b^*b|_1 \leq |b^*|_1 |b|_1$. It follows that $|b|_1 = |b^*|_1$ and hence that $|b|_1^2 \leq |b^*b|_1 \leq |b|_1^2$.

Finally, suppose that A is an A*-algebra with norm $||\cdot||$ and auxiliary norm $|\cdot|$. By the first part of the proof there is an auxiliary norm $|\cdot|_1$ on B corresponding to $|\cdot|$; it therefore suffices to define a norm on B under which it becomes a Banach *-algebra. However, if we set $||L_a + \lambda I|| = ||a|| + |\lambda|$, it is easily verified that B is a complete normed *-algebra and the *-isomorphism $\Phi$ preserves both the auxiliary norm and the complete norm. □

Let B be a Banach algebra with norm $|\cdot|$ and A a subalgebra of B which is a Banach algebra under a second norm $||\cdot||$. If $x \in A$, then $\sigma_B(x) \subseteq \sigma_A(x) \cup \{0\}$ and it follows from (B.4.12) that

$$\nu_B(x) = |x|_{\sigma_B} \leq |x|_{\sigma_A} = \nu_A(x).$$

This observation will be useful for the proof of the next lemma.

(23.2) LEMMA. *Let* A *be an A*-algebra with norm* $||\cdot||$ *and auxiliary norm* $|\cdot|$. *Then* $|x|^2 \leq |x^*x|_\sigma$ *for all* $x \in A$. *In particular,* $|x^*x|_\sigma = 0$ *implies that* $x = 0$.

*Proof*. Let B denote the completion of A with respect to the norm $|\cdot|$, and note by the observation preceding the lemma that $\nu_B(x) \leq \nu_A(x)$ for each $x \in A$. If h is any hermitian element of A, $|h|^2 = |h^*h| = |h^2|$ and, by iteration, $|h| = |h^{2^n}|^{1/2^n}$ for every n. Then

$$|h| = \lim_{n\to\infty} |h^{2^n}|^{1/2^n} = \nu_B(h) \leq \nu_A(h).$$

In particular, for $x \in A$, $|x|^2 = |x^*x| \leq \nu_A(x^*x) = |x^*x|_\sigma$. □

(23.3) LEMMA. *Let* A *be a Banach algebra and* B *any subalgebra. If* $x \in \mathrm{Rad}(B)$, *then* $\nu_A(x) = 0$.

*Proof*. If $x \in \mathrm{Rad}(B)$, then, as in the proof of (B.5.17), (c), $\sigma_B(x) = \{0\}$. Since $\sigma_A(x) \subseteq \sigma_B(x)$ the lemma follows from (B.4.12).

It follows from the next proposition that every A*-algebra is semi-simple.

(23.4) PROPOSITION. *Every *-subalgebra of an A*-algebra* A *is semi-simple.*

*Proof*. Let B be a *-subalgebra of A and suppose that $x \in \mathrm{Rad}(B)$.

Then $x^*x \in \mathrm{Rad}(B)$ and, by (23.3) $\nu_A(x^*x) = 0$. Then $x = 0$ by (23.2) and so $\mathrm{Rad}(B) = \{0\}$, i.e., B is semisimple. □

(23.5) PROPOSITION. *The involution in an A*-algebra is continuous with respect to both norms.*

*Proof.* Since the auxiliary norm satisfies the condition $|x^*| = |x|$, the involution is clearly continuous with respect to it. Since $x \to x^*$ is a linear mapping of a real Banach space onto itself, it suffices, by the closed graph theorem, to show that $x_n \to a$ and $x_n^* \to b$ implies $b = a^*$. Now, $|b - a^*| \le |b - x_n^*| + |x_n^* - a^*|$, and by (23.2), $|x|^2 \le \nu_A(x^*x) \le ||x^*x|| \le ||x^*||\cdot||x||$. Hence,

$$|b - x_n^*|^2 \le ||b - x_n^*||\cdot||b^* - x_n||$$

and

$$|x_n^* - a^*|^2 \le ||x_n^* - a^*||\cdot||x_n - a||.$$

Observe that on the right-hand side of each of these inequalities one factor is bounded while the other converges to zero as $n \to \infty$. Thus both $|b - x_n^*| \to 0$ and $|x_n^* - a^*| \to 0$. This implies that $|b - a^*| = 0$, as required. □

(23.6) COROLLARY. *Let* A *be an A*-algebra with norm* $||\cdot||$ *and auxiliary norm* $|\cdot|$. *Then there exists a constant* $\beta$ *such that* $|x| \le \beta||x||$ *for all* $x \in A$.

*Proof.* By (23.5) there exists a constant $\alpha \ge 0$ such that $||x^*|| \le \alpha||x||$ for all $x \in A$. Then by (23.2), $|x|^2 \le \nu_A(x^*x) \le ||x^*x|| \le ||x^*||\cdot||x|| \le \alpha||x||^2$, and, setting $\beta = \alpha^{1/2}$, we have $|x| \le \beta||x||$. □

(23.7) PROPOSITION. *Let* A *be an A*-algebra with norm* $||\cdot||$ *and auxiliary norm* $|\cdot|$. *Let* B *be a *-subalgebra of* A *which is a Banach algebra under a norm* $||\cdot||_1$. *Then there exists a constant* $\alpha$ *such that* $||x|| \le \alpha||x||_1$ *for all* $x \in B$; *that is, the inclusion mapping of* B *into* A *is continuous.*

*Proof.* By (23.6) there exists $\beta$ such that $|x| \le \beta||x||$ for all $x \in A$. Also, since B is an A*-algebra in its own right, there exists

$\beta'$ such that $|x| \leq \beta' ||x||_1$ for all $x \in B$. The proposition will be proved if we show that the injection mapping of B into A is continuous. Again we shall apply the closed graph theorem. Let $\{x_n\}$ be a sequence in B which converges to an element $b \in B$ in the norm of B and also to an element $a \in A$ in the norm of A. It then follows from the relation

$$|a - b| \leq |a - x_n| + |x_n - b| \leq \beta||a - x_n|| + \beta'||x_n - b||_1$$

that $a = b$. Hence, the injection of B into A has closed graph and therefore is continuous. □

(23.8) COROLLARY. *Let* A *be a *-algebra which admits an auxiliary norm. Then all norms which make* A *into a Banach algebra are equivalent. In particular, every A*-algebra has a unique complete norm topology.*

(23.9) LEMMA. *Let* $\phi$ *be a homomorphism of a Banach algebra* B *onto an algebra* A, *and let* K *denote the kernel of* $\phi$. *Then* $\phi(\overline{K}) \subseteq \text{Rad}(A)$.

*Proof.* If $\text{Rad}(A) = A$ the lemma is obvious. Otherwise, let $M$ denote the set of all maximal modular left ideals in A and recall that Rad(A) is equal to the intersection of all ideals in $M$ (see (B.5.12 )). For each $J \in M$, the inverse image $\phi^{-1}(J)$ is a maximal modular left ideal in B and is therefore closed. Further, $K \subseteq \phi^{-1}(J)$ and, since $\phi^{-1}(J)$ is closed, then $\overline{K} \subseteq \phi^{-1}(J)$. Hence, $\phi(\overline{K}) \subseteq \phi(\phi^{-1}(J)) \subseteq J$ for each $J \in M$, and so $\phi(\overline{K}) \subseteq \text{Rad}(A)$. □

(23.10) COROLLARY. *If* B *is a semisimple algebra, then the kernel of any homomorphism of a Banach algebra onto* B *is necessarily closed.*

*Proof.* Let K denote the kernel of a homomorphism of a Banach algebra onto B. By (23.9) and the semisimplicity of B, $\phi(\overline{K}) = \{0\}$. Hence, if $x \in \overline{K}$, then $\phi(x) = 0$ and so $x \in K$, i.e., $\overline{K} \subseteq K$ and K is closed. □

We shall show next that under certain conditions a homomorphism of one Banach algebra into another is automatically continuous.

(23.11) THEOREM. *Let* $\phi$ *be a homomorphism of a Banach algebra* B *into an A*-algebra* A. *If* $\phi(B)$ *is a *-subalgebra of* A, *then* $\phi$ *is continuous.*

*Proof.* Since $\phi(B)$ is a *-subalgebra of A, it is semisimple by (23.4). Hence, by (23.10), the kernel K of $\phi$ is a closed ideal of B. It follows that $\phi(B)$, which is isomorphic to B/K, is a Banach algebra under the norm

$$||\phi(x)||_o = \inf\{||y||: \phi(y) = \phi(x)\}.$$

Since $||\phi(x)||_o \leq ||x||$, the theorem follows from (23.7). □

## §24. *Homomorphisms and quotients of C*-algebras.*

We return now to C*-algebras, and look at some of the basic properties of *-homomorphisms and quotients. We shall assume that the norm in a C*-algebra satisfies $||x^*x|| = ||x||^2$ for all x. It follows from results on A*-algebras that every C*-algebra is semisimple and that *-homomorphisms of Banach *-algebras into C*-algebras are continuous. We shall, however, give simple direct proofs of these facts for C*-algebras.

(24.1) PROPOSITION. *Let* x *be an element in a C*-algebra* A. *Then:*
*(a)* $||x|| = |x|_\sigma$ *if* x *is normal.*
*(b)* $||x|| = |x^*x|_\sigma^{1/2}$.
*(c)* A *is semisimple.*

*Proof.* (a) Since $xx^* = x^*x$ we have

$$||x^2||^2 = ||(x^2)^*x^2|| = ||(x^*x)^*x^*x|| = ||x^*x||^2 = ||x||^4.$$

Hence $||x^2|| = ||x||^2$. Iterating we obtain $||x|| = ||x^{2^n}||^{1/2^n}$, and by ( B.4.12 ), $||x|| = |x|_\sigma$.

(b) Applying (a) to the hermitian element $x^*x$ we have $||x||^2 = ||x^*x|| = |x^*x|_\sigma$.

(c) If $x \in \mathrm{Rad}(A)$, then $x^*x \in \mathrm{Rad}(A)$ and so $|x^*x|_\sigma = 0$ by ( B.5.17 ). Applying (b), $||x||^2 = |x^*x|_\sigma = 0$. □

(24.2) PROPOSITION. *Let* A *be a *-Banach algebra and* B *a C*-algebra. If* $f: A \to B$ *is a *-homomorphism, then* $||f(x)|| \leq ||x||$ *for all* $x \in A$. *In particular,* f *is continuous with* $||f|| \leq 1$.

*Proof*. Let $x \in A$. We first show $\sigma_B(f(x)) \subseteq \sigma_A(x) \cup \{0\}$. Assume $\lambda \notin \sigma_A(x) \cup \{0\}$. Then $x/\lambda$ has a quasi-inverse in $A$ and hence, $f(x)/\lambda$ a quasi-inverse in $B$, i.e., $\lambda \notin \sigma_B(f(x))$.

Since $\sigma_B(f(x)) \subseteq \sigma_A(x) \cup \{0\}$, it follows that

$$|f(x)|_\sigma \leq |x|_\sigma. \tag{1}$$

By (1) and (24.1), (a) we have

$$\begin{aligned} ||f(x)||^2 &= ||f(x)^*f(x)|| = |f(x)^*f(x)|_\sigma \\ &= |f(x^*x)|_\sigma \leq |x^*x|_\sigma \\ &\leq ||x^*x|| \leq ||x||^2 \end{aligned} \tag{2}$$

and so $||f(x)|| \leq ||x||$; that is, $||f|| \leq 1$. □

Of course the continuity of $f$ in (24.2) could also be obtained by quoting (23.11) since every $C^*$-algebra is an $A^*$-algebra. However, we would not get the nice estimate on the norm.

An immediate consequence of (24.2), obtained by considering the identity map, is that there is at most one norm on a $^*$-algebra under which it is a $C^*$-algebra.

(24.3) PROPOSITION. *Let* $A$ *be a* $C^*$*-algebra,* $B$ *a* $^*$*-normed algebra, and* $f: A \to B$ *an injective* $^*$*-homomorphism of* $A$ *into* $B$. *Then* $||f(x)|| \geq ||x||$ *for each* $x \in A$.

*Proof*. We begin by making several reductions. Let $x \in A$. First, it suffices to show that $||f(x^*x)|| \geq ||x^*x||$; for then

$$||x||^2 = ||x^*x|| \leq ||f(x^*x)|| = ||f(x)^*f(x)|| \leq ||f(x)||^2.$$

Thus, we may assume that $x$ is hermitian. By restricting $f$ to the $C^*$-subalgebra of $A$ generated by $x$ we may assume that $A$ is commutative. Replacing $B$ by $f(A)$ we may assume that $B$ is commutative, and by replacing $B$ with its completion, we may assume $B$ is complete. Finally, we may also adjoin identity elements to $A$ and $B$, respectively. Summarizing, it is enough to consider the case when $A$ and $B$ are commutative, complete, and have identity elements.

Consider the compact Hausdorff structure spaces $\hat{A}$ and $\hat{B}$ of A and B, respectively. For each $\psi \in \hat{B}$, $\psi \circ f$ is an MLF on A, i.e., $\psi \circ f \in \hat{A}$. Hence, we may define a mapping $\Phi_f : \hat{B} \to \hat{A}$ by

$$\Phi_f(\psi) = \psi \circ f.$$

If $x \in A$, $\psi \in \hat{B}$, and $\{\psi_\alpha\}$ is a net in $\hat{B}$ converging to $\psi$, then $\psi_\alpha(f(x)) \to \psi(f(x))$; thus $\Phi_f$ is continuous on $\hat{B}$, and $\Phi_f(\hat{B})$ is a compact subset of $\hat{A}$.

We assert that $\Phi_f(\hat{B}) = \hat{A}$. Suppose that $\Phi_f(\hat{B}) \neq \hat{A}$ and choose, by Urysohn's lemma, continuous complex-valued functions $g, h \in C(\hat{A})$ such that $g \neq 0$, $h = 1$ on $\Phi_f(\hat{B})$, and $gh = 0$. By (7.1) there exist elements $x$ and $y$ in A which correspond to $g$ and $h$ under the Gelfand representation such that $x \neq 0$, $\psi(f(y)) = 1$ for all $\psi \in \hat{B}$, and $xy = 0$. Then $f(y)$ is invertible in B by (B.6.6), $f(x)f(y) = 0$, and since f is injective, $f(x) \neq 0$; but this is clearly impossible. Therefore, $\Phi_f(\hat{B}) = \hat{A}$. It then follows, for all $x \in A$, that

$$||x|| = \sup\{|\phi(x)| : \phi \in \hat{A}\} = \sup\{|\psi(f(x))| : \psi \in \hat{B}\} \leq ||f(x)||,$$

since $|\psi(f(x))| \leq ||\psi|| \cdot ||f(x)|| = ||f(x)||$. □

(24.4) COROLLARY. *Let* $f: A \to B$ *be an injective* *-*homomorphism of a* C*-*algebra* A *into a* C*-*algebra* B. *Then* f *is an isometry; in particular,* f(A) *is closed in* B.

*Proof.* Quote (24.2) and (24.3). □

The next result shows that a quotient of a C*-algebra by a closed two-sided ideal is again a C*-algebra.

(24.5) PROPOSITION. *Let* A *be a* C*-*algebra and* I *a closed two-sided ideal of* A. *Then* I *is a* *-*ideal and* A/I *is a* C*-*algebra.*

*Proof.* In view of (6.1) we may assume that A has an identity e. By (13.2) there is a net $\{e_\alpha\}$ in $I \cap A^+$ such that $||e_\alpha|| \leq 1$ for all $\alpha$, and $||e_\alpha x - x|| \to 0$ for all $x \in I$. So, if $x \in I$ then $x^*e_\alpha \in I$ and $||x^*e_\alpha - x^*|| = ||e_\alpha x - x|| \to 0$; hence $x^* \in \bar{I} = I$ and I is a *-ideal.

Since $A/I$ is a Banach *-algebra, it remains to prove the C*-norm condition, and for this it is enough to show that the quotient norm satisfies $||x||^2 \leq ||xx^*||$. To this end we show, for $x + I \in A/I$, that

$$||x + I|| = \lim_{\alpha} ||x - e_\alpha x||. \tag{1}$$

Indeed, if $y \in I$ is arbitrary, then $y - e_\alpha y \to 0$, and therefore

$$\begin{aligned}\overline{\lim_{\alpha}} ||x - e_\alpha x|| &= \overline{\lim_{\alpha}} ||(x - e_\alpha x) + (y - e_\alpha y)|| \\ &= \overline{\lim_{\alpha}} ||(e - e_\alpha)(x + y)|| \\ &\leq ||x + y||\cdot\overline{\lim_{\alpha}} ||e - e_\alpha|| \\ &\leq ||x + y||, \end{aligned} \tag{2}$$

where the inequality $\overline{\lim} ||e - e_\alpha|| \leq 1$ follows from (12.4), (b). Then Then, by (2),

$$\begin{aligned}||x + I|| &= \inf\{||x + y||: y \in I\} \geq \overline{\lim} ||x - e_\alpha x|| \\ &\geq \underline{\lim} ||x - e_\alpha x|| \geq \inf\{||x + y||: y \in I\} \\ &= ||x + I||,\end{aligned}$$

which proves (1).

It now follows from (1) and $||e - e_\alpha|| \leq 1$ that

$$\begin{aligned}||x + I||^2 &= \lim_{\alpha} ||x - e_\alpha x||^2 = \lim_{\alpha} ||(x - e_\alpha x)(x - e_\alpha x)^*|| \\ &= \lim_{\alpha} ||(e - e_\alpha)(xx^*)(e - e_\alpha)|| \\ &\leq \overline{\lim_{\alpha}} ||(e - e_\alpha)xx^*||\cdot||e - e_\alpha|| \\ &\leq \lim_{\alpha} ||xx^* - e_\alpha xx^*|| = ||xx^* + I|| \\ &= ||(x + I)(x + I)^*||,\end{aligned}$$

as required. □

(24.6) PROPOSITION. *Let* $I$ *be a closed left ideal in a* $C^*$*-algebra* $A$ *and let* $x \in A$. *Then* $x \in I$ *if and only if* $x^*x \in I$.

*Proof.* If $x \in I$ then it is clear that $x^*x \in I$. Conversely suppose that $y = x^*x \in I$ and let $B$ denote the closed commutative $*$-subalgebra of $A$ generated by the hermitian element $y$. Clearly $B \subseteq I$. Let $\lambda > 0$ be fixed. By (12.6) $y$ is positive; thus the element $-\lambda y$ has a (hermitian) quasi-inverse $(-\lambda y)'$ in $B$, hence in $I$. Define $z_\lambda = x - x(-\lambda y)'$. Since $(-\lambda y)' \in I$ then $z_\lambda - x \in I$. Hence, since $I$ is closed, to prove that $x \in I$ it will suffice to prove that $z_\lambda \to 0$ as $\lambda \to \infty$.

We adjoin an identity $e$ to $A$ if one is not already present. Then we have

$$||z_\lambda||^2 = ||z_\lambda^* z_\lambda|| = ||(e - (-\lambda y)')x^*x(e - (-\lambda y)')||$$

$$= ||(e - (-\lambda y)')y(e - (-\lambda y)')||.$$

Passing to a functional representation of the commutative algebra $B$ and noting that $e - (-\lambda y)' = (e + \lambda y)^{-1}$ we obtain

$$||(e - (-\lambda y)')y(e - (-\lambda y)')|| = ||(e + \lambda y)^{-2}y||$$

$$\leq \sup\{t/(1 + \lambda t)^2 : t \geq 0, t \in R\} = 1/4\lambda.$$

Hence $||z_\lambda|| \leq 1/2\lambda^{1/2}$, and thus $z_\lambda \to 0$ as $\lambda \to \infty$. □

(24.7) COROLLARY. *Let* $I$ *be a closed right ideal in a* $C^*$*-algebra* $A$ *and let* $x \in A$. *Then* $x \in I$ *if and only if* $xx^* \in I$.

*Proof.* Apply (24.6) to the closed left ideal $I^*$ in $A$. □

(24.8) PROPOSITION. *If* $f$ *is a* $*$*-homomorphism of a* $C^*$*-algebra* $A$ *into a* $C^*$*-algebra* $B$, *then* $f(A)$ *is closed in* $B$. *If* $K = \ker(f)$, *then the* $C^*$*-algebras* $A/K$ *and* $f(A)$ *are (canonically) isometrically* $*$*-isomorphic:*

$$A/K \simeq f(A).$$

*Proof.* By (24.2) $f$ is continuous and hence the kernel $K$ of $f$ is a closed two-sided *-ideal of $A$. Then $A/K$ is a C*-algebra by (24.5), and the map $g: A/K \to B$ defined by $g(x + K) = f(x)$ is a *-isomorphism of $A/K$ into $B$. By (24.4) $g$ is an isometry, and since $f$ and $g$ have the same range, $f(A)$ is closed in $B$. □

(24.9) COROLLARY. *Let* $f$ *be a *-homomorphism of a C*-algebra* $A$ *into a C*-algebra* $B$ *and set* $K = \ker(f)$. *If* $I$ *is any closed two-sided ideal in* $A$ *which contains* $K$, *then* $A/I$ *is isometrically *-isomorphic to the C*-algebra* $f(A)/f(I)$:

$$A/I \simeq f(A)/f(I).$$

*Proof.* By (24.8) $C = f(A)$ is a C*-subalgebra of $B$, and $f(I)$ is a closed two-sided ideal in $C$. Let $\tau: C \to C/f(I)$ denote the natural homomorphism. Then $\ker(\tau) = f(I)$, and $\tau \circ f$ is a *-homomorphism of $A$ onto $C/f(I)$. Since $x \in \ker(\tau \circ f)$ if and only if $f(x) \in \ker(\tau) = f(I)$ if and only if $x \in f^{-1}(f(I)) = I$, we have $\ker(\tau \circ f) = I$. Applying (24.8) to $\tau \circ f$, it follows that $A/I$ is isometrically *-isomorphic to $(\tau \circ f)(A) = f(A)/f(I)$. □

(24.10) COROLLARY. *If* $K$ *and* $I$ *are closed two-sided ideals in a C*-algebra* $A$ *such that* $K \subseteq I$, *then the C*-algebras* $A/I$ *and* $(A/K)/(I/K)$ *are isometrically *-isomorphic*:

$$A/I \simeq (A/K)/(I/K).$$

*Proof.* Let $B = A/K$ and $\tau: A \to B$ the quotient map. Then $\ker(\tau) = K$, and also $\tau(I) = I/K$. By (24.9), $A/I \simeq \tau(A)/\tau(I) = (A/K)/(I/K)$. □

(24.11) COROLLARY. *Let* $A$ *be a C*-algebra,* $B$ *a C*-subalgebra of* $A$, *and* $I$ *a closed two-sided ideal in* $A$. *Then* $B + I$ *is a C*-subalgebra of* $A$, *and the C*-algebras* $(B + I)/I$ *and* $B/(B \cap I)$ *are isometrically *-isomorphic:*

$$(B + I)/I \simeq B/(B \cap I).$$

*Proof.* Let $\tau: A \to A/I$ denote the natural homomorphism, and let $f = \tau|_B$. Then $f$ is a *-homomorphism from $B$ into $A/I$, and $\ker(f) =$

$B \cap \ker(\tau) = B \cap I$. Moreover, $f(B) = (B + I)/I$; indeed, if $y \in f(B)$ then there is an $x \in B$ such that $y = f(x) = x + I = (x + 0) + I \in (B + I)/I$. On the other hand, if $x + I \in (B + I)/I$, then $x = y + z$ with $y \in B$, $z \in I$; hence $x + I = (y + z) + I = y + I = f(y) \in f(B)$.

It follows from (24.8) that $(B + I)/I$ is closed in $A/I$, and hence that $B + I$ is closed in $A$; hence $B + I$ is a $C^*$-algebra. Finally, by (24.8) we have $B/(B \cap I) \simeq f(B) = (B + I)/I$. □

EXERCISES

(IV.1) Verify that all of the examples (1) - (9) in §21 are *-algebras.

(IV.2) Let $A$ be a *-algebra and let $x \in A$. Show in the decomposition $x = h + ik$ that the hermitian elements $h$ and $k$ are unique.

(IV.3) Show that every commutative *-subalgebra of a *-algebra $A$ is contained in a maximal commutative *-subalgebra of $A$.

(IV.4) Let $A$ and $B$ be *-algebras and $f: A \to B$ a *-homomorphism. Prove that:

(a) if $M$ is a *-subalgebra of $A$, then $f(M)$ is a *-subalgebra of $B$;

(b) if $N$ is a *-subalgebra of $B$, then $f^{-1}(N)$ is a *-subalgebra of $A$;

(c) $\ker(f) = f^{-1}(\{0\})$ is a *-ideal of $A$;

(d) $A/\ker(f)$ is *-isomorphic to $f(A)$;

(e) if $I$ is a *-ideal of $A$ and $\ker(f) \subseteq I$, then $A/I$ is *-isomorphic to $f(A)/f(I)$;

(f) if $J$ is any *-ideal of $B$, then $A/f^{-1}(J)$ is *-isomorphic to $f(A)/(J \cap f(A))$.

(IV.5) If $I$ and $J$ are *-ideals in a *-algebra $A$ with $J \subseteq I$, prove that $I/J$ is a *-ideal of $A/J$ and that $(A/J)/(I/J)$ is *-isomorphic to $A/I$.

(IV.6) If $I$ and $J$ are *-ideals in a *-algebra $A$, prove that $I/(I \cap J)$ is *-isomorphic to $(I + J)/J$.

(IV.7) Let $f_1$ and $f_2$ be *-homomorphisms from a *-algebra $A$ onto *-algebras $B_1$ and $B_2$, respectively. If $\ker(f_1) \subseteq \ker(f_2)$, prove that there exists a unique *-homomorphism $g: B_1 \to B_2$ such that $f_2 = g \circ f_1$.

(IV.8) Let $I$ be a $^*$-ideal of a $^*$-algebra $A$ and suppose $B_1$ and $B_2$ are $^*$-subalgebras of $A$ which contain $I$. If $\tau: A \to A/I$ is the quotient mapping, prove that:

(a) $B_1 \subseteq B_2$ iff $\tau(B_1) \subseteq \tau(B_2)$;

(b) $\tau(B_1 \cap B_2) = \tau(B_1) \cap \tau(B_2)$.

(IV.9) Let $B$ be a $^*$-subalgebra of a $^*$-algebra $A$. If $I$ is a $^*$-ideal of $A$ such that $B \cap I = \{0\}$, prove that $B$ is a $^*$-isomorphic to a $^*$-subalgebra of the $^*$-algebra $A/I$.

(IV.10) Let $A$ be a $^*$-algebra with identity and $M_n(A)$ the $^*$-algebra of $n \times n$ matrices over $A$. Prove:

(a) if $I$ is a $^*$-ideal of $A$, then $M_n(I)$ is a $^*$-ideal in the matrix algebra $M_n(A)$.

(b) every $^*$-ideal in $M_n(A)$ is of the form $M_n(I)$ for some $^*$-ideal $I$ in $A$.

(IV.11) Show that a linear functional $f$ on a $^*$-algebra $A$ is hermitian iff $f(x)$ is real for each hermitian element $x$ in $A$.

(IV.12) Let $A$ be a $^*$-algebra with identity $e$. If $y$ is a hermitian invertible element of $A$, show that the mapping $x \to x'$, where $x' = y^{-1}x^*y$, is an involution on $A$.

(IV.13) Let $A$ be a $^*$-algebra with identity $e$. Prove that an element $x$ in $A$ is invertible iff both $x^*x$ and $xx^*$ are invertible.

(IV.14) An involution $x \to x^*$ in an algebra $A$ is called *proper* if $x^*x = 0$ implies $x = 0$. The involution is called *quasi-proper* if $xx^* = 0$ implies $x^*x = 0$.

(a) Prove that every proper involution is quasi-proper;

(b) Give an example of a quasi-proper involution which is not proper;

(c) Give an example of an involution which is not quasi-proper.

(IV.15) Let $A$ be a $^*$-algebra with identity $e$ and assume that the involution is proper. Prove that $x^*x = -e$ implies $xx^* = -e$.

(IV.16) Prove that a $^*$-algebra with proper involution has no nonzero left or right nilpotent ideals. (An ideal $I$ is called *nilpotent* if $I^n = \{0\}$ for some positive integer $n$, where $I^n$ denotes the set of all finite sums of products of $n$ elements taken from $I$.)

(IV.17) Assume that A is an algebra with identity e and proper involution $x \to x^*$. If u is a normal element of A, prove that:

(a) $(e - u)x = 0$ iff $x^*(e - u) = 0$ for all $x \in A$.

(b) If u is right or left quasi-regular, then u is quasi-regular.

(IV.18) Let A be a *-algebra with identity e. An element x in A satisfying $x^* = x^n$ for some integer $n > 1$ will be called a *hermitian element of order* n or an $H_n$*-element*. Prove that:

(a) Every $H_n$-element is normal.

(b) If x is an $H_n$-element, then $x = x^{n^2}$.

(c) If x is an $H_n$-element, then x is invertible iff $x^{n^2-1} = e$.

(d) If x and y are $H_n$-elements for which $xy = yx$, then xy is an $H_n$-element.

(IV.19) Let A be a *-algebra with identity e. If $n > 1$, define $H_{1/n}$-*elements* by $(x^*)^n = x$. Prove every $H_{1/n}$-element is an $H_n$-element.

(IV.20) Let A be a *-algebra and p a positive functional on A. For $a \in A$, define $p_a: A \to C$ by $p_a(x) = p(ax)$. Show that

$$\bigcap_{a \in A} \{x \in A: p_a(x) = 0\} = \{x \in A: p(x^*x) = 0\}.$$

(IV.21) Let A be a *-algebra with identity e and $F$ the set of all positive functionals p on A such that $p(e) = 1$. Assume that $F$ is nonempty and define

$$|x| = \sup\{p(x^*x)^{1/2}: p \in F\}.$$

Prove that $|\cdot|$ has the following properties:

(a) $|x + y| \leq |x| + |y|$;

(b) $|\lambda x| = |\lambda| \cdot |x|$, $\lambda \in C$

(c) $|xy| \leq |x| \cdot |y|$

(d) $|x^*| = |x|$.

(IV.22) Let A be a *-algebra with identity e and p a positive functional on A such that $p(e) = 1$ and $p(x^2) = p(x)^2$ for all hermitian $x \in A$. Prove that:

(a) ker(p) is a *-ideal of A;

(b) $p(xy) = p(x)p(y)$ for all $x, y \in A$.

(IV.23) Construct a Banach algebra with discontinuous involution which is distinct from those in §21.

(IV.24) Show that if the hypothesis of semisimplicity is dropped in (22.4) the conclusion may fail.

(IV.25) Give an example of a *-Banach algebra without identity which admits no nonzero positive functionals.

(IV.26) Give an example of a normed *-algebra A satisfying $||x^*x|| = ||x||^2$ for all x in A and a positive functional p on A which is discontinuous.

(IV.27) Let A and B be *-algebras with identities $e_A$ and $e_B$, respectively. A *-linear map $\phi$ of A into B with $\phi(e_A) = e_B$ is called a C*-*homomorphism* if $\phi(x^2) = \phi(x)^2$ for all hermitian $x \in A$. Prove that:

(a) a *-linear map $\phi: A \to B$ is a C*-homomorphism iff $\phi(xy + yx) = \phi(x)\phi(y) + \phi(y)\phi(x)$ for all $x, y \in A$.

(b) a *-linear map $\phi: A \to B$ is a C*-homomorphism iff $\phi(x^n) = \phi(x)^n$ for all $x \in A$ and each positive integer n.

(IV.28) The involution $x \to x^*$ of a Banach *-algebra A is *locally continuous* if it is continuous on every maximal commutative *-subalgebra of A. Prove that if $x \to x^*$ is locally continuous and A is spanned by a finite number of its maximal commutative *-subalgebras (such algebras are said to be *full*), then $x \to x^*$ is continuous.

(IV.29) Give an example of an infinite-dimensional, noncommutative, Banach *-algebra which is full.

(IV.30) Let A be a Banach *-algebra. Prove that the involution $x \to x^*$ is continuous iff the set F of all continuous hermitian linear functionals on A is total in A* (i.e., for each nonzero $x \in A$ there corresponds $f \in F$ such that $f(x) \neq 0$).

(IV.31) Let A be a Banach *-algebra. The involution $x \to x^*$ is said to be *regular* if $x \in H(A)$ and $\nu(x) = 0$ imply $x = 0$. The involution is *proper* if $xx^* = 0$ implies $x = 0$.

(a) If * is proper on A, prove that it is proper on $A_e$.

(b) If * is regular on A, prove that it is regular on $A_e$.

(c) Prove that $x \to x^*$ is regular iff every maximal commutative *-subalgebra of A is semisimple.

(IV.32) Let A be a unital *-Banach algebra. If p is a positive functional on A prove, for $x_i, y_i \in A$, $i = 1,\ldots,n$. that:

(a) $|p(\Sigma_{i=1}^{n} y_i^* x_i)|^2 \leq p(\Sigma_{i=1}^{n} y_i^* y_i) p(\Sigma_{i=1}^{n} x_i^* x_i)$.

(b) $p((\Sigma_{i=1}^{n} y_i^* x_i)^* (\Sigma_{i=1}^{n} y_i^* x_i)) \leq p(\Sigma_{i=1}^{n} x_i^* x_i) \nu(\Sigma_{i=1}^{n} y_i^* y_i)$.

(IV.33) Let $A$ be a *-algebra with identity $e_A$ and $B$ a commutative $C^*$-algebra with identity $e_B$. Suppose that $\phi$ is a *-linear map of $A$ into $B$ with $\phi(e_A) = e_B$ and $\phi(x^*x) \geq 0$ for each $x \in A$. Prove that $\phi(x^2) = \phi(x)^2$ for each hermitian element $x$ in $A$ iff $\phi$ is a *-homomorphism.

(IV.34) Let $A$ be a $C^*$-algebra, $B$ a commutative $C^*$-algebra and $f: A \to B$ a positive linear map. Prove that

$$|f(xy)|^2 \leq f(xx^*) f(y^*y), \quad x, y \in A$$

where $|x|$ denotes the unique positive square root of $x^*x$.

(IV.35) Find a $C^*$-algebra with identity $e$ and an element $x$ in $A$ such that $xx^* = e$ but such that $x$ is not invertible in $A$.

(IV.36) Let $A$ be a $C^*$-algebra with identity and let $x$ be invertible in $A$. If $||x|| = ||x^{-1}|| = 1$, prove that $x$ is unitary.

(IV.37) Let $A$ and $B$ be unital $C^*$-algebras and let $\phi: A \to B$ be a homomorphism which preserves the identities. Prove that if $||\phi(x)|| \leq ||x||$ for all $x \in A$ then $\phi$ is a *-homomorphism.

(IV.38) Let $A$ be a $C^*$-algebra and $f$ and $g$ positive functionals on $A$. Prove that $||f + g|| = ||f|| + ||g||$.

(IV.39) Let $A$ be a $C^*$-algebra. Show that for $x, y \in A$:

(a) $xy = 0$ iff $x^*xy = 0$;

(b) $xx^*x = x$ iff $x^*x$ is a projection;

(c) if $x^*x = xx^*$ and $x^2 = x$, then $x^* = x$.

(IV.40) Let $A$ be a $C^*$-algebra. If $x, y \in A$ and if $zx = zy$ for all $z \in A$, prove that $x = y$.

(IV.41) Let $A$ be a $C^*$-algebra with identity. An element $x$ in $A$ is *relatively invertible* if there is $u \in A$ such that $xux = x$. If $u$ is a projection in $A$, and $x$ is any element of $A$ such that $||x - u|| < 1$, prove that $xu$ is relatively invertible in $A$.

(IV.42) Let $A$ be a $C^*$-algebra. If $I$ is a nonzero left or right ideal of $A$, prove that $I^2$ is nonzero.

(IV.43) Let A be a C*-algebra. An element $s \in A$ is called *single* if, whenever $asb = 0$ for some a, b in A, then at least one of as, sb is zero. Let $S$ denote the set of single elements of A. Prove that:

(a) if $s \in S$, then $s^* \in S$;

(b) if $s \in S$ and $x \in A$, then $xs, sx \in S$;

(c) if $s \in S$ and s is normal, then $s^2 = \lambda s$ for some complex number $\lambda$;

(d) if $s \in S$ and s is normal, then $s = \lambda e$, where $e \in S$ and e is a projection;

(e) if $s \in S$, then there exist hermitian idempotents e and f, both in $S$, such that $s = fse$.

(IV.44) Verify, in the proof of (22.18), that $u_\lambda^* = u_\lambda^{-1}$.

(IV.45) Let A be the C*-algebra C([0,1]) and let $I = \{f \in A: f(0) = 0\}$. Show that I is a *-ideal and describe the quotient algebra A/I.

(IV.46) Let B be a C*-subalgebra of a C*-algebra A. Show that for each positive functional p on B there is a norm-preserving extension of p to a positive functional on A.

(IV.47) Let I be a closed two-sided ideal of a C*-algebra A and let x be an hermitian element of A. Show that there is an element $y \in I$ satisfying $||x + y|| = ||x + I||$.

# 5

# *-Representations on a Hilbert Space: A Closer Look

## §25. *Introduction.*

We now wish to study *-representations and positive functionals in detail. Basic properties of *-representations will be established. The GNS construction will be considered once again. This time the construction will be given for Banach *-algebras with arbitrary, possibly discontinuous, involutions. Further, we will see that the identity element can be dispensed with. Irreducible *-representations and pure states will be studied next. A bijective correspondence will be established between these two concepts which will enable us to show that every C*-algebra has enough irreducible *-representations to separate the points of the algebra. Finally, the *-radical will be defined and studied. Applications of these results will be given later.

## §26. **-representations on a Hilbert space.*

Many notions concerning *-representations will be defined and studied in this section. For the most part the results given here lie near the surface.

(26.1) DEFINITION. *Let* A *be a *-algebra. A *-<u>representation</u> of* A *on a Hilbert space* H *is a *-homomorphism* $\pi$ *of* A *into the *-algebra* B(H). *If* A *has an identity* e *and* $\pi(e) = I$, *the identity operator on* H, *then* $\pi$ *is called <u>unital</u>. The Hilbert space* H *is called the <u>space</u> of* $\pi$, *sometimes denoted* $H_\pi$. *The <u>dimension</u> of the Hilbert space* H *is called the <u>dimension</u> of* $\pi$, *denoted* dim $\pi$. *A linear subspace* M *of* H *is said to be <u>invariant</u> under* $\pi$ *if*

$\pi(x)(M) \subseteq M$ *for all* $x \in A$. *If* $\ker(\pi) = \{0\}$, *the representation* $\pi$ *is said to be* *faithful*.

If $\pi$ is a $^*$-representation of $A$ on $H$, then $\{0\}$ and $H$ are trivial invariant subspaces for $\pi$. The next result shows that if $M$ is an invariant subspace, then two others are always at hand.

(26.2) PROPOSITION. *Let* $\pi$ *be a* $^*$*-representation of a* $^*$*-algebra* $A$ *on a Hilbert space* $H$. *If* $M$ *is a linear subspace of* $H$ *invariant under* $\pi$, *then the orthogonal complement* $M^{\perp}$ *and the closure* $\overline{M}$ *of* $M$ *are also invariant under* $\pi$.

*Proof.* Let $\xi \in M^{\perp}$, $\eta \in M$, and $x \in A$. Then

$$(\pi(x)\xi|\eta) = (\xi|\pi(x)^*\eta) = (\xi|\pi(x^*)\eta) = 0$$

since $\pi(x^*)\eta \in M$. Hence, $\pi(x)\xi$ is orthogonal to every $\eta \in M$ and is therefore in $M^{\perp}$. Thus, $M^{\perp}$ is invariant under $\pi$. Since $\overline{M} = (M^{\perp})^{\perp}$, it also follows that $\overline{M}$ is invariant. □

The following characterization of closed invariant subspaces in terms of their associated projections is frequently useful.

(26.3) PROPOSITION. *Let* $A$ *be a* $^*$*-algebra and* $\pi$ *a* $^*$*-representation of* $A$ *on a Hilbert space* $H$. *Let* $M$ *be a closed subspace of* $H$ *and let* $P$ *be the orthogonal projection which maps* $H$ *onto* $M$. *Then* $M$ *is invariant under* $\pi$ *iff* $P\pi(x) = \pi(x)P$ *for all* $x \in A$.

*Proof.* Assume that $M$ is invariant under $\pi$; if $x \in A$ and $\xi \in H$ then $P\xi \in M$, $\pi(x)P\xi \in M$, and so $P\pi(x)P\xi = \pi(x)P\xi$. Therefore, for $x \in A$,

$$\pi(x)P = P\pi(x)P. \tag{1}$$

Taking adjoints we have $P\pi(x)^* = P\pi(x)^*P$ or

$$P\pi(x^*) = P\pi(x^*)P \tag{2}$$

for all $x \in A$. Replacing $x$ by $x^*$ in (2) gives

$$P\pi(x) = P\pi(x)P. \tag{3}$$

Combining (1) and (3) we have $P\pi(x) = \pi(x)P$.

Conversely, suppose that $P\pi(x) = \pi(x)P$ for all $x \in A$. If $\xi \in M$ and $x \in A$ we have

$$P\pi(x)\xi = \pi(x)P\xi = \pi(x)\xi$$

so that $\pi(x)\xi \in M$. Hence, M is invariant under $\pi$. □

We briefly recall the definition of the direct sum of a family of Hilbert spaces. Let $\{H_\alpha\}_{\alpha\in\Gamma}$ be such a family. The direct sum $\bigoplus_{\alpha\in\Gamma} H_\alpha$ of the family is the set of all functions $\xi: \Gamma \to \bigcup_{\alpha\in\Gamma} H_\alpha$ with $\xi(\alpha) \in H_\alpha$ for all $\alpha \in \Gamma$ and such that $\sum_\alpha ||\xi(\alpha)||_\alpha^2 < \infty$. The operations in $\bigoplus_{\alpha\in\Gamma} H_\alpha$ are defined coordinatewise and the inner product is defined by

$$(\xi|\eta) = \sum_\alpha (\xi(\alpha)|\eta(\alpha))_\alpha.$$

One can then verify that $\bigoplus_{\alpha\in\Gamma} H_\alpha$ is a Hilbert space.

(26.4) DEFINITION. *Let* A *be a *-algebra, and let* $\{\pi_\alpha\}_{\alpha\in\Gamma}$ *be a family of *-representations of* A *on Hilbert spaces* $H_\alpha$. *Assume, for each* $x \in A$, *that there is a nonnegative real number* $k_x$ *such that*

$$||\pi_\alpha(x)|| \leq k_x \quad \textit{for all} \quad \alpha \in \Gamma. \tag{4}$$

*The* <u>*direct sum*</u> $\pi = \bigoplus_{\alpha\in\Gamma} \pi_\alpha$ *of* $\{\pi_\alpha\}_{\alpha\in\Gamma}$ *is defined to be the *-representation of* A *which acts on the Hilbert space direct sum* $H = \bigoplus_{\alpha\in\Gamma} H_\alpha$ *as follows: if* $\xi \in H$, $x \in A$, *and* $\eta(\alpha) = \pi_\alpha(x)\xi(\alpha)$ *for all* $\alpha \in \Gamma$, *then*

$$\pi(x)\xi = \eta.$$

*Because of* (4) *each* $\pi(x)$ *is a bounded linear operator on* H; *indeed,*

$$||\pi(x)\xi||^2 = \sum_\alpha ||\pi_\alpha(x)\xi(\alpha)||_\alpha^2 \leq \sum_\alpha ||\pi_\alpha(x)||^2||\xi(\alpha)||_\alpha^2 \leq k_x^2||\xi||^2.$$

The boundedness condition (4) automatically holds if $\Gamma$ is finite, or if A is a *-Banach algebra (see (24.2)). The Hilbert spaces $H_\alpha$ in the direct sum $H = \bigoplus_{\alpha\in\Gamma} H_\alpha$ can be regarded (in an obvious way) as closed linear subspaces of H. As such they are invariant subspaces of the direct sum representation $\pi$. Thus, H is a direct sum of invariant subspaces of $\pi = \bigoplus_{\alpha\in\Gamma} \pi_\alpha$.

(26.5) DEFINITION. *Let* A *be a* *-*algebra and* $\pi$ *a* *-*representation of* A *on a Hilbert space* H. *The* <u>*null space*</u> *of* $\pi$ *is the set*

$$N(\pi) = \{\xi \in H: \pi(x)\xi = 0 \ \text{for all} \ x \in A\}.$$

$N(\pi)$ *is a closed invariant subspace of* H *on which* $\pi$ *is trivial. The orthogonal complement* $N(\pi)^\perp$ *of* $N(\pi)$ *in* H *is called the* <u>*essential space*</u> *of* $\pi$. *If* $N(\pi) = \{0\}$, $\pi$ *is said to be* <u>*nondegenerate*</u>.

One verifies easily that a *-representation $\pi$ of a *-algebra A is nondegenerate iff $\pi(A)H = \{\pi(x)\xi: x \in A, \xi \in H\}$ spans a dense subspace of H. If A has an identity, then $\pi$ is nondegenerate iff $\pi$ is unital.

Any *-representation $\pi$ of A is the direct sum $\pi_1 \oplus \pi_2$ of a nondegenerate representation $\pi_1$ and a trivial representation $\pi_2$; indeed, set $\pi_1 = \pi|_{N(\pi)^\perp}$ and $\pi_2 = \pi|_{N(\pi)}$. Finally, the direct sum of a family of nondegenerate *-representations is again nondegenerate.

(26.6) DEFINITION. *Let* $\pi$ *be a* *-*representation of a* *-*algebra* A *on a Hilbert space* H. *If* K *is a closed invariant subspace of* H, *then the mapping* $x \to \pi(x)|_K$ ($\pi(x)$ *restricted to* K) *defines a* *-*representation of* A *with space* K. *This representation, denoted* $\pi_K$, *is called the* <u>*subrepresentation of*</u> $\pi$ <u>*acting on*</u> K.

Any subrepresentation of a nondegenerate *-representation is nondegenerate.

Recall that a linear isometry of one Hilbert space onto another is called a *unitary transformation.*

(26.7) DEFINITION. *Two* *-*representations* $\pi$ *and* $\pi'$ *of a* *-*algebra* A *on Hilbert spaces* H *and* H', *respectively, are said to be* <u>*unitarily equivalent*</u> *(or simply* <u>*equivalent*</u>*), denoted* $\pi \simeq \pi'$, *if there exists a unitary transformation* U *mapping* H *onto* H' *such that* $\pi(x) = U\pi'(x)U^{-1}$ *for all* $x \in A$. *The unitary transformation* U *is said to* <u>*implement*</u> *the equivalence between* $\pi$ *and* $\pi'$.

Unitary equivalence is obviously an equivalence relation on the set of all *-representations of A. In general, we shall only be interested in the classification of representations up to equivalence.

We also mention another notion along these lines. Let H and H' be Hilbert spaces. Suppose that A and B are *-subalgebras of B(H) and B(H') respectively. If there is a *-isomorphism $\phi: A \to B$ and a unitary transformation U of H onto H' such that $\phi(S) = USU^{-1}$ for all S in A, then A and B are said to be *spatially isomorphic*, and $\phi$ is called a *spatial isomorphism*.

(26.8) DEFINITION. *Let* $\pi$ *and* $\pi'$ *be two *-representations of a *-algebra* A. *If* $\pi'$ *is unitarily equivalent to a subrepresentation of* $\pi$, *then* $\pi'$ *is said to be* <u>*contained in*</u> $\pi$, *and* $\pi$ *is said to* <u>*contain*</u> $\pi'$. *If* $\pi'$ *is contained in* $\pi$, *we shall write* $\pi' \leq \pi$.

If $\pi$ is the direct sum of a family $\{\pi_\alpha\}_{\alpha\in\Gamma}$ of *-representations, then clearly $\pi_\alpha \leq \pi$ for all $\alpha \in \Gamma$.

(26.9) DEFINITION. *A *-representation* $\pi$ *of a *-algebra* A *on a Hilbert space* H *is said to be* <u>*cyclic*</u> *if there is a vector* $\xi$ *in* H *such that the linear subspace* $\pi(A)\xi = \{\pi(x)\xi: x \in A\}$ *is dense in* H. *The vector* $\xi$ *is called a* <u>*cyclic vector*</u> *for* $\pi$.

Every cyclic *-representation is obviously nondegenerate. On the other hand, the following proposition essentially reduces the study of *-representations to that of cyclic *-representations.

(26.10) PROPOSITION. *Let* $\pi$ *be a nondegenerate *-representation of a *-algebra* A *on a Hilbert space* H. *Then* $\pi$ *is the direct sum of cyclic *-representations.*

*Proof*. Let $\xi_0$ be a nonzero vector in H. Since $\pi$ is nondegenerate, $\pi(x)\xi_0 \neq 0$ for some $x \in A$. Set $H_0 = \{\pi(x)\xi_0: x \in A\}^-$; then $H_0$ is a nonzero closed invariant subspace of H with cyclic vector $\xi_0$. If $H_0 = H$, we are done; otherwise, $H_0^\perp$ is a nonzero invariant subspace of H (26.2). Repeating this argument gives a nonzero closed invariant subspace $H_1$ on which $\pi$ is cyclic and such that $H_1 \perp H_0$.

Now, let T denote the set whose elements are collections of nonzero mutually orthogonal closed invariant cyclic subspaces of H. (A closed invariant subspace M is called cyclic if $\pi|M$ is a cyclic representation.) The set T is nonempty; for example, the pair $\{H_0,H_1\}$

constructed above is an element in $T$. Partially order $T$ by inclusion; a straightforward application of Zorn's lemma yields a maximal element $\{H_\alpha\}_{\alpha\in\Gamma}$ in $T$.

Let $K = \bigoplus_{\alpha\in\Gamma} H_\alpha$. If we can show that $K = H$, then $\pi$ will be the direct sum of the cyclic representations $\pi_\alpha = \pi|_{H_\alpha}$. Assume, to the contrary, that $K \neq H$. Then $K^\perp$ is a nonzero closed invariant subspace of $H$. Let $\xi$ be a nonzero vector in $K^\perp$. Since $\pi$ is nondegenerate, $H' = \{\pi(x)\xi: x \in A\}^-$ is a nonzero closed invariant cyclic subspace of $H$. Since $H'$ is orthogonal to all of the $H_\alpha$, then $\{H_\alpha\}_{\alpha\in\Gamma} \cup \{H'\}$ belongs to $T$, and it properly contains $\{H_\alpha\}_{\alpha\in\Gamma}$, a contradiction of the maximality of $\{H_\alpha\}_{\alpha\in\Gamma}$ in $T$. □

The next proposition gives necessary and sufficient conditions for two *-representations to be unitarily equivalent.

(26.11) PROPOSITION. *Let* $A$ *be a *-algebra.*

*(a) If two *-representations of* $A$ *are unitarily equivalent and one is cyclic, then so is the other.*

*(b) Two cyclic *-representations* $\pi$ *and* $\pi'$ *of* $A$ *on spaces* $H$ *and* $H'$, *respectively, are equivalent iff there exist cyclic vectors* $\xi \in H$ *and* $\xi' \in H'$ *such that for all* $x \in A$,

$$(\pi(x)\xi|\xi) = (\pi'(x)\xi'|\xi'). \qquad (1)$$

*Proof.* (a) If a vector $\xi$ in $H$ is cyclic for $\pi$ on $H$, and $\pi$ is unitarily equivalent to $\pi'$ on $H'$ under a unitary transformation $U$ of $H$ onto $H'$, $\pi'(x) = U\pi(x)U^{-1}$, $x \in A$, then $U\xi$ is a cyclic vector for $\pi'$ on $H'$.

(b) The necessity follows from the proof of (a) with $\xi' = U\xi$. Conversely, assume there are cyclic vectors $\xi$, $\xi'$ such that (1) is true for all $x$ in $A$. If $x, y \in A$, then

$$(\pi(x)\xi|\pi(y)\xi) = (\pi(y^*x)\xi|\xi)$$

$$= (\pi'(y^*x)\xi'|\xi') = (\pi'(x)\xi'|\pi'(y)\xi').$$

Setting $y = x$ gives $||\pi(x)\xi|| = ||\pi'(x)\xi'||$ for all $x$ in $A$. Define a mapping $U_0: \pi(A)\xi \to \pi'(A)\xi'$ by

$$U_0(\pi(x)\xi) = \pi'(x)\xi' \quad (x \in A).$$

Then $U_0$ is a linear isometry of a dense subspace of $H$ onto a dense subspace of $H'$ and hence can be extended to a unitary mapping $U$ of $H$ onto $H'$. To show that $\pi'(x) = U\pi(x)U^{-1}$, $x \in A$, we need only show that it holds on the dense subspace $\{\pi'(y)\xi' : y \in A\}$ of $H'$. But, for $x \in A$,

$$\begin{aligned} U\pi(x)U^{-1}(\pi'(y)\xi') &= U\pi(x)\pi(y)\xi \\ &= U\pi(xy)\xi = \pi'(xy)\xi' \\ &= \pi'(x)(\pi'(y)\xi'). \end{aligned}$$

Therefore, $\pi$ and $\pi'$ are equivalent. □

At this time we introduce a function which will play a particularly important role in our study of hermitian and symmetric *-algebras.

(26.12) DEFINITION. *Let* $A$ *be a Banach* *-*algebra. The* *__Pták function__* $\rho: A \to [0,\infty)$ *is the function defined by the formula* $\rho(x) = |x^*x|_\sigma^{1/2}$, $x \in A$. *If it is necessary to specify the algebra in which* $\rho$ *is being evaluated, we shall write* $\rho_A$.

The final result of this section concerns automatic continuity of *-representations on Banach *-algebras.

(26.13) THEOREM. *Let* $A$ *be a Banach* *-*algebra and let* $\pi$ *be a* *-*representation of* $A$ *on a Hilbert space* $H$. *Then:*

*(a)* $\pi$ *is continuous.*

*(b)* $||\pi(x)|| \leq \rho(x)$ *for all* $x \in A$.

*(c)* $||\pi|| \leq 1$ *if the involution in* $A$ *is isometric.*

*Proof.* (a) Since $B(H)$ is an $A^*$-algebra, $\pi$ is continuous by (23.11).

(b) Let $x \in A$. By equation (2) in the proof of (24.2), $||\pi(x)||^2 = |\pi(x^*x)|_\sigma \leq |x^*x|_\sigma$, i.e., $||\pi(x)|| \leq \rho(x)$.

(c) follows immediately from (24.2). □

## §27. *The GNS-construction revisited.*

The beautiful construction of Gelfand-Naimark and Segal considered in §17 will be examined in detail in this section. If $\pi$ is a $^*$-representation of a $^*$-algebra $A$ on a Hilbert space $H$ and $\xi$ is a fixed vector in $H$, then the mapping $p: A \to C$ defined by

$$p(x) = (\pi(x)\xi|\xi) \qquad (x \in A)$$

is a positive functional on $A$. We shall show that if $A$ is a Banach $^*$-algebra, then every extendable positive functional on $A$ arises in the above manner for some cyclic $^*$-representation and cyclic vector. The result will be established first for unital algebras and then extended to the general case. Once again we emphasize that no continuity assumptions will be made on the involutions in our algebras. One consequence of the results in this section will be that every positive functional on a Banach $^*$-algebra with bounded approximate identity is necessarily continuous, proving anew Varopoulos' theorem (22.12).

In the next definition we single out an important class of functionals.

(27.1) DEFINITION. *Let* $A$ *be a unital Banach* $^*$*-algebra with identity* $e$. *A* <u>*state*</u> *on* $A$ *is a positive functional* $p$ *on* $A$ *such that* $p(e) = 1$. *The set of all states on* $A$ *will be denoted by* $S(A)$.

If $p$ is a positive functional on a unital $^*$-Banach algebra, then $||p|| = p(e)$ by (22.11). Hence $p$ is a state iff $||p|| = 1$. When studying algebras without identity it is common to define a state as a positive functional $p$ such that $||p|| = 1$. One reason for introducing states is that the set of such functionals will lie in the closed unit ball of the dual space $A^*$ of $A$. This allows one to apply powerful compactness and duality theorems from functional analysis. The term "state" originated from connections with physics.

If $p$ is a state on a unital Banach $^*$-algebra we summarize for the reader's convenience properties which have already been established. For $x, y \in A$:

(1°) $p$ is hermitian (21.6), (a);

(2°) $|p(y^*x)|^2 \leq p(x^*x)p(y^*y)$ (21.5), (b);

(3°) $|p(x)| \leq p(x^*x)^{1/2}$ (21.5), (b), $y = e$;

(4°) $|p(h)| \leq |h|_\sigma$, $h \in H(A)$ (22.7), $x = e$;

(5°) $|p(x)| \leq \rho(x)$ (22.8), (b);

(6°) $p(x^*y^*yx) \leq |y^*y|_\sigma p(x^*x)$ (22.7), $h = y^*y$.

Let A be a unital Banach *-algebra and $\pi$ a *-representation of A on a Hilbert space H. If $\xi$ is a unit vector in H and p is defined by $p(x) = (\pi(x)\xi|\xi)$, $x \in A$, then p is a state on A. Conversely:

(27.2) THEOREM. *(Gelfand-Naimark-Segal). Let* A *be a unital Banach *-algebra and* p *a state on* A. *Then there exists a cyclic *-representation* $\pi$ *of* A *on a Hilbert space* H, *with cyclic vector* $\xi$ *of norm* 1 *in* H, *such that*

$$p(x) = (\pi(x)\xi|\xi) \quad (x \in A).$$

*Further, the representation* $\pi$ *is unique up to unitary equivalence.*

*Proof.* This proof is similar to the one in §17; however, it differs in two respects. First, a few more details will be supplied, and secondly it does not require the involution to be continuous.

For $x, y \in A$ set

$$(x|y) = p(y^*x).$$

This inner product on A is linear in x, conjugate-linear in y and $(x|x) \geq 0$. In general this inner product is degenerate, i.e., $(x|x) = 0$ for some $x \neq 0$, so a reduction is necessary to obtain nondegeneracy. To this end we define the associated null ideal

$$\begin{aligned} I &= \{x \in A\colon p(x^*x) = 0\} \\ &= \{x \in A\colon p(y^*x) = 0 \text{ for all } y \in A\}. \end{aligned} \tag{1}$$

The equality of these two sets follows easily from the Cauchy-Schwarz inequality 2° in the above list. Hence I is a left ideal in A. Then the quotient space $X = A/I$ is a pre-Hilbert space with respect to the induced inner product

$$(x + I|y + I) = p(y^*x).$$

To see that this product is well defined, assume that $x_1 + I = x_2 + I$ and $y_1 + I = y_2 + I$. Then $x_1 - x_2 \in I$ and $y_1 - y_2 \in I$; so by (1) and (1°) we have

$$p(y_1^*x_1) - p(y_2^*x_2) = p(y_1^*(x_1 - x_2)) + p((y_1 - y_2)^*x_2) = 0,$$

that is, $(x_1 + I|y_1 + I) = (x_2 + I|y_2 + I)$. Further, this inner product is definite since $(x + I|x + I) = 0$ iff $p(x^*x) = 0$ iff $x \in I$.

For each $x \in A$ define a linear operator $\theta(x)$ on $X$ by

$$\theta(x)(y + I) = xy + I.$$

Since $I$ is a left ideal, each $\theta(x)$ is well defined. Further, the map $x \to \theta(x)$ has the following easily verified properties: For $x, y \in A$ and complex $\lambda$, $\theta(x + y) = \theta(x) + \theta(y)$, $\theta(\lambda x) = \lambda\theta(x)$, $\theta(xy) = \theta(x)\theta(y)$, and $\theta(e)$ is the identity operator on $X$. Also, for $x \in A$, $\theta(x^*) = \theta(x)^*$ since

$$\begin{aligned}(\theta(x)(y + I)|z + I) &= (xy + I|z + I) = p(z^*(xy)) \\ &= p((x^*z)^*y) = (y + I|x^*z + I) \\ &= (y + I|\theta(x^*)(z + I)).\end{aligned}$$

Hence $x \to \theta(x)$ is a *-representation of $A$ on the pre-Hilbert space $X$.

We wish to show that $\theta(x)$ is a bounded linear operator on $X$ for each $x \in A$. Indeed for any $y$ not in $I$ we have by 6° that

$$\frac{||\theta(x)(y + I)||^2}{||y + I||^2} = \frac{||xy + I||^2}{||y + I||^2} = \frac{p(y^*x^*xy)}{p(y^*y)} \leq |x^*x|_\sigma.$$

It follows that $||\theta(x)|| \leq \rho(x)$ for each $x \in A$.

Let $H$ be the Hilbert space completion of $X$, and let $\pi(x)$ denote the unique extension of $\theta(x)$ to a bounded linear operator on $H$. Since $X$ is dense in $H$, the algebraic properties of $\theta(x)$ extend to $\pi(x)$ and hence $x \to \pi(x)$ is a *-representation of $A$ on $H$. Setting $\xi = e + I \in X$, we have $||\xi|| = 1$, and for all $x \in A$,

$$p(x) = p(e^*x) = (x + I|e + I) = (\pi(x)\xi|\xi).$$

Also, since $X$ is dense in $H$, $\xi$ is a cyclic vector for $\pi$.

We make two final observations to be recorded below. First, $\ker(\pi) = \ker(\theta)$ and

$$\ker(\pi) = \{x \in A\colon xy \in I \text{ for all } y \in A\} = (I:A).$$

Second, for each $x \in A$,

$$p(x^*x) = ||x + I||^2 = ||\theta(x)(e + I)||^2 = ||\pi(x)\xi||^2 \le ||\pi(x)||^2;$$

hence $p(x^*x)^{1/2} \le ||\pi(x)||$ for each $x \in A$. Uniqueness follows from (26.11). □

(27.3) REMARK. Given a state $p$ on the unital Banach *-algebra $A$, the triple $(\pi, H, \xi)$ constructed in (27.2) is often denoted by $(\pi_p, H_p, \xi_p)$ to show dependence on $p$, and is called the GNS-*triple associated with* $p$. Replacing $\pi$, $H$, $\xi$ and $I$ by $\pi_p$, $H_p$, $\xi_p$ and $I_p$, respectively, and letting $\tau_p\colon A \to A/I_p$ denote the quotient mapping $\tau_p(x) = x + I_p$, we note that the following properties were established in the proof of (27.2). For all $x, y \in A$:

(a) $\tau_p$ is a continuous linear map of $A$ onto a dense linear subspace of $H_p$;

(b) $(\tau_p(x)|\tau_p(y)) = p(y^*x)$;

(c) $\pi_p(a)\tau_p(x) = \tau_p(ax)$;

(d) $\xi_p = \tau_p(e)$, so $\tau_p(x) = \pi_p(x)\xi_p$;

(e) $p(x^*x)^{1/2} \le ||\pi_p(x)|| \le \rho(x)$;

(f) $\ker(\pi_p) = (I_p:A) = \{x \in A\colon xy \in I_p \text{ for all } y \in A\}$.

With (27.2) established, we give a second proof of (22.12), which states that *every* positive functional on a Banach *-algebra with bounded approximate identity is continuous.

(27.4) THEOREM. *Let* $p$ *be a positive functional on a Banach *-algebra* $A$ *with bounded approximate identity. Then* $p$ *is continuous. In particular, if* $A$ *is unital, every state on* $A$ *is continuous.*

*Proof*. Suppose first that $A$ is unital and that $p$ is a state on

A. By (27.2) p can be written in the form $p(x) = (\pi(x)\xi|\xi)$, $x \in A$, where $\pi$ is a *-representation of A on a Hilbert space H and $\xi \in H$. By (26.13) $\pi$ is continuous; hence there exists $\gamma > 0$ such that $||\pi(x)|| \leq \gamma||x||$ for all $x \in A$. Therefore

$$|p(x)| = |(\pi(x)\xi|\xi)| \leq ||\pi(x)||\cdot||\xi||^2 \leq \gamma||\xi||^2||x||$$

for all $x \in A$. Thus, if p is a state it is continuous.

Now, still assuming that A is a unital Banach *-algebra, let p be any positive functional on A. If $p \equiv 0$, it is certainly continuous. If $p \not\equiv 0$, then there is a state q on A and $\lambda > 0$ such that $p = \lambda q$. Indeed, set $\lambda = p(e)$ and let $q(x) = p(e)^{-1}p(x)$ ($\lambda > 0$ by (21.6), (b)). Then q is continuous from the above and, hence, p is continuous.

Finally, assume that A has a bounded approximate identity and that p is any positive functional on A. For fixed $a \in A$ the linear functional ${}_{a^*}p_a(x) = p(a^*xa)$ is positive and extendable by (21.8), hence is continuous by the preceding step. The remainder of the proof is identical to that of (22.12). □

(27.5) COROLLARY. *Let* A *be a Banach *-algebra with bounded approximate identity. Then every positive functional on* A *is continuous, hermitian, and extendable.*

*Proof.* Apply (27.4) to obtain continuity and then observe that the proof of (22.15), (a) applies with $||p||$ replaced by $M\cdot||p||$, where $M = \sup||e_\alpha^* e_\alpha||$. □

We next extend (27.2) to arbitrary Banach *-algebras.

(27.6) THEOREM. *Let* A *be a Banach *-algebra. Let* $\pi$ *be a *-representation of* A *on a Hilbert space* H, *and let* $\xi \in H$. *Then the function* $p(x) = (\pi(x)\xi|\xi)$, $x \in A$, *is an extendable positive linear functional on* A.

*Conversely, if* p *is an extendable positive linear functional on* A, *then there is a cyclic *-representation* $\pi$ *of* A *on a Hilbert space* H, *unique up to unitary equivalence, and a cyclic vector* $\xi \in H$ *such that*

$$p(x) = (\pi(x)\xi|\xi) \quad (x \in A).$$

*In particular, if* A *has a bounded approximate identity, there is a one-*

*to-one correspondence between positive functionals on* A *and unitary equivalence classes of cyclic *-representations of* A.

*Proof*. Let $\pi$ be a *-representation of A on a Hilbert space H. If $A_e$ is the unitization of A, then $\pi$ can be extended to a *-representation $\pi'$ of $A_e$ by setting $\pi'(x + \lambda e) = \pi(x) + \lambda I$, where I denotes the identity operator on H. The corresponding positive functional $p'(x + \lambda e) = (\pi'(x + \lambda e)\xi|\xi)$, $x + \lambda e \in A_e$, is clearly a positive extension of p.

Conversely, assume that p is extendable. By (21.7), (b) there is $k \geq 0$ such that $|p(x)|^2 \leq kp(x^*x)$ for all $x \in A$. Assume k is the least such nonnegative number. In the proof of (21.7), (b) it was shown that $p'(x + \lambda e) = p(x) + k\lambda$ is a positive functional on $A_e$. By (27.2) there is a cyclic *-representation $\pi'$ of $A_e$ on a Hilbert space H corresponding to p', with cyclic vector $\xi \in H$. It is clear that $\pi'$ restricted to A is a *-representation on H which corresponds to p. It remains to show that this restriction $\pi$ is still cyclic.

Recall that H is the completion of $A_e/I'$, where I' is a left ideal in $A_e$, and $(x' + I'|y' + I') = p'((y')^*(x'))$ for $x', y' \in A_e$. Since k is minimal, we may select a sequence $\{x_n\}$ in A such that $p(x_n^*x_n) = 1$ and $|p(x_n)|^2 > k - 1/n$. Set $y_n = \overline{p(x_n)}x_n$. Then we have

$$\begin{aligned} ||(e + I') - (y_n + I')||^2 &= ||(e - y_n) + I'||^2 = p'((e - y_n)^*(e - y_n)) \\ &= p'(e - y_n^* - y_n + y_n^*y_n) \\ &= k - 2|p(x_n)|^2 + |p(x_n)|^2 p(x_n^*x_n) \\ &= k - |p(x_n)|^2 < 1/n. \end{aligned}$$

This shows that as an element of H, $\xi \equiv e + I' = \lim_n (y_n + I')$. Hence, for each $x \in A$ and $\lambda \in C$,

$$\begin{aligned} \pi'(x + \lambda e)\xi &= (x + \lambda e) + I' = \lim_n (x + \lambda y_n + I') \\ &= \lim_n \pi'(x + \lambda y_n)\xi. \end{aligned}$$

Since $x + \lambda y_n \in A$ for each $n$, this shows that $\{\pi'(x)\xi : x \in A\}^- =$ $\{\pi'(x + \lambda e)\xi : x + \lambda e \in A_e\}^- = H$, and thus $\pi = \pi'|A$ is cyclic with cyclic vector $\xi$. The uniqueness of $\pi$ follows from (26.11). The final statement is a consequence of (27.5). □

Let $H$ be a Hilbert space and $A$ a *-subalgebra of $B(H)$. Given $\xi \in H$ define a positive functional $\omega_\xi$ on $A$ by setting

$$\omega_\xi(S) = (S\xi|\xi) \qquad (S \in A).$$

Any positive functional $p$ on $A$ such that $p = \omega_\xi$ for some $\xi \in H$ is called a *vector functional*. If $A$ contains the identity operator on $H$ and $\xi$ is a unit vector, then $\omega_\xi$ is a state on $A$, called the *vector state defined by* $\xi$.

If $\pi$ is a *-representation of a *-algebra $A$ on a Hilbert space $H$ and $\xi \in H$, then the positive functional $p(x) = (\pi(x)\xi|\xi)$, $x \in A$, associated with $\pi$ and $\xi$ can be written simply as $p = \omega_\xi \circ \pi$.

## §28. *Irreducible *-representations.*

Let $A$ be a nonempty set of bounded linear operators on a Hilbert space $H$. If $\{0\}$ and $H$ are the only closed linear subspaces of $H$ which are invariant under all operators $T$ in $A$, then $A$ is said to be *topologically irreducible*. Otherwise, $A$ is said to be *reducible*, and a proper closed linear subspace $M$ of $H$ invariant under all $T$ in $A$ is said to *reduce* $A$.

(28.1) DEFINITION. *A *-representation* $\pi$ *of a *-algebra* $A$ *on a Hilbert space* $H$ *is said to be* <u>*topologically irreducible*</u> *if the set* $\pi(A) = \{\pi(x) : x \in A\}$ *of operators on* $H$ *is topologically irreducible.*

Thus, a *-representation $\pi$ of $A$ on $H$ is irreducible iff the only closed invariant subspaces of $H$ for $\pi$ are $\{0\}$ and $H$. Recall that irreducibility of $\pi$ in the algebraic sense (see ( B.5.6 )) means that the only linear subspaces of $H$ invariant under $\pi(A)$ are $\{0\}$ and $H$. In general, if $\dim \pi$ is infinite, algebraic irreducibility is much more restrictive than topological irreducibility. However, if

A is a C*-algebra it is a remarkable theorem of R. Kadison (see Dixmier [5, p. 53]) that these two notions of irreducibility coincide.

We shall be interested in topological irreducibility in this section (and in most of the remainder of our study of *-representations) so unless stated otherwise "irreducible" will mean "topologically irreducible."

Clearly, every *-representation of a *-algebra on a one-dimensional Hilbert space is irreducible. Of course, a one-dimensional *-representation of a *-algebra A can be regarded as a multiplicative linear functional on A.

Given a *-algebra A, one of the central problems in the theory is to find *all* of its irreducible *-representations up to unitary equivalence. It may be the case that no irreducible *-representations exist for a given *-algebra. However, we shall see later (30.7) that each C*-algebra has enough irreducible *-representations to separate its points.

(28.2) THEOREM. *Let* A *be a *-algebra and* $\pi$ *a non-trivial *-representation of* A *on a Hilbert space* H. *The following conditions on* $\pi$ *are equivalent:*

*(a)* $\pi$ *is irreducible;*

*(b) Every nonzero vector in* H *is a cyclic vector for* $\pi$*;*

*(c) The only bounded operators commuting with all* $\pi(x)$, $x \in A$, *are of the form* $\lambda I$ *for* $\lambda \in \mathbb{C}$.

*Proof.* (a) implies (b): Assume that $\pi$ is irreducible, and let $K = \{\xi \in H: \pi(x)\xi = 0$ for all $x \in A\}$, i.e., K is the null space of $\pi$ (26.5). Then K is a closed invariant subspace of $\pi$; hence either $K = \{0\}$ or $K = H$. The latter is ruled out because $\pi$ is non-trivial; thus $K = \{0\}$.

Now, let $\xi \in H$, $\xi \neq 0$. Then $M = \pi(A)\xi$ is a linear subspace of H invariant under $\pi$, and $\{0\} \neq M$. Since $\overline{M}$ is invariant under $\pi$ by (26.2), $\overline{M} = H$; that is, $\xi$ is a cyclic vector for $\pi$.

(b) implies (a): Suppose that every nonzero vector in H is cyclic and let M be a nonzero closed invariant subspace of H. Let $\xi \in M$, $\xi \neq 0$. Then $H = \{\pi(x)\xi: x \in A\}^- \subseteq M$ and hence $M = H$. Thus $\pi$ is irreducible.

(c) implies (a): Assume that (c) holds. By (26.3) a closed subspace

M of H is invariant iff the projection P of H onto M commutes with every $\pi(x)$. Thus, by hypothesis $P = \lambda I$ for some $\lambda \in C$. Since P is a projection, either $P = 0$ or $P = I$. Thus, $M = \{0\}$ or $M = H$.

(a) implies (c): Assume that $\pi$ is irreducible, and let S be a nonzero hermitian operator on H commuting with all $\pi(x)$. Let B denote the commutative C*-subalgebra of B(H) generated by S and the identity operator. By (8.1), (a) the spectrum $\sigma(S)$ of S is real. We assert that $\sigma(S)$ consists of one point. Indeed, if $\lambda_1$ and $\lambda_2$ are distinct points in $\sigma(S)$ choose, by Uryshohn's lemma, nonnegative continuous functions $f_1$, $f_2$ in $C(\sigma(S))$ such that $f_1 f_2 = 0$ and $f_1(\lambda_1) = 1 = f_2(\lambda_2)$. By (7.1) there exist operators $f_1(S)$, $f_2(S)$ in B corresponding to $f_1$, $f_2$ such that $f_1(S)f_2(S) = 0$ and $f_1(S) \neq 0$, $f_2(S) \neq 0$. Since S commutes with all $\pi(x)$, then $f_1(S)$, $f_2(S)$ also commute with all $\pi(x)$. Hence $M_1 = \{f_1(S)\xi\colon \xi \in H\}^-$ and $M_2 = \{f_2(S)\xi\colon \xi \in H\}^-$ are closed invariant subspaces for $\pi$. Neither of these subspaces is $\{0\}$ since $f_1(S) \neq 0$, $f_2(S) \neq 0$. Further, since $f_1(S)f_2(S) = 0$ it follows that both $M_1$ and $M_2$ are proper subspaces of H. This contradicts the irreducibility of $\pi$, and proves that $\sigma(S)$ has one element. A functional representation of S now shows that $S = \lambda I$ for some $\lambda \in C$.

Finally, if T is any bounded operator on H commuting with all $\pi(x)$, then $T^*\pi(x) = (\pi(x)^*T)^* = (T\pi(x)^*)^* = \pi(x)T^*$, so that $T^*$ likewise commutes with all $\pi(x)$. Hence $T_1 = (T + T^*)/2$ and $T_2 = (T - T^*)/2i$ are hermitian operators commuting with all $\pi(x)$, and so each has the form $\lambda I$ by the preceding paragraph. Therefore, $T = T_1 + iT_2$ also has this form. □

(28.3) PROPOSITION. *Every finite-dimensional *-representation* $\pi$ *of a *-algebra* A *is the direct sum of a finite number of irreducible *-representations.*

*Proof*. Let H be the representation space of $\pi$. The proposition is trivially true if $\dim H = 1$. Assume it is true for all representations $\pi$ on H such that $\dim H < n$. If $\dim H = n$ and $\pi$ is reducible, let $H_1$, $0 < \dim H_1 < n$, be an invariant subspace of $\pi$. Then $H_1^\perp$ is also invariant (26.2), and $0 < \dim H_1^\perp < n$. By the inductive hypothesis $\pi$ is the direct sum of irreducible *-representations on $H_1$ and $H_1^\perp$. □

(28.4) PROPOSITION. *Let* $A$ *be a separable Banach *-algebra and suppose that* $\pi$ *is an irreducible *-representation of* $A$ *on a Hilbert space* $H$. *Then* $H$ *is separable.*

*Proof.* If $H$ is zero-dimensional it is clearly separable. Otherwise, let $\xi$ be a nonzero vector in $H$. Then by (28.2), (b) $\xi$ is a cyclic vector for $\pi$ and hence $\{\pi(x)\xi: x \in A\}^{-} = H$. Since $A$ is separable there is a countable dense subset $D$ of $A$; then it is easy to check that $\{\pi(x)\xi: x \in D\}$ is a countable dense subset of $H$. □

§29. *Pure states and irreducible *-representations.*

The results of §27 show how to construct a *-representation from a given positive functional on a Banach *-algebra. We now concern ourselves with the problem of determining when this procedure leads to irreducible *-representations.

(29.1) DEFINITION. *Let* $p$ *and* $q$ *be positive functionals on a *-algebra* $A$. *If* $p(x^*x) \geq q(x^*x)$ *for all* $x$ *in* $A$, *then* $p$ *is said to* *__dominate__* $q$ *and we write* $p > q$ *or* $q < p$.

When $A$ is a $C^*$-algebra, the relation $<$ is a partial order on the positive functionals of $A$. Indeed, if $p > 0$ and $p < 0$, then $p$ is zero on $A^+$ and hence is zero on $H(A)$. Since each element of $A$ is a linear combination of hermitian elements, $p$ is zero.

(29.2) THEOREM. *Let* $A$ *be a Banach *-algebra,* $\pi$ *a cyclic *-representation of* $A$ *on a Hilbert space* $H$, $\xi$ *in* $H$ *a cyclic vector for* $\pi$, *and*

$$p(x) = (\pi(x)\xi|\xi) \quad (x \in A).$$

*If* $q$ *is an extendable positive functional on* $A$ *which is dominated by* $p$, *then there is a unique positive operator* $T$ *on* $H$, *such that*

$$q(x) = (T\pi(x)\xi|\xi) \quad (x \in A). \tag{1}$$

*Moreover,* $T$ *commutes with each* $\pi(x)$, $x \in A$, *and* $||T|| \leq 1$. *Conversely, any such operator* $T$ *defines, by relation* (1), *an extendable positive functional on* $A$ *which is dominated by* $p$.

*Proof*. Assume that $q$ is an extendable positive functional on $A$ which is dominated by $p$. Let $K = \{\pi(x)\xi : x \in A\}$. Since $\xi$ is cyclic, $K$ is a dense linear subspace of $H$.

For each $x, y$ in $A$,

$$|q(y^*x)|^2 \leq q(x^*x)q(y^*y) \leq p(x^*x)p(y^*y)$$
$$= ||\pi(x)\xi||^2||\pi(y)\xi||^2.$$

It follows from this that $q(y^*x) = 0$ if either $\pi(x)\xi = 0$ or $\pi(y)\xi = 0$. Hence, if we let

$$\Phi(\pi(x)\xi, \pi(y)\xi) = q(y^*x),$$

then $\Phi$ is a well-defined continuous sesquilinear functional on $K \times K$ which can be uniquely extended to a continuous sesquilinear functional on $H \times H$. Hence by (A.11) there is a unique bounded operator $T$ on $H$, $||T|| \leq 1$, such that

$$q(y^*x) = (T\pi(x)\xi|\pi(y)\xi) \quad (x, y \in A). \tag{2}$$

Since $q$ is positive, we have $(T\pi(x)\xi|\pi(x)\xi) = q(x^*x) \geq 0$ for all $x$ in $A$; since $K$ is dense in $H$, it follows that $T$ is a positive operator on $H$.

To see that $T$ commutes with each $\pi(x)$, let $x, y, z \in A$. Then

$$(T\pi(x)\pi(y)\xi|\pi(z)\xi) = q(z^*(xy)) = q((x^*z)^*y)$$

$$= (T\pi(y)\xi|\pi(x^*)\pi(z)\xi)$$

$$= (\pi(x)T\pi(y)\xi|\pi(z)\xi)$$

and again using the denseness of $K$ in $H$ we obtain $\pi(x)T = T\pi(x)$ for all $x$ in $A$.

We next show that $q(x) = (T\pi(x)\xi|\xi)$. If $A$ is unital, this follows from (2) by setting $y = e$. Otherwise we argue as follows. Since $q$ is extendable, there exists $k \geq 0$ such that

$$|q(x)|^2 \leq kq(x^*x) \leq kp(x^*x) = k||\pi(x)\xi||^2$$

for all $x$ in $A$. Hence, $q$ may be regarded as a continuous linear functional on $K$ which may be extended uniquely to $H$. By the Riesz-Fréchet theorem (A.9) there exists a vector $\eta \in H$ such that $q(x) = (\pi(x)\xi|\eta)$, $x \in A$.

For each $x, y$ in $A$ we then have

$$(\pi(x)\xi|T\pi(y)\xi) = (T\pi(x)\xi|\pi(y)\xi) = q(y^*x)$$

$$= (\pi(y^*x)\xi|\eta) = (\pi(x)\xi|\pi(y)\eta)$$

from which it follows that $T\pi(y)\xi = \pi(y)\eta$ for all $y$ in $A$. Therefore,

$$(T\pi(x)\xi|\xi) = (\pi(x)\eta|\xi) = (\eta|\pi(x^*)\xi)$$

$$= \overline{(\pi(x^*)\xi|\eta)} = \overline{q(x^*)} = q(x).$$

To show that $T$ is unique, suppose that $S$ is a positive operator on $H$ which also represents $q$ in the sense of (1), that is,

$$(S\pi(x)\xi|\xi) = (T\pi(x)\xi|\xi) \quad (x \in A).$$

Since $S$ and $T$ are hermitian,

$$(\pi(x)\xi|S\xi) = (\pi(x)\xi|T\xi)$$

for all $x \in A$, so that $S\xi = T\xi$. Since $\xi$ is a cyclic vector for $\pi$, it is a separating vector for the commutant of $\pi(A)$ (see B.6.19) and hence $S = T$.

Conversely, assume that $q(x) = (T\pi(x)\xi|\xi)$, where $T \geq 0$, $||T|| \leq 1$, and $T\pi(x) = \pi(x)T$ for all $x$ in $A$. Then, for $x$ in $A$

$$q(x^*x) = (T\pi(x^*x)\xi|\xi) = (T\pi(x)\xi|\pi(x)\xi) \geq 0$$

and

$$q(x^*) = (T\pi(x^*)\xi|\xi) = (\xi|T\pi(x)\xi) = \overline{q(x)};$$

hence $q$ is positive and hermitian. If $S$ denotes the unique positive square root of $T$, then $S$ commutes with each $\pi(x)$ (by applying the functional calculus to the commutant algebra $\pi(A)'$) and so

$$q(x) = (S^2\pi(x)\xi|\xi) = (\pi(x)S\xi|S\xi)$$

for all $x$ in $A$; hence $q$ is extendable. (See proof of (27.6).) Finally, since $||T|| \leq 1$, we have $q(x^*x) = (T\pi(x)\xi|\pi(x)\xi) \leq ||T||\cdot||\pi(x)\xi||^2 \leq p(x^*x)$, and we see that $p$ dominates $q$. □

(29.3) COROLLARY. *Let* $p$ *and* $q$ *be states on a Banach *-algebra with identity. Then the following conditions are equivalent:*

*(a)* $p$ *dominates* $q$*;*

*(b) There exists a *-representation* $\pi$ *of* $A$ *on a Hilbert space* $H$*, a vector* $\xi$ *in* $H$ *and a bounded positive operator* $T$ *on* $H$ *with* $||T|| \leq 1$*, such that* $T$ *commutes with every* $\pi(x)$ *and*

$$p(x) = (\pi(x)\xi|\xi),$$

$$q(x) = (T\pi(x)\xi|\xi).$$

(29.4) DEFINITION. *A positive functional* $p$ *on a *-algebra* $A$ *is said to be* <u>*pure*</u> *(or* <u>*indecomposable*</u>*) if every extendable positive functional on* $A$ *which is dominated by* $p$ *is of the form* $\lambda p$ *for some* $\lambda$*,* $0 \leq \lambda \leq 1$.

The significance of positive functionals which are pure can be seen from the following theorem.

(29.5) THEOREM. *Let* $A$ *be a *-algebra,* $\pi$ *a cyclic *-representation of* $A$ *on a Hilbert space* $H$*,* $\xi \in H$ *a cyclic vector for* $\pi$*, and* $p(x) = (\pi(x)\xi|\xi)$ *for all* $x$ *in* $A$*. Then* $\pi$ *is irreducible iff* $p$ *is pure.*

*Proof.* Suppose that $\pi$ is irreducible, and let $q$ be any extendable positive functional on $A$ which is dominated by $p$. By (29.2) there is a positive operator $T$ on $H$, $||T|| \leq 1$, which commutes with all $\pi(x)$, $x \in A$, such that $q(x) = (T\pi(x)\xi|\xi)$, $x \in A$. Since $\pi$ is irreducible (28.2) implies that $T = \lambda I$ for some $\lambda \in C$, where $I$ is the identity operator on $H$. Since $T \geq 0$ and $||T|| \leq 1$ it follows that $0 \leq \lambda \leq 1$. Hence, we have

$$q(x) = (T\pi(x)\xi|\xi) = \lambda(\pi(x)\xi|\xi) = \lambda p(x)$$

for all $x$ in $A$, with $0 \leq \lambda \leq 1$; i.e., $p$ is pure.

Conversely, assume that $p$ is pure and let $Q$ be a projection operator on $H$ which commutes with all $\pi(x)$, $x \in A$. Then $q(x) = (Q\pi(x)\xi|\xi) = (Q^2\pi(x)\xi|\xi) = (\pi(x)Q\xi|Q\xi)$ is an extendable (see the proof of (27.6)) positive functional on $A$ which is dominated by $p$ (29.2); because $p$ is pure there is a real $\lambda$, $0 \leq \lambda \leq 1$, such that $q = \lambda p$. By the uniqueness statement made in (29.2) $Q = \lambda I$, and since $Q$ is a projection, $Q = 0$ or $Q = I$. It now follows from (26.3) that $\pi$ is irreducible. □

(29.6) PROPOSITION. *Let* $A$ *be a *-Banach algebra with an approximate identity bounded by* 1. *Let* $B$ *denote the set of all positive functionals* $p$ *on* $A$ *with* $||p|| \leq 1$. *Then*

*(a)* $B$ *is a weak *-compact convex subset of the dual of* $A$.

*(b)* $0$ *is an extreme point of* $B$.

*(c) a nonzero* $p$ *in* $B$ *is an extreme point of* $B$ *iff* $||p|| = 1$ *and* $p$ *is pure.*

*(d)* $B$ *is the weak *-closed convex hull of its extreme points.*

*Proof.* (a) The set $B$ is clearly convex. Since the weak *-topology on $A^*$ is the topology of pointwise convergence it follows that $B$ is a weak *-closed subset of the unit ball of $A^*$; since this ball is weak *-compact (Alaoglu theorem), so is $B$.

(b) If $0 \in B$ were not an extreme point of $B$, there would exist elements $p, q \in B$, $p \neq q$, such that $0 = \frac{1}{2}(p + q)$; that is, $q = -p$. Hence there would be a nonzero $p$ in $B$ with $-p \in B$ also. Therefore $p(x^*x) \geq 0$ and $-p(x^*x) \geq 0$ for all $x$ in $A$, i.e., $p(x^*x) = 0$ for all $x$ in $A$. By (22.15), (b), $|p(x)|^2 \leq ||p||p(x^*x) = 0$ for all $x$ in $A$, i.e., $p = 0 = q$, a contradiction.

(c) Let $p$ be a nonzero extreme point of $B$. Then $||p|| = 1$; indeed, if $0 < ||p|| < 1$, then $p = ||p||(p/||p||) + (1 - ||p||)\cdot 0$ where $p/||p||$ and $0$ belong to $B$ and $p/||p|| \neq 0$, contradicting the fact that $p$ is an extreme point of $B$. Now let $p_1$ be any nonzero element of $B$ which is dominated by $p$. We must show that $p_1 = \lambda p$ for some $\lambda$, $0 \leq \lambda \leq 1$. If $p = p_1$, we may take $\lambda = 1$; so assume that $p \neq p_1$ and let $p_2 = p - p_1$. Then $p_2$ is positive, and since $1 = ||p|| = ||p_1 + p_2|| = ||p_1|| + ||p_2||$ (see (22.16), (b)), we see that $||p_2|| \leq 1$, i.e., $p_2$ is a nonzero element of $B$. Let $\lambda = ||p_1||$. Then $0 < \lambda < 1$ and $||p_2|| = 1 - \lambda$. Set $q_1 = \lambda^{-1}p_1$ and $q_2 = (1 - \lambda)^{-1}p_2$.

Then $p = \lambda q_1 + (1 - \lambda)q_2$, with $q_1, q_2 \in B$. Since $p$ is an extreme point of $B$, we have $p = q_1 = q_2$. Hence $p_1 = \lambda p$ with $0 < \lambda < 1$, i.e., $p$ is pure.

Conversely, assume that $||p|| = 1$ and that $p$ is pure. Suppose that $p = \lambda p_1 + (1 - \lambda)p_2$ with $0 < \lambda < 1$ and $p_1, p_2 \in B$. Then $\lambda p_1$ is dominated by $p$; hence $\lambda p_1 = \mu p$, where $0 \leq \mu \leq 1$. Since $||p|| = 1$, we have $1 = ||p|| = \lambda||p_1|| + (1 - \lambda)||p_2||$ by (22.16), (b). This equality combined with $||p_1|| \leq 1$ and $||p_2|| \leq 1$ implies that $||p_1|| = ||p_2|| = 1$. Therefore, $\lambda = \lambda||p_1|| = ||\lambda p_1|| = ||\mu p|| = \mu||p|| = \mu$ so that $p_1 = p = p_2$; thus $p$ is an extreme point of $B$.

(d) This is an immediate consequence of the Krein-Milman Theorem (A.6). □

Let $A$ be a unital Banach *-algebra with identity $e$ (no continuity assumptions are made on the involution). If $p$ is a state on $A$, then $p$ is hermitian (21.6) and therefore real valued on $H(A)$, the real vector space of hermitian elements of $A$. It follows (27.4) that the restriction of $p$ to $H(A)$, denoted $p_H$, is an element of $H(A)^*$, the dual of $H(A)$. The map $p \to p_H$ is obviously an affine map (i.e., it preserves convex combinations) and is continuous relative to the weak *-topologies. Moreover, since each $x$ in $A$ can be written uniquely as $x = h + ik$ with $h, k \in H(A)$, it follows that $p$ is determined by $p_H$. This establishes an affine homeomorphism of $S(A)$ onto $S_H(A) = \{p_H: p \in S(A)\}$. We can now use this natural correspondence to establish:

(29.7) PROPOSITION. *Let* $A$ *be a unital Banach *-algebra. Then* $S(A)$ *is a weak *-compact convex subset of the dual of* $A$. *Further, the *-representation* $\pi_p$ *associated with a state* $p$ *on* $A$ *is irreducible iff* $p$ *is an extreme point of* $S(A)$.

*Proof.* Let $U_1$ be the closed unit ball of the dual of $H(A)$. If $p \in S(A)$ and $h \in H(A)$, we have by property 4° following (27.1) that $|p(h)| \leq |h|_\sigma \leq ||h||$; hence $p_H \in U_1$. Further, $p \in H(A)^*$ lies in $S_H(A)$ iff $p(x^*x) \geq 0$ for all $x \in A$ and $p(e) = 1$. Therefore $S_H(A)$ is a weak *-closed convex subset of $U_1$. Since $U_1$ is weak *-compact, $S_H(A)$ is also. The same must hold for $S(A)$ in view of the affine homemorphism between $S(A)$ and $S_H(A)$.

To prove the second assertion it suffices by (29.5) to show that a state $p$ is an extreme point of $S(A)$ iff $p$ is pure. First assume $p$ is a pure state and $p = \frac{1}{2}(p_1 + p_2)$ where $p_1, p_2 \in S(A)$. Since $p(x^*x) = \frac{1}{2}p_1(x^*x) + \frac{1}{2}p_2(x^*x)$, we have $p \geq \frac{1}{2}p_1$ and $p \geq \frac{1}{2}p_2$; so there exist $\lambda_1, \lambda_2, 0 \leq \lambda_i \leq 1$ $(i = 1, 2)$, such that $\frac{1}{2}p_1 = \lambda_1 p$ and $\frac{1}{2}p_2 = \lambda_2 p$. Evaluating at $e$ we conclude that $\lambda_1 = \lambda_2 = \frac{1}{2}$ so that $p_1 = p_2 = p$. For the converse suppose that $p$ is an extreme point of $S(A)$ and $q \leq p$. If $q(e) = 0$, then $q = 0$ by the Cauchy-Schwarz inequality, so $q = 0 \cdot p$. If $q(e) > 0$, let $q_1 = q(e)^{-1}q$ and choose $\lambda$, $0 < \lambda < q(e)$. Then $q_1 \in S(A)$, $h = (1 - \lambda)^{-1}(p - \lambda q_1) \in S(A)$, and $p = \lambda q_1 + (1 - \lambda)h$ where $0 < \lambda < 1$. Since $p$ is an extreme point of $S(A)$, $q_1 = p$; i.e., $q = q(e) \cdot p$ where $0 < q(e) \leq p(e) = 1$. □

(29.8) PROPOSITION. *Let* $A$ *be a *-algebra,* $\pi$ *a nontrivial irreducible *-representation of* $A$, $\xi_1, \xi_2$ *two nonzero vectors in the space* $H_\pi$ *of* $\pi$, *and* $p_1, p_2$ *the positive functionals defined by* $(\pi, \xi_1)$ *and* $(\pi, \xi_2)$. *Then* $p_1 = p_2$ *iff there exists a complex number* $\lambda$ *with* $|\lambda| = 1$, *such that* $\xi_2 = \lambda\xi_1$.

*Proof.* Assume that $\xi_2 = \lambda\xi_1$ for a complex number $\lambda$ with $|\lambda| = 1$. Then $p_2(x) = (\pi(x)\xi_2|\xi_2) = (\pi(x)(\lambda\xi_1)|(\lambda\xi_1)) = \lambda\bar{\lambda}(\pi(x)\xi_1|\xi_1) = |\lambda|^2(\pi(x)\xi_1|\xi_1) = p_1(x)$ for all $x$ in $A$, i.e., $p_1 = p_2$.

Conversely, assume that $p_1 = p_2$. Since $\xi_1$ and $\xi_2$ are both cyclic vectors for $\pi$ (by 28.2)), there exists a unitary operator $U: H_\pi \to H_\pi$ which commutes with each $\pi(x)$, such that $U\xi_1 = \xi_2$ (see (26.11)). Hence, again by (28.2), $U = \lambda I$ for some complex $\lambda$. Hence, $\xi_2 = \lambda\xi_1$ and since $||U|| = 1 = ||U^{-1}||$, it follows that $|\lambda| = 1$. □

Irreducible *-representations may not exist for a given *-algebra. When they do, the key to their existence is the availability of certain positive functionals. The next two results are concerned with these ideas.

(29.9) THEOREM. *Let* $A$ *be a *-Banach algebra with approximate identity bounded by* 1, *and let* $x \in A$. *Let* $B$ *denote the set of all* *positive functionals* $p$ *on* $A$ *with* $||p|| \leq 1$. *Then the following are equivalent:*

(a) *there is* $p \in B$ *with* $p(x) \neq 0$.

(b) *there is* $p \in B$ *with* $p(x^*x) > 0$.

(c) *there is an irreducible *-representation* $\pi$ *of* $A$ *such that* $\pi(x) \neq 0$.

(d) *there is a *-representation* $\psi$ *of* $A$ *such that* $\psi(x) \neq 0$.

*Proof*. (a) implies (b): This is an immediate consequence of the inequality $|p(x)|^2 \leq ||p||\cdot p(x^*x) \leq p(x^*x)$, if $p \in B$ (see (22.15).)

(b) implies (c): Since $B$ is the weak *-closed convex hull of its extreme points, the Krein-Milman theorem implies that there is an extreme point $p$ of $B$ such that $p(x^*x) > 0$. Let $\pi$ and $\xi$ be the cyclic *-representation and cyclic vector defined by $p$. Then $\pi$ is irreducible (by (29.6), (c) and (29.5)) and $||\pi(x)\xi||^2 = p(x^*x) > 0$; hence $\pi(x) \neq 0$.

(c) implies (d) is obvious.

(d) implies (a): Assume that $\psi: A \to B(H)$ is a *-representation of $A$ such that $\psi(x) \neq 0$. By polarization it follows that there is a vector $\xi$ in $H$ such that $(\psi(x)\xi|\xi) \neq 0$. Normalizing we may assume that $||\xi|| = 1$. It then follows that the functional $p(x) = (\psi(x)\xi|\xi)$ is an element of $B$ (26.13, (c)) and $p(x) \neq 0$. □

For unital Banach *-algebras with arbitrary (possibly discontinuous) involutions we have:

(29.10) THEOREM. *Let* $A$ *be a unital Banach *-algebra. Then the following are equivalent:*

1° *the set* $S(A)$ *is nonempty.*

2° *the set of nonzero irreducible *-representations is nonempty.*

3° *the set of nonzero *-representations of* $A$ *is nonempty.*

*Proof*. 1° implies 2°. If $S(A)$ is nonempty there exists an extreme point $p$ of $S(A)$. By (29.7) the *-representation $\pi_p$ associated with $p$ is irreducible. 2° implies 3° is clear. That 3° implies 1° follows since the positive functional defined by a *-representation with a unit vector is a state (cf. the remark preceding (27.2).) □

§30. *The *-radical.*

We turn our attention now to the so-called *-radical of a Banach *-algebra. This useful concept was introduced in 1948 by Naimark [1], who called it the reducing ideal. The definition can be given for any *-algebra.

(30.1) DEFINITION. *Let* A *be a *-algebra. The *-radical of* A, *denoted by* $R^*(A)$, *is the intersection of the kernels of all (topologically) irreducible *-representations of* A *on Hilbert spaces. If* $R^*(A) = \{0\}$, *then* A *is said to be *-semisimple.*

The following result contains first properties of the *-radical.

(30.2) PROPOSITION. *Let* A *be a Banach *-algebra.*

*(i)* $R^*(A)$ *is a closed *-ideal.*

*(ii)* $\mathrm{Rad}(A) \subseteq \rho^{-1}(\{0\}) \subseteq R^*(A)$.

*(iii)* $A/R^*(A)$ *is *-semisimple.*

*Proof.* (i) By (26.13), (a) every *-representation of A is continuous. Hence the kernel of such a representation is a closed *-ideal, from which (i) follows.

(ii) If $x \in \mathrm{Rad}(A)$, then $x^*x \in \mathrm{Rad}(A)$ and so $\rho(x) = |x^*x|_\sigma^{1/2} = 0$. If $\rho(x) = 0$, then $\pi(x) = 0$ for every *-representation $\pi$ of A by (26.13), (b).

(iii) Since $R^*(A)$ is a closed *-ideal, $A/R^*(A)$ is a Banach *-algebra. Let $\tau: A \to A/R^*(A)$ be the natural homomorphism. Consider an irreducible *-representation $\pi$ of A on a Hilbert space and let $K = \ker(\pi)$. Then $R^*(A) \subseteq K$. For each $\tau(x) \in A/R^*(A)$, define $\tilde{\pi}(\tau(x)) = \pi(x)$. Since $\tau(x) = \tau(y)$ implies $y - x \in R^*(A)$, it follows that $\pi(x) = \pi(y)$, so that $\tilde{\pi}(\tau(x))$ is well-defined. It is clear that $\tau(x) \to \tilde{\pi}(\tau(x))$ is an irreducible *-representation of $A/R^*(A)$ on a Hilbert space. Since $x \notin K$ implies $\tilde{\pi}(\tau(x)) \neq 0$ and $R^*(A)$ is the intersection of all such K, it follows that $A/R^*(A)$ is *-semisimple. □

It was shown in the proof of (ii) of the above result that Rad(A) is actually contained in the intersection of the kernels of *all* *-representations of A on Hilbert spaces. We next observe the stronger result that the *-radical is *equal* to the intersection of the kernels of all *-representations of A on Hilbert spaces. Although the result is true for arbitrary Banach *-algebras (see Rickart [ 1 , p. 225], Bonsall-Duncan [ 5 , p. 223]) we restrict our attention here (and in the remainder of this section) to unital Banach *-algebras since this is the only case we shall need.

(30.3) THEOREM. *Let* A *be a unital Banach *-algebra. Then the *-radical of* A *is equal to the intersection of the kernels of all *-representations of* A *on Hilbert spaces.*

*Proof.* Let J denote the intersection of the kernels of all *-representations of A on Hilbert spaces. Then it is clear that $J \subseteq R^*(A)$. To obtain the opposite inclusion, consider $x \in A$ with $x \notin J$. We wish to show that $x \notin R^*(A)$. Now $x \notin J$ implies that there exists a *-representation $\phi$ of A on a Hilbert space H such that $\phi(x) \neq 0$. Since $\phi(x^*) = \phi(x)^* \neq 0$, there is a unit vector $\xi$ in H such that $\phi(x^*)\xi \neq 0$. Define

$$q(a) = (\phi(a)\xi|\xi) \quad (a \in A).$$

Then q is a state on A and $q(xx^*) = ||\phi(x^*)\xi||^2 > 0$. By the Krein-Milman theorem (A.6) there is a state p on A such that p is an extreme point of S(A) (see (29.7)) and $p(xx^*) \neq 0$. In fact, since S(A) is weak *-compact, the map $\tilde{p} \to \tilde{p}(xx^*)$ assumes its maximum $\gamma_0 > 0$ on S(A). Then

$$K = \{\tilde{p} \in S(A): \tilde{p}(xx^*) = \gamma_0\}$$

is a nonempty compact convex set and any extreme point of K is also an extreme point of S(A).

Let $\pi$ be the *-representation of A associated with p. By (29.7) $\pi$ is irreducible. Set

$$I_p = \{a \in A: p(a^*a) = 0\}$$

(see (27.3)). Then $x^* \notin I_p$. If $xx^* \in I_p$, then from 3° following (27.1) $p(xx^*)^2 \leq p(xx^*xx^*) = 0$, which contradicts the fact that $p(xx^*) \neq 0$. Hence $xx^* \notin I_p$ and therefore $x \notin (I_p: A) = \ker(\pi)$. Since x is not in the kernel of the irreducible *-representation $\pi$, it follows that $x \notin R^*(A)$. □

Various characterizations of the *-radical are given in the following.

(30.4) PROPOSITION. *Let* A *be a unital Banach *-algebra. Then* $R^*(A) \neq A$ *iff* $S(A) \neq \emptyset$. *When* $S(A)$ *is nonempty, then:*

*(a)* $R^*(A) = \cap \{(I_p: A): p \in S(A)\}$.

*(b)* $R^*(A) = \cap \{I_p: p \in S(A)\}$.

*(c)* $R^*(A) = \cap \{p^{-1}(\{0\}): p \in S(A)\}$.

*Proof.* Suppose that $R^*(A) \neq A$. Then there is a nontrivial *-representation $\pi$ of A on a Hilbert space H. For any unit vector $\xi$ in H the functional $p(x) = (\pi(x)\xi|\xi)$ belongs to $S(A)$. On the other hand, assume $S(A) \neq \emptyset$. If $p \in S(A)$, then (30.3) and (27.3) give $R^*(A) \subseteq (I_p: A) = \ker \pi_p \subsetneq A$. Since $|p(x)|^2 \leq p(x^*x)$ for all x in A, we also have $I_p \subseteq p^{-1}(\{0\})$. Hence

$$R^*(A) \subseteq (I_p: A) \subseteq I_p \subseteq p^{-1}(\{0\}).$$

To complete the proof it suffices to show that $\cap \{p^{-1}(\{0\}): p \in S(A)\} \subseteq R^*(A)$. To this end assume that $x \in A \setminus R^*(A)$. Then there is a *-representation $\pi$ of A on a Hilbert space H with $\pi(x) \neq 0$. Hence there is a unit vector $\xi$ in H such that $(\pi(x)\xi|\xi) \neq 0$. Define a functional p on A by setting $p(a) = (\pi(a)\xi|\xi)$ as in the first step of the proof. Then $p(x) \neq 0$ and $p \in S(A)$. □

On a Banach *-algebra there is a very useful seminorm, called the *Gelfand-Naimark seminorm*, which is closely related to the *-radical. Its existence and some of its properties are the subject of the next proposition.

(30.5) PROPOSITION. *Let* A *be a unital Banach *-algebra. Then there exists an algebra seminorm* $\gamma$ *on* A *such that, for each* $x \in A$:

*(a)* $\gamma(x^*x) = \gamma(x)^2$.

*(b)* $R^*(A) = \{x \in A: \gamma(x) = 0\}$.

*(c) If* p *is a state on* A, $|p(x)| \leq \gamma(x)$.

*(d) If* $\pi$ *is a *-representation of* A, $||\pi(x)|| \leq \gamma(x)$.

*(e) If* $R^*(A) \neq A$, *then*

$$\begin{aligned}\gamma(x)^2 &= \sup\{p(x^*x): p \in S(A)\} \\ &= \sup\{p(x^*x): p \in E(S(A))\}\end{aligned}$$

$$= \sup\{||\pi_p(x)||^2 : p \in S(A)\}$$

$$= \sup\{||\pi(x)||^2 : \pi \text{ } a \text{ *-representation of } A\}.$$

*(f)* $\gamma(x) \leq \rho(x)$.

*Proof.* If the *-radical coincides with A, set $\gamma \equiv 0$. Then (a), (b), (e), and (f) hold. Properties (c) and (d) also hold since there are no nontrivial *-representations, hence no states, on A.

Suppose $R^*(A) \neq A$. By (30.4) the set $S(A)$ is nonempty. Define $\gamma(x) = \sup\{p(x^*x)^{1/2} : p \in S(A)\}$ so that

$$\gamma(x)^2 = \sup\{p(x^*x) : p \in S(A)\}.$$

By the Krein-Milman theorem the supremum may be taken over $E(S(A))$, the set of extreme points of $S(A)$. If $\pi$ is a *-representation of A on a Hilbert space H and $\xi$ a unit vector in H, then $p(x) = (\pi(x)\xi|\xi)$ defines a state p on A. Further,

$$\begin{aligned} ||\pi(x)\xi||^2 &= (\pi(x)\xi|\pi(x)\xi) \\ &= (\pi(x^*x)\xi|\xi) \\ &= p(x^*x) \leq \gamma(x)^2, \end{aligned}$$

which proves part (d).

If $p \in S(A)$, then by (21.6), (b), we have $|p(x)|^2 \leq p(x^*x) \leq \gamma(x)^2$ from which (c) follows.

To prove (e), note that since

$$\begin{aligned} ||\pi_p(x)||^2 &= \sup\{||\pi_p(x)\xi||^2 : ||\xi|| = 1\} \\ &\geq ||\pi_p(x)\xi_p||^2 \\ &= (\pi_p(x)\xi_p|\pi_p(x)\xi_p) = p(x^*x) \end{aligned}$$

it follows that

$$\begin{aligned} \gamma(x)^2 &= \sup\{p(x^*x) : p \in S(A)\} \\ &\leq \sup\{||\pi_p(x)||^2 : p \in S(A)\} \end{aligned}$$

$$\leq \sup\{||\pi(x)||^2 : \pi \text{ a } {}^*\text{-representation of } A\}$$

$$\leq \gamma(x)^2, \text{ by part (d).}$$

Part (b) is immediate from (e) and (30.3).

By property 5° of the list following (27.1) we have for each $p \in S(A)$ and $x \in A$ that $p(x^*x) \leq \rho(x^*x) = \rho(x)^2$, which proves (f).

Finally, utilizing the formulas in part (e) we can prove that (a) holds. Indeed, since $\gamma(x) = \sup\{||\pi(x)|| : \pi$ is a $^*$-representation of $A\}$, $\gamma$ is an *algebra seminorm* such that $\gamma(x^*) = \gamma(x)$ and $\gamma(x^*x) = \gamma(x)^2$. □

Every Banach $^*$-algebra with nonempty state space is contained as a dense $^*$-subalgebra of a certain $C^*$-algebra. The details are given in the following theorem.

(30.6) THEOREM. *Let* $A$ *be a unital Banach* $^*$*-algebra which admits at least one state. Then there exists a* $C^*$*-algebra* $B$ *and a* $^*$*-representation* $\pi$ *of* $A$ *onto a dense* $^*$*-subalgebra of* $B$ *such that* $\pi$ *is an isometry relative to the Gelfand-Naimark seminorm* $\gamma$ *on* $A$. *Further, the map* $\Phi: S(B) \to S(A)$ *defined by* $\Phi(q) = q \circ \pi$ *bijectively pairs the sets* $S(B)$ *and* $S(A)$.

*Proof*. For each state $p \in S(A)$, consider the associated $^*$-representation $\pi_p$ of $A$ on a Hilbert space $H_p$ and set $\pi = \bigoplus_{p \in S(A)} \pi_p$, $H = \bigoplus_{p \in S(A)} H_p$. This can be done because of (26.4) and (26.13). We wish to show that $\gamma(x) = ||\pi(x)||$ for each $x$ in $A$. To see this, it suffices to note that

$$||\pi(x)|| = \sup\{||\pi_p(x)|| : p \in S(A)\} \qquad \text{(cf. (26.4))}$$

$$= \gamma(x), \text{ by (30.5), (e).}$$

By (30.4) and (30.5) we have

$$\ker(\pi) = \cap\,\{I_p : p \in S(A)\} = R^*(A) = \gamma^{-1}(\{0\}).$$

Therefore $A/R^*(A)$ with norm $\gamma$ is isometric with a $^*$-subalgebra $C$ of

$B(H)$. Let $B$ denote the closure of $C$. Then $B$ is a $C^*$-algebra which contains $\pi(A)$ isometrically (with respect to $\gamma$) as a dense $^*$-subalgebra.

It remains only to verify the last statement of the theorem. If $q \in S(B)$, then the function $p$ on $A$ defined by $p(x) = q(\pi(x))$ is obviously a state on $A$. For the converse, let $p \in S(A)$ and let $\xi$ be the element of $H$ whose $p^{th}$ coordinate is $\xi_p$ and whose remaining coordinates are zero. Then the function $q$ defined on $B$ by $q(S) = (S\xi|\xi)$ is a state on $B$ and $p(x) = (\pi_p(x)\xi_p|\xi_p) = (\pi(x)\xi|\xi) = q(\pi(x))$ for all $x \in A$, as required. □

The $C^*$-algebra $B$ constructed in (30.6) is called the *enveloping $C^*$-algebra* of the Banach $^*$-algebra $A$. The $^*$-representation

$$\pi = \bigoplus_{p \in S(A)} \pi_p$$

constructed in the proof is called the *universal representation* of $A$.

We close this section with the following important "completeness theorem" for $C^*$-algebras.

(30.7) THEOREM. *(Gelfand-Naimark). Every $C^*$-algebra $A$ has enough irreducible $^*$-representations to separate its points.*

*Proof.* We may assume, by (6.1), that $A$ has an identity. Given a nonzero $x$ in $A$, we must show that there exists an irreducible $^*$-representation $\pi$ of $A$ such that $\pi(x) \neq 0$. However, by (18.1), there exists a positive functional $p$ on $A$ such that $p(x) \neq 0$. It then follows from (29.9) that there is an irreducible $^*$-representation $\pi$ of $A$ such that $\pi(x) \neq 0$. □

EXERCISES

(V.1) Show that a $^*$-representation $\pi$ of a $^*$-algebra $A$ on a Hilbert space $H$ is nondegenerate iff $\{\pi(x)\xi: x \in A, \xi \in H\}$ spans a dense subspace of $H$.

(V.2) Show that a subrepresentation of a nondegenerate $^*$-representation of a $^*$-algebra $A$ on a Hilbert space $H$ is nondegenerate.

(V.3) Give an example of a $^*$-algebra with no nonzero irreducible $^*$-representations.

(V.4) Let $A$ be a commutative *-algebra and $\pi$ an irreducible *-representation of $A$ on a nonzero Hilbert space $H$. Prove that $H$ is one-dimensional.

(V.5) Show that the decomposition in (26.10) of a nondegenerate *-representation $\pi$ of a *-algebra $A$ into a direct sum of cyclic *-representations is not unique.

(V.6) Prove that if $A$ is a commutative unital *-Banach algebra, then each extreme point of $S(A)$ is multiplicative.

(V.7) Let $A$ be a separable $C^*$-algebra and $I = \{x \in A: p(x^*x) = 0\}$, for a nonzero positive functional $p$. Prove that the pre-Hilbert space $A/I_p$ in the GNS-construction is separable.

(V.8) Let $\pi$ be a *-representation of a $C^*$-algebra $A$. Prove that the following are equivalent, where $x > 0$ means $\sigma_A(x) \subseteq (0,\infty)$ and $x = x^*$:

(a) $\pi$ is faithful;

(b) $||\pi(x)|| = ||x||$ for all $x \in A$;

(c) $\pi(x) > 0$ for all $x \in A$ with $x > 0$.

(V.9) Let $A$ be a Banach *-algebra with bounded approximate identity $\{e_\alpha\}$ and $\pi$ a nondegenerate *-representation of $A$ on a Hilbert space $H$. Prove that $\pi(e_\alpha)\xi \to \xi$ for each $\xi \in H$.

(V.10) Let $A$ be a $C^*$-algebra and $\pi$ a nondegenerate *-representation of $A$ on a Hilbert space $H$.

(a) If $\xi \in H$ and $p(x) = (\pi(x)\xi|\xi)$, $x \in A$, show that $||p|| = ||\xi||^2$.

(b) Show that part (a) remains true if $A$ is a *-Banach algebra with approximate identity bounded by 1.

(V.11) Let $A$ be a Banach *-algebra with isometric involution and approximate identity bounded by one. If $\pi$ is a *-representation of $A$ on a Hilbert space, prove that

$$||\pi(x)||^2 = \sup p(x^*x), \quad x \in A,$$

where $p$ varies over the set of positive functionals on $A$ associated with $\pi$ (see §27) such that $||p|| \leq 1$.

(V.12) Let $A$ be a Banach *-algebra with isometric involution and approximate identity bounded by one. Prove that there exists a seminorm $\nu$

on $A$ such that, for all $x, y \in A$, we have:

(a) $\nu(xy) \leq \nu(x)\nu(y)$;

(b) $\nu(x^*) = \nu(x)$;

(c) $\nu(x^*x) = \nu(x)^2$;

(d) $\nu(x) \leq ||x||$.

(V.13) Let $A$ be a $C^*$-algebra and $I$ a closed two-sided ideal in $A$. Show that if $\pi$ is an irreducible $*$-representation of $A$, then the restriction of $\pi$ to $I$ is either the zero representation or is irreducible.

(V.14) Let $A$ be a $C^*$-algebra. An element $x$ in $A$ is *strictly positive* if $p(x) > 0$ for each nonzero positive linear functional $p$ on $A$. If $x \in A$ is strictly positive and $\pi$ is a nondegenerate $*$-representation of $A$ on a Hilbert space $H$, prove that the set $\{\pi(x)\xi: \xi \in H\}$ is dense in $H$.

(V.15) Let $A$ be a $C^*$-algebra with identity. If $x$ is a normal element of $A$, show there exists a state $p$ on $A$ such that $|p(x)| = ||x||$.

(V.16) Let $A$ be a $C^*$-algebra and $B$ a closed $*$-subalgebra of $A$. Prove that every state on $B$ extends to a state on $A$.

(V.17) Let $A$ and $B$ be $C^*$-algebras with identities, and let $\phi$ be a surjective $*$-homomorphism of $A$ onto $B$. If $f$ is a linear functional on $B$, prove that:

(a) $f\circ\phi$ is a state on $A$ iff $f$ is a state on $B$;

(b) $f\circ\phi$ is a pure state on $A$ iff $f$ is a pure state on $B$.

(V.18) Let $A$ and $B$ be $C^*$-algebras with identities $e_A$ and $e_B$, respectively, and let $\phi: A \to B$ be a $*$-homomorphism. Let $\phi^*$ be the dual map of $\phi$ defined by

$$\phi^*(f)(x) = f(\phi(x)) \quad \text{for} \quad f \in B^*,\ x \in A.$$

Prove that:

(a) $\phi^*$ is a linear map which is weak$^*$-continuous;

(b) $\phi^*$ maps the state space of $B$ into the state space of $A$;

(c) If $\phi$ is onto, then the pure states on $B$ are mapped into the pure states on $A$ by $\phi^*$.

(V.19) Let $I$ and $J$ be closed two-sided ideals in a $C^*$-algebra $A$. Show that the ideal $I + J$ is closed.

(V.20) Let $A$ be a C*-algebra with identity, and let $x \in A$. Prove that $x$ is hermitian iff $f(x)$ is real for every state $f$ on $A$.

(V.21) Let $A$ be a Banach *-algebra and $f$ a positive functional on $A$. If the involution is isometric, prove that $f(x^*) = \overline{f(x)}$ for all $x \in A$ if there exists a real constant $k \geq 0$ such that $|f(x)|^2 \leq kf(x^*x)$ for all $x \in A$. (Cf. (21.7).)

# 6
# Hermitian and Symmetric*-Algebras

## §31. *Introduction.*

The study of Banach algebras with involution originated with Gelfand and Naimark's celebrated 1943 paper (see §3) on the characterization of $C^*$-algebras. To obtain their main results they assumed the involution to be symmetric, i.e., each element of the form $e + x^*x$ is invertible, where $e$ is the identity in the given algebra. As pointed out in §3 they suspected that this assumption was not necessary. In 1947 I. Kaplansky introduced a closely related condition on the involution which he conjectured was equivalent to symmetry. This new condition, called hermiticity, required each hermitian element in the algebra to have real spectrum. Kaplansky's conjecture remained unresolved for many years; then in 1970, S. Shirali and J. W. M. Ford [ 1 ] confirmed the conjecture was true for *Banach* *-algebras. The question was not decided for arbitrary *-algebras until 1973 when J. Wichmann [ 1 ] constructed examples of hermitian *-algebras which are *not* symmetric. In this chapter we shall study properties of *-algebras, with and without norm, which are symmetric and hermitian.

By utilizing properties of the spectral radius and the function $\rho(x) = |x^*x|_\sigma^{1/2}$, V. Pták [ 1 ] showed in 1970 that an elegant theory for Banach *-algebras arises from the inequality

$$|x|_\sigma \leq \rho(x).$$

This inequality characterizes hermitian (and symmetric) Banach *-algebras and also reveals, in an extremely simple way, that many properties of $C^*$-algebras persist under the purely algebraic assumption of symmetry.

Further characterizations of $C^*$-algebras follow as a natural result of Pták's theory, and the whole subject is unified and simplified by it.

We shall look first at the basic properties of hermitian and symmetric algebras in the context of *-algebras without norm. Then the theory will be developed for Banach *-algebras with applications to characterizations of $C^*$-algebras.

## §32. *Definitions and basic properties.*

In this section we consider arbitrary *-algebras. Recall that an element $x$ in a *-algebra $A$ is quasi-regular if there exists an element $y$ in $A$ such that $x \circ y = 0 = y \circ x$, where $x \circ y = x + y - xy$.

(32.1) DEFINITION. *The involution in a *-algebra* $A$ *is said to be* ___symmetric___ *if every element of the form* $-x^*x$ *is quasi-regular in* $A$. *When the involution in a *-algebra* $A$ *is symmetric, then* $A$ *is called a* ___symmetric algebra___.

If a *-algebra $A$ has an identity $e$, then $A$ is symmetric iff $e + x^*x$ is invertible for all $x$ in $A$. This follows immediately from the relation $e - (x \circ y) = (e - x)(e - y)$.

Many *-algebras are symmetric; for instance all $C^*$-algebras are symmetric, as are the group algebras of abelian or compact groups. The field $C(x_1, \ldots, x_n)$ of fractions of complex polynomials in $x_1, \ldots, x_n$ with involution defined by $x_i^* = x_i$, $i = 1, 2, \ldots, n$, is an example of a nonnormable *-algebra which is symmetric.

On the other hand, there are many *-algebras which are not symmetric. Perhaps the simplest example is $C^2$ furnished with the involution $(a,b)^* = (\bar{b}, \bar{a})$. The disk algebra $A(D)$, i.e., the algebra of continuous complex functions on the disk $D = \{\lambda : |\lambda| \le 1\}$ analytic on the interior with the involution $f^*(\lambda) = \overline{f(\bar{\lambda})}$ is also not symmetric. In general, the group algebra of an arbitrary locally compact group need not be symmetric; in fact, it is an important question to determine which groups have symmetric group algebras.

(32.2) DEFINITION. *The involution in a *-algebra* $A$ *is said to be* ___hermitian___ *if every hermitian element in* $A$ *has real spectrum. When the involution in a *-algebra* $A$ *is hermitian then* $A$ *is called a* ___hermitian algebra___.

Our first two propositions give elementary characterizations of hermitian and symmetric $^*$-algebras.

(32.3) PROPOSITION. *A $^*$-algebra* $A$ *is hermitian iff every element of the form* $-h^2$, *where* $h^* = h$, *is quasi-regular.*

*Proof.* If $A$ is hermitian and $h^* = h$, then $\sigma_A(h)$ is real and, since $\sigma_A(h^2) = \sigma_A(h)^2$, it follows that $\sigma_A(h^2) \geq 0$. In particular, $-1 \notin \sigma_A(h^2)$, so that $-h^2$ is quasi-regular.

Conversely, assume that each element of the form $-h^2$, where $h^* = h$, is quasi-regular. Let $h$ be hermitian and suppose, to the contrary, that $\lambda + i\mu$ is a complex number in $\sigma_A(h)$, where $\mu \neq 0$. Consider the polynomial

$$p(t) = \mu^{-1}(\lambda^2 + \mu^2)^{-1}(\lambda t^2 + (\mu^2 - \lambda^2)t)$$

for $t$ complex, and set $k = p(h)$. Then $k$ is clearly hermitian and by the spectral mapping theorem

$$i = p(\lambda + i\mu) \in p(\sigma_A(h)) = \sigma_A(p(h)) = \sigma_A(k).$$

Hence, $-1 \in \sigma_A(k^2)$ and so $-k^2$ is quasi-singular, a contradiction of the hypothesis. □

(32.4) COROLLARY. *Every symmetric $^*$-algebra is hermitian.*

*Proof.* Let $h \in H(A)$. Since $A$ is symmetric, $-h^2 = -h^*h$ is quasi-regular. Now apply (32.3). □

(32.5) PROPOSITION. *A $^*$-algebra* $A$ *is symmetric iff* $\sigma_A(x^*x) \geq 0$ *for all* $x$ *in* $A$.

*Proof.* Assume that $A$ is symmetric. Since $x^*x$ is hermitian, $\sigma_A(x^*x)$ is a subset of the reals by (32.4). Suppose that $\alpha$ is a negative real number in $\sigma_A(x^*x)$ and set $\beta = (-\alpha)^{-1/2}$. Then the element $\alpha^{-1}x^*x = -(\beta x)^*(\beta x)$ is quasi-singular, contradicting the assumption that $A$ is symmetric.

Conversely, assume $\sigma_A(x^*x) \geq 0$ for all $x \in A$. Since $-1 \notin \sigma_A(x^*x)$ for all $x$ in $A$, then each element of the form $-x^*x$ is quasi-regular, i.e., $A$ is symmetric. □

Although symmetric *-algebras are hermitian (32.4), these two classes of algebras do not coincide, as the following examples, due to J. Wichmann [ 1], show.

(32.6) *Examples.* Let $n$ be any positive integer. The ring $C(x_1,\ldots,x_n)[x_{n+1}]$ of polynomials in $x_{n+1}$, over the field $C(x_1,\ldots,x_n)$ of fractions of complex polynomials in $x_1,\ldots,x_n$, with involution defined by $x_i^* = x_i$ $(i = 1,\ldots,n+1)$, is a commutative *-algebra with identity.

The hermitian polynomial $(1 + x_1^2 + \cdots + x_n^2) + x_{n+1}^2$ is an irreducible polynomial in $x_{n+1}$ over $C(x_1,\ldots,x_n)$. Therefore, the ideal generated by this polynomial is a maximal *-ideal and

$$B_n = C(x_1,\ldots,x_n)[x_{n+1}]/(1 + x_1^2 + \cdots + x_n^2 + x_{n+1}^2)$$

is a complex *-algebra (which is even a field!)

The elements of $B_n$ can be represented uniquely in the form $p + qy$, with $p$ and $q$ in $C(x_1,\ldots,x_n)$ such that $y^2 = -(1 + x_1^2 + \cdots + x_n^2)$ and $(p + qy)^* = p^* + q^*y$.

Let $h = h^* \in B_n$. Then $1 + ih \neq 0$ (otherwise, $1 = -ih$ and 1 would not be hermitian.) Hence, $1 + h^2 = (1 + ih)(1 - ih) \neq 0$ and is therefore invertible in the field $B_n$. Thus, each $B_n$ is a hermitian *-algebra.

However $B_1$ is not symmetric since

$$1 + (x_1 + iy)^*(x_1 + iy) = 1 + x_1^2 + y^2 = 0.$$

Also, the algebra $B_2$ is not symmetric since

$$1 + [(1 + ix_1)^{-1}(x_2 + iy)]^*[(1 + ix_1)^{-1}(x_2 + iy)]$$

$$= (1 + x_1^2)^{-1}(1 + x_1^2 + x_2^2 + y^2) = 0.$$

Consequently $B_1$ and $B_2$ are examples of hermitian *-algebras which are not symmetric. □

(32.7) PROPOSITION. *Every maximal commutative *-subalgebra* B *of a hermitian (resp. symmetric) *-algebra* A *is hermitian (resp. symmetric).*

*Proof.* Suppose that A is hermitian and let x be a hermitian element of B. Then $\sigma_A(x)$ is real and, by (B.4.3), (b), $\sigma_B(x) = \sigma_A(x)$.

Next, assume that A is symmetric and let x belong to B. The symmetry of A implies that the quasi-inverse $y = (-x^*x)'$ of $-x^*x$ exists in A. Since y commutes with any element which commutes with $-x^*x$, y commutes with elements in B; and since B is maximal commutative we have $y \in B$, i.e., B is symmetric. □

If a hermitian or symmetric *-algebra does not have an identity element, we may always adjoin one and retain these properties. This was first proved for symmetric Banach *-algebras in 1959 by P. Civin and B. Yood [ 1,p.421] thus answering a question posed in 1949 by I. Kaplansky [ 1]. The result for arbitrary symmetric *-algebras was proved in 1972 by R. S. Doran [ 3 ].

(32.8) PROPOSITION. *If* A *is a hermitian (resp. symmetric)* *-*algebra, then the unitization* $A_e$ *of* A *is also hermitian (resp. symmetric).*

*Proof.* Assume that A is hermitian and let $x + \lambda e$ be a hermitian element of $A_e$. Then x is a hermitian element of A and $\lambda$ is real. Since $\sigma_{A_e}(x + \lambda e) = \sigma_A(x) + \lambda$, it follows that $\sigma_{A_e}(x + \lambda e)$ is real and hence that $A_e$ is hermitian.

Next, assume that A is symmetric. Let $y = x + \lambda e \in A_e$ and assume that $-y^*y$ is quasi-singular in $A_e$; then $e + y^*y$ is singular in $A_e$. Set $h = \bar{\lambda}x + \lambda x^* + x^*x$, $k = (1 + |\lambda|^2)e$, and $z = -k^{-1}h = -(1 + |\lambda|^2)^{-1}h$. Then $e + y^*y = k + h = k(e - z)$. Clearly, $z = z^*$; since A is an ideal in $A_e$, $z \in A$, and hence $yz \in A$. Since $e - z = k^{-1}(e + y^*y)$ is singular in $A_e$, $1 \in \sigma_{A_e}(z)$. Consider the polynomial

$$p(t) = 1 + |\lambda|^2t^2 - (1 + |\lambda|^2)t^3,$$

and note that $p(z) = e + (yz)^*(yz)$. Since $0 = p(1) \in p(\sigma_{A_e}(z)) = \sigma_{A_e}(p(z))$, it follows that $e + (yz)^*(yz)$ is singular in $A_e$; that is, $-(yz)^*(yz)$ is quasi-singular in A. This contradicts the symmetry of A. □

Recalling that a *-algebra A lies in its unitization $A_e$ as a maximal *-ideal, and noting that $A_e/A$ is isomorphic to the complex numbers (a symmetric *-algebra), it follows that the next result, proved by J. Wichmann [ 4 ] in 1978, generalizes (32.8).

(32.9) PROPOSITION. *Let* A *be a* *-*algebra and* I *a* *-*ideal of* A. *Then* A *is hermitian (resp. symmetric) iff* I *and* A/I *are hermitian (resp. symmetric).*

*Proof.* We shall give the proof for symmetric algebras, leaving the hermitian case for the reader. Clearly, if A is symmetric, then so are I and A/I.

Conversely, suppose that I and A/I are symmetric. Let x be any element in A. Since the quotient A/I is symmetric, there exists an element y in A such that $z = e - (e - y)(e + x^*x) \in I$. Then

$$h = z + z^* - z^*z = e - (e + x^*x)(e - y)^*(e - y)(e + x^*x)$$

is a hermitian element in I. Since I is a symmetric *-ideal, the element $-(xh)^*(xh)$ is quasi-regular. Hence the simple algebraic identity

$$\begin{aligned} e + (xh)^*(xh) &= e + hx^*xh \\ &= h(e + x^*x) + (e - hx^*x)(e - h) \\ &= h(e + x^*x) + (e - hx^*x)(e + x^*x)(e - y)^*(e - y)(e + x^*x) \\ &= [h + (e - hx^*x)(e + x^*x)(e - y)^*(e - y)](e + x^*x) \end{aligned}$$

and the similarly proven one

$$e + (xh)^*(xh) = (e + x^*x)[h + (e - y)^*(e - y)(e + x^*x)(e - x^*xh)]$$

show that $-x^*x$ is quasi-regular. □

Before proceeding to the next section we make a few remarks (without proof) about symmetry in arbitrary *-algebras. For details see Wichmann [3].

(1) Let A be a *-algebra and n a positive integer. Then A is symmetric iff the *-algebra $A_n$ of $n \times n$ matrices over A is symmetric.

(2) If A is a symmetric *-algebra, then the *-algebra $A_\infty$ of all infinite matrices $a = (a_{ij})$ with $a_{ij} \in A$, $a_{ij} = 0$ for almost all i and j, is also symmetric.

(3) The product $\Pi A_i$ of a family of *-algebras $A_i$ is a symmetric *-algebra (relative to the involution $(a_i)^* = (a_i^*)$) iff each $A_i$ is symmetric.

## §33. *Hermitian Banach *-algebras.*

Many properties of hermitian *Banach* *-algebras will be established in this section. We will show, in particular, that they coincide with the class of symmetric Banach *-algebras (the Shirali-Ford theorem). A key result is the following fundamental inequality which is valid in any hermitian algebra A:

$$|x|_\sigma^2 \leq |x^*x|_\sigma \quad (x \in A). \tag{1}$$

Armed with this tool, it is relatively easy to show that the function $\rho(x) = |x^*x|_\sigma^{1/2}$ is an algebra seminorm on A which coincides with $|x|_\sigma$ on normal elements, and $\rho^{-1}(\{0\})$ equals the Jacobson radical of A. Once these facts have been established, many additional results follow easily.

We mention that inequality (1) is an algebraic analogue of the $C^*$-condition

$$||x^*x|| = ||x^*||\cdot||x||. \tag{2}$$

Indeed, the relation in (2) is, by submultiplicativity of the norm, equivalent to the inequality $||x^*||\cdot||x|| \leq ||x^*x||$. Replacing the norms by the spectral radii and using the fact that $|x^*|_\sigma = |x|_\sigma$, $x \in A$, gives inequality (1). It turns out that this inequality actually characterizes hermitian Banach *-algebras.

Since we may always adjoin an identity to a symmetric or hermitian *-algebra and retain these properties (see (32.8)), we shall assume unless stated otherwise throughout this section that all algebras possess an identity element, denoted by e or 1, of norm 1.

(33.1) THEOREM. *(Pták). Let* A *be a hermitian Banach *-algebra. Then:*

*(a)* $|x|_\sigma \leq \rho(x)$ *for* $x \in A$;

*(b)* $|x|_\sigma = \rho(x)$ *for normal* $x \in A$;

*(c)* $|hk|_\sigma \leq |h|_\sigma |k|_\sigma$ *for hermitian* $h, k \in A$;

*(d)* $\rho(xy) \leq \rho(x)\rho(y)$ *for* $x, y \in A$;

*(e)* $|xy|_\sigma \leq |x|_\sigma |y|_\sigma$ *for normal* $x, y \in A$;

*(f)* $\mathrm{Rad}(A) = \{x \in A: \rho(x) = 0\} = \rho^{-1}(\{0\})$;

*(g)* *If* $x \geq 0$, $y \geq 0$, *then* $x + y \geq 0$;

*(h)* $|h + k|_\sigma \leq |h|_\sigma + |k|_\sigma$ *for hermitian* $h, k \in A$;

*(i)* $|\frac{1}{2}(x^* + x)|_\sigma \leq \rho(x)$ *for* $x \in A$;

*(j)* $\rho(x + y) \leq \rho(x) + \rho(y)$ *for* $x, y \in A$;

*(k)* $\rho(x^*x) = \rho(x)^2$ *for* $x \in A$.

*Proof*. (a) The definition of the spectral radius implies that the inequality in (a) holds iff given $z \in A$ and $\lambda \in C$ with $|\lambda| > \rho(z)$, the element $\lambda e - z$ is invertible. By considering the element $x = \lambda^{-1}z$, it suffices to prove that $\rho(x) < 1$ implies $e - x$ is invertible.

Suppose, then, that $\rho(x) < 1$. It follows that $|x^*x|_\sigma < 1$ and $e - x^*x$ is positive, so that by (22.5), $e - x^*x$ can be written as $a^2$ for a positive invertible $a$ in A. Consider the identity

$$\begin{aligned}(e + x^*)(e - x) &= e + x^* - x - x^*x \\ &= a^2 + x^* - x \\ &= a(e + a^{-1}(x^* - x)a^{-1})a.\end{aligned}$$

Both $a$ and $e + a^{-1}(x^* - x)a^{-1}$ are invertible, the latter since $a^{-1}(x^* - x)a^{-1}$, being skew hermitian, has purely imaginary spectrum. The element $e - x$ is now clearly left invertible. Since $\rho(x^*) = \rho(x) < 1$, the corresponding identity for $(e - x)(e + x^*)$ shows that $e - x$ is also right invertible.

(b) If $x$ is normal, then $\rho(x)^2 = |x^*x|_\sigma \leq |x^*|_\sigma |x|_\sigma = |x|_\sigma^2 \leq \rho(x)^2$ by part (a) and the submultiplicativity of $|\cdot|_\sigma$ on commuting elements.

(c) If $h$, $k$ are hermitian, then by (a), we have

$$|hk|_\sigma \leq \rho(hk) = |khhk|_\sigma^{1/2} = |h^2k^2|_\sigma^{1/2}.$$

Hence by induction

$$|hk|_\sigma \leq |h^n k^n|_\sigma^{1/n} \leq ||h^n||^{1/n}||k^n||^{1/n}$$

where $n$ has the form $n = 2^m$. Passing to the limit as $m \to \infty$ gives (c).

(d) Since $x^*x$ and $yy^*$ are hermitian, we have from (c) that

$$\rho(xy) = |y^*x^*xy|_\sigma^{1/2} = |x^*xyy^*|_\sigma^{1/2}$$

$$\leq |x^*x|_\sigma^{1/2}|yy^*|_\sigma^{1/2} = \rho(x)\rho(y).$$

(e) This is immediate from (a), (b), and (d).

(f) Let $x \in \mathrm{Rad}(A)$. Then $x^*x \in \mathrm{Rad}(A)$ and hence $\rho(x) = |x^*x|_\sigma^{1/2} = 0$. On the other hand, if $x \in \ker\rho$, then $\rho(x) = 0$ and by (a) and (d) we have $|ax|_\sigma \leq \rho(ax) \leq \rho(a)\rho(x)$ for $a \in A$. Hence $|ax|_\sigma = 0$ for all $a \in A$ and so $x \in \mathrm{Rad}(A)$ by (B.5.17), (d).

(g) Suppose $x \geq 0$, $y \geq 0$; i.e., $x, y \in H(A)$ and each has nonnegative spectrum. To prove that $x + y \geq 0$ it suffices to prove that $-1 \notin \sigma_A(x + y)$, i.e., that $e + x + y$ is invertible. Both $e + x$ and $e + y$ are invertible and hence we can write

$$e + x + y = (e + x)(e + y) - xy$$

$$= (e + x)(e - ab)(e + y),$$

where $a = (e + x)^{-1}x$ and $b = y(e + y)^{-1}$. Since $|a|_\sigma < 1$ and $|b|_\sigma < 1$ (functional calculus), part (c) implies that $|ab|_\sigma < 1$. Hence $e - ab$ is invertible, and so $e + x + y$ is invertible, being the product of three invertible elements.

(h) We have $|h|_\sigma \pm h \geq 0$ and $|k|_\sigma \pm k \geq 0$; hence according to (g), $|h|_\sigma + |k|_\sigma \pm (h + k) \geq 0$, i.e., $|\lambda| \leq |h|_\sigma + |k|_\sigma$ for $\lambda \in \sigma_A(h + k)$. Therefore $|h + k|_\sigma \leq |h|_\sigma + |k|_\sigma$.

(i) Let $x = h + ik$ with $h, k \in H(A)$. Then $x^*x + xx^* = 2(h^2 + k^2)$. It is clear that $|h^2 + k^2|_\sigma - (h^2 + k^2) \geq 0$ and $k^2 \geq 0$. Hence, by (g),

$$|h^2 + k^2|_\sigma - h^2 = (|h^2 + k^2|_\sigma - h^2 - k^2) + k^2 \geq 0.$$

Thus, by (h),

$$|\tfrac{1}{2}(x^* + x)|_\sigma^2 = |h|_\sigma^2 \le \tfrac{1}{2}|x^*x + xx^*|_\sigma$$

$$\le \tfrac{1}{2}|x^*x|_\sigma + \tfrac{1}{2}|xx^*|_\sigma = |x^*x|_\sigma = \rho(x)^2,$$

which proves (i).

(j) Let $x, y \in A$. Then, by (h),

$$\rho(x + y)^2 = |(x^* + y^*)(x + y)|_\sigma$$

$$\le |x^*x|_\sigma^2 + |y^*y|_\sigma^2 + |x^*y + y^*x|_\sigma.$$

Further (i) and (d) yield

$$|x^*y + y^*x|_\sigma \le 2\rho(y^*x) \le 2\rho(y^*)\rho(x)$$

$$= 2\rho(x)\rho(y).$$

Therefore, $\rho(x + y)^2 \le [\rho(x) + \rho(y)]^2$.

(k) For $x \in A$, we have $\rho(x^*x)^2 = |x^*xx^*x|_\sigma = |x^*x|_\sigma^2 = \rho(x)^4$ and hence $\rho(x^*x) = \rho(x)^2$. □

We saw in (32.6) that, for arbitrary *-algebras, the notions of symmetry and hermiticity are not the same. We are now prepared to prove the Shirali-Ford Theorem which states that these two notions do coincide for *Banach* *-algebras. The result was first observed by S. Shirali [ 1 ], but his proof contained an error. It was corrected in a joint publication by S. Shirali and J. W. M. Ford [ 1 ]. N. Suzuki [ 2 ] gave an independent proof a little later. Alternative proofs have been given by T. W. Palmer [ 6 ], V. Pták [ 2 ], and L. A. Harris [ 1 ]. The proof given here is a slight simplification of Harris' proof due to A. W. Tullo.

(33.2) THEOREM. *(Shirali-Ford). A Banach *-algebra* A *is symmetric iff it is hermitian.*

*Proof.* If A is symmetric then it is hermitian by (32.4). To prove the converse, assume that A is hermitian and let $\delta = \sup\{-\mu : \mu \in \sigma_A(x^*x), \rho(x) \le 1\}$. It suffices to show that $\delta \le 0$. Assuming $\delta > 0$, there exists $x \in A$ and $\lambda \in \sigma_A(x^*x)$ with $-\lambda > \frac{1}{4}\delta$ and $\rho(x) < 1$. Let $y = 2x(e + x^*x)^{-1}$; then $e - y^*y = (e - x^*x)^2(e + x^*x)^{-2}$ and by Gelfand representation theory applied to a maximal abelian *-subalgebra containing $x^*x$ we have

$$\sigma_A(y^*y) = \{1 - f(t)^2 : t \in \sigma_A(x^*x)\},$$

where $f(t) = (1 - t)/(1 + t)$. Thus $\sigma_A(y^*y) \subseteq (-\infty,1)$. Set $y = h + ik$ with $h, k \in H(A)$. Then

$$yy^* = 2h^2 + 2k^2 - y^*y.$$

By (33.1), (g) $2h^2 + 2k^2 + (e - y^*y) \geq 0$ and so $\sigma_A(yy^*) \subseteq [-1,\infty)$. But $\sigma_A(y^*y) \cup \{0\} = \sigma_A(yy^*) \cup \{0\}$ and hence $\sigma_A(y^*y) \subset [-1,1)$ from which it follows that $\rho(y) \leq 1$. According to the definition of $\delta$ we have $-(1 - f(\lambda)^2) \leq \delta$, whence $f(\lambda) \leq (1 + \delta)^{1/2}$. Since $f(f(t)) = t$ and $f$ is decreasing on $(-1,\infty)$, we also have $\lambda = f(f(\lambda)) \geq f((1 + \delta)^{1/2})$ and therefore

$$-\lambda \leq \frac{(1 + \delta)^{1/2} - 1}{(1 + \delta)^{1/2} + 1} \leq \frac{\delta/2}{2} = \frac{\delta}{4},$$

which contradicts the choice of $\lambda$. □

We shall need the following lemma.

(33.3) LEMMA. *Let* A *be a Banach algebra with identity* 1 *and suppose* $x \in A$. *The following properties are equivalent:*

*(a)* $\sigma_A(x) \subseteq R$.

*(b)* $|e^{\lambda ix}|_\sigma = 1$ *for all* $\lambda \in R$.

*(c)* *there exists* $c > 0$ *such that* $|e^{\lambda ix}|_\sigma \leq c$ *for all* $\lambda \in R$.

*(d)* $\displaystyle\lim_{\lambda\to 0} \frac{|1 + i\lambda x|_\sigma - 1}{|\lambda|} = 0, \quad (\lambda \in R).$

*Proof.* (a) iff (b): By the holomorphic functional calculus we have $\sigma_A(e^{\lambda ix}) = \{e^{i\lambda\alpha} : \alpha \in \sigma_A(x)\}$ for all $\lambda \in R$. Hence (a) implies (b). Conversely, if $|e^{i\lambda x}|_\sigma = 1$ for all $\lambda \in R$, then $|e^{ix}|_\sigma = |e^{-ix}|_\sigma = 1$, whence $\sigma_A(e^{ix}) \subseteq \{z \in C: |z| = 1\}$. Thus $\sigma_A(x) \subseteq R$ follows from the equality in the second line of the proof.

(b) iff (c): Clearly (b) implies (c). In the other direction, if $n$ is any positive integer, we have $|e^{inx}|_\sigma \leq c$ and $|e^{-inx}|_\sigma \leq c$ and hence $|e^{\pm ix}|_\sigma \leq \lim_{n\to\infty} c^{1/n} = 1$. If $|e^{ix}|_\sigma < 1$ or $|e^{-ix}|_\sigma < 1$, then the inequality $1 = |e^{ix}e^{-ix}|_\sigma \leq |e^{ix}|_\sigma |e^{-ix}|_\sigma$ implies a contradiction.

(a) iff (d): Since $\sigma_A(1 + \lambda ix) = \{1 + \lambda i\alpha: \alpha \in \sigma_A(x)\}$, we have $|1 + \lambda ix|_\sigma = (1 + \lambda^2|x|_\sigma^2)^{1/2}$ if $\sigma_A(x) \subseteq R$; hence

$$\lim_{\lambda\to 0} \frac{|1 + \lambda ix|_\sigma - 1}{|\lambda|} = 0$$

where $\lambda$ is real. Conversely, suppose that $\lim_{\lambda\to 0} \frac{|1 + \lambda ix|_\sigma - 1}{|\lambda|} = 0$ with, for example $a + ib \in \sigma_A(x)$, where $a, b \in R$ and $b \leq 0$. Since $1 - \lambda b + i\lambda a$ is in $\sigma_A(1 + \lambda ix)$, we obtain for $\lambda > 0$ the inequalities:

$$0 \leq \frac{(1 - 2\lambda b + \lambda^2 b^2 + \lambda^2 a^2)^{1/2} - 1}{\lambda} \leq \frac{|1 + \lambda ix|_\sigma - 1}{\lambda}$$

and letting $\lambda$ tend to 0, this forces $b = 0$. The case $b \geq 0$ works in an analogous way by taking $\lambda < 0$. □

In 1971 B. Aupetit [ 1 ] announced, without proof, the characterizations of symmetric Banach *-algebras given in the next theorem. V. Pták [1] published the first proof. The argument given below for (c) implies (a) can be found in Rickart [1, p.190]. The reader will note that in parts (a), (b), and (c) it is not assumed that A contains an identity.

(33.4) THEOREM. *Let* A *be a Banach *-algebra. The following properties are equivalent:*

*(a)* A *is symmetric;*

*(b)* $|x^*x|_\sigma \geq |x|_\sigma^2$ *for all* $x \in A$*;*

*(c) there exists a constant* $c > 0$ *such that* $|x^*x|_\sigma \geq c|x|_\sigma^2$ *for all normal* $x \in A$*;*

*In case* A *has an identity, these properties are also equivalent to:*

*(d) there exists a constant* $k > 0$ *such that* $|e^{ih}|_\sigma \leq k$ *for all hermitian* $h \in A$.

*Proof.* (a) implies (b) is a consequence of (32.4) and (33.1), (a). Also, (b) implies (c) is clear.

(c) implies (a): We use a variation of Arens' argument (see (7.1)). Suppose that $h \in H(A)$ with $\sigma_A(h) \not\subseteq R$. Then we may assume that there is $a \in R$ such that $a + i \in \sigma_A(h)$. Indeed, since $\sigma_A(h)$ is not contained in R, there exist $\alpha$ and $\beta$ real, $\beta \neq 0$, such that $\alpha + i\beta \in \sigma_A(h)$. Set $a = \beta^{-1}\alpha$ and redefine $h$ to be the element $\beta^{-1}h$. Then $h$ is still hermitian and $a + i \in \sigma_A(h)$. Let B be a (closed) maximal commutative *-subalgebra of A containing h. Set $v = (h - a + ni)^m h \in B$, where m and n are positive integers. By (21.3) $\sigma_A(h) = \sigma_B(h)$ so there is a multiplicative linear functional $\phi$ on B such that $\phi(h) = a + i \neq 0$. Then $\phi(v) = (n + 1)^m i^m(a + i)$, whence $|v|_\sigma \geq (n + 1)^m(1 + a^2)^{1/2}$. But $v^*v = ((h - a)^2 + n^2)^m h^2$, so that $|v^*v|_\sigma \leq |h|_\sigma^2[(|h|_\sigma + |a|)^2 + n^2]^m$ which with the hypothesis gives $c(n + 1)^{2m}(1 + a^2) \leq |h|_\sigma^2[(|h|_\sigma + |a|)^2 + n^2]^m$ so that

$$c^{1/m}(n + 1)^2(1 + a^2)^{1/m} \leq |h|_\sigma^{2/m}[(|h|_\sigma + |a|)^2 + n^2].$$

Letting $m \to \infty$, we obtain $(n + 1)^2 \leq (|h|_\sigma + |a|)^2 + n^2$, which is false for $2n + 1 > (|h|_\sigma + |a|)^2$.

(a) iff (d) follows from (33.2) and (33.3). □

(33.5) COROLLARY. *(Aupetit). Let* A *be a Banach *-algebra. Then* A/Rad A *is symmetric and commutative iff* $|x^*x|_\sigma = |x|_\sigma^2$ *for all* $x \in A$.

*Proof.* If B = A/Rad A is commutative and symmetric, we have $|x|_\sigma^2 \leq |x^*x|_\sigma \leq |x^*|_\sigma |x|_\sigma = |x|_\sigma^2$ by (33.1), (a) together with the fact that $|x + \text{Rad } A|_\sigma = |x|_\sigma$ (see (B.5.16)). Conversely, if $|x^*x|_\sigma = |x|_\sigma^2$ for all x then, by (33.4), B is symmetric. Then by (33.1) there exists a seminorm p on B such that $p(x + \text{Rad } A) = |x + \text{Rad } A|_\sigma$ for all $x \in A$. Hence by (B.6.17), the algebra B is commutative. □

When A has an identity, the preceding result can be improved in a local way. Indeed, B. Aupetit [3, p. 118] has shown that if A is a Banach *-algebra with identity, then A/Rad A is symmetric and commutative iff $|x^*x|_\sigma = |x|_\sigma^2$ for all x in a neighborhood of the identity of A.

If A is a symmetric normed *-algebra with isometric involution, is the completion $\tilde{A}$ symmetric? This question was posed by J. Wichmann [3] in 1976. The answer was shown to be negative by P. G. Dixon [4], who adapted an example he had constructed earlier for another purpose. We present his example next.

(33.6) *Example.* Let $A_o$ be the free complex associative algebra on generators $e_n$ $(n = 1,2,3,\ldots)$ with the relations $d = 0$ for every monomial $d = e_{i_1}e_{i_2}\cdots e_{i_r}$ containing more than $n$ occurrences of $e_n$ where $n = \max\{i_1,\ldots,i_r\}$. Hence $A_o$ is the algebra generated by $\{e_n\colon n = 1,2,3,\ldots\}$ with these relations. The countable set $M = \{d_i\colon i = 1,2,3,\ldots\}$ of non-zero monomials is a vector space basis for $A_o$, so we can define a norm by

$$||\Sigma_{i=1}^{n}\lambda_i d_i|| = \sum_{i=1}^{n}|\lambda_i| \qquad (\lambda_i \in C,\ d_i \in M).$$

One checks easily that this is an *algebra* norm. The involution is defined on $A_o$ setting $e_n^* = e_n$ $(n = 1,2,3,\ldots)$; it is clearly isometric.

We wish to show that the algebra $A_o$ is nil (every element is nilpotent). First one can see inductively that if $d = e_{i_1}e_{i_2}\cdots e_{i_r}$ and $r \geq (n+1)!$ where $n = \max\{i_1,\ldots,i_r\}$, then $d = 0$. The bound is precise in the sense that there is a (unique) non-zero monomial of length $(n+1)! - 1$. To see this, define inductively a sequence $\{g_n\}$ of non-zero monomials by $g_1 = e_1$ and $g_n = g_{n-1}e_n g_{n-1}e_n\cdots e_n g_{n-1}$, an alternating product in which $e_n$ occurs $n$ times. Then each $g_n$ is non-zero and has length $(n+1)! - 1$. Further, an inductive argument shows that no non-zero monomial whose largest index is $n$ can be longer than $g_n$. Now $(\sum_{i=1}^{n}\lambda_i d_i)^k$ is a sum of terms of the form $\lambda\cdot f_1\cdot f_2\cdots f_k$ where $f_i$ is some $d_j$, $j = 1,2,\ldots,n$. If $k = (p+1)!$ where $p = \max\{j\colon e_j$ occurs in some $d_i$, $i = 1,\ldots,n\}$, then every $\lambda\cdot f_1\cdots f_k$ must be zero.

Since every nilpotent element is quasi-regular, it follows that $A_o$ is symmetric. Let $A$ denote the completion of $A_o$, so that each element of $A$ has the unique representation $\sum_{i=1}^{\infty}\lambda_i d_i$ $(\lambda_i \in C,\ d_i \in M)$ with $\sum_{i=1}^{\infty}|\lambda_i| < \infty$. Consider the hermitian element $h = \sum_{n=1}^{\infty}2^{-n}e_n$ in $A$. We show that the spectrum of $h$ in $A$ contains an open disk about zero, from which it follows that $A$ is not hermitian, hence not symmetric.

Whenever $\lambda h$ has a quasi-inverse $y = y(\lambda)$ it is expressible as an absolutely convergent series in the $d_i$, say as $\sum_{i=1}^{\infty}\gamma_i(\lambda)d_i$. The relation $\lambda h + y = (\lambda h)y$ can then be written as

$$\sum_{n=1}^{\infty} \lambda 2^{-n} e_n + \sum_{i=1}^{\infty} \gamma_i(\lambda) d_i = \left(\sum_{m=1}^{\infty} \lambda 2^{-m} e_m\right)\left(\sum_{j=1}^{\infty} \gamma_j(\lambda) d_j\right). \qquad (*)$$

We have $\{e_1, e_2, \ldots\} \subseteq \{d_1, d_2, \ldots\}$. Let, say, $e_n = d_{i_n}$, $n \in N$. Since no single $e_n$'s occur in the expanded version of the right side of (*), we infer from (*) that

$$\lambda 2^{-n} + \gamma_{i_n}(\lambda) = 0 \quad \text{for all } n. \qquad (1)$$

If $i \notin \{i_1, i_2, \ldots\}$, then for some pair $(m,j)$ we have $d_i = e_m d_j$. Hence (*) shows that

$$\gamma_i(\lambda) = \lambda 2^{-m} \gamma_j(\lambda). \qquad (2)$$

We now show by induction on $r$ that

$$\text{if } d_i = e_{j_1} \cdots e_{j_r}, \text{ then } \gamma_i(\lambda) = -\lambda^r 2^{-(j_1 + \cdots + j_r)}. \qquad (**)$$

The case $r = 1$ is contained in (1). If (**) is true for $r$, consider $d_i = e_{j_1} \cdots e_{j_r} e_{j_{r+1}}$. Write

$$d_j = e_{j_2} \cdots e_{j_r} e_{j_{r+1}} \qquad (3)$$

so that $d_i = e_{j_1} d_j$. Then by (2) we have

$$\gamma_i(\lambda) = \lambda 2^{-j_1} \gamma_j(\lambda) \qquad (4)$$

and by the induction hypothesis and (3), $\gamma_j(\lambda) = -\lambda^r 2^{-(j_2 + \cdots + j_r + j_{r+1})}$. Therefore (4) yields $\gamma_i(\lambda) = -\lambda^{r+1} 2^{-(j_1 + j_2 + \cdots + j_r + j_{r+1})}$, and completes the inductive proof.

We now consider a certain sequence $\{b_n\}_{n=1}^{\infty}$ in $M$ whose coefficients in the expansion of $y(\lambda)$ can be usefully estimated. Define $j_n = \max\{j: 2^{j-1} \text{ divides } n\}$ and set

$$b_n = e_{j_1} \cdots e_{j_n} = e_1 e_2 e_1 e_3 e_1 e_2 e_1 e_4 e_1 e_2 e_1 e_3 e_1 \cdots \quad (n \text{ factors}).$$

The fact that $b_n \neq 0$ is a consequence of two elementary properties of the sequence $\{j_n\}$: (1°) if $j_p = j_r$ for $p < r$ then there is a $q$, $p < q < r$, for which $j_q > j_r$; (2°) for any sequence of consecutive terms $j_n, j_{n+1}, \ldots j_{n+p}$ the maximum value occurs exactly once. Property (2°), which follows easily from (1°), shows that $e_{j_1} e_{j_2} \cdots e_{j_n}$ cannot contain a zero factor. With $b_n \neq 0$ confirmed, we see that $b_n \in M$; that is, $b_n$ is a $d_i$. Hence by what was shown in (**), the coefficient of $b_n$ in $y(\lambda)$ is $\lambda^n/2^{t_n}$ where $t_n = j_1 + j_2 + \cdots + j_n$. We would like to show that $t_n \leq 2n$. This will follow easily upon verifying

$$t_n = \sum_{i=0}^{s}(2^{i+1} - 1)\alpha_i \tag{1}$$

where the $\alpha_i$'s are the coefficients of the binary expansion of $n$, $n = \alpha_s 2^s + \cdots + \alpha_1 \cdot 2 + \alpha_o$. Two further properties of the sequence $\{j_n\}$ enter here: (3°) $j_{2^k+p} = j_p$ for $1 \leq p < 2^k$ and (4°) $j_{2^{k+1}} = j_{2^k} + 1$. Using (3°) we see that

$$t_{2^k+p} = t_{2^k} + t_p \quad \text{for} \quad 1 \leq p < 2^k. \tag{2}$$

Then by induction using (4°) we find that

$$t_{2^k} = 2^{k+1} - 1, \quad k = 1,2,3,\ldots . \tag{3}$$

Now applying (2) and (3) inductively to $n$ written in its binary expansion, we obtain (1). So $t_n = \sum_{i=0}^{s}(2^{i+1} - 1)\alpha_i \leq \sum_{i=0}^{s} 2^{i+1}\alpha_i = 2n$ as desired.

Since $||y|| = \sum_{i=1}^{\infty}|\gamma_i(\lambda)| \geq \sum_{n=1}^{\infty}|\lambda^n|e^{-t_n} \geq \sum_{n=1}^{\infty}|\lambda|^n 2^{-2n} = \sum_{n=1}^{\infty}|\lambda/4|^n$, and $||y||$ is finite, it follows that $|\lambda| < 4$. Thus $\lambda h$ is quasi-regular in $A$ only if $|\lambda| < 4$; that is, $\lambda^{-1}h$ is quasi-singular whenever $|\lambda| \leq 1/4$. Hence $\{\lambda \in \mathbb{C}: |\lambda| \leq 1/4\} \subseteq \sigma_A(h)$, as claimed, and so $A$ is not symmetric. □

The next theorem gives several characterizations of hermitian Banach *-algebras.

(33.7) THEOREM. *(Pták). Let* A *be a Banach *-algebra. The following are equivalent:*

*1)* A *is hermitian.*

*2)* $|x|_\sigma^2 \leq |x^*x|_\sigma$ *for all* $x \in A$.

*3)* $|x|_\sigma \leq \rho(x)$ *for all* $x \in A$.

*4)* $|x|_\sigma \leq \rho(x)$ *for all* $x \in N(A)$.

*5)* $|x|_\sigma^2 = |x^*x|_\sigma$ *for all* $x \in N(A)$.

*6)* $|x^* + x|_\sigma \leq 2|x^*x|_\sigma^{1/2}$ *for all* $x \in A$.

*7)* $\rho(x + y) \leq \rho(x) + \rho(y)$ *for all* $x, y \in A$.

*8)* $|u|_\sigma = 1$ *for all* $u \in U(A)$.

*9)* $|u|_\sigma \leq 1$ *for all* $u \in U(A)$.

*10)* $|u|_\sigma \le \beta$ *for all* $u \in U(A)$ *and some finite* $\beta$.

*11)* *For each* $x \in A$, *the real part of every* $\lambda \in \sigma_A(x^*x)$ *is nonnegative.*

*12)* $\sigma_A(x^*x) \ge 0$ *for each* $x \in A$.

*13)* A *is symmetric.*

*Proof.* (1) implies (2) follows from (33.1), (a) and (2) => (3) => (4) are clear.

(4) implies (5) is a consequence of the fact that, in any Banach algebra, the spectral radius is submultiplicative on commuting elements (see (B. 6.20)).

To show that (5) implies (1) we use, once again, an Arens' type argument. Assume there is an $h$ in $H(A)$ such that $\lambda + i\mu \in \sigma_A(h)$ with $\lambda, \mu \in R$ and $\mu \neq 0$. Then $k = \mu^{-1}(h - \lambda e)$ is hermitian, and $i \in \sigma_A(k)$. For each $\beta > 0$, the element $k + \beta ie$ is normal and $(\beta + 1)i \in \sigma_A(k + \beta ie)$. From condition (5) we have

$$|(\beta + 1)i|^2 \le |k + \beta ie|_\sigma^2 = |(k + \beta ie)^*(k + \beta ie)|_\sigma$$
$$= |k^2 + \beta^2 e|_\sigma \le |k^2|_\sigma + \beta^2$$

from which it follows that $2\beta + 1 \le |k^2|_\sigma$ for all $\beta > 0$; a contradiction. Hence (1) is true, and conditions (1) through (5) are equivalent.

Note that implications (11) => (12) => (13) are obvious. It was shown in (32.4) that (13) implies (1). Hence (1), (11), (12), and (13) are equivalent.

The implication (1) implies (7) is given in (33.1), (j). We now observe that if $\rho$ is subadditive, then (6) follows. Indeed, given $x \in A$, we have

$$|\tfrac{1}{2}(x^* + x)|_\sigma = \rho(\tfrac{1}{2}(x^* + x))$$
$$\le \tfrac{1}{2}\rho(x^*) + \tfrac{1}{2}\rho(x) = \rho(x).$$

Hence, we have shown (1) => (7) => (6).

Let us show next that (6) implies (4): Let $x \in N(A)$ and write $x = h + ik$ with $h, k \in H(A)$. Then, since $hk = kh$, (B.6.20) gives

$$|x|_\sigma \leq |h|_\sigma + |k|_\sigma.$$

From (6) it follows that

$$|x|_\sigma \leq |h|_\sigma + |k|_\sigma \leq \rho(x) + \rho(ix) = 2\rho(x).$$

Hence

$$\begin{aligned} |x|_\sigma^n = |x^n|_\sigma \leq 2\rho(x^n) &= 2|(x^n)^*x^n|_\sigma^{1/2} \\ &= 2|(x^*)^n x^n|_\sigma^{1/2} = 2|(x^*x)^n|_\sigma^{1/2} \\ &= 2\rho(x)^n \end{aligned}$$

so that $|x|_\sigma \leq 2^{1/n}\rho(x)$ for each $n = 1,2,\ldots$ ; therefore (4) follows.

The implications (5) => (8) => (9) => (10) are all obvious. To complete the proof, it suffices to show that (10) implies (1). Let $u \in U(A)$ and note that $u^n \in U(A)$ for each $n = 1,2,3,\ldots$; hence $|u|_\sigma = |u^n|_\sigma^{1/n} \leq \beta^{1/n}$ for $n = 1,2,3,\ldots$, so that $|u|_\sigma \leq 1$. Because $u^{-1} \in U(A)$ also, $|u^{-1}|_\sigma \leq 1$, and it follows that each point in $\sigma_A(u)$ has modulus one. In particular, if $\lambda \in \sigma_A(u)$, then $\lambda^{-1} = \bar{\lambda}$.

Now assume that $h \in H(A)$ satisfies $|h|_\sigma < 1$, and let $B$ be a maximal commutative *-subalgebra of $A$ containing $h$. Since

$$|e - (e - h^2)|_\sigma = |h^2|_\sigma < 1,$$

and $B$ is closed there exists, by (22.5), an element $k \in H(A) \cap B$ such that $k^2 = e - h^2$. Let $u = h + ik$: then $u^* = h - ik$ and $u \in U(A)$. For $\phi \in \hat{B}$, $\phi(h) + i\phi(k) = \phi(u) \in \sigma_A(u)$ and $\phi(h) - i\phi(k) = \phi(u^*) \in \sigma_A(u^*)$. Hence, $|\phi(h) \pm i\phi(k)| = 1$. However, since we also have $\phi(h)^2 + \phi(k)^2 = \phi(h^2 + k^2) = \phi(e) = 1$, it follows that $\phi(h)$ is real. Therefore, $A$ is hermitian and the proof is complete. □

Let $A$ be a *-algebra with identity $e$. A linear functional $f$ on $A$ is said to be *unital* if $f(e) = 1$, and *weakly positive* if $f(h^2) \geq 0$ for all $h$ in $H(A)$.

It is clear that every positive functional is weakly positive. A few basic properties of weakly positive functionals are given in the following proposition.

(33.8) PROPOSITION. *Let* $f$ *be a unital weakly positive functional on a unital Banach *-algebra* $A$. *Then:*

*(a)* $f$ *is hermitian, i.e.,* $f(h)$ *is real for each* $h \in H(A)$.

*(b)* $|f(h)| \leq |h|_\sigma$ *for each* $h \in H(A)$.

*(c)* $f(h)^2 \leq f(h^2)$ *for each* $h \in H(A)$.

*(d)* $|f(x)| \leq \rho(x)$ *for each* $x \in A$.

*Proof.* (a) Let $h \in H(A)$ with $|h|_\sigma < 1$. Then $|e - (e - h)|_\sigma < 1$, so, by (22.5), there is an element $x \in H(A)$ such that $e - h = x^2$. Therefore $1 - f(h) = f(x^2) \geq 0$ from which it follows that $f(h)$ is real.

(b) The proof in part (a) shows that $f(k)$ is real for all $k \in H(A)$ and that $f(k) \leq 1$ if $|k|_\sigma < 1$. Replacing $k$ by $-k$, we obtain $|f(k)| \leq 1$. Hence, $|k|_\sigma < 1$ implies that $|f(k)| \leq 1$, from which it follows that $|f(h)| \leq |h|_\sigma$ for each $h \in H(A)$.

(c) For $h \in H(A)$ and $\alpha \in R$ we have $f(h^2) + 2\alpha f(h) + \alpha^2 = f((h + \alpha e)^2) \geq 0$. By (a), $f(h)$ is real. Setting $\alpha = -f(h)$ gives $f(h)^2 \leq f(h^2)$.

(d) Let $x \in A$ and write $x = h + ik$, with $h$ and $k$ hermitian. Since $f(h)$ and $f(k)$ are real by (a), it follows from (b) and (c) that

$$\begin{aligned} |f(x)|^2 &= f(h)^2 + f(k)^2 \leq f(h^2) + f(k^2) \\ &= f(h^2 + k^2) \\ &= f(\tfrac{1}{2}(x^*x + xx^*)) \\ &= \tfrac{1}{2}f(x^*x) + \tfrac{1}{2}f(xx^*) \\ &\leq \tfrac{1}{2}|x^*x|_\sigma + \tfrac{1}{2}|xx^*|_\sigma = \rho(x)^2. \quad \square \end{aligned}$$

The reader will recognize that the proof of the next result is another slight variant of the Arens' argument, which has been used several times.

(33.9) PROPOSITION. *Let* $A$ *be a unital Banach *-algebra and* $f$ *a* *unital linear functional on* A. *If* $|f(x)| \leq \rho(x)$ *for all* $x \in H(A) + iR$, *then* f *is hermitian and* $|f(h)| \leq |h|_\sigma$ *for* $h \in H(A)$.

*Proof.* For $h \in H(A)$, set $f(h) = \mu + i\nu$, with $\mu, \nu \in R$. Then $k = h - \mu e \in H(A)$ and $f(k) = i\nu$. Let $\alpha$ be a real number. Then

$$|\alpha + \nu| = |i\nu + i\alpha| = |f(k + i\alpha e)|$$

$$\leq \rho(k + i\alpha e) = |k^2 + \alpha^2 e|_\sigma^{1/2}.$$

Because $\sigma_A(k^2 + \alpha^2 e) = \sigma_A(k^2) + \alpha^2$, we have $|k^2 + \alpha^2 e|_\sigma \leq |k^2|_\sigma + \alpha^2$. Hence

$$|\alpha + \nu| \leq (|k^2|_\sigma + \alpha^2)^{1/2},$$

or

$$\alpha^2 + 2\alpha\nu + \nu^2 \leq |k^2|_\sigma + \alpha^2$$

for each real $\alpha$. Therefore $\nu = 0$ and $f(h)$ is real. Finally, $|f(h)| \leq \rho(h) = |h|_\sigma$ for each h in $H(A)$. □

(33.10) LEMMA. *Let* A *be a unital hermitian Banach *-algebra, and* f *a unital hermitian linear functional on* A *such that* $|f(h)| \leq |h|_\sigma$ *for* $h \in H(A)$. *Then*

*1)* $f(h) \in \text{conv}\,\sigma_A(h)$ *for each* $h \in H(A)$, *and*

*2)* f *is weakly positive.*

*Proof.* (1) Let $h \in H(A)$. Since A is hermitian, $\sigma_A(h)$ is a compact subset of the real line. Hence, there exist real numbers $\alpha$ and $\beta$ such that $\alpha \leq \sigma_A(h) \leq \beta$. Setting $\mu = \frac{1}{2}(\alpha + \beta)$ and $\nu = \frac{1}{2}(\beta - \alpha)$, we have $|h - \mu e|_\sigma \leq \nu$. Therefore

$$|f(h) - \mu| = |f(h - \mu e)| \leq |h - \mu e|_\sigma \leq \nu,$$

and it follows that $\alpha = \mu - \nu \leq f(h) \leq \mu + \nu = \beta$. Since $\sigma_A(h)$ is closed, $f(h) \in \text{conv}\,\sigma_A(h)$ for h in $H(A)$.

(2) We have, by part (1), that $f(h^2) \in \text{conv}\,\sigma_A(h^2) = \text{conv}\,(\sigma_A(h))^2 \geq 0$. Hence, f is weakly positive. □

The next proposition shows, among other things, that the notions of positivity and weak positivity coincide on hermitian Banach *-algebras.

(33.11) PROPOSITION. *Let* A *be a unital hermitian Banach *-algebra, and* f *a unital linear functional on* A. *Then the following are equivalent:*

*(a)* f *is hermitian and* $|f(h)| \leq |h|_\sigma$ *for* $h \in H(A)$.

*(b)* f *is a state on* A.

*(c)* f *is weakly positive.*

*(d)* $|f(x)| \leq \rho(x)$ *for all* $x \in A$.

*(e)* $|f(x)| \leq \rho(x)$ *for all* $x \in N(A)$.

*(f)* $|f(x)| \leq \rho(x)$ *for all* $x \in H(A) + iR$.

*Proof.* (a) implies (b): By (a) and (33.10), we have

$$f(x^*x) \in \operatorname{conv} \sigma_A(x^*x) \geq 0, \quad (x \in A)$$

(the inequality follows from (32.5) and (33.2)) and hence f is a state on A. (b) implies (c) is obvious. (c) implies (d) follows from (33.8), (d). The implications (d) $\Rightarrow$ (e) $\Rightarrow$ (f) are immediate. Finally, (f) implies (a) is the content of (33.9). □

REMARK. Only the implication (a) implies (b) in (33.11) required that A be hermitian. Recall that the set of states on a unital Banach *-algebra is denoted by S(A) (see (27.1)).

The next result characterizes hermitian algebras in terms of their states and the Gelfand-Naimark seminorm $\gamma$ (see (30.5)).

(33.12) THEOREM. *Let* A *be a unital Banach *-algebra. Then the following are equivalent:*

*(a)* A *is hermitian.*

*(b)* S(A) *is nonempty and* $\rho(x) = \gamma(x)$ *for all* $x \in A$.

*(c)* S(A) *is nonempty and* $|x^*x|_\sigma = \sup\{p(x^*x)\colon p \in S(A)\}$ *for all* $x \in A$.

*Proof.* Let e be the identity of A. (a) implies (b). Let $x \in A$ and let B be a maximal commutative *-subalgebra containing $x^*x$ and e; B is closed, as noted after (22.1). By (21.3), $\sigma_B(x^*x) = \sigma_A(x^*x) \geq 0$ (see (32.5) and (33.2)), and since the spectrum is a compact subset of R, the number $\lambda = |x^*x|_\sigma$ belongs to $\sigma_A(x^*x)$. Therefore, there is an element

$\psi \in \hat{B}$ with $\psi(x^*x) = \lambda$. Since A is hermitian, so is B by (32.7); hence every multiplicative linear functional on B assumes real values on hermitian elements. This is true, in particular, for $\psi$ and therefore, for each $b \in B$,

$$|\psi(b)|^2 = \overline{\psi(b)}\cdot\psi(b) = \psi(b^*)\psi(b)$$

$$= \psi(b^*b) \leq |b^*b|_\sigma = \rho(b)^2.$$

Now according to (33.1), (j), $\rho$ is a seminorm on A. Therefore (A.1) furnishes an extension p of $\psi$ from B to A such that

$$|p(a)| \leq \rho(a) \quad (a \in A).$$

Since $\psi$ is multiplicative and p extends $\psi$, we have $p(e) = 1$. Hence, p is a state by (33.11), and by (30.5), (e), $\gamma(x) \geq p(x^*x)^{1/2} = \lambda^{1/2} = \rho(x)$. It was also shown in (30.5) that $\gamma(x) \leq \rho(x)$ for all x in A. When combined with the above, this shows that $\gamma(x) = \rho(x)$.

(b) implies (c) is obvious in view of (30.5).

(c) implies (a): Let $x \in A$, set $\mu = |x^*x|_\sigma$, and consider the element $u = \mu e - x^*x$. If $p \in S(A)$, then

$$\begin{aligned} p(u^2) &= p(\mu^2 e - 2\mu x^*x + x^*(xx^*)x) \\ &= \mu^2 - 2\mu p(x^*x) + p(x^*(xx^*)x). \end{aligned} \tag{1}$$

Since $p \in S(A)$, we have, by (27.1), 6°, that

$$p(x^*a^*ax) \leq |a^*a|_\sigma p(x^*x), \quad (a \in A).$$

Hence, setting $a = x^*$ and using $|x^*x|_\sigma = |xx^*|_\sigma$, gives

$$p(x^*xx^*x) \leq |x^*x|_\sigma p(x^*x) = \mu p(x^*x). \tag{2}$$

From (1) and (2) we have

$$p(u^2) \leq \mu^2 - \mu p(x^*x) \leq \mu^2. \tag{3}$$

Since $p \in S(A)$ was arbitrary, it follows from (c) and (3) that

$$|u^2|_\sigma = |uu^*|_\sigma = \sup\{p(uu^*)\colon p \in S(A)\}$$
$$= \sup\{p(u^2)\colon p \in S(A)\}$$
$$\leq \mu^2.$$

Therefore $|u|_\sigma \leq \mu$, so that $|\mu e - x^*x|_\sigma \leq \mu$. It follows that the real part of each $\lambda$ in $\sigma_A(x^*x)$ is nonnegative, and hence A is hermitian by (11) of (33.7). □

By combining (30.5), (30.6), (33.1) (f), and (33.12) we obtain the following important corollary relating the Jacobson and *-radicals:

(33.13) COROLLARY. *Let* A *be a unital hermitian Banach *-algebra. Then*

$$\mathrm{Rad}(A) = R^*(A) = \rho^{-1}(\{0\}).$$

*Furthermore, there exists a Hilbert space* H *and a *-representation* $\pi$ *of* A *on* H *such that:*

*1°)* $\pi^{-1}(\{0\}) = \mathrm{Rad}(A)$.

*2°)* $||\pi(a)|| = \rho(a) = \gamma(a)$ *for each* $a \in A$.

*If* A *is semisimple, then* $\rho$ *is an algebra norm on* A *which satisfies the* C**-condition, and hence* A *is an* A**-algebra.*

(33.14) REMARKS. We close this section by making a few general remarks about *-ideals in hermitian and symmetric algebras. In 1949, I. Kaplansky [1, Theorem 4.4] proved that every primitive ideal in a hermitian *-algebra is a *-ideal. Recently, in a preprint, J. Wichmann has shown the converse to be false. He also shows that a *commutative* *-algebra A is hermitian iff each maximal modular ideal of A is a *-ideal, thus generalizing a classical result known previously to hold for commutative *Banach* *-algebras.

In another direction, Wichmann [4, Corollary, p. 86] proved that the closure of a symmetric *-ideal in a Banach *-algebra is again symmetric. This should be contrasted with the situation described in (33.6) where an example was given of a nonsymmetric Banach *-algebra with a dense symmetric *-subalgebra.

If a *-ideal I of a Banach *-algebra A contains the Jacobson radical of A, then by Johnson's uniqueness of the norm theorem (B.5.35), the closure of I in A is again a $^*$-ideal of A. This is not true for all $^*$-ideals as the following example, due to Wichmann [4], shows.

(33.15) *Example.* Let A be the Banach algebra of all bounded sequences $\{x_n\}$ with the norm $||\{x_n\}|| = \sup |x_n|$ and trivial multiplication, that is, $AA = \{0\}$. Let $e_n$ be the sequence having as only nonzero entry the number one in the n-th coordinate. Extend the linearly independent set $e_1, e_2, \ldots$, $\{1/n\}$, $\{1,1,\ldots\}$ to a Hamel basis of A. Define an involution on A by setting $\{1/n\}^* = \{1,1,\ldots\}$, $\{1,1,\ldots\}^* = \{1/n\}$, and $v^* = v$ for all other basis elements v. Then the set I of all sequences $\{x_n\}$ with $x_n = 0$ for almost all n is a $^*$-ideal of the symmetric Banach $^*$-algebra A. The closure of I is an ideal of A which contains $\{1/n\}$, but not $\{1/n\}^*$. □

## §34. *Equivalent C*-norms.*

What conditions can be imposed on a Banach $^*$-algebra A which force A to admit an equivalent C*-norm? I. Kaplansky [1, p. 405] conjectured in 1949 that if A is symmetric and $\alpha > 0$ is a scalar such that $|h|_\sigma \geq \alpha||h||$ for all hermitian h, then A admits such a norm. He also conjectured that a Banach $^*$-algebra satisfying $||x^*x|| \geq \alpha||x^*||\cdot||x||$ for some $\alpha > 0$ and all x admits an equivalent C*-norm. After considerable effort B. Yood [2, 4] obtained these results in their full generality. They, as well as two additional properties due to Pták, are shown in this section to be equivalent to an algebra's admitting an equivalent C*-norm. We shall state our results in terms of symmetric Banach $^*$-algebras but freely use the equivalence of this hypothesis with the hypothesis that A is hermitian.

(34.1) LEMMA. *Let* A *be a symmetric Banach $^*$-algebra.*
1° *If* A *is semisimple, then the involution is continuous.*
2° *The Pták function* $\rho$ *is continuous on* A.

*Proof.* 1°. Assume $\text{Rad}\, A = \{0\}$. We shall show that $x \to x^*$ has a closed graph. Indeed, suppose that $x_n \to 0$ and $x_n^* \to y$. Then

$$\rho(y) \leq \rho(y - x_n^*) + \rho(x_n^*), \text{ by (33.1), (j),}$$

$$\leq ||y^* - x_n||^{1/2}||y - x_n^*||^{1/2} + ||x_n||^{1/2}||x_n^*||^{1/2}.$$

Since $x_n \to 0$, $y - x_n^* \to 0$ and the other factors are bounded, $\rho(y) = 0$ and hence $y \in \text{Rad } A$ by (33.1), (f). Therefore $y = 0$.

2°. Let $R = \text{Rad } A$ and $\tau\colon A \to A/R$ the quotient map. By 1° the involution on $A/R$ is continuous; thus there exists $\gamma > 0$ satisfying $||\tau(x^*)|| \leq \gamma||\tau(x)||$ for all $x \in A$. Since $|x|_\sigma = |\tau(x)|_\sigma$ for each $x \in A$ by (B.5.16), it follows that $|x^*x|_\sigma = |\tau(x^*x)|_\sigma = |\tau(x^*)\tau(x)|_\sigma \leq ||\tau(x^*)\tau(x)|| \leq ||\tau(x^*)||\cdot||\tau(x)|| \leq \gamma||x||^2$. Because p is a seminorm, this proves continuity. □

(34.2) LEMMA. *Let* A *be a Banach *-algebra. If* $\alpha > 0$ *is a scalar such that*

$$\alpha||h||^2 \leq ||h^2|| \qquad (h \in H(A)) \tag{1}$$

*then* $\alpha||h|| \leq |h|_\sigma$ *for each* $h \in H(A)$.

*Proof*. Multiplying inequality (1) by $\alpha$, we have $\alpha^2||h||^2 \leq \alpha||h^2||$, i.e., $\alpha||h|| \leq (\alpha||h^2||)^{1/2}$. By induction we obtain $\alpha||h|| \leq (\alpha||h^{2^n}||)^{1/2^n}$ for $n = 1,2,3,\ldots$ . Letting $n \to \infty$, we have $\alpha||h|| \leq |h|_\sigma$. □

The main result of this section is the following:

(34.3) THEOREM. *Let* A *be a Banach *-algebra with identity* 1. *The following are equivalent.*

*(a)* A *admits an equivalent C*-norm.*

*(b) there exists* $\alpha > 0$ *such that* $||x^*x|| \geq \alpha||x^*||\cdot||x||$ *for each* $x \in A$.

*(c) the set* $E = \{e^{ih}\colon h \in H(A)\}$ *is bounded.*

*(d) the set* U(A) *of unitaries is bounded.*

*(e)* A *is symmetric and there exists finite* $\beta > 0$ *such that* $||h|| \leq \beta|h|_\sigma$ *for each* $h \in H(A)$.

*Proof*. It is clear that (a) implies (b), (c), and (d). Let us show that each of these conditions implies (e). Finally, we will show that (e) implies (a) to complete the proof.

(b) implies (e): Condition (b) implies that $\alpha||h||^2 = \alpha||h^*||\cdot||h|| \leq ||h^*h|| = ||h^2||$ for $h \in H(A)$ and hence $\alpha||h|| \leq |h|_\sigma$ for $h \in H(A)$ by (34.2). To show A is symmetric, we prove that $|x|_\sigma^2 \leq |x^*x|_\sigma$ for all normal $x$ and apply (33.4). Suppose $x$ is normal. Now $x^*x$ is hermitian and from $\alpha||h|| \le |h|_\sigma$, with $h = x^*x$, we have $\alpha||x^*x|| \le |x^*x|_\sigma$. Combining this with (b) gives $\alpha^2||x||\cdot||x^*|| \le \alpha||x^*x|| \le |x^*x|_\sigma$. Replacing $x$ by $x^n$ and using normality yields

$$\alpha^2||x^n||\cdot||(x^*)^n|| = \alpha^2||x^n||\cdot||(x^n)^*|| \le |(x^n)^*x^n|_\sigma$$

$$= |(x^*)^n x^n|_\sigma = |(x^*x)^n|_\sigma = |x^*x|_\sigma^n.$$

It follows that $|x|_\sigma^2 = |x|_\sigma|x^*|_\sigma \le |x^*x|_\sigma$ by taking $n^{th}$ roots and then letting $n$ tend to infinity.

(c) implies (e): Suppose there exists $\beta > 0$ such that $||e^{ih}|| \le \beta$ for all hermitian $h$. Let $h \in H(A)$ and assume that $\lambda + i\mu \in \sigma_A(h)$, $\lambda$, $\mu$ real where $\mu \neq 0$. Set $z = \mu^{-1}(h - \lambda)$. Then $z \in H(A)$ and $i \in \sigma_A(z)$. Now, for $\gamma > 0$, the element $e^{-i\gamma z}$ has $e^\gamma$ in its spectrum. Then $e^\gamma \le \beta$ for all $\gamma > 0$, which is impossible. Hence $A$ is hermitian and therefore by (33.2) symmetric. To prove that $||h|| \le \beta|h|_\sigma$, let $h$ be any element of $H(A)$, $||h|| = 1$, and set $\delta = ||h^2||^{1/3}$. Then $||h^n|| \le \delta^n$ for all $n \ge 2$. Given $s > 0$ we have

$$ish = e^{ish} - 1 - \sum_{n=2}^{\infty} \frac{(ish)^n}{n!}$$

and it follows that for all $s > 0$,

$$s = ||ish|| \le \beta + 1 + \sum_{n=2}^{\infty} \frac{(s\delta)^n}{n!} = \beta + e^{s\delta} - s\delta.$$

Thus $\delta > 0$. If $s$ is chosen to be $\beta/\delta$, then $\delta \ge \beta\cdot e^{-\beta}$. Therefore $||h^2|| = \delta^3 > (\beta\cdot e^{-\beta})^3$, and then by normalizing, $||k^2|| \ge (\beta\cdot e^{-\beta})^3||k||^2$ for each $k \in H(A)$. By (34.2) we have $|k|_\sigma \ge (\beta\cdot e^{-\beta})^3||k||$ for each $k \in H(A)$.

(d) implies (e): By (d) there is $\beta > 0$ with $||u|| \le \beta$, so that $|u|_\sigma \le \beta$, for all $u \in U(A)$. By (33.2) and (33.7) $A$ is symmetric. If $h$ hermitian and $|h|_\sigma < 1$, then $1 - h^2$ is positive and there exists, by (22.5), an element $k \in H(A)$ which commutes with $h$ such that $k^2 =$

$1 - h^2$. Then $u = h + ik \in U(A)$ and $h = (u + u^*)/2$. Hence $||h|| \leq \frac{1}{2}(||u|| + ||u^*||) \leq \beta$. Now, if $h \in H(A)$ is arbitrary, the inequality $||h|| \leq \beta|h|_\sigma$ follows by considering the element $h/(|h|_\sigma + \varepsilon)$, $\varepsilon > 0$.

Finally, we prove (e) implies (a). Let $x \in A$ and write $x = h + ik$ with $h, k \in H(A)$. The inequality in (e) gives us

$$||x|| \leq ||h|| + ||k|| \leq \beta(|h|_\sigma + |k|_\sigma)$$

$$= \beta\left|\frac{1}{2}(x^* + x)\right|_\sigma + \beta\left|\frac{1}{2}((ix)^* + ix)\right|_\sigma.$$

Since $A$ is symmetric, it follows from parts (d), (j), and (k) of (33.1) that $\rho$ is a seminorm satisfying the $C^*$-condition; and by (i), that $||x|| \leq 2\beta\rho(x)$. Therefore, the norm $||\cdot||$ on $A$ is equivalent to the $C^*$-norm $\rho$ by (23.6). □

## §35. *The Russo-Dye theorem in symmetric algebras.*

We return to the Russo-Dye theorem (see (22.19)) and obtain some extensions and refinements involving the exponential $e^{ih}$, $h \in H(A)$, in symmetric Banach *-algebras. An application of the results will be made in the next section. The results presented here are due primarily to T. W. Palmer [1], L. A. Harris [1], and V. Pták [2].

Let $A$ be a symmetric Banach *-algebra with identity 1. For $x \in A$ with $\rho(x) < 1$ define $T_x$ by

$$T_x(y) = (1 - xx^*)^{-1/2}(y + x)(1 + x^*y)^{-1}(1 - x^*x)^{1/2}$$

for all $y \in A$ for which $|x^*y|_\sigma < 1$. The reader will note that $T_x(y)$ is defined, in particular, for all $y \in A$ such that $\rho(x)\rho(y) < 1$, thanks to (33.1), (a) and (d). Throughout this section $A$ will denote a symmetric Banach *-algebra with identity 1.

(35.1) LEMMA. *For* $x \in A$ *with* $\rho(x) < 1$ *the function* $f(\lambda) = T_x(\lambda)$ *is defined and holomorphic in a neighborhood of* $D = \{\lambda \in \mathbb{C}: |\lambda| \leq 1\}$ *and maps the unit circle* $\partial D$ *into the component of* $U(A)$ *containing* 1.

*Furthermore,* $f(0) = x$, *and if* $x = tu$, $0 < t < 1$, $u \in U(A)$, *then for each* $\lambda \in \partial D$ *the point* $-\lambda$ *does not belong to the spectrum of* $T_x(\lambda)$.

*Proof.* From the definition of $T_x$ we have

$$f(\lambda) = (1 - xx^*)^{-1/2}(\lambda + x)(1 + \lambda x^*)^{-1}(1 - x^*x)^{1/2}.$$

Since $\rho(x) < 1$ and $A$ is symmetric, we have by (33.7), (3), $|x^*|_\sigma = |x|_\sigma < 1$; therefore $(1 + \lambda x^*)^{-1}$ exists in a neighborhood of $D$. Since the mapping $y \to y^{-1}$ is holomorphic on $A^{-1}$, $f(\lambda)$ is defined and holomorphic in a neighborhood of $D$. If $|\lambda| = 1$, then the inverse $(\lambda + x)^{-1}$ clearly exists, and we write $u$ for $f(\lambda)$, i.e.,

$$u = \overline{\lambda}(1 - xx^*)^{-1/2}(\lambda + x)(\overline{\lambda} + x^*)^{-1}(1 - x^*x)^{1/2}.$$

The same calculations given in (22.19) show that $u \in U(A)$, i.e., $f(\lambda) \in U(A)$ for $|\lambda| = 1$. Also, $f(0) = (1 - xx^*)^{-1/2}x(1 - x^*x)^{1/2} = x$.

Now consider the mapping $g$ defined on $[0,1]$ by

$$g(t) = (1 - t^2xx^*)^{-1/2}(1 + tx)(1 + tx^*)^{-1}(1 - t^2x^*x)^{1/2}.$$

Clearly $g$ is continuous and $g(0) = 1$, $g(1) = f(1)$. Further, since $g(t) = T_{tx}(1)$, then the range of $g$ lies in $U(A)$. Since $f(\lambda) = \lambda T_{\overline{\lambda}x}(1)$ for $|\lambda| = 1$, we have a continuous curve connecting 1 to any element of the form $f(\lambda)$ with $|\lambda| = 1$. This shows that $f(\lambda)$ lies in the component of $U(A)$ containing the identity whenever $|\lambda| = 1$.

To prove the last statement, let $x$ be of the form $x = tu$, $0 < t < 1$, $u \in U(A)$. Then $\rho(x) = t < 1$; now if $|\lambda| = 1$, then $T_x(\lambda) = (\lambda + x)(1 + \lambda x^*)^{-1}$ and hence

$$\lambda + T_x(\lambda) = (\lambda(1 + \lambda x^*) + (\lambda + x))(1 + \lambda x^*)^{-1}$$

$$= \lambda(2 + \lambda x^* + \overline{\lambda}x)(1 + \lambda x^*)^{-1}.$$

Therefore to show that $\lambda + T_x(\lambda)$ is invertible, it is enough to show that $2 + \lambda x^* + \overline{\lambda}x$ is invertible. However, since $A$ is symmetric (33.1), (i) gives $|\frac{1}{2}(\lambda x^* + \overline{\lambda}x)|_\sigma \le \rho(\overline{\lambda}x) = \rho(\overline{\lambda}tu) = t < 1$, which implies that $1 + \frac{1}{2}(\lambda x^* + \overline{\lambda}x)$ is invertible. □

(35.2) LEMMA. *Let* A *be a Banach* $^*$*-algebra with identity and continuous involution. If* $u \in U(A)$ *and* $\sigma_A(u)$ *is a proper subset of the unit circle, then* $u = e^{ih}$ *for some* $h \in H(A)$.

*Proof.* Let $V = C \setminus (-\infty,0]$, $W = R \times (-\pi,\pi)$, E the exponential function restricted to W, L the principal branch of the logarithm in V. Thus:

$$E(W) = V \quad \text{and} \quad L(V) = W \tag{1}$$

$$E(L(z)) = z \quad \text{for all} \quad z \in V \tag{2}$$

$$L(E(w)) = w \quad \text{for all} \quad w \in W. \tag{3}$$

We can assume, by rotating u if necessary, that $-1 \notin \sigma_A(u)$, that is, $\sigma_A(u) \subseteq T \setminus \{-1\} \subseteq V$ where $T = \{z \in C: |z| = 1\}$. Then we can form $k = L(u)$. By (B.8.1), (vi)

$$\sigma_A(k) = L(\sigma_A(u)) \subseteq L(T \setminus \{-1\}) = \{it: -\pi < t < \pi\}. \tag{4}$$

In the open disk of radius $\pi$ centered at 0 the function E is given by its power series. Therefore (5) shows that we may form E(k) and in fact $E(k) = e^k$. Then

$$e^k = E(k) = E(L(u)) = u \quad \text{by (2) and (B.8.3).}$$

By the continuity of *

$$u^* = (e^k)^* = \Big(\sum_{n=0}^{\infty} \frac{k^n}{n!}\Big)^* = \sum_{n=0}^{\infty} \frac{(k^*)^n}{n!} = e^{k^*},$$

whence

$$u = (u^*)^{-1} = e^{-k^*}.$$

From (4) we have

$$\sigma_A(-k^*) \subseteq \{t: -\pi < t < \pi\} \subseteq \{z: |z| < \pi\}.$$

Hence we may form $E(-k^*)$ and in fact $E(-k^*) = e^{-k^*}$. Then

$$L(e^{-k^*}) = L(E(-k^*)) = -k^* \quad \text{by (3).}$$

Therefore

$$k = L(u) = L(e^{-k^*}) = -k^*.$$

Then $h = -ik$ is the required element. □

(35.3) THEOREM. *Let* A *be a symmetric Banach *-algebra with identity* 1. *Then for each* $x \in A$ *with* $\rho(x) < 1$ *and each* $\varepsilon > 0$ *there exists a positive integer* N *(depending on* x *and* ε*) such that for all* $n \geq N$

$$||x - \frac{1}{n}\sum_{k=1}^{n} T_x(\exp k\frac{2\pi i}{n})|| < \varepsilon$$

*Proof.* Since A is symmetric, we have the inequality $|x|_\sigma \leq \rho(x)$ (see (33.1)). Since $\rho(x) = |x^*x|_\sigma^{1/2}$ and $\rho(x) < 1$, then $|x|_\sigma < 1$ and $|x^*x|_\sigma < 1$. We may now apply the arguments given in (22.19) and (22.20) to complete the proof. □

(35.4) COROLLARY. *Let* A *be a symmetric Banach *-algebra with identity* 1. *Then the set* $U(\rho) = \{x \in A: \rho(x) \leq 1\}$ *is* $\overline{\text{conv}}\, U(A)$, *the closed convex hull of* $U(A)$.

*Proof.* Let $S = \overline{\text{conv}}\, U(A)$. Note that $U(\rho)$ is closed by (34.1), 2°. Since $U(A) \subseteq U(\rho)$ and $\rho$ is a seminorm, $S \subseteq U(\rho)$. Assume that $x \in U(\rho)$. For any $t$, $0 < t < 1$, $\rho(tx) < 1$; so by (35.3) and (35.1) $tx \in S$. Since $tx \to x$ as $t \uparrow 1$ and S is closed, then $x \in S$ also. □

(35.5) COROLLARY. *Let* A *be a symmetric Banach *-algebra with identity* 1. *Suppose* $x \in A$ *and* $\rho(x) \neq 0$. *Then for each* $\varepsilon > 0$, *there exists a positive integer* n, *a sequence of positive numbers* $\lambda_1, \ldots, \lambda_n$ *and unitary elements* $u_1, \ldots, u_n$ *such that* $\sum_{k=1}^{n} \lambda_k = \rho(x)$ *and* $||x - \sum_{k=1}^{n} \lambda_k u_k|| < \varepsilon$.

*Proof.* Let $x \in A$ with $\rho(x) \neq 0$. Define $y = \frac{x}{\rho(x)}$. Then $y \in U(\rho)$. By (35.4), y belongs to the closed convex hull of $U(A)$. Hence, given $\varepsilon > 0$ there exists $w \in \text{conv}\, U(A)$ such that $||y - w|| < \frac{\varepsilon}{\rho(x)}$. Therefore $||x - \rho(x)w|| < \varepsilon$. □

(35.6) THEOREM. *Let* A *be a semisimple, symmetric, Banach *-algebra with identity. Let* $E = \{e^{ih}: h \in H(A)\}$. *Then* $E \subseteq U(A)$ *and the closed convex hull of* $U(A)$ *coincides with the closed convex hull of* E.

*Proof.* Since A is semisimple and symmetric, $x \to x^*$ is continuous by (34.1). Hence $E \subseteq U(A)$. Now, let $u \in U(A)$ and $\varepsilon > 0$. Set $\delta = \min\{1, \varepsilon/2\}$ and $x = u/(1 + \delta)$. Then $0 < \rho(x) = 1/(1 + \delta) < 1$. By (35.5) there exists a positive integer n, positive numbers $\lambda_1, \ldots, \lambda_n$ and complex numbers $\mu_1, \ldots, \mu_n$ on the unit circle such that $\sum_{k=1}^{n} \lambda_k = \rho(x)$ and $||x - \sum_{k=1}^{n} \lambda_k T_x(\mu_k)|| < \delta$. By (35.1) and (35.2) we have $T_x(\mu_k) = \exp(ih_k)$ for suitable $h_k \in H(A)$, $k = 1,2,\ldots,n$. It follows that

$$||u - (1 + \delta)\sum_{k=1}^{n} \lambda_k \exp(ih_k)|| < (1 + \delta)\delta \leq \varepsilon.$$

Therefore $E \subseteq U(A) \subseteq \overline{\text{conv}}\, E$, the closed convex hull of E. □

## §36. *Further characterizations of C*-algebras.*

In this section we summarize and extend our list of necessary and sufficient conditions on a Banach *-algebra in order that it be isometrically *-isomorphic to a (concrete) C*-algebra. The theorem in the form given below is due to V. Pták [2]; however, it is the result of the combined efforts of many workers in the field. Among these E. Berkson [1], B. W. Glickfeld [1], T. W. Palmer and I. Vidav deserve special mention. We shall discuss the relevance of their work in more detail in the chapter on numerical range (see Chapter VIII).

(36.1) THEOREM. *Let* A *be a unital Banach *-algebra. The following conditions are equivalent:*

*(a)* $||x^*x|| = ||x||^2$ *for all* $x \in A$;

*(b)* $||x|| = \rho(x)$ *for all* $x \in A$;

*(c)* $||x|| \leq \rho(x)$ *for all* $x \in A$;

*(d)* $||x^*||\cdot||x|| \leq |x^*x|_\sigma$ *for all* $x \in A$;

*(e)* $||x^*||\cdot||x|| \leq ||x^*x||$ *for all* $x \in A$;

*(f)* $||x^*||\cdot||x|| = ||x^*x||$ *for all* $x \in A$;

*(g)* $||x^*||\cdot||x|| = ||x^*x||$ *for all normal* $x \in A$;

*(h)* $||u|| = 1$ *for all unitary* $u \in A$;

*(i)* $||\exp(ih)|| = 1$ *for all hermitian* $h \in A$.

*Proof.* (a) implies (b): If $h \in H(A)$ then $||h^2|| = ||h||^2$, from which it follows that $|h|_\sigma = ||h||$. In particular, $||x||^2 = ||x^*x|| = |x^*x|_\sigma$, which is (b).

(b) implies (c) is clear. We show next that (c) implies (a), which will establish the equivalence of the first three conditions. Now, condition (c) gives $||x||^2 \leq \rho(x)^2 = |x^*x|_\sigma \leq ||x^*x|| \leq ||x^*||\cdot||x||$ for $x \in A$, which implies that $||x^*|| = ||x||$. It follows that $||x||^2 \leq ||x^*x|| \leq ||x^*||\cdot||x|| = ||x||^2$ for all $x \in A$.

(c) implies (d): From (c) we have $||x|| \leq \rho(x)$ and $||x^*|| \leq \rho(x^*) = \rho(x)$; hence $||x^*||\cdot||x|| \leq \rho(x)^2 = |x^*x|_\sigma$, which is (d). That (d) implies (e) is clear, and that (e) implies (f) is immediate from the submultiplicativity of the norm. Also (f) implies (g) is obvious.

(g) implies (h): Assume (g) holds. If $u$ is unitary in A, then $||u^{*n}||\cdot||u^n|| = 1$ for all $n$. Hence $|u|_\sigma^2 = |u^*|_\sigma|u|_\sigma = 1$ so that $1 = |u|_\sigma \leq ||u||$ and $1 \leq |u^*|_\sigma \leq ||u^*||$. It follows that $||u|| = ||u^*|| = 1$, which proves (h).

(h) implies (i): Suppose (h) is true. By the proof of (34.4) the function $\rho$ is an equivalent norm on A, so there exists a scalar $\beta > 0$ such that $||x|| \leq \beta\rho(x)$ for all $x \in A$. Now $||x||^2 \leq \beta^2\rho(x)^2 = \beta^2|x^*x|_\sigma \leq \beta^2||x^*||\cdot||x||$ from which it follows that $||x^*|| \leq \beta^2||x||$ for each $x \in A$, i.e., $x \to x^*$ is continuous. Therefore, elements of the form $\exp(ih)$ are unitary, which proves that $||\exp(ih)|| = 1$ for all hermitian $h$.

The theorem will be proved if we show that (i) implies (c). If $||e^{ih}|| = 1$ for each $h \in H(A)$, it follows from the proof of (34.4) that $\rho$ is an equivalent norm on A. In particular, A is symmetric and semisimple so that by (35.6) the set $U(\rho) = \{x \in A: \rho(x) \leq 1\}$ coincides with the closed convex hull of the set $E = \{e^{ih}: h \in H(A)\}$. If $B_1$ denotes the closed unit ball of A, then assumption (i) implies $E \subseteq B_1$, hence $U(\rho) \subseteq B_1$ also. Consequently $||x|| \leq \rho(x)$ for all $x \in A$. □

The presence of an identity element in (36.1) is necessary only for parts (h) and (i). The equivalence of (a), (e), and (f) was shown in Chapter III for Banach *-algebras in which an identity need not be present (see (16.1)). The assumption that an identity be present in part (g) was shown to be unnecessary in 1970 by G. A. Elliott [1]. A result of J. F. Aarnes and R. V. Kadison [1] on the existence of an approximate identity commuting with a given strictly positive element in a C*-algebra $A$ enabled him to extend the norm on $A$ to $A_e$ so that the algebra $A_e$ still satisfied the C*-condition on normal elements. We mention here that Jacob Feldman [1] was the first to observe that, for algebras with identity, it was sufficient to assume the C*-condition on normal elements.

We turn our attention next to a refinement of (36.1), part (i). It will be shown that this part of the theorem can be sharpened in the following way: we replace the submultiplicativity of the norm in $A$ by the weaker assumption

$$||a^*a|| \leq ||a^*||\cdot||a|| \quad \text{for all} \quad a \in A.$$

Observe that under this assumption, even the existence of exp(ih) is not entirely obvious; but it will be confirmed below. We remark that in the next chapter we shall dispense with the submultiplicativity of the norm altogether when dealing with the C*-axioms.

The results presented in (36.2) and (36.3) are due to Zoltan Magyar [1].

(36.2) LEMMA. *Let* $A$ *be a* *-*algebra with identity. Let* $p$ *be a complete linear space norm on* $A$ *such that the following hold for a suitable finite constant* $\beta$*:*

*(i)* $p(a^*a) \leq \beta\cdot p(a^*)\cdot p(a)$ *for all* $a \in A$*;*

*(ii)* $p(\exp(ih)) \leq \beta$ *if* $h \in H(A)$ *and* $\exp(ih)$ *exists.*

*Then there exists a norm* $||\cdot||_c$ *on* $A$*, equivalent to* $p$*, such that* $(A, ||\cdot||_c)$ *is a* C*-*algebra.*

*Proof.* Utilizing the polarization identity

$$\begin{aligned}4xy &= (y + x^*)^*(y + x^*) - (y - x^*)^*(y - x^*) \\ &\quad + i(y + ix^*)^*(y + ix^*) - i(y - ix^*)^*(y - ix^*),\end{aligned} \tag{1}$$

which holds in the *-algebra $A$, and applying (i) we obtain, for all $x, y \in A$:

$$4p(xy) \le 4\beta\cdot(p(y^*) + p(x))\cdot(p(y) + p(x^*)). \tag{2}$$

Let $u, v \in A$. Setting $x = p(v^*)^{1/2}\cdot p(v)^{1/2}\cdot u$ and $y = p(u^*)^{1/2}\cdot p(u)^{1/2}\cdot v$ in (2), we obtain

$$p(uv) \le \beta\cdot(p(u^*)^{1/2}\cdot p(v^*)^{1/2} + p(u)^{1/2}\cdot p(v)^{1/2})^2. \tag{3}$$

Now, define a new norm on $A$ by setting

$$||a|| = 4\beta\cdot\max\{p(a^*), p(a)\}. \tag{4}$$

We then have, by (3), that

$$||ab|| \le ||a||\cdot||b||;$$

$$||a^*|| = ||a||; \tag{5}$$

$$p(a) \le \frac{1}{4\beta}\cdot||a||$$

for all $a, b \in A$.

Let $B$ denote the completion of $(A, ||\cdot||)$. It follows from (5) that the algebra operations and the norm $p$ have unique continuous extensions to $B$. Moreover, $(B, ||\cdot||)$ is a *-normed algebra, $p$ is a continuous seminorm on it, and (i), (4) and (5) are also valid in $B$.

Because $(B, ||\cdot||)$ is a Banach algebra with identity, given $a \in B$, we can define $\exp_B(a) = \sum_{n=0}^{\infty} a^n/n!$ with respect to $||\cdot||$. For $a \in A$, the series $\sum_{n=0}^{\infty} a^n/n!$ is Cauchy in $(A, ||\cdot||)$ and hence, by (5), in $(A, p)$ also. However, $p$ is a complete norm on $A$, and thus there exists a unique element $\exp_A(a) = \sum_{n=0}^{\infty} a^n/n!$ in $A$, with respect to $p$. Since $p$ is continuous relative to $||\cdot||$, we also have $p(\exp_A(a) - \exp_B(a)) = 0$ for all $a \in A$. Consequently, from hypothesis (ii) we have

$$p(\exp_B(ih)) \le \beta \quad \text{if} \quad h \in H(A). \tag{6}$$

Since the involution $x \to x^*$ is continuous with respect to $||\cdot||$, $(\exp_B(a))^* = \exp_B(a^*)$ for all $a \in B$; in particular, we have $(\exp_B(ih))^* = \exp_B(-ih)$ for $h \in H(A)$. Hence, by (6) and (4), we see that

$$||\exp_B(ih)|| \le 4\beta^2 \quad \text{if} \quad h \in H(A). \tag{7}$$

Since the hermitian part of $A$ is dense in that of $B$, (7) is true even when $h \in H(B)$. It follows that

$$||a||_c = |a^*a|_\sigma^{1/2}$$

defines a $C^*$-norm on $B$, equivalent to the norm $||\cdot||$ (see (34.4)).

Therefore, there exist positive constants $\lambda$, $\mu$ such that

$$\lambda\cdot||a||_c \leq ||a|| \leq \mu\cdot||a||_c \quad \text{for all} \quad a \in B.$$

Setting $\gamma = \lambda(4\beta)^{-1}$, $\omega = \mu(4\beta)^{-1}$, we have by (4) that

$$p(a) \leq \omega\cdot||a||_c \quad \text{for all} \quad a \in B$$

and (8)

$$p(h) \geq \gamma\cdot||h||_c \quad \text{if} \quad h \in H(B).$$

Hence, by (i), we have

$$\gamma\cdot||a||_c^2 = \gamma\cdot||a^*a||_c \leq p(a^*a) \leq \beta\cdot p(a^*)\cdot p(a).$$

This result and (8) imply that

$$p(a) \geq \gamma\cdot(\beta\omega)^{-1}\cdot||a||_c \quad \text{for all} \quad a \in B,$$

and hence $p$ and $||\cdot||_c$ are equivalent. □

The reader should note that the first part of the proof of (36.2) shows that $\exp(a)$ exists in $A$ for all $a \in A$.

(36.3) THEOREM. *(Magyar). If the assumptions of* (36.2) *hold with* $\beta = 1$, *then* $p = ||\cdot||_c$; *that is,* $(A,p)$ *is a* $C^*$*-algebra.*

*Proof.* Because $|a|_\sigma = \lim_{n\to\infty} ||a^n||_c^{1/n}$, it follows from (36.2) that

$$|a|_\sigma = \lim_{n\to\infty} p(a^n)^{1/n} \quad \text{for all} \quad a \in A. \tag{9}$$

Applying assumption (i) to $a = h^{2^n}$, where $h$ is hermitian, we obtain $p(h^{2^n}) \leq p(h)^{2^n}$ for all $n = 1,2,3,\cdots$, and hence, by (9), we have

$$|h|_\sigma \leq p(h) \quad \text{if} \quad h \in H(A). \tag{10}$$

But $|a^*a|_\sigma = ||a||_c^2$ for all $a \in A$, and consequently

$$||a||_c^2 \leq p(a^*a) \quad \text{for all} \quad a \in A. \tag{11}$$

Now, the Palmer-Russo-Dye theorem (see the refinements in §35) states that the closed unit ball in a $C^*$-algebra is the closed convex hull of the elements of the form exp(ih), where $h \in H(A)$.

Utilizing hypothesis (ii), we see that $p(a) \leq 1$ if $a$ is a convex combination of elements of the form exp(ih) with $h \in H(A)$. Moreover, $p$ is continuous relative to $||\cdot||_c$ and therefore we obtain

$$p(a) \leq ||a||_c \quad \text{for all} \quad a \in A. \tag{12}$$

Comparing (11), (12) and (i) we get

$$||a||_c^2 \leq p(a^*a) \leq p(a^*)\cdot p(a) \leq ||a^*||_c\cdot||a||_c = ||a||_c^2;$$

that is, $||a||_c^2 = p(a^*)\cdot p(a)$ for all $a \in A$. This and (12) show that $p = ||\cdot||_c$, which completes the proof. □

We remark that the completeness of the norm in (36.2) and (36.3) is not essential. Indeed, we may drop it and replace (ii) by:

(iii) $\lim_{k\to\infty} p(\sum_{n=0}^{k} \frac{(ih)^n}{n!}) \leq \beta$ if $h \in H(A)$ and the limit exists.

Then the conclusion must be modified so that the completion of $(A,p)$ is an equivalent $C^*$-algebra (resp., a $C^*$-algebra if $\beta = 1$). The proof is unchanged.

## EXERCISES

(VI.1) Let $A$ be a commutative Banach $^*$-algebra with radical $R$. Let $I$ be a closed $^*$-ideal in $A$ such that $I \subseteq R$. Prove that if $A/I$ is symmetric, then $A$ is symmetric.

(VI.2) Let $x \to x^*$ and $x \to x'$ be two involutions on an algebra $A$. Prove that the map $x \to x^\#$, where $x^\# = x'^{*'}$, is an involution on $A$ which is symmetric iff $x \to x^*$ is symmetric.

(VI.3) Let $A$ be a Banach $^*$-algebra. The involution is *regular* if $x \in H(A)$ and $\nu(x) = 0$ imply $x = 0$. The involution is *proper* if $xx^* = 0$

implies $x = 0$. Prove that if $A$ is a symmetric Banach $^*$-algebra with regular involution, then $*$ is proper.

(VI.4) Let $I$ be a proper two-sided ideal in a symmetric Banach $^*$-algebra $A$ with identity. Prove that:

(a) $I + I^*$ is a proper two-sided ideal in $A$.

(b) Each maximal two-sided ideal of $A$ is a $^*$-ideal of $A$.

(VI.5) Let $A$ be a symmetric Banach $^*$-algebra with identity $e$. If $I$ is any proper left ideal in $A$, prove that the closure of $I + I^*$ does not contain $e$.

(VI.6) If $A$ is a Banach $^*$-algebra with radical $R$, then $A/R$ is symmetric (resp. hermitian) if $A$ is symmetric (resp. hermitian).

(VI.7) Let $A$ be a symmetric Banach $^*$-algebra. If $x$ is an idempotent element in $A$ with $(x^*x)^n = 0$ for some positive integer $n$, prove that $x = 0$.

(VI.8) Give an example of a commutative semisimple Banach $^*$-algebra with proper involution which is not symmetric.

(VI.9) Let $A$ be a symmetric $^*$-algebra with identity $e$ and suppose that $f$ is a linear functional on $A$ which is multiplicative on each subalgebra generated by an hermitian element and $e$. Prove that $f$ is multiplicative on $A$. (Cf. IV.22.)

(VI.10) Give an example of a weakly positive functional on a $^*$-algebra $A$ which is not positive.

(VI.11) Let $A$ be a unital symmetric Banach $^*$-algebra. If $h \in H(A)$, prove that

$$\sigma_A(\exp(ih)^* - \exp(-ih)) = \{0\}.$$

(VI.12) Let $A$ be a unital symmetric Banach $^*$-algebra. If $x$ is normal and invertible in $A$ and satisfies $\sigma_A(x^* - x^{-1}) = \{0\}$, prove that $|x|_\sigma \leq 1$.

(VI.13) Let $A$ be a unital symmetric Banach $^*$-algebra. Prove that if $h \in H(A)$, then $|\exp(ih)|_\sigma \leq 1$.

(VI.14) Let $A$ be a unital Banach $^*$-algebra. If there exists a real $\beta > 0$ such that $|\exp(ih)|_\sigma \leq \beta$ for each $h \in H(A)$, prove that $A$ is symmetric.

(VI.15) Give an example of a Banach *-algebra $A$ with hermitian involution which satisfies the condition $||x^*x|| = ||x||^2$ for all *hermitian* $x$ in $A$, but fails to satisfy this equality *for all* $x$ in $A$.

# 7
# A Further Weakening of the C*-Axioms

## §37. *Introduction.*

We have now seen that a $C^*$-algebra can be defined in several equivalent ways. Can the axioms describing a $C^*$-algebra be reduced in number? With respect to this question the first named author puzzled with whether the condition of submultiplicativity $||xy|| \leq ||x||\cdot||y||$ is necessary in the $C^*$-axioms. During an informal conversation in 1972 the problem was brought to the attention of Huzihiro Araki. Some months later, Araki and G. A. Elliott [1] proved the following two results.

THEOREM 1. *(Araki-Elliott). Let* $A$ *be a* $^*$*-algebra with a complete linear space norm such that* $||x^*x|| = ||x||^2$ *for all* $x \in A$. *Then* $A$ *is a* $C^*$*-algebra.*

THEOREM 2. *(Araki-Elliott). Let* $A$ *be a* $^*$*-algebra with a complete linear space norm and continuous involution such that* $||x^*x|| = ||x^*||\cdot||x||$ *for all* $x \in A$. *Then* $A$ *is a* $C^*$*-algebra.*

It is natural to ask if it suffices in Theorem 1 or 2 to assume $||x^*x|| = ||x||^2$, respectively $||x^*x|| = ||x^*||\cdot||x||$, only for all *normal* $x$. However, the following counterexample shows that the answer to both questions is negative. Let $B(H)$ be the $^*$-algebra of all bounded operators on a Hilbert space $H$ of dimension $\geq 2$. The numerical radius of an operator $x$ on $H$ is defined by

$$||x||_1 = \sup\{|(x\xi,\xi)|: \xi \in H, ||\xi|| = 1\}.$$

By Halmos [1, Chapter 17], $||\cdot||_1$ is a complete linear space norm on

B(H) with $\frac{1}{2}||x|| \leq ||x||_1 \leq ||x||$ for all $x \in B(H)$ and $||x||_1 = ||x||$ for all normal $x \in B(H)$, where $||\cdot||$ is the usual operator norm. The norm $||\cdot||_1$ has the following properties.

$$||x^*||_1 = ||x||_1 \quad \text{for all} \quad x \in B(H),$$

$$||x^*x|| \geq ||x||_1^2 = ||x^*||_1||x||_1 \quad \text{for all} \quad x \in B(H);$$

and

$$||x^*x||_1 = ||x||_1^2 = ||x^*||_1||x||_1$$

for all normal $x \in B(H)$ but not all $x \in B(H)$.

Even though Theorems 1 and 2 do not hold for $x$ restricted to normal elements Zoltan Sebestyén [1] was able to prove the following general characterization of C*-algebras.

THEOREM 3. *(Sebestyén). Let* A *be a *-algebra with complete linear space norm such that*

$$||x^*x|| \leq ||x||^2 \quad \textit{for all} \quad x \in A$$

*and*

$$||x^*x|| = ||x||^2 \quad \textit{for all normal} \quad x \in A.$$

*Then* A *is a* C*-*algebra.*

In a later paper, Sebestyén [2] claimed to prove that continuity of the involution could be dropped from Theorem 2 above. However, G. Elliott pointed out an error on line four of page 212 of Sebestyén's paper. Indeed, the series displayed there, although convergent, is not shown to converge to the quasi-inverse of $\lambda^{-1}x$. In a recent paper Z. Magyar and Sebestyén [1] have given a new proof which circumvents the above difficulty and establishes Theorem 2 without the continuity assumption on the involution.

In this chapter we confine our attention to proving a generalization of Theorem 1, due to Sebestyén [ 7 ], which shows that every C*-seminorm is automatically submultiplicative. Applications of this result to extensions of a $C^*$-seminorm or a $^*$-representation on a Hilbert space from a $^*$-ideal to the whole algebra will be given in §39.

§38. *Every $C^*$-seminorm is automatically submultiplicative.*

Let $A$ be a $^*$-algebra and $p$ a seminorm on $A$; that is, $p$ is a real-valued function on $A$ such that for $a, b \in A$ and $\lambda \in C$:

1° $p(a) \geq 0$,

2° $p(\lambda a) = |\lambda| p(a)$,

3° $p(a + b) \leq p(a) + p(b)$.

A $C^*$-*seminorm* is a seminorm $p$ on $A$ such that

4° $p(a^*a) = p(a)^2 \quad (a \in A)$.

It is important to note that we *do not* require $p$ to be submultiplicative:

5° $p(ab) \leq p(a)p(b) \quad (a, b \in A)$.

The main result of this section is:

(38.1) THEOREM. *(Sebestyén). Every $C^*$-seminorm $p$ on a $^*$-algebra $A$ is submultiplicative.*

*Proof.* We are assuming conditions 1° - 4° and must prove 5°. Utilizing the polarization identity

$$4uv = \sum_{n=0}^{3} i^n (v + i^n u^*)^*(v + i^n u^*)$$

valid for all $u, v \in A$ and 2° - 4°, we obtain

$$\begin{aligned} 4p(uv) &\leq [p(v + u^*)]^2 + [p(v + iu^*)]^2 \\ &\quad + [p(v - u^*)]^2 + [p(v - iu^*)]^2 \\ &\leq 4[p(u^*) + p(v)]^2. \end{aligned}$$

Hence, for any positive integer $n$, the substitutions $u \to (p(u^*) + 1/n)^{-1}u$, $v \to (p(v) + 1/n)^{-1}v$ give us

$$\begin{aligned} p(uv) &\leq (p(u^*) + 1/n)(p(v) + 1/n)[(p(u^*) + 1/n)^{-1}p(u^*) + (p(v) + 1/n)^{-1}p(v)]^2 \\ &\leq 4(p(u^*) + 1/n)(p(v) + 1/n). \end{aligned}$$

Since $n$ was arbitrary, we obtain:

$$p(uv) \leq 4p(u^*)p(v) \quad (u, v \in A). \tag{1}$$

The C*-property 4° implies, by induction, for any hermitian $h \in A$ that:

$$p(h^{2^n}) = [p(h)]^{2^n}, \quad n = 1,2,\ldots . \tag{2}$$

Applying (1) and (2) we obtain, for any $a \in A$ and natural number $n$:

$$[p(a^*a)]^{2^n} = p((a^*a)^{2^n}) = p(a^*(aa^*)^{2^n-1}a)$$

$$\leq 4p(a)p((aa^*)^{2^n-1}a)$$

$$\leq 4^2[p(a)]^2 p((aa^*)^{2^n-1}).$$

Now, considering $p((aa^*)^{2^n-1})$, we have by repeated use of (1) and (2):

$$p((aa^*)^{2^n-1}) = p((aa^*)^{1+2+\cdots+2^{n-1}})$$

$$\leq 4^{n-1}p(aa^*)p((aa^*)^2)\cdot \ \ldots \ \cdot p((aa^*)^{2^{n-1}}) = 4^{n-1}p(aa^*)^{2^n-1}$$

for all $n$. So, taking $2^n$-th roots and letting $n \to \infty$, we obtain

$$p(a^*a) \leq p(aa^*) \quad (a \in A).$$

The substitution $a \to a^*$ proves the reverse inequality and thus the identity

$$p(a^*a) = p(aa^*) \quad (a \in A).$$

By 4° this yields the isometry of the involution with respect to $p$:

6° $p(a) = p(a^*) \quad (a \in A)$.

By (1) and 6° we have

$$p(ab) \leq 4p(a)p(b) \quad (a, b \in A). \tag{3}$$

Therefore, the set

$$J_p = \{a \in A: p(a) = 0\}$$

is a *-ideal in A, and the quotient algebra $A_p = A/J_p$ is a *-algebra under the involution

$$a + J_p \to a^* + J_p \quad (a \in A).$$

Further, the norm $|\cdot|$ on $A_p$ defined by

$$|a + J_p| = p(a) \quad (a \in A)$$

inherits the properties of p.

Let B denote the completion of $A_p$ with respect to the norm $|\cdot|$. Then for any $a, b \in B$ the relations

$$\begin{aligned} |a^*a| &= |a|^2 \\ |a^*| &= |a| \\ |ab| &\le 4|a|\cdot|b| \end{aligned} \tag{4}$$

are valid. Consequently, defining as usual

$$||a|| = \sup\{|ab|: b \in B, |b| \le 1\} \quad (a \in B), \tag{5}$$

we obtain an algebra norm on B such that

$$|a| \le ||a|| \le 4|a| \quad (a \in B).$$

The spectral radius $r(\cdot)$ in the Banach *-algebra $(B, ||\cdot||)$ satisfies for any normal element $a \in B$:

$$\begin{aligned} r(a)^2 &= \lim_{n\to\infty} ||a^{2^n}||^{2^{-n+1}} = \lim_{n\to\infty} |a^{2^n}|^{2^{-n+1}} \\ &= \lim_{n\to\infty} |(a^*a)^{2^n}|^{2^{-n}} = \lim_{n\to\infty} ||(a^*a)^{2^n}||^{2^{-n}} \\ &= r(a^*a), \end{aligned} \tag{6}$$

and, in particular, utilizing (2) and (4) we obtain

$$r(a) = \lim_{n\to\infty} |(a^*a)^{2^n}|^{2^{-n-1}} = |a^*a|^{1/2} = |a|. \tag{7}$$

We next prove that the spectrum of any hermitian element in $B$ is real. Suppose to the contrary, and let $t + i$ be a point of the spectrum of $h = h^* \in B$ for some real $t$. Then there exists a multiplicative linear functional $\phi$ on some maximal commutative *-subalgebra of $B$ containing $h$ such that $\phi(h) = t + i$. Let $b$ in this subalgebra be such that $\phi(b) = 1$ and let $a = (h - t + ni)^m b$ for natural numbers $m$ and $n$. (Here $a \in B$, but the factorization may be possible only in the unitization $B_e$.) Then, by (6),

$$(1 + n)^{2m} = |\phi(a)|^2 \leq (r(a))^2 = r(a^*a)$$

$$= r((h - t)^2 + n^2)^m b^*b)$$

$$\leq [(r(h) + |t|)^2 + n^2]^m r(b^*b).$$

Taking $m^{th}$ roots and letting $m \to \infty$, we have

$$(1 + n)^2 \leq [r(h) + |t|]^2 + n^2,$$

or, $1 + 2n \leq [r(h) + |t|]^2$. Letting $n \to \infty$ gives a contradiction.

We are going to show next that

$$r(a)^2 \leq r(a^*a) \quad (a \in B). \tag{8}$$

To verify (8), it is convenient to work in the unitization $B_e$ of $B$. Suppose $r(a^*a) < |\lambda|^2$ for some $a \in B$ and $\lambda \in C$. The hermitian element

$$h = \sum_{n=0}^{\infty} \binom{1/2}{n} (-a^*a)^n / |\lambda|^{2n}$$

lies in $B_e$, is invertible, and satisfies the identity

$$h^2 = e - (a^*a)/|\lambda|^2.$$

Moreover, we have the identities

$$(e + a^*/\bar{\lambda})(e - a/\lambda) = e - a^*a/|\lambda|^2 + a^*/\bar{\lambda} - a/\lambda$$

$$= h^2 + a^*/\bar{\lambda} - a/\lambda$$

$$= -ih[ie + ih^{-1}(a^*/\bar{\lambda} - a/\lambda)h^{-1}]h$$

$$= -ih(ie + k)h,$$

where

$$k = ih^{-1}(a^*/\bar{\lambda} - a/\lambda)h^{-1}$$

is a hermitian element in $B$ and, consequently, has real spectrum. Hence

$$(e + a^*/\bar{\lambda})(e - a/\lambda)$$

is invertible, so that $e - a/\lambda$ is left invertible. It is right invertible by a similar argument. Therefore, $\lambda e - a$ is invertible, and (8) is proved.

Now let $a, b \in B$. Then by (B.4.8) and (8)

$$r(b^*(a^*a)b) = r(a^*abb^*) \leq [r((a^*abb^*)^*(a^*abb^*))]^{1/2}$$

$$(\alpha) \qquad = [r(bb^*(a^*a)^2bb^*)]^{1/2}.$$

In particular, if $a = k \in H(B)$, $b = h \in H(B)$, we have $r(hk^2h) \leq [r(h^2k^4h^2)]^{1/2}$. Induction on this and (7) gives, for any natural number n,

$$(\beta) \qquad r(hk^2h) \leq [r(h^{2^{n-1}}k^{2^n}h^{2^{n-1}})]^{2^{-n+1}} = |h^{2^{n-1}}k^{2^n}h^{2^{n-1}}|^{2^{-n+1}}.$$

Consider again any $a, b \in B$. By (4) and (7) we have

$$|ab|^2 = |(ab)^*(ab)| = r((ab^*)(ab)) = r(b^*(a^*a)b)$$

$$\leq [r(bb^*(a^*a)^2bb^*)]^{1/2} \quad \text{by } (\alpha)$$

$$\leq |(bb^*)^{2^{n-1}}(a^*a)^{2^n}(bb^*)^{2^{n-1}}|^{2^{-n}} \quad \text{by } (\beta). \qquad (9)$$

It then follows by applying (9) and (4) that

$$|ab|^2 \leq 16^{2^{-n}}|(a^*a)^{2^n}|^{2^{-n}}|(bb^*)^{2^{n-1}}|^{2^{-n+1}}$$

$$= 16^{2^{-n}}|a^*a|\cdot|bb^*| = 16^{2^{-n}}|a|^2|b|^2$$

for all natural numbers n. Hence

$$|ab| \leq |a|\cdot|b| \quad (a, b \in B).$$

Therefore

$$p(ab) = |ab + J_p| \leq |a + J_p| \cdot |b + J_p| = p(a)p(b)$$

for all $a, b \in A$. □

§39. *Some applications.*

A few applications of Theorem (38.1) will be given in this section.

(39.1) PROPOSITION. *Let* $p$ *be a* $C^*$*-seminorm on a complex* $*$*-algebra* $A$. *Then there exists a* $*$*-representation* $\pi$ *of* $A$ *on a Hilbert space such that* $p(a) = ||\pi(a)||$ *for all* $a \in A$.

*Proof*. We observed in the proof of (38.1) that if $B$ denotes the completion of $A/J_p$ with the norm $|a + J_p| = p(a)$, then $B$ is a $C^*$-algebra. The Gelfand-Naimark theorem (19.1) provides a universal $*$-representation $\hat{\pi}$ of $B$ such that $||\hat{\pi}(a + J_p)|| = |a + J_p|$ for all $a \in A$. If $\tau: A \to A/J_p$ denotes the quotient map, then the composition $\pi = \hat{\pi} \circ \tau$ is a $*$-representation of $A$ with the required properties. □

The next result gives a necessary and sufficient condition that a $C^*$-seminorm can be extended from a $*$-ideal to the whole algebra.

(39.2) PROPOSITION. *Let* $A$ *be a* $*$*-algebra and* $J$ *a* $*$*-ideal in* $A$. *If* $p$ *is a* $C^*$*-seminorm on* $J$, *then there exists a* $C^*$*-seminorm on* $A$ *whose restriction to* $J$ *equals* $p$ *iff*

$$\sup\{p(ab): b \in J,\ p(b) \leq 1\} < \infty \tag{10}$$

*holds for all* $a \in A$.

*Proof*. Assume first that there exists a $C^*$-seminorm $q$ on $A$ such that $q(b) = p(b)$ for all $b \in J$. Then for all $a \in A$, $b \in J$ with $p(b) \leq 1$ we have

$$p(ab) = q(ab) \leq q(a)q(b) = q(a)p(b) \leq q(a),$$

since $q$ is an algebra seminorm by (38.1).

Conversely, assume (10) holds and define a seminorm $q$ on $A$ by

$$q(a) = \sup\{p(ab): b \in J,\ p(b) \leq 1\} \quad (a \in A).$$

We first show that $q$ is a $C^*$-seminorm, i.e., $q(a^*a) = [q(a)]^2$ for all $a \in A$. Let $a \in A$, $b \in J$ with $p(b) \leq 1$. Then

$$p(ab)^2 = p(b^*a^*ab) \leq p(b^*)p(a^*ab)$$

$$\leq p(a^*ab) \leq q(a^*a).$$

Therefore

$$(q(a))^2 \leq q(a^*a) \quad (a \in A). \tag{11}$$

To prove the opposite inequality, let $a \in A$, $q(a^*a) > 0$ and $b \in J$, $p(b) \leq 1$. We consider two cases. If $p(ab) = 0$, then

$$(p(a^*ab))^2 = p(b^*a^*aa^*ab) \leq p(b^*a^*aa^*)p(ab) = 0.$$

If, on the other hand, $p(ab) > 0$, then

$$p(a^*ab) = p(a^*\frac{ab}{p(ab)})p(ab) \leq q(a^*)q(a).$$

Hence,

$$q(a^*a) \leq q(a^*)q(a). \tag{12}$$

From (10) and (11) we get $q(a) \leq q(a^*)$, and the substitution $a \to a^*$ yields $q(a^*) = q(a)$. By (12), this implies the desired inequality

$$q(a^*a) \leq (q(a))^2.$$

It remains to prove that $q$ is an extension of $p$. For $a \in J$ the relation $q(a) \leq p(a)$ is obvious. On the other hand, if $p(a) > 0$, then

$$p(a) = p(a^*) = p(a\frac{a^*}{p(a^*)}) \leq q(a),$$

which completes the proof. □

(39.3) PROPOSITION. *Let* $p$ *be a* $C^*$*-seminorm on a* $*$*-algebra* $A$. *Then there exists a* $C^*$*-seminorm on the unitization* $A_e$ *of* $A$ *which extends* $p$.

*Proof.* Since for all $a, b \in A$ and complex $\lambda$ we have

$$p((\lambda e + a)b) \leq p(\lambda b) + p(ab) \leq (|\lambda| + p(a))p(b),$$

it follows that

$$\sup\{p((\lambda e + a)b) : b \in A,\ p(b) \leq 1\} \leq |\lambda| + p(a),$$

and (39.2) can be applied. □

The preceding result is a considerable generalization of theorem (6.1) concerning the extension of the norm of a C*-algebra to the unitization.

As a final application, we prove the following:

(39.4) PROPOSITION. *Let* $A$ *be a *-algebra and* $\pi$ *a *-representation of a *-ideal* $J$ *in* $A$ *on a Hilbert space. Then there exists a *-representation* $\hat{\pi}$ *of* $A$ *on some Hilbert space with* $||\pi(b)|| = ||\hat{\pi}(b)||$ *for all* $b \in J$ *iff, for any* $a \in A$,

$$q(a) = \sup\{||\pi(ab)|| : b \in J,\ ||\pi(b)|| \leq 1\} < \infty \qquad (13)$$

*is satisfied.*

*Proof.* The existence of such a *-representation $\hat{\pi}$ ensures, for any $a \in A$, that

$$q(a) \leq ||\hat{\pi}(a)||,$$

since for arbitrary $b \in J$ we have

$$||\pi(ab)|| = ||\hat{\pi}(ab)|| \leq ||\hat{\pi}(a)||\cdot||\hat{\pi}(b)|| = ||\hat{\pi}(a)||\cdot||\pi(b)||.$$

Conversely, if we assume (13), then applying (39.2) to the C*-seminorm

$$p(b) = ||\pi(b)|| \qquad (b \in J),$$

we find that $q$ is a C*-seminorm on $A$ such that $q$ equals $p$ on $J$. Now (39.1) provides the desired *-representation $\hat{\pi}$ on some Hilbert space. □

## EXERCISES

(VII.1) Let $H$ be a Hilbert space with $\dim H \geq 2$. For $T \in B(H)$ define

$$||T||_1 = \sup\{|(T\xi|\xi)|: \xi \in H, ||\xi|| = 1\}.$$

Let $||\cdot||$ denote the usual operator norm on $B(H)$. For $T \in B(H)$ prove that:

(a) $||\cdot||_1$ is a complete linear space norm on $B(H)$;

(b) $\frac{1}{2}||T|| \leq ||T||_1 \leq ||T||$;

(c) $||T||_1 = ||T||$ if $T$ is normal;

(d) $||T^*||_1 = ||T||_1$;

(e) $||T^*T|| \geq ||T||_1^2$;

(f) $||T||_1^2 = ||T^*||_1||T||_1$;

(g) $||T^*T||_1 = ||T||_1^2 = ||T^*||_1||T||_1$ if $T$ is normal;

(h) Give an example of a Hilbert space $H$ and $T \in B(H)$ such that $||T^*T||_1 \neq ||T||_1^2$.

(VII.2) Let $H$ be a Hilbert space and consider the norm $||\cdot||_1$ defined above on $B(H)$. If $T \in B(H)$ is idempotent and $||T||_1 \leq 1$, prove that $T$ is a projection.

(VII.3) Let $H$ be a Hilbert space. If $T \in B(H)$ with $||T|| \leq 1$ and $\xi \in H$ is such that $T\xi = \xi$, prove that $T^*\xi = \xi$.

(VII.4) Let $H$ be a Hilbert space and $N$ the set of normal operators in $B(H)$. Prove that the adjoint operation $T \to T^*$ on $B(H)$ is strong operator continuous when restricted to $N$.

(VII.5) Let $A$ be a Banach *-algebra for which $||x^*x|| = ||x^*||\cdot||x||$ for all normal $x$ in $A$. Show that $||x^*|| = ||x||$ for all normal $x$ in $A$.

(VII.6) Let $A$ be a *-algebra. Verify, for $x, y \in A$ and complex scalars $\lambda$, $\mu$, that:

$$4\lambda\mu xy = \Sigma_{n=0}^{3} i^n(\bar{\mu}y^* + (-i)^n\lambda x)(\mu y + i^n\bar{\lambda}x^*).$$

(VII.7) Let $A$ be a *-algebra with a linear space norm satisfying $||x^*x|| = ||x||^2$ for all $x$ in $A$. Show, by direct computation, that:

(a) $||xy|| \leq 4||x^*||\cdot||y||$ for $x, y \in A$.

(b) $||xy|| \leq 16||x||\cdot||y||$ for $x, y \in A$.

(VII.8) Let $A$ be a *-algebra with a complete linear space norm satisfying $||x^*x|| = ||x||^2$ for all $x$ in $A$. Prove, using the preceding exercise, that any norm-closed commutative *-subalgebra of $A$ is a C*-algebra.

(VII.9) Let $A$ be a *-algebra with a linear space norm satisfying $||x^*x|| = ||x^*||\cdot||x||$ for all $x$ in $A$. Given $x, y \in A$, prove that

$$||xy|| \leq (||x||^{1/2}||y||^{1/2} + ||x^*||^{1/2}||y^*||^{1/2})^2.$$

(VII.10) Let $A$ be a *-algebra with a linear space norm satisfying $||x^*x|| = ||x^*||\cdot||x||$ for all $x$ in $A$ and assume $x \to x^*$ is continuous. Prove that:

(a) there exist real numbers $\alpha \geq 1$ and $\beta \geq 2$ such that $||x^*|| \leq \alpha||x||$ and $||x + x^*|| \leq \beta||x^*x||^{1/2}$ for all $x \in A$;

(b) $||xy|| \leq (1 + \gamma)^2||x||\cdot||y||$ for some $\gamma \geq 1$ and all $x, y \in A$;

(c) if $x$ and $y$ are commuting hermitian elements of $A$, then $||xy|| \leq ||x||\cdot||y||$ and $||x^n|| = ||x||^n$ for $n = 1,2,\ldots$

(d) if $x$ is a normal element of $A$, $||x + x^*|| \leq 2||x^*x||^{1/2}$.

# 8
# Geometrical Characterizations of C*-Algebras

## §40. *Introduction.*

The first step toward a geometrical characterization of $C^*$-algebras among complex Banach algebras was taken in 1956 by Ivan Vidav [1]. To state his result in an appropriate form, let us collect some basic ideas and results. Proofs of many of the results stated in this introduction will be given in subsequent sections of this chapter. Throughout the chapter all algebras will be assumed to lie over the complex field unless explicitly stated otherwise.

Let $A$ be a unital Banach algebra, i.e., a Banach algebra with identity $e$ of norm 1. A continuous linear functional $f$ on $A$ is called a *state* if $||f|| = 1 = f(e)$. This definition exploits an earlier, involution-independent, geometrical characterization of the positive functionals on a $C^*$-algebra due to H. Frederic Bohnenblust and Samuel Karlin [1]: a continuous linear functional $f$ on a unital $C^*$-algebra is positive iff $||f|| = f(e)$ (see (22.18)). Gunter Lumer [1] made strikingly successful use of states to define hermitian elements in an arbitrary unital Banach algebra $A$. An element $x$ in $A$ is called *hermitian* if $f(x)$ is real for every state $f$ on $A$. In the special case where $A$ is a $C^*$-algebra, an element $x$ is hermitian iff $x^* = x$. Further, it turns out that the following conditions for an element $x$ of a unital Banach algebra are equivalent:

1. $f(x)$ is real for every state $f$ on $A$;
2. $||e + i\alpha x|| = 1 + o(\alpha)$ ($\alpha$ real);
3. $||\exp(i\alpha x)|| = 1$ ($\alpha$ real).

In fact, Vidav [1] used the second condition to define hermitian elements in unital Banach algebras. Obviously in the algebra of complex numbers we have

$$|1 + i\alpha x| = 1 + o(\alpha) \quad (\alpha \text{ real})$$

if and only if $x$ is a real number. In the $C^*$-algebra of all bounded operators on a Hilbert space the *hermitian operators* (the operators $x$ with $x^* = x$) play the same role as the real numbers in the algebra of complex numbers. Motivated by this observation, Vidav—as he pointed out in a private letter to Josef Wichmann—asked if the hermitian operators could be characterized in a similar way. And, indeed, he was able to show quite easily that an element $x$ in a concrete $C^*$-algebra is hermitian if and only if

$$||e + i\alpha x|| = 1 + o(\alpha) \quad (\alpha \text{ real}). \tag{1}$$

Here is his argument. Let $x$ be any bounded operator on a Hilbert space, and write $x = h + ik$, where $h$ and $k$ are hermitian. For all real $\alpha$ we have:

$$||e + i\alpha x||^2 = \sup||\xi + i\alpha x\xi||^2,$$

where the supremum is taken over all vectors $\xi$ of norm 1. We can write:

$$||\xi + i\alpha x\xi||^2 = (\xi|\xi) - 2\alpha(k\xi|\xi) +$$

$$+ \alpha^2[||h\xi||^2 + ||k\xi||^2 + i(k\xi|h\xi) - i(h\xi|k\xi)].$$

Hence, if $||\xi|| = 1$, then

$$||\xi + i\alpha x\xi||^2 = 1 - 2\alpha(k\xi|\xi) + O(\alpha^2),$$

i.e., there exists a bounded function $M$ such that

$$||\xi + i\alpha x\xi||^2 = 1 - 2\alpha(k\xi|\xi) + \alpha^2 M(\xi),$$

where $|M(\xi)| \leq (||h|| + ||k||)^2$. Now if (1) holds, then we also have $||e + i\alpha x||^2 = 1 + o(\alpha)$ from which it follows that $(k\xi|\xi) = 0$ for every vector $\xi$. To see this, suppose, for example, that $(k\xi_o|\xi_o) < 0$ for some (unit) vector $\xi_o$, and set

$$L = \sup\{-2(k\xi|\xi) : ||\xi|| = 1\} > 0.$$

Then,

$$||e + i\alpha x||^2 = 1 + \sup\{-2\alpha(k\xi|\xi) + \alpha^2 M(\xi)\},$$

where the supremum is taken over all unit vectors $\xi$. Hence, for $\alpha > 0$, the number

$$||e + i\alpha x||^2 - 1 = \sup\{-2\alpha(k\xi|\xi) + \alpha^2 M(\xi)\}$$

lies between $\alpha L - \alpha^2(||h|| + ||k||)^2$ and $\alpha L + \alpha^2(||h|| + ||k||)^2$. Consequently, as $\alpha \to 0^+$, we have

$$\frac{||e + i\alpha x||^2 - 1}{\alpha} \to L > 0,$$

contradicting the assumption that $||e + i\alpha x||^2 = 1 + o(\alpha)$. This implies that $k = 0$, i.e., $x = h$ is hermitian.

Conversely, if $x = h$ is hermitian, then

$$||\xi + i\alpha h\xi||^2 = (\xi|\xi) + \alpha^2||h\xi||^2$$

and so

$$||e + i\alpha h||^2 = 1 + \alpha^2||h||^2,$$

which implies that

$$||e + i\alpha h|| = 1 + o(\alpha).$$

Thus, an element $x$ in a unital $C^*$-algebra is hermitian if and only if

$$||e + i\alpha x|| = 1 + o(\alpha) \quad (\alpha \text{ real}).$$

Further investigations of the set $H(A)$ of hermitian elements in a unital Banach algebra $A$ led Vidav [1] to a rather deep geometrical characterization of $C^*$-algebras.

THEOREM. *(Vidav). Let* $A$ *be a unital Banach algebra such that:*

*i)* $A = H(A) + iH(A)$;

*ii) if* $h \in H(A)$ *then* $h^2 = a + ib$ *for some* $a, b \in H(A)$ *with* $ab = ba$.

*Then the algebra* A *has the following properties:*

*1. The decomposition* $x = h + ik$, *with* $h, k \in H(A)$, *is unique.*

*2. Setting* $x^* = h - ik$ *if* $x = h + ik$, *the map* $x \to x^*$ *is an involution on* A. *Furthermore, for* $h \in H(A)$ *we have* $||h^2|| = ||h||^2$.

*3.* $||x||_o = ||x^*x||^{1/2}$ *defines a* $C^*$*-norm on* A *which is equivalent to the original norm.*

Nearly ten years later Barnett W. Glickfeld [1] and Earl Berkson [1] showed independently that A is actually a C*-algebra under its *original* norm. Their proofs in the commutative case are quite different. Berkson utilized the notion of semi-inner-product space introduced by Lumer [1], and the theory of scalar type operators as developed by N. Dunford (see Dunford and Schwartz [1], [2]). Glickfeld recognized the importance of the exponential function and obtained the commutative theorem via the hermiticity condition $||\exp(i\alpha x)|| = 1$ ($\alpha$ real) for $x \in A$. A simplification of his proof was pointed out by Robert B. Burckel [1]. The extension to arbitrary (possibly noncommutative) unital Banach algebras is a consequence of the Russo-Dye theorem (22.19). Based on a refinement of this theorem Theodore W. Palmer [1] finally showed that condition (ii) in Vidav's theorem is unnecessary and he also gave the first simple proof that A is already a C*-algebra under its original norm. Thus, in 1968, the following elegant characterization of C*-algebras was established.

THEOREM. *A unital Banach algebra* A *admits an involution with respect to which it is a* $C^*$*-algebra iff* $A = H(A) + iH(A)$.

This result of Vidav-Palmer will be proved in §45.

A few years later (1971) Robert T. Moore [4] gave deep duality characterizations of C*-algebras. He defined *hermitian functionals* on an arbitrary unital Banach algebra A to be those in the real span H(A*) of the states on A. He showed that every functional f in the dual A* of A can be decomposed as $f = h + ik$, where h and k are hermitian functionals. Moore's proof uses the usual decomposition of measures. Independently, Allan M. Sinclair [1] gave an interesting direct proof in which the measure theory is replaced by convexity and Hahn-Banach separation arguments. Their result is a useful strengthening of the Bohnenblust-Karlin vertex theorem [1], which asserts that the states on a unital Banach algebra separate points in A (cf. (43.2)). Substantial simplifications of the proofs of Moore and Sinclair have been given by L. A. Asimow and A. J. Ellis [1].

Clearly, in the special case where A is a C*-algebra, a continuous linear functional f on A is hermitian iff $f(x^*) = \overline{f(x)}$ for all $x \in A$. Moreover, every hermitian functional on a C*-algebra is the difference of two positive functionals (see Corollary 2.6.4 of Dixmier [5]). We have seen that C*-algebras are characterized among unital Banach algebras as those for which there are *enough* hermitian elements. Moore's duality characterization shows that they may also be characterized as those for which there are *not too many* hermitian functionals.

THEOREM. *(Moore). A unital Banach algebra* A *admits an involution with respect to which it is a C*-algebra iff the dual* $A^*$ *decomposes as a real direct sum* $A^* = H(A^*) + iH(A^*)$*; or, equivalently, iff the hermitian elements in* A *separate points in* $A^*$.

This reduces an important property of a Banach algebra to properties of its dual space and may play a crucial role in further investigations.

## §41. *The numerical range of an element in a normed algebra.*

Let A denote a unital normed algebra with identity e. In this section we define and study elementary properties of the numerical range of an element of A. Once again, we remind the reader that an element f in the dual space $A^*$ of A is called a *state* on A provided that $f(e) = ||f|| = 1$. The set of all states on A is denoted by S(A), i.e.,

$$S(A) = \{f \in A^*: f \text{ is a state on } A\}.$$

(41.1) DEFINITION. *Let* A *be a unital normed algebra. Given an element* $a \in A$, *the* <u>*numerical range*</u> *of* a *is the set of scalars*

$$V(a) = \{f(a): f \in S(A)\}.$$

*When it is necessary to show dependence on the particular algebra* A, *we shall write* V(A;a) *for* V(a).

If B is a subalgebra of A containing the identity of A, then

$V(B;b) = V(A;b)$ for all $b \in B$. Indeed, by the Hahn-Banach theorem, the restriction map $f \to f|B$ sends $S(A)$ onto $S(B)$. Hence, $V(A;a)$ is independent of the choice of the algebra $A$. In particular, $V(A;a)$ is unchanged when $A$ is replaced by its completion.

Since $S(A)$ is the intersection of the weak*-compact convex subset $\{f \in A^*: ||f|| \leq 1\}$ of $A^*$ with the weak*-closed convex set $\{f \in A^*: f(e) = 1\}$, then $S(A)$ is a nonempty weak*-compact convex subset of $A^*$. (It is nonempty by the Hahn-Banach theorem.) It follows that $V(a)$ is a nonempty compact convex subset of scalars; this is a consequence of the linearity and weak*-continuity of the map $f \to f(a)$, $f \in A^*$, $a \in A$. We mention here that the above properties of $V(a)$ remain true if $A$ is a real or complex normed linear space with distinguished element $e$ of norm 1, and $B$ is any linear subspace of $A$. Many of the results we shall prove remain true in this generality. While we shall largely restrict our attention to complex unital normed algebras, it is often convenient, in proving results about a single element $a$, to take advantage of this larger setting by working only in the *subspace* spanned by $e$ and $a$. (Cf. the proof of (42.1).)

(41.2) PROPOSITION. *Let* $A$ *be a unital normed algebra with identity* $e$, *and let* $a, b \in A$, $\lambda, \mu \in C$. *Then:*

*(a)* $V(a)$ *is a nonempty compact convex subset of scalars which is independent of* $A$;

*(b)* $V(\lambda e + \mu a) = \lambda + \mu V(a)$;

*(c)* $V(a + b) \subseteq V(a) + V(b)$;

*(d)* $|\alpha| \leq ||a||$ *for all* $\alpha \in V(a)$;

*(e)* $V(a) = \bigcap_{\alpha \in C} E(\alpha, ||\alpha e - a||)$, *where* $E(\alpha,\beta) = \{\lambda \in C: |\lambda - \alpha| \leq \beta\}$.

*Proof.* (a) was proved in the discussion preceding the statement of the proposition, and parts (b), (c), and (d) follow immediately from the definitions.

To prove (e), let $\lambda \in V(a)$. Then $\lambda = f(a)$ for some $f \in S(A)$, and, for all $\alpha \in C$, we have

$$|\lambda - \alpha| = |f(a - \alpha e)| \leq ||a - \alpha e||,$$

that is,

$$\lambda \in E(\alpha, ||\alpha e - a||) \text{ for all scalars } \alpha. \tag{1}$$

Conversely, suppose (1) holds. If $a = \beta e$ for some scalar $\beta$, then $||\alpha e - a|| = |\alpha - \beta|$, and setting $\alpha = \beta$ we have $\lambda \in E(\beta,0)$, i.e., $\lambda = \beta$. However, when $a = \beta e$, $V(a) = \{\beta\}$. Now, assume that $e$ and $a$ are linearly independent, and define $f_o$ on their linear span by:

$$f_o(\alpha e + \beta a) = \alpha + \beta\lambda, \quad (\alpha,\beta \text{ scalars}).$$

Since $\lambda \in E(\alpha,||\alpha e - a||)$ for all $\alpha$, we see that $||f_o|| \leq 1$ and

$$|f_o(\alpha e + \beta a)| \leq ||\alpha e + \beta a||.$$

Extending $f_o$ to $f \in A^*$ with $||f|| \leq 1$, we obtain an element $f \in S(A)$ such that $f(a) = f_o(a) = \lambda$. Hence, $\lambda \in V(a)$. □

Part (e) of (41.2) reveals that $V(a)$ can be expressed as an intersection of closed disks.

(41.3) PROPOSITION. *Let* $A$ *be a unital Banach algebra. Then* $\sigma_A(a) \subseteq V(a)$ *for all* $a \in A$.

*Proof.* Suppose that $\lambda \in C \setminus V(a)$. By (41.2), (d), there is a complex number $\alpha$ such that $|\alpha - \lambda| > ||\alpha e - a||$. Hence $||(\alpha - \lambda)^{-1}(\alpha e - a)|| < 1$, and by (B.3.3), the element $e - (\alpha - \lambda)^{-1}(\alpha e - a)$ is invertible. It follows easily that $\lambda e - a$ is invertible, and so $\lambda \in C \setminus \sigma_A(a)$. □

A direct proof of (41.3) which does not depend on part (e) of (41.2) can be given as follows: Let $\lambda \in \sigma_A(a)$. Then $\lambda e - a$ is singular in A. If $\lambda e - a$ has no left inverse, then $J = \{x(\lambda e - a): x \in A\}$ is a proper left ideal of A. Since A is a unital Banach algebra, then $||e - x|| \geq 1$ for all $x \in J$. By the Hahn-Banach theorem, there exists $f \in A^*$ such that $f(e) = ||f|| = 1$ and $f(J) = \{0\}$. Then $f \in S(A)$ and $f(\lambda e - a) = 0$; that is, $\lambda = f(a) \in V(a)$. If $\lambda e - a$ has no right inverse, a parallel argument can be given in terms of right ideals.

(41.4) PROPOSITION. *If* $A$ *and* $B$ *are unital normed algebras, with identities* $e$ *and* $e'$ *respectively, and* $\phi: A \to B$ *is a homomorphism such that* $||\phi|| \leq 1$ *and* $\phi(e) = e'$, *then* $V(B;\phi(a)) \subseteq V(A;a)$ *for all* $a \in A$.

*Proof.* Let $\lambda \in V(B;\phi(a))$. Then there exists $p \in S(B)$ such that

$\lambda = p(\phi(a))$. Define a linear functional $f$ on $A$ by $f(x) = p(\phi(x))$, $(x \in A)$. Then $f(e) = 1$, $|f(x)| \leq ||p||\cdot||\phi||\cdot||x||$, $(x \in A)$; hence $f \in S(A)$ and $\lambda = f(a) \in V(A;a)$ as required. □

If $I$ is a closed two-sided ideal of $A$, and $\phi: A \to A/I$ denotes the canonical homomorphism, then, by (41.4), $V(A/I;\phi(a)) \subseteq V(A;a)$ for all $a \in A$. The next result shows, in this case, that even more can be said.

(41.5) PROPOSITION. *Let* $A$ *be a unital normed algebra, and let* $I$ *be a closed two-sided ideal of* $A$. *Then*

$$V(A/I;a + I) = \bigcap_{x\in I} V(a + x).$$

*Proof.* From the definition of the quotient norm on $A/I$ we have, for $\alpha, \lambda \in C$, that $|\alpha - \lambda| \leq ||\alpha e - (a + I)||$ iff

$$|\alpha - \lambda| \leq ||\alpha e - (a + x)|| \quad (x \in I).$$

Hence, $E(\alpha,||\alpha e - (a + I)||) = \bigcap_{x\in I} E(\alpha,||\alpha e - (a + x)||)$. The proposition now follows from (41.2), (e). □

We conclude this section with the following proposition:

(41.6) PROPOSITION. *Let* $A$ *be a unital normed algebra, and let* $a \in A$. *Then:*

*1°)* $V(a) = \cup \{V_x(a): x \in A, ||x|| \leq 1\}$, *where* $V_x(a) = \bigcap_{\lambda\in C} E(\lambda,||(\lambda e - a)x||)$.

*2°)* $\inf\{\mathrm{Re}\lambda: \lambda \in V(a)\} \leq \inf\{||ax||: x \in A, ||x|| = 1\}$.

*Proof. 1°)* Suppose $||x|| \leq 1$, $x \in A$. Then $||(\lambda e - a)x|| \leq ||\lambda e - a|| = ||(\lambda e - a)e||$, and (41.2), (e) implies that $V_x(a) \subseteq V(a) = V_e(a)$.

*2°)* Once again let $x \in A$ with $||x|| \leq 1$. Then $V_x(a) \subseteq E(0,||ax||)$, and so, by *1°)* we have $\inf\{\mathrm{Re}\lambda: \lambda \in V(a)\} \leq \inf\{\mathrm{Re}\lambda: \lambda \in V_x(a)\} \leq ||ax||$. □

§42. *Two numerical range formulas.*

The purpose of this section is to prove the following theorem.

(42.1) THEOREM. *If* $A$ *is a unital normed algebra with identity* $e$, *then, for* $a \in A$:

*(a)* $$\sup \operatorname{Re} V(a) = \inf_{\alpha>0} \frac{||\alpha a + e|| - 1}{\alpha} = \lim_{\alpha\to 0+} \frac{||\alpha a + e|| - 1}{\alpha}$$

*(b)* $$\sup \operatorname{Re} V(a) = \sup_{\alpha>0} \frac{1}{\alpha} \log||\exp(\alpha a)|| = \lim_{\alpha\to 0+} \frac{1}{\alpha} \log||\exp(\alpha a)||.$$

This theorem plays an important role in the theory of numerical ranges; it is the foundation, in particular, on which our proof of the Vidav-Palmer characterization theorem (cf. (45.1)) rests.

*Proof.* (a) Assume first that $A$ is a real normed algebra, and let $A_o$ be the two-dimensional subspace

$$A_o = \{\alpha e + \beta a\colon \alpha, \beta \in R\}.$$

Since the numerical range of $a$ is independent of the containing subspace, we may replace $A$ by $A_o$. Given $f \in A_o^*$ with $f(e) = 1$, we have that $f \in S(A_o)$, i.e., $||f|| = 1$, iff, for all $\alpha \in R$,

$$|f(\alpha e + a)| \leq ||\alpha e + a||,$$

or equivalently,

$$-||\alpha e + a|| - \alpha \leq f(a) \leq ||\alpha e + a|| - \alpha. \tag{1}$$

Let $\beta = \inf_\alpha\{||\alpha e + a|| - \alpha\}$. Since $\sup_\alpha\{-||\alpha e + a|| - \alpha\} \leq \beta$, the linear function $f_o$ on $A_o$ defined by $f_o(e) = 1$ and $f_o(a) = \beta$ satisfies $f_o(a) \leq ||\alpha e + a|| - \alpha$ for all $\alpha \in R$. By the Hahn-Banach theorem $f_o$ extends to a state $f$ on $A$. Since any state on $A$ satisfies (1), we have $\sup V(a) = \beta$.

The function $g\colon R \to R$ defined by $g(\alpha) = ||\alpha e + a|| - \alpha$ is decreasing since, if $\alpha \leq \gamma$, then $\gamma = \alpha + \delta$ for some $\delta \geq 0$, and

$$\begin{aligned} g(\gamma) &= ||(\alpha + \delta)e + a|| - (\alpha + \delta) \\ &\leq ||\alpha e + a|| + \delta - (\alpha + \delta) \\ &= ||\alpha e + a|| - \alpha = g(\alpha). \end{aligned}$$

Hence,

$$\sup V(a) = \beta = \inf_{\alpha}\{||\alpha e + a|| - \alpha\} = \lim_{\alpha\to+\infty}\{||\alpha e + a|| - \alpha\}$$

$$= \lim_{\alpha\to 0+}\{||\frac{1}{\alpha}e + a|| - \frac{1}{\alpha}\} = \lim_{\alpha\to 0+}\frac{||\alpha a + e|| - 1}{\alpha}$$

$$= \inf_{\alpha>0}\frac{||\alpha a + e|| - 1}{\alpha}.$$

If $A$ is a complex normed algebra, let $A_R$ be the underlying real space. Then the map $f \to \text{Re } f$ is an isometry of $A^*$ onto $(A_R)^*$. To see this, let $f \in A^*$. Then, clearly, $\text{Re } f \in (A_R)^*$ and $||\text{Re } f|| \le ||f||$. On the other hand, given $\lambda \in C$ with $|\lambda| = 1$, we have

$$|\text{Re}(\lambda f(x))| = |\text{Re } f(\lambda x)| \le ||\text{Re } f||\cdot||\lambda x|| = ||\text{Re } f||\cdot||x||.$$

Since there exists a complex number $\lambda$ with $\text{Re}(\lambda f(x)) = |f(x)|$, we have $|f(x)| \le ||\text{Re } f||\cdot||x||$, and, therefore, $||f|| = ||\text{Re } f||$. Now, since $f \to \text{Re } f$ is an isometry, it restricts to a map of $S(A)$ onto $S(A_R)$, and the complex form of (a) is an immediate consequence of the real result.

(b) Set $\phi(\alpha) = \log ||\exp(\alpha a)||$. Since the last expression in part (b) is a right derivative (of $\phi$ at 0), we have from the chain rule that

$$\lim_{\alpha\to 0+}\frac{1}{\alpha}\log ||\exp(\alpha a)|| = \lim_{\alpha\to 0+}\frac{\phi(\alpha) - \phi(0)}{\alpha - 0}$$

$$= \frac{1}{||\exp(\alpha a)||}\Bigg|_{\alpha=0}\cdot\lim_{\alpha\to 0+}\frac{||\exp(\alpha a)|| - 1}{\alpha}$$

$$= \lim_{\alpha\to 0+}\frac{||\alpha a + e|| - 1}{\alpha}$$

$$= \sup \text{Re } V(a), \quad \text{(by (a))}$$

where we have used the fact that $\exp(\alpha a) = e + \alpha a + 0(\alpha^2)$.

The function $\phi$ is subadditive, i.e.,

$$\phi(\alpha_1 + \alpha_2) \leq \phi(\alpha_1) + \phi(\alpha_2),$$

and we have $\lim_{\alpha \to 0+} \phi(\alpha) = 0$. Further, we claim that

$$\sup_{\alpha>0} \frac{\phi(\alpha)}{\alpha} = \lim_{\alpha\to 0+} \frac{\phi(\alpha)}{\alpha} .$$

To see this, we note that the quotient $\frac{\phi(\alpha)}{\alpha}$ is bounded on (0,1], since $\lim_{\alpha\to 0+} \frac{\phi(\alpha)}{\alpha} \equiv L$ is a real number and $\frac{\phi(\alpha)}{\alpha}$ is a continuous function of $\alpha$. On the interval $[1,+\infty)$, $\frac{\phi(\alpha)}{\alpha}$ is bounded above because $\phi(\alpha)$ is. Let $K = \sup_{\alpha>0} \frac{\phi(\alpha)}{\alpha}$. Clearly, $L \leq K$. For any $\varepsilon > 0$, choose $\alpha > 0$ with $K - \varepsilon < \frac{\phi(\alpha)}{\alpha}$. For any $h$, $0 < h < \alpha$, write $\alpha = nh + \delta$, where $n$ is a positive integer and $0 \leq \delta < h$. By the subadditivity of $\phi$,

$$K - \varepsilon < \frac{\phi(\alpha)}{\alpha} \leq (\frac{nh}{\alpha})(\frac{(h)}{h}) + \frac{\phi(\delta)}{\alpha} .$$

Thus as $h \to 0^+$, $K - \varepsilon \leq 1 \cdot L + 0$, and hence $K \leq L$. □

The first inequality in (42.1) is due essentially to G. Lumer [1] who proved it in the context of "semi-inner product spaces." The theorem is discussed in detail in Bonsall and Duncan [3], [4], [5]. The presentation we have given has been influenced by Effros [1].

## §43. *The numerical radius.*

Let a be an element of a unital normed algebra. We define the *numerical radius of* a by the formula

$$v(a) = \sup\{|\lambda| : \lambda \in V(a)\}.$$

It follows from (41.2), parts (b) and (c) that $v(\cdot)$ is a seminorm on A. It will be shown below that for complex algebras $v(\cdot)$ is, in fact, a norm on A which is equivalent to the original norm. Bohnenblust and Karlin [1] were the first to recognize and prove this useful fact.

(43.1) THEOREM. *(Bohnenblust and Karlin). Let* A *be a complex unital normed algebra. Then, for all* $a \in A$,

$$||a|| \geq v(a) \geq \frac{1}{e}||a||,$$

*where* e *denotes the base of the natural logarithm.*

*Proof.* The first inequality follows from (41.2), (d). To establish the second, we may assume, without loss of generality, that A is complete. The proof utilizes (42.1) and term-by-term integration of an A-valued power series in the unit disk. Indeed, let $a \in A$ and suppose that $v(a) \leq 1$. If $|\lambda| = 1$, we have, by (42.1), (b), that

$$||\exp(\lambda a)|| \leq e.$$

Since

$$a = \frac{1}{2\pi i} \int_{C(0,1)} \exp(\lambda a)\frac{d\lambda}{\lambda^2},$$

where $C(0,1) = \{\lambda \in C: |\lambda| = 1\}$, it follows that $||a|| \leq e$. □

Another proof of (43.1), which depends on n-th roots of unity, can be found in Bonsall and Duncan [3, p. 34]. The argument given there is, in certain respects, more elementary than the above, but is also more technical. We also mention in passing that Bonsall and Duncan [3, p. 36] have given a simple example which illustrates that (43.1) may be false for *real* normed algebras. Indeed, letting A be the complex numbers viewed as a real Banach algebra with modulus norm, let $a = i$. If f is a state on A, then for some real number q,

$$f(x + iy) = x + qy \quad (x, y \in R).$$

Since $||f|| = 1$, then $1 + qt \leq (1 + t^2)^{1/2}$ for all $t \in R$. Hence $q = 0$ from which we see that $V(a) = \{0\}$.

(43.2) COROLLARY. *Let* A *be a complex unital normed algebra. Then the set* S(A) *of states on* A *separates the points of* A.

*Proof.* If a is a nonzero element of A, then $v(a) \neq 0$ by (43.1), and hence there exists $f \in S(A)$ such that $f(a) \neq 0$. □

§44. *Hermitian elements in a unital normed algebra.*

Recall that an element $a$ in a unital normed algebra $A$ is *hermitian* if $f(a)$ is real for each state $f$ on $A$, i.e., if $V(a) \subseteq R$. The set of hermitian elements in $A$ is denoted by $H(A)$. In this section we establish basic properties of the hermitian elements which will be needed to prove the Vidav-Palmer theorem. The reader wishing to pursue the subject further should consult Bonsall-Duncan [3], [4].

(44.1) PROPOSITION. *Let* $A$ *be a unital normed algebra with identity* $e$*, and let* $x \in A$*. Then the following statements are equivalent:*

*(a)* $x \in H(A)$

*(b)* $\lim_{\alpha\to 0} \frac{||e + i\alpha x|| - 1}{\alpha} = 0 \quad (\alpha \text{ real})$

*(c)* $||\exp(i\alpha x)|| = 1 \quad (\alpha \text{ real})$.

*Proof.* We have $x \in H(A)$ iff

$$\inf\{\text{Im } \lambda: \lambda \in V(x)\} = 0 = \sup\{\text{Im } \lambda: \lambda \in V(x)\}. \tag{1}$$

By (42.1)

$$\inf\{\text{Im } \lambda: \lambda \in V(x)\} = -\sup\{\text{Re } \lambda: \lambda \in V(ix)\} = \tag{2}$$

$$-\sup\{\frac{1}{\alpha} \log ||\exp(i\alpha x)||: \alpha > 0\}$$

and

$$\sup\{\text{Im } \lambda: \lambda \in V(x)\} = \sup\{\text{Re } \lambda: \lambda \in V(-ix)\} = \tag{3}$$

$$\sup\{\frac{1}{\alpha} \log ||\exp(-i\alpha x)||: \alpha > 0\}.$$

Therefore $x \in H(A)$ iff

$$\text{both } ||\exp(i\alpha x)|| \leq 1 \text{ and } ||\exp(-i\alpha x)|| \leq 1 \text{ for all } \alpha > 0. \tag{4}$$

Since

$$||\exp(i\alpha x)||\cdot||\exp(-i\alpha x)|| \geq ||\exp(i\alpha x)\cdot\exp(-i\alpha x)|| = 1,$$

(4) can prevail iff the norms there equal 1 for all positive $\alpha$, hence for all real $\alpha$. This establishes the equivalence of (a) and (c).

If (c) holds, sup Re $V(ix)$ = sup Re $V(-ix) = 0$ by (2) and (3). Therefore by (42.1), (a)

$$\lim_{\alpha\to 0^+} \frac{||e + i\alpha x|| - 1}{\alpha} = 0 = \lim_{\alpha\to 0^+} \frac{||e - i\alpha x|| - 1}{\alpha} = \lim_{\alpha\to 0^-} \frac{||e + i\alpha x|| - 1}{\alpha}.$$

Therefore (b) holds. Assuming (b), we likewise obtain from (42.1) that $\sup \operatorname{Re} V(ix) = \sup \operatorname{Re} V(-ix) = 0$ so that by (2) and (3) we obtain (1), i.e., $x \in H(A)$. □

Proposition (44.1) allows us to show that if $A$ is a unital Banach *-algebra satisfying

$$||x^*x|| = ||x||^2 \quad (x \in A),$$

then an element $h$ in $A$ belongs to $H(A)$ iff $h^* = h$. Indeed, let $h^* = h \in A$ and suppose that $\alpha$ is real. Then

$$||e + \alpha^2h^2|| = ||(e + i\alpha h)(e - i\alpha h)|| = ||e + i\alpha h||^2,$$

so that

$$\lim_{\alpha\to 0} \frac{1}{\alpha}\{||e + i\alpha h|| - 1\} = \lim_{\alpha\to 0} \frac{1}{\alpha}\{||e + \alpha^2h^2||^{1/2} - 1\} = 0$$

Hence $h \in H(A)$. To establish the opposite inclusion, let $x \in H(A)$. Then $x = h + ik$ with $h^* = h$ and $k^* = k$. Further,

$$V(x) = \{f(h) + if(k): f \in S(A)\},$$

and since $x, h, k \in H(A)$, it follows that $f(k) = 0$ for all $f \in S(A)$. Whence, $v(k) = 0$ and so $k = 0$, by (43.2).

Returning to the unital normed algebra setting, we wish to establish that $|h|_\sigma = ||h||$ for each $h \in H(A)$. For this we shall utilize the *arcsine function* defined, for $|\lambda| \leq 1$, by

$$\arcsin \lambda = \lambda + \frac{1}{2}\cdot\frac{1}{3}\lambda^3 + \frac{1}{2}\cdot\frac{3}{4}\cdot\frac{1}{5}\lambda^5 + \ldots = \sum_{n=1}^{\infty} \gamma_n\lambda^n.$$

When $A$ is complete and $x$ is an element whose spectrum is contained in the open disk, then

$$\arcsin x = \sum_{k=1}^{\infty} \gamma_k x^k$$

is a well-defined element of $A$. Further, for all $x$ in $A$, we define

$$\sin x = x - \frac{1}{3!}x^3 + \frac{1}{5!}x^5 - \ldots$$

and note that it also exists in A and satisfies

$$\sin x = [\exp(ix) - \exp(-ix)]/2i. \tag{1}$$

The following lemma establishes that the arcsine function and the sine function, as defined above, in a unital Banach algebra satisfy the expected inverse relationship.

(44.2) LEMMA. *Let* A *be a unital Banach algebra, and suppose that* h *is an element of* A *such that* $\sigma_A(h) \subseteq (-\frac{\pi}{2}, \frac{\pi}{2})$. *Then* $\sigma_A(\sin h) \subseteq \{\lambda \in C: |\lambda| < 1\}$, *and* $\arcsin(\sin h) = h$.

*Proof.* Since by (B.8.1), (vi), $\sigma_A(\sin h) = \{\sin \lambda: \lambda \in \sigma_A(h)\}$, then $\sigma_A(\sin h) \subseteq (-1,1) \subseteq \{\lambda \in C: |\lambda| < 1\}$. Hence,

$$\arcsin(\sin h) = \sum_{k=1}^{\infty} \gamma_k (\sin h)^k.$$

Let N be an open neighborhood of $\sigma_A(h)$ such that $\sin \lambda$ belongs to the open unit disk for all $\lambda \in N$. Then $\arcsin(\sin \lambda)$ is an analytic function of $\lambda$ for all $\lambda \in N$, and, since $\arcsin(\sin \alpha) = \alpha$ for all $\alpha \in [-\frac{\pi}{2}, \frac{\pi}{2}]$, we obtain $\arcsin(\sin \lambda) = \lambda$ for all $\lambda \in N$. Set $t_n(\lambda) = \sum_{k=1}^{n} \gamma_k (\sin \lambda)^k$. Then $t_n(\lambda) \to \lambda$ uniformly on a neighborhood of $\sigma_A(h)$. It follows from the continuity of the functional calculus (B.8.1), (v), that $\lim_{n\to\infty} t_n(h) = h$. □

(44.3) THEOREM. *(Sinclair). Let* A *be a unital normed algebra. If* $h \in H(A)$, *then* $|h|_\sigma = ||h||$.

*Proof.* Since $|h|_\sigma \leq ||h||$ is always true, we need only to establish the reverse inequality, and for this we may assume that A is complete, and that $|h|_\sigma < \pi/2$. It suffices to show that $||h|| \leq \pi/2$; indeed, if this is so, then for any $t$, $0 < t < \pi/2$, we have $|[t/(|h|_\sigma + (\pi/2 - t))]h|_\sigma < \pi/2$ and hence $||[t/(|h|_\sigma + (\pi/2 - t))]h|| \leq \pi/2$. Factoring out scalars and letting $t \to (\pi/2)^-$, we obtain $||h|| \leq |h|_\sigma$.

Now we have by (41.3), that $\sigma_A(h) \subseteq V(h) \subseteq R$. Hence $\sigma_A(h) \subseteq (-\pi/2, \pi/2)$, and by (44.2) we have

$$h = \arcsin(\sin h) = \sum_{k=1}^{\infty} \gamma_k (\sin h)^k \tag{2}$$

By (44.1), (c), $||\exp(\pm ih)|| = 1$, and therefore $||\sin h|| \leq 1$ by (1). Since $\gamma_k \geq 0$ for all $k = 1,2,3,\ldots$ it follows from (2) that $||h|| \leq \sum_{k=1}^{\infty} \gamma_k = \pi/2$. □

(44.4) PROPOSITION. *Let* A *be a unital Banach algebra. Then* H(A) *is a real Banach space, and* $i(hk - kh) \in H(A)$ *whenever* $h, k \in H(A)$.

*Proof.* It is straightforward to verify that H(A) is a closed real linear subspace of A. Suppose that $h, k \in H(A)$, let $\alpha \in R$, and set

$$a = \exp(i\alpha h)\exp(i\alpha k)\exp(-i\alpha h)\exp(-i\alpha k).$$

Then $||a|| \leq 1$ and $||a^{-1}|| \leq 1$; whence $||a|| = 1$, and upon expanding the exponentials we obtain

$$||e - \alpha^2(hk - kh)|| = 1 + O(\alpha^3) \quad (\alpha \to 0).$$

Interchanging h and k we obtain

$$||e + \alpha^2(hk - kh)|| = 1 + O(\alpha^3) \quad (\alpha \to 0),$$

and so

$$||e + \alpha(hk - kh)|| = 1 + o(\alpha) \quad (\alpha \to 0).$$

Hence $\lim_{\alpha\to 0} \frac{1}{\alpha}\{||e + i\alpha x|| - 1\} = 0$, where $x = i(hk - kh)$, and the result follows from (44.1). □

(44.5) PROPOSITION. *Let* A *be a unital Banach algebra. If* $h \in H(A)$, *then* $V(h) = \mathrm{conv}\ \sigma_A(h)$, *the convex hull of* $\sigma_A(h)$.

*Proof.* By (41.2) and (41.3), $\mathrm{conv}\ \sigma_A(h) \subseteq V(h) \subseteq R$. On the other hand, since $|h - \alpha e|_\sigma \leq v(h - \alpha e) \leq ||h - \alpha e||$, (44.3) implies that $|h - \alpha e|_\sigma = v(h - \alpha e)$ for all $\alpha \in R$. To see that $V(h) \subseteq \mathrm{conv}\ \sigma_A(h)$, using this equality, we proceed as follows. Let $V(h) = [\alpha,\beta]$ for real numbers $\alpha$, $\beta$ with $\alpha < \beta$. We will show that $\alpha, \beta \in \sigma_A(h)$. Indeed, $V(h - \alpha e) = [0,\beta-\alpha]$, so $v(h - \alpha e) = \beta - \alpha = |h - \alpha e|_\sigma$. Since $\sigma_A(h - \alpha e) \subseteq V(h - \alpha e) \subseteq [0,+\infty)$, then $|h - \alpha e|_\sigma \in \sigma_A(h - \alpha e)$. Therefore $\beta - \alpha \in \sigma_A(h - \alpha e)$ and so $\beta \in \sigma_A(h)$. Similarly, $V(h - \beta e) = [\alpha-\beta,0]$ so $v(h - \beta e) = \beta - \alpha = |h - \beta e|_\sigma$. Since $\sigma_A(h - \beta e) \subseteq V(h - \beta e) \subseteq (-\infty,0]$, then $-|h - \beta e|_\sigma \in \sigma_A(h - \beta e)$. Therefore $\alpha - \beta \in \sigma_A(h - \beta e)$ so that $\alpha \in \sigma_A(h)$. □

(44.6) PROPOSITION. *Let* A *be a unital Banach algebra, and let* $x \in A$. *If* $x = h + ik$, *where* $h, k \in H(A)$ *and* $hk = kh$, *then* $V(x) = \operatorname{conv} \sigma_A(x)$.

*Proof.* We begin by making several reductions. Indeed, since the hypotheses of the proposition hold for $x$ replaced by $\lambda e + \mu x$, $\lambda, \mu \in C$, since $\operatorname{conv} \sigma_A(a)$ is the intersection of the closed half-planes that contain $\sigma_A(a)$, and since $V(x)$ is compact and convex, it suffices to show that $\max \operatorname{Re} V(x) = \max \operatorname{Re} \sigma_A(x)$.

To this end, let B be a maximal commutative subalgebra of A containing h and k. By (B.4.3) we have

$$\sigma_A(x) = \sigma_B(x) = \{\phi(h) + i\phi(k) : \phi \in \hat{B}\}.$$

Since $h, k \in H(A)$ if follows that

$$\max \operatorname{Re} \sigma_A(x) = \max \sigma_B(h) = \max \sigma_A(h)$$

and

$$\max \operatorname{Re} V(x) = \max V(h).$$

An application of (44.5) completes the proof. □

Let A be a unital Banach algebra. An element x in A is said to be *positive* if $V(x)$ is a subset of the nonnegative reals. We denote by $A^+$ the set of all positive elements of A. Since $x \in A^+$ iff $f(x) \geq 0$ for all $f \in S(A)$, it is clear that $A^+$ is a closed cone in $H(A)$. Further, the identity e is an interior point of $A^+$ in $H(A)$. Indeed, if $x \in H(A)$ with $||e - x|| < 1$, then $v(e - x) < 1$ and so $V(x)$ is a subset of the nonnegative reals. Hence $x \in A^+$ and e is an interior point. Finally we note, by (44.5), that an element x in A belongs to $A^+$ iff $x \in H(A)$ and $\sigma_A(x) \geq 0$.

(44.7) DEFINITION. *A* <u>*Vidav algebra*</u> *is a unital Banach algebra* A *such that* $A = H(A) + iH(A)$, i.e., *each element* x *in* A *can be written in the form* $x = h + ik$ *with* $h, k \in H(A)$.

A Vidav algebra $A$ does not *a priori* come with an involution defined on it. However, we shall show (Vidav-Palmer theorem) that a Vidav algebra, with involution defined by

$$(h + ik)^* = h - ik,$$

is, in fact, a C*-algebra. The following result is an important step towards establishing this result.

(44.8) PROPOSITION. *Let* $A$ *be a Vidav algebra. Then the mapping* $*: A \to A$ *defined by* $(h + ik)^* = h - ik$ *is a continuous involution on* $A$.

*Proof.* We first show that each element $a \in A$ has a *unique* representation of the form $h + ik$, with $h, k \in H(A)$. Indeed, if $h, k \in H(A)$, and $h + ik = 0$, then $V(h) = -iV(k) \subseteq R \cap iR = \{0\}$. By (43.1), $h = k = 0$. Uniqueness follows immediately from this. Using this uniqueness it is routine to verify that $*$ is conjugate-linear of period two.

To verify $(ab)^* = b^*a^*$ we observe a few properties of hermitian elements and their squares. Let $h \in H(A)$ and write $h^2 = p + iq$, where $p, q \in H(A)$. Then $h(p + iq) = (p + iq)h$, and so $hp - ph = i(qh - hq)$. Setting $z = hp - ph$, we have $z, iz \in H(A)$ by (44.4); hence $V(z) = \{0\}$, and $z = 0$ by (43.1). Therefore $hp = ph$, $h^2p = ph^2$, and so $pq = qp$. Applying (44.6) we obtain $V(h^2) = \text{conv}\ \sigma_A(h^2) = \text{conv}\ \sigma_A(h)^2 \subseteq R$, i.e., $h^2 \in H(A)$. Now, given $h, k \in H(A)$, we have $h^2, k^2, (h + k)^2 \in H(A)$, and hence $hk + kh \in H(A)$.

We are now in a position to show that $*$ is an algebra involution. Indeed, let $a, b \in A$, where $a = h + ik$ and $b = p + iq$. By (44.4) we have

$$\begin{aligned} ab + b^*a^* = (hp + ph) - (kq + qk) + i(kp - pk) \\ + i(hq - qh) \in H(A), \end{aligned}$$

$$\begin{aligned} i(ab - b^*a^*) = i(hp - ph) - i(kq - qk) - (kp + pk) \\ - (hq + qh) \in H(A). \end{aligned}$$

Therefore in the representation

$$ab = [\tfrac{1}{2}(ab + b^*a^*)] + i[\tfrac{1}{2i}(ab - b^*a^*)]$$

each square-bracketed term belongs to $H(A)$ and so by definition of $*$

$$(ab)^* = [\tfrac{1}{2}(ab + b^*a^*)] - i[\tfrac{1}{2i}(ab - b^*a^*)] = b^*a^*.$$

Finally, we show that $*$ is continuous. Let $h, k \in H(A)$. By (44.5) and (44.3),

$$\begin{aligned} ||h|| &= \max\{|f(h)|: f \in S(A)\} \\ &\le \max\{|f(h) + if(k)|: f \in S(A)\} \\ &\le ||h + ik|| \end{aligned}$$

and similarly $||k|| \le ||h + ik||$. Hence we have

$$||(h + ik)^*|| = ||h - ik|| \le ||h|| + ||k|| \le 2||h + ik||. \quad \square$$

The next proposition shows that every positive element in a Vidav algebra admits a positive square root. Although a proof based on the power series expansion of $(1 - t)^{1/2}$ can be given (similar to (22.5)), we prefer to give a well-known recursive argument.

(44.9) PROPOSITION. *Let* $A$ *be a Vidav algebra. Given* $k \in A^+$, *let* $B$ *denote the smallest closed subalgebra of* $A$ *containing the identity* $e$ *and* $k$. *Then there exists* $u \in B \cap A^+$ *such that* $u^2 = k$.

*Proof.* By considering $k/(|k|_\sigma + 1)$, we may clearly assume that $|k|_\sigma \le 1$. Set $a = e - k$ so that $|a|_\sigma \le 1$, and let $P$ denote the set of all polynomials in $a$ with coefficients in $R^+$. Let $x_0 = 0$ and set

$$x_n = \tfrac{1}{2}(a + x_{n-1}^2) \quad (n = 1,2,\ldots).$$

By induction, $x_n \in P$, and since

$$x_{n+1} - x_n = \tfrac{1}{2}(x_n + x_{n-1})(x_n - x_{n-1}),$$

it follows that $x_n - x_{n-1} \in P$ $(n = 1,2,\ldots)$. Also, since $ax_n = x_na$, an induction argument gives $|x_n|_\sigma \le 1$ $(n = 1,2,\ldots)$.

Since $\sigma_B(a) \subseteq [0,1]$, the sequence $\{\hat{x}_n\}$ of Gelfand transforms on $\hat{B}$ is an increasing sequence of nonnegative functions bounded above by 1. It therefore converges pointwise to a function $\psi$ with $0 \le \psi \le 1$ on $\hat{B}$. Since $\psi = \frac{1}{2}(\hat{a} + \psi^2)$ and $\hat{a}$ is continuous, $\psi = 1 + (1 - \hat{a})^{1/2}$ is continuous; applying Dini's theorem, $\hat{x}_n \to \psi$ uniformly on $\hat{B}$. Hence $\{x_n\}$ is a

Cauchy sequence relative to $|\cdot|_\sigma$, and thus, by (44.3), relative to $||\cdot||$. Consequently, $x_n \to x \in H(A)$, where $\hat{x} = \psi$, and $(1 - \hat{x})^2 = 1 - \hat{a} = \hat{k}$. Let $u = e - x$. Then $(u^2 - k)\hat{} = 0$, and, since $u^2 - k \in H(A)$, $||u^2 - k|| = |u^2 - k|_\sigma = 0$, so $u^2 = k$. Since $\sigma_A(u) \subseteq \sigma_B(u) \subseteq [0,1]$, then $u \in A^+$ by (44.5). □

(44.10) COROLLARY. *Let* A *be a Vidav algebra. Then, for each* $x \in H(A)$, *there exist elements* $x^+, x^- \in A^+$ *such that* $x = x^+ - x^-$ *and* $x^+x^- = x^-x^+ = 0$.

*Proof.* Let B be a maximal commutative subalgebra of A containing x. By (44.9) there is an element $u \in B^+ = B \cap A^+$ such that $u^2 = x^2$. Then

$$\sigma_A(u \pm x) \subseteq \sigma_B(u \pm x) = \{\phi(u) \pm \phi(x): \phi \in \hat{B}\}.$$

Given $\phi \in \hat{B}$, we have

$$\phi(x)^2 = \phi(x^2) = \phi(u^2) = \phi(u)^2,$$

and therefore $\phi(x) = \pm\phi(u)$. Since $u \in B^+$, then $\sigma_B(u \pm x) \subseteq R^+$, and so $u \pm x \in A^+$ by (44.5). Set $x^+ = \frac{1}{2}(u + x)$ and $x^- = \frac{1}{2}(u - x)$. □

The next result corresponds to (12.6). Except for a few minor changes, due to numerical range considerations, the proof is the same as before.

(44.11) PROPOSITION. *Let* A *be a Vidav algebra and let* $x \in A$. *Then* $x^*x \in A^+$.

*Proof.* Let $y \in A$ be such that $-y^*y \in A^+$. Since $\sigma_A(y^*y)\backslash\{0\} = \sigma_A(yy^*)\backslash\{0\}$, we have by (44.5), that $-yy^* \in A^+$. Set $y = h + ik$, where $h, k \in H(A)$. Then $y^*y = 2h^2 + 2k^2 - yy^* \in A^+$, since $A^+$ is a cone. Hence $V(y^*y) = \{0\}$, and so $y^*y = 0$ by (43.1). By an analogous argument $yy^* = 0$, and so $h^2 + k^2 = 0$. Hence $V(h^2) = V(k^2) = \{0\}$, $h = k = 0$, and so $y = 0$.

Now suppose that x is any element of A. Then $x^*x \in H(A)$, and so, by (44.10), there exist $p, q \in A^+$ such that $x^*x = p - q$, $pq = qp = 0$. Let $y = xq$. Then $-y^*y = -qx^*xq = q^3 \in A^+$, and so $y = 0$ by the previous paragraph. It follows that $x^*xq = 0$, $q^2 = 0$, $q = 0$ and hence $x^*x = p \in A^+$. □

## 45. *The Vidav-Palmer theorem and applications.*

We are now prepared to prove the Vidav-Palmer theorem. A few applications of the theorem will be made to $C^*$-algebras.

(45.1) THEOREM. *(Vidav-Palmer). Let* $A$ *be a Vidav algebra. Then* $A$, *under the involution* $*$ *defined by* $(h + ik)^* = h - ik$ $(h, k \in H(A))$, *is isometrically *-isomorphic to a concrete* $C^*$*-algebra. In particular,* $||x^*x|| = ||x||^2$ *for all* $x \in A$.

*Proof*. If $f \in S(A)$, then $f(x^*x) \geq 0$ by (44.11), and hence each element of $S(A)$ is a positive linear functional. By (27.2) and the proof of (44.8) there exists, for each $f \in S(A)$, a Hilbert space $H_f$ and a $*$-representation $x \to \pi_f(x)$ from $A$ into $B(H_f)$ such that

$$||\pi_f(h)|| \leq ||h|| \quad (h \in H(A)),$$

$$||\pi_f(x)|| \leq \sqrt{2}||x|| \quad (x \in A).$$

Let $H$ and $\pi$ denote the direct sum of the Hilbert spaces $H_f$ and $*$-representations $\pi_f$, respectively. Then

$$||\pi(h)|| \leq ||h|| \quad (h \in H(A)),$$

$$||\pi(x)|| \leq \sqrt{2}||x|| \quad (x \in A),$$

and, by (44.3) and (44.5), we have

$$\begin{aligned} ||\pi(x)||^2 &\geq \sup\{f(x^*x) : f \in S(A)\} \\ &= v(x^*x) = ||x^*x|| \quad (x \in A). \end{aligned} \tag{1}$$

By (21.6), $|f(x)|^2 \leq f(x^*x)$ $(x \in A)$ and hence $v(x)^2 \leq ||x^*x||$. Thus, by (1),

$$||\pi(x)||^2 \geq ||x^*x|| \geq v(x)^2 \quad (x \in A)$$

and, by (43.1), we obtain

$$||\pi(x)|| \geq v(x) \geq \frac{1}{e}||x|| \quad (x \in A) \tag{2}$$

It follows that $x \to \pi(x)$ is injective, has a $C^*$-algebra as range, and, by (44.3), satisfies $||\pi(h)|| = ||h||$ for all $h \in H(A)$. Identifying $A$ with $\pi(A)$, $A$ is a $C^*$-algebra under the norm $|\cdot|$ defined by

$$|x| = ||\pi(x)|| \quad (x \in A).$$

The proof will be complete if we can show that this norm coincides with the original norm on $A$. To this end, let $E = \{\exp(ih): h \in H(A)\}$, and let $x$ be an element of $A$ such that $|x| = 1$. By (35.6) and (15.2) there exists a sequence $\{y_n\}$ in conv(E) such that $|y_n - x| \to 0$. Hence, by (2), $||y_n - x|| \to 0$, and so $||y_n|| \to ||x||$. By (44.1),

$$||\exp(ih)|| = 1 \quad (h \in H(A)),$$

hence $||y_n|| \leq 1$ for $n = 1,2,\ldots$, and so $||x|| \leq 1$. It follows that

$$||x|| \leq |x| \quad (x \in A).$$

If for some $a \in A$ we have $||a|| < |a|$, then

$$||a^*a|| \leq ||a^*||\cdot||a|| < |a^*|\cdot|a| = |a^*a| = ||a^*a||.$$

This contradiction shows that $||a|| = |a|$ for all $a \in A$, as desired. □

The geometrical nature of the Vidav-Palmer theorem derives from its dependence on the underlying Banach space structure. Indeed, suppose that $A$ is any Banach space and $e$ is an element of $A$ such that $||e|| = 1$. The Vidav-Palmer theorem shows that if $A = H(A) + iH(A)$, then any product $\cdot$ on $A$ under which $(A,\cdot)$ is a Banach algebra with $e$ as identity necessarily also makes $(A,\cdot)$ into a $C^*$-algebra.

As a first application of (45.1) we re-prove (16.1).

(45.2) THEOREM. *Let* $A$ *be a Banach* $^*$*-algebra. The following two conditions are equivalent:*

*(a)* $||x^*x|| = ||x||^2 \quad (x \in A)$.

*(b)* $||x^*x|| = ||x^*||\cdot||x|| \quad (x \in A)$.

*Proof*. (a) clearly implies (b). To show that (b) implies (a) we may assume, by (14.1), that $A$ has an identity element $e$ of norm 1. By the Vidav-Palmer theorem it is sufficient to show that $A = H(A) + iH(A)$, and to prove this we need only show that each $x \in A$ satisfying $x^* = x$

belongs to $H(A)$. Let $x = x^* \in A$ and set $\mu = \max \operatorname{Re} V(ix)$, $\lambda = \min \operatorname{Re} V(ix)$. Given $\alpha > 0$, (42.1), (a) implies that

$$||e + i\alpha x|| = 1 + \alpha\mu + o(\alpha), \text{ as } \alpha \to 0^+$$

$$||e - i\alpha x|| = 1 - \alpha\lambda + o(\alpha), \text{ as } \alpha \to 0^+.$$

Therefore

$$\begin{aligned} 1 + \alpha(\mu - \lambda) + o(\alpha) &= ||e + i\alpha x||\cdot||(e + i\alpha x)^*|| \\ &= ||e + \alpha^2 x^2|| \\ &= 1 + O(\alpha^2). \end{aligned}$$

Hence $\lambda = \mu$, $V(ix) \subseteq \lambda + iR$, $V(x + i\lambda e) \subseteq R$, and thus there exists $h \in H(A)$ such that $x = h - i\lambda e$. Since $\sigma_A(x^*) = \overline{\sigma_A(x)}$, and since $\sigma_A(h) \subseteq R$, it follows that $\lambda = 0$ and so $x \in H(A)$. □

Recall that a linear functional $f$ on a $^*$-algebra $A$ is hermitian provided that $f(x^*) = \overline{f(x)}$ for all $x \in A$.

(45.3) THEOREM. *Let* $A$ *be a unital Banach* $^*$*-algebra. If each* $f$ *in* $S(A)$ *is hermitian, then* $A$ *is a C*-algebra.*

*Proof.* Let $h \in A$ with $h^* = h$. Then

$$V(h) = \{f(h): f \in S(A)\} \subseteq R.$$

The result now follows from the Vidav-Palmer theorem. □

(45.4) THEOREM. *Let* $A$ *be a unital Banach* $^*$*-algebra such that:*

*(a)* $\sigma_A(h) \subseteq R$ *when* $h^* = h$;

*(b)* $|e + \lambda h|_\sigma = ||e + \lambda h||$ *when* $h^* = h$, $\lambda \in C$.

*Then* $A$ *is a C*-algebra.*

*Proof.* Let $h = h^*$. The condition in (b), together with the characterization of conv $\sigma_A(h)$ as the intersection of the closed disks that contain $\sigma_A(h)$, shows that $V(h) \subseteq \text{conv}\, \sigma_A(h)$. By part (a) we then have $V(h) \subseteq R$, and the Vidav-Palmer theorem can be applied. □

The Vidav-Palmer theorem can be used to give another proof of (24.5).

(45.5) PROPOSITION. *Let* A *be a* $C^*$*-algebra and* I *a closed two-sided ideal of* A. *Then* $I^* = I$, *and* A/I *is a* $C^*$*-algebra.*

*Proof.* We assume first that A is unital with identity e. Then A/I is a unital Banach algebra with identity e + I. If $\phi \in S(A/I)$, define $\tilde{\phi}$ on A by

$$\tilde{\phi}(a) = \phi(a + I) \quad (a \in A).$$

Then $\tilde{\phi} \in S(A)$. Now let $h \in H(A)$. Then $\phi(h + I) = \tilde{\phi}(h) \in \mathbb{R}$ and hence $h + I \in H(A/I)$. Because

$$A/I = \{(h + I) + i(k + I): h, k \in H(A)\},$$

the Vidav-Palmer theorem implies that A/I is a $C^*$-algebra under the quotient norm and involution

$$[(h + I) + i(k + I)]^* = (h + I) - i(k + I).$$

Consequently $a \to a + I$ is a *-homomorphism and so $I^* = I$.

If A does not have an identity, we can adjoin one to obtain a unital $C^*$-algebra B. Then I is a closed two-sided ideal of B, so that $I^* = I$ and B/I is a $C^*$-algebra by the preceding paragraph. Since A/I is a closed *-subalgebra of B/I, the result follows. □

## EXERCISES

(VIII.1) Verify parts (b), (c), and (d) of Proposition (41.2).

(VIII.2) Verify that H(A) is a closed real linear subspace of A in the proof of Proposition (44.4).

(VIII.3) Show that $||e - \alpha^2(hk - kh)|| = 1 + O(\alpha^3)$ in the proof of Proposition (44.4).

(VIII.4) Let $E = C^2$ with norm

$$||(\lambda,\mu)|| = \max\{|\lambda|, |\mu|, \frac{1}{\sqrt{2}}|\lambda - i\mu|\}.$$

Let A denote the unital Banach algebra B(E) and consider the element

$$x = \begin{pmatrix} 1 & 0 \\ 0 & 0 \end{pmatrix}.$$

(a) Show that $x \notin H(A)$.

(b) Find the spectrum $\sigma_A(x)$.

(c) Find the numerical range $V(x)$.

(d) Show that $V(x) \neq \text{conv}\,\sigma_A(x)$.

(VIII.5) Let A denote the algebra $C([0,1])$ of continuous functions on $[0,1]$ with norm

$$||f|| = \sup\{\tfrac{1}{2}|f(s) + f(t)| + \tfrac{1}{2}|f(s) - f(t)|: s, t \in [0,1]\}.$$

Prove that A is a unital Banach algebra and that $||\cdot||$ coincides with the supremum norm $||\cdot||_\infty$ on *real* functions in A but, in general, $||\cdot||$ is not equal to $||\cdot||_\infty$. (This shows that (45.4) fails if, in part (b), complex scalars are replaced by real scalars).

(VIII.6) A *pre-$C^*$-algebra* is a complex normed $^*$-algebra A such that $||x^*x|| = ||x||^2$ for all x in A. Prove that if A is a unital normed $^*$-algebra such that $A = H(A) + iH(A)$, then A is a pre-$C^*$-algebra.

(VIII.7) Give an example to show that the numerical radius $v$ defined in §43 is not necessarily submultiplicative.

# 9
# Locally C*-Equivalent Algebras

## §46. *Introduction.*

In 1972 and 1973 Bruce Barnes [ 1 ], [ 3 ] attempted to characterize C*-algebras in terms of their commutative closed *-subalgebras. He specifically asked the following question: if A is a Banach *-algebra such that each of its commutative closed *-subalgebras is a C*-algebra in an equivalent norm, is A a C*-algebra in an equivalent norm? Although unable to give an answer in general, Barnes was able to answer the question affirmatively in several important special cases. A few years later, in 1976, Joachim Cuntz [ 1 ], making ingenious use of a restricted form of uniform C*-equivalence, answered Barnes' question affirmatively for general Banach *-algebras. The purpose of the present chapter is to study locally equivalent C*-algebras and to prove Cuntz's theorem.

## §47. *Locally* C*-*equivalent algebras.*

(47.1) DEFINITION. *A Banach* *-*algebra* A *is called* <u>C*-*equivalent*</u> *if there is a norm on* A, *equivalent to the given norm, which makes* A *into a* C*-*algebra.*

It should be noted that by (23.8) a Banach *-algebra A is C*-equivalent iff it admits a C*-norm (i.e., a norm under which A is a C*-algebra). Looking at it another way, A is C*-equivalent iff there exists an algebraic *-isomorphism of A onto some C*-algebra B. If such an isomorphism exists, it is necessarily continuous by (23.11). Then, by the open mapping theorem, the norm on B can be carried back to a C*-norm on A that is equivalent to its original norm. Since any *-isomorphism between A and B is a homeomorphism, A and B have the same algebraic and topological structure, but not necessarily the same geometrical structure.

(47.2) DEFINITION. *A Banach *-algebra* A *is called* <u>*locally*</u> <u>C**-equivalent*</u> *if, for each hermitian* h *in* A, *the closed *-subalgebra* C(h) *of* A *generated by* h *is* C**-equivalent.*

Thus, a Banach *-algebra A is locally C*-equivalent if, for every hermitian h in A, the closed *-subalgebra of A generated by h is *-isomorphic to a C*-algebra. If A has an identity, we do not require C(h) to contain it.

If the Banach *-algebra A has an identity e, we denote by C'(h) the closed *-subalgebra of A generated by the hermitian element h and e.

Since C'(h) is hermitian, $\sigma_{C'(h)} \subseteq R$ so that $\sigma_{C'(h)}(h)$ has no interior points. By exercise (B.65) the spectrum of h in A is the same as the spectrum of h in C'(h). It will be denoted by $\sigma_A(h)$.

Throughout this chapter we shall freely identify C'(h) with the Banach algebra of all continuous complex-valued functions on $\sigma_A(h)$. The norm $|\cdot|$ (given in (47.4)) then coincides on C'(h) with the uniform norm. If h is invertible in A, then C(h) = C'(h); indeed, since h is invertible in A, $0 \notin \sigma_A(h) = \sigma_{C'(h)}(h)$ and hence h is invertible in C'(h). Therefore, there exist polynomials $p_n$ and complex numbers $c_n$ such that $p_n(h) + c_n \cdot e \to h^{-1}$. Thus, the element $h \cdot p_n(h) + c_n h$ in C(h) converges to the identity e and so $e \in C(h)$. It follows that C(h) and C'(h) coincide.

If h is not invertible, then C(h) consists of those functions in C'(h) which vanish at 0.

The following lemma permits us to restrict our attention to algebras with identity. Recall that the norm on the unitization $A_e$ is $||x + \lambda e|| = ||x|| + |\lambda|$.

(47.3) LEMMA. *Let* A *be a Banach *-algebra and* $A_e$ *its unitization. Then:*

*(a)* A *is* C**-equivalent iff* $A_e$ *is* C**-equivalent.*

*(b)* A *is locally* C**-equivalent iff* $A_e$ *is locally* C**-equivalent.*

*Proof.* Part (a) follows immediately from (23.1). Since for any $\lambda \in C$, $C(h + \lambda e) \subseteq C(h)_e = C'(h)$ for $h = h^* \in A$, part (b) follows from (a). □

Several useful properties of locally C*-equivalent algebras are collected in the next lemma.

(47.4) LEMMA. *Let* A *be a locally* C*-*equivalent algebra. Then:*

*(a)* A *is hermitian.*

*(b)* A *is semisimple.*

*(c)* A *is* *-*semisimple.*

*(d) there is a unique, not necessarily complete, algebra norm* $|\cdot|$ *on* A *satisfying* $|x^*x| = |x|^2$ *for all* x *in* A.

*Proof.* We may assume that A has an identity.

(a) For each hermitian h in A, the spectrum of h in C(h) is real by (8.1), (a). Then $\sigma_A(h)$ is real, being contained in $\sigma_{C(h)}(h)$.

(b) If h is a hermitian element in Rad(A), then $h^*h \in$ Rad(A) and so $|h^*h|_\sigma = 0$. Since h is contained in a C*-equivalent algebra, then $h = 0$ by (23.2). Recalling that the radical is *-closed (see paragraph preceding (21.1)) and writing $x = h + ik$, where $h, k \in \mathrm{Rad}(A) \cap H(A)$, proves (b).

(c) Since for unital hermitian algebras, the *-radical coincides with Rad(A) (see (33.13)), it follows from (a) and (b) that A is *-semisimple.

(d) By (33.13) there is a (not necessarily complete) norm $|\cdot|$ on A with the C*-property $|a^*a| = |a|^2$ for all a in A (this norm is the Gelfand-Naimark seminorm). To show uniqueness, let $|\cdot|_1$ be another norm satisfying the condition in (d). Now let $|||\cdot|||$ be a C*-norm in $C(a^*a)$ which is equivalent to the original norm in A (exists by hypothesis). It is complete so we can apply (24.4) to the injection of $(C(a^*a), |||\cdot|||)$ into the $(A, |\cdot|)$-completion and again to the injection of $(C(a^*a), |||\cdot|||)$ into the $(A, |\cdot|_1)$-completion. It follows that $|a|^2 = |a^*a| = |||a^*a||| = |a^*a|_1 = |a|_1^2$ for all $a \in A$ as desired. □

We next prove that *commutative* locally C*-equivalent algebras are C*-equivalent, a result first established by Barnes [ 1 ]. Clearly, any closed *-subalgebra of a locally C*-equivalent algebra is locally C*-equivalent. Therefore, the proposition implies that every maximal commutative *-subalgebra of a locally C*-equivalent algebra is C*-equivalent. The simple proof given here is due to Josef Wichmann [ 2 ].

(47.5) PROPOSITION. *Let* A *be a commutative locally* C*-*equivalent algebra. Then* A *is* C*-*equivalent.*

*Proof.* Again we may assume that A has an identity. Let h be a hermitian element of A. Since A is locally C*-equivalent, (34.4) implies that there is a real number $M > 0$, depending on h, such that $||e^{ik}|| \leq M$ for each hermitian k in C(h). By (47.4), A is hermitian and semisimple, and hence the involution is continuous by (34.1) and (33.2). Therefore, the set H(A) of hermitian elements is closed in A. For $n = 1,2,3,\cdots$, set

$$H_n = \{h \in H(A): ||e^{ith}|| \leq n \quad \text{for all} \quad t \in R\}.$$

It is clear that each $H_n$ is closed and that H(A) is the union of the $H_n$. By the Baire Category theorem applied in H(A) there exists a positive integer m such that the interior of $H_m$ is not empty, i.e., it contains an open ball of radius r in H(A) centered at a hermitian $h_o$. If $h \in H(A)$ with $||h|| < r$, then by (B.3.4)

$$||e^{ith}|| \leq ||e^{it(h+h_o)}||\cdot||e^{-ith_o}|| \leq m^2$$

for every real t; in particular, $||e^{ith}|| \leq m^2$ for all hermitian h and real t. It follows from (34.4) that A is C*-equivalent. □

The following result shows that the property of local C*-equivalence is preserved under continuous *-homomorphisms.

(47.6) PROPOSITION. *If* A *is a locally* C*-*equivalent algebra and* I *is a closed* *-*ideal of* A, *then* A/I *is locally* C*-*equivalent.*

*Proof.* Assume that $s + I$ is a hermitian element of A/I, where $s \in A$. Set $t = (s + s^*)/2$. Since $s - s^* \in I$, then $s - t = (s - s^*)/2 \in I$. Hence $s + I = t + I$, where $t \in H(A)$. Now, set $B = C(t) + I$. Define a map $\phi : C(t)/(I \cap C(t)) \to B/I$ by

$$\phi(a + I \cap C(t)) = a + I \quad (a \in C(t)).$$

Then $\phi$ is a *-isomorphism of $C(t)/(I \cap C(t))$ onto B/I. Now B/I is complete; indeed, if $\{b_n + I\}$ is Cauchy in B/I, then there exists $a_n \in b_n + I \subseteq B$ with $a_n$ Cauchy in A. Hence $\{a_n + I \cap C(t)\}$ is

Cauchy in $B/(I \cap C(t)) = C(t)/(I \cap C(t))$ which implies that $a_n + I \cap C(t) \to a + I \cap C(t)$ for some $a \in C(t)$. Therefore $b_n + I = a_n + I \to a + I$ in $B/I$. It follows that $B/I$ is C*-equivalent. Then $s + I \in B/I$, which is a closed *-subalgebra of $A/I$. Therefore $C(s + I) \subseteq B/I$, and since $B/I$ is C*-equivalent, then $C(s + I)$ is C*-equivalent. It follows that $A/I$ is locally C*-equivalent. □

(47.7) PROPOSITION. *Let* A *be a locally* C**-equivalent algebra,* B *a *-subalgebra of* A*, and* I *a closed *-ideal of* A *such that* $I \subseteq B$. *If* I *is* C**-equivalent and* B/I *is* C**-equivalent, then* B *is* C**-equivalent (and hence closed in* A*).*

*Proof.* By (47.4), (d) there is a unique norm $|\cdot|$ on A with the C*-property. We prove that $|\cdot|$ is a complete norm on B. Let $\tilde{B}$ denote the completion of B with respect to $|\cdot|$. The ideal I is complete in the norm $|\cdot|$ via (24.4), so I is a closed *-ideal of $\tilde{B}$. Consider the usual quotient norm

$$|a + I|' = \inf\{|a - b| : b \in I\}$$

on $\tilde{B}/I$. By (24.5), $|\cdot|'$ is a C*-norm on $\tilde{B}/I$. Since $B/I$ is a C*-algebra in some norm, $B/I$ is complete in the norm $|\cdot|'$ (recall that any two C*-norms on a C*-algebra coincide by (24.4) again).

Suppose now that $\{b_n\} \subseteq B$ and $|b_n - b_m| \to 0$. Then $|(b_n - b_m) + I|' \to 0$. Therefore there exists $b \in B$ such that $|(b_n - b) + I|' \to 0$, and hence we can choose $\{a_n\} \subseteq I$ such that $|(b_n - b) - a_n| \to 0$. Then $|a_n - a_m| \to 0$, and, since $|\cdot|$ is complete on I, there exists $a \in I$ such that $|a_n - a| \to 0$. Finally, $|b_n - (b + a)| \to 0$, so that $|\cdot|$ is complete on B. □

## §48. *Local* C**-equivalence implies* C**-equivalence.*

We turn now to the main result of the chapter. Our goal is to prove the following theorem:

(48.1) THEOREM. *(Cuntz). Every locally* C**-equivalent Banach *-algebra is* C**-equivalent.*

Utilizing the Gelfand-Naimark theorem for commutative C*-algebras (see (7.1)), the theorem can be given the following equivalent formulation:

(48.2) THEOREM. *Let* A *be a Banach *-algebra. If, for every hermitian element* h *in* A, *there is a locally compact Hausdorff space* X *such that* C(h) *is *-isomorphic to the algebra* $C_o(X)$ *of all continuous complex-valued functions vanishing at infinity on* X, *then* A *is* C*-*equivalent.*

These theorems show that, generally speaking, every characterization of commutative C*-equivalent algebras (or equivalently, every characterization of $C_o(X)$ among its Banach *-subalgebras) may be generalized to noncommutative Banach *-algebras.

PROOF OF THEOREM (48.1)

The proof of (48.1) will be developed in a sequence of lemmas. To begin, we make a few observations: Since A is a locally C*-equivalent algebra, then A is equipped with its original norm $||\cdot||$ and, by (47.4), (d), with a unique norm $|\cdot|$ with the C*-property (the Pták function $\rho$). We will show that these two norms are equivalent on A.

By (23.6) we have $|x| \leq \beta||x||$ for some positive $\beta$, hence it suffices to show that there is a positive real number M such that $||x|| \leq M\cdot|x|$ for all $x \in A$. We may assume, by (47.3), that A has an identity element which we shall denote by $\mathbb{1}$.

In order to state several of our lemmas concisely and to facilitate the proof it will be helpful to introduce the following terminology: Let K be a real positive constant. A locally C*-equivalent algebra A is said to be K-*indecomposable* if given a normal element x in A and a non-C*-equivalent closed *-subalgebra B of A such that $x\cdot B = \{0\} = B\cdot x$, then $||x|| \leq K\cdot|x|$. Here $x\cdot B = \{xy: y \in B\}$ and $B\cdot x = \{yx: y \in B\}$.

The proof of (48.1) is indirect, i.e., *reductio ad absurdum;* our first lemma contains an important reduction of the problem. In essence it states that if a Banach *-algebra which is locally C*-equivalent but is not C*-equivalent exists, then there is a real number $K > 0$ for which such a K-indecomposable algebra exists. Hence, we may restrict our attention to K-indecomposable locally equivalent C*-algebras.

(48.3) LEMMA. *Let* A *be a locally* C*-*equivalent algebra with* $\mathbb{1}$, *such that* A *is not* C*-*equivalent. Then there exists a positive real number* K *and a Banach* *-*subalgebra* A' *of* A *containing* $\mathbb{1}$ *such that* A' *is locally* C*-*equivalent but not* C*-*equivalent and* A' *is* K-*indecomposable.*

*Proof.* Assume the contrary. Since, then, A is not 1-indecomposable, there is a normal $x_1 \in A$ and a closed non-C*-equivalent *-subalgebra $B_1$ of A such that $x_1 \cdot B_1 = \{0\} = B_1 \cdot x_1$ and $||x_1|| > |x_1|$. In this proof, let us write $\tilde{B}_1$ for the algebra obtained by adjoining $\mathbb{1}$ to $B_1$. Then $x_1$ commutes with $\tilde{B}_1$.

Suppose that normal elements $x_1, \cdots, x_{n-1}$ of A commuting with each other and with a closed *-subalgebra $\tilde{B}_{n-1}$ of A that is not C*-equivalent have been constructed such that $||x_i|| > i|x_i|$ and $\mathbb{1} \in \tilde{B}_i$ for $1 \leq i \leq n - 1$.

By assumption, $\tilde{B}_{n-1}$ is not n-indecomposable. Consequently there is a normal $x_n \in \tilde{B}_{n-1}$ and a closed non-C*-equivalent *-subalgebra $B_n$ of $\tilde{B}_{n-1}$ such that $x_n \cdot B_n = \{0\} = B_n \cdot x_n$ and $||x_n|| > n|x_n|$. Further, $x_1, \cdots, x_n$ commute with $\tilde{B}_n$.

Hence, by induction, there is a commutative subset $\{x_n : n = 1,2,3,\cdots\}$ of normal elements in A such that $||x_n|| > n|x_n|$. However, this is impossible since the commutative $|\cdot|$-closed (and therefore $||\cdot||$-closed) *-subalgebra generated by $\{x_n : n = 1,2,3,\cdots\}$ must be C*-equivalent by (47.5), the commutative case. □

In all that follows let A be a fixed locally C*-equivalent but not C*-equivalent Banach *-algebra with identity $\mathbb{1}$ and let A be K-indecomposable for a fixed $K > 0$. In view of (48.3), Theorem 48.1 will be proved when we succeed in deducing a contradiction from this assumption.

We shall say that the norms $||\cdot||$ and $|\cdot|$ on A are *equivalent on a subset* X of A, if there exists a real number $M > 0$ such that $||x|| \leq M \cdot |x|$ for all $x \in X$. The next lemma will be used in (48.7) and (48.8).

(48.4) LEMMA. *Let* x *be a non-zero element of* A *and let* X *be a subset of* A *such that* $x \cdot X = X \cdot x = x \cdot X^* = X^* \cdot x = \{0\}$.

*(a) If* $x$ *is normal and* $||x||/|x| > K$*, then* $||\cdot||$ *and* $|\cdot|$ *are equivalent on* $X$.

*(b) If* $||x||/|x| > 2K$*, then* $||\cdot||$ *and* $|\cdot|$ *are equivalent on* $X$.

*Proof.* (a) If $B$ is the $|\cdot|$-closed (and therefore $||\cdot||$-closed) *-subalgebra of $A$ generated by $X$, then $x\cdot B = \{0\} = B\cdot x$. Since $A$ is K-indecomposable, it follows that $B$ is C*-equivalent.

(b) Assume that $|x| = 1$ and write $x = x_1 + ix_2$, where $x_1 = (x + x^*)/2$ and $x_2 = (x - x^*)/2i$ are hermitian. Then necessarily $|x_1| \leq 1$ and $|x_2| \leq 1$. On the other hand, the hypothesis that $||x||/|x| > 2K$ and the triangle inequality imply that

$$\text{either } ||x_1|| > K \text{ or } ||x_2|| > K.$$

Thus, either $x_1$ or $x_2$ satisfies the conditions of part (a), and it follows that $||\cdot||$ and $|\cdot|$ are equivalent on $X$. □

The reader will observe that since $A$ is not C*-equivalent $A$ contains hermitian elements $h$ for which $||h||/|h|$ is arbitrarily large. Indeed, if a positive real constant $M$ were to exist such that $||h|| \leq M\cdot|h|$ for all $h \in H(A)$, then, given any $x \in A$, $x = h + ik$ with $h, k \in H(A)$, it would follow that $||x|| \leq ||h|| + ||k|| \leq M\cdot|h| + M|k| \leq M\cdot|x| + M\cdot|x| = 2M\cdot|x|$, which would imply that $||\cdot||$ and $|\cdot|$ are equivalent.

Recall that, given a hermitian $h$ in $A$, $C'(h)$ denotes the closed *-subalgebra of $A$ generated by $h$ and the identity $\mathbb{1}$. Also recall that the spectrum of $h$ in $A$ coincides with the spectrum of $h$ in $C'(h)$ and is denoted by $\sigma_A(h)$. In the following lemma we freely identify $C'(h)$ with the Banach algebra of continuous complex-valued functions on $\sigma_A(h)$. It is important to keep in mind that $C(h) = C'(h)$ when $h$ is invertible and, for a non-invertible $h$, that $C(h)$ consists of those functions in $C'(h)$ which vanish at 0.

(48.5) LEMMA. *Let* $h$ *be a hermitian element of* $A$ *such that* $|h| = 1$ *and* $||h|| > 4K + ||\mathbb{1}||$*. Then the spectrum of* $h$ *contains two points* $\xi_1$ *and* $\xi_2$ *with the following properties:*

*1°) If* $f \in C'(h)$ *and* $f(\xi_1) = f(\xi_2) = 0$*, then* $||f|| \leq K\cdot|f|$.

*2°) If* $f \in C'(h)$ *with* $0 \leq f \leq 1$, $f(\xi_1) = 1$, *and* $f(\xi_2) = 0$, *then* $||f|| > K|f|$.

*3°) If* $f \in C'(h)$ *with* $0 \leq f \leq 1$, $f(\xi_1) = 0$, *and* $f(\xi_2) = 1$, *then* $||f|| > K|f|$.

*Proof.* The hypotheses imply that the spectrum $\sigma_A(h)$ is a topological subspace of the closed interval $[-1,1]$. Hence, the restriction to $\sigma_A(h)$ of any continuous complex-valued function $g$ on $[-1,1]$ defines an element $\tilde{g}$ in $C'(h)$.

Let $N$ denote the natural numbers. By induction we shall construct two sequences $\{g_n\}_{n\in N}$ and $\{g_n'\}_{n\in N}$ of non-negative continuous functions on $[-1,1]$ such that the supports of $g_n$ and $g_n'$ are intervals of length less than $3/n$, such that

$$\mathrm{supp}(g_n) \subseteq \mathrm{supp}(g_{n-1})$$

$$\mathrm{supp}(g_n') \subseteq \mathrm{supp}(g_{n-1}'),$$

and such that $||\cdot||$ and $|\cdot|$ are not equivalent on the product $\tilde{g}_n \cdot A \cdot \tilde{g}_n'$.

To begin, define a function $\mathbb{1}_o$ on $[-1,1]$ by $\mathbb{1}_o(\xi) = 1$ for all $\xi \in [-1,1]$, so that $\mathbb{1}_o = \mathbb{1}$, the identity element of $A$. Then set $g_1 = g_1' = \mathbb{1}_o$.

Suppose that $g_1,\ldots,g_{n-1}$ and $g_1',\ldots,g_{n-1}'$ with the desired properties have been constructed. We choose, then, continuous non-negative functions $k_1,\ldots,k_n$ on $[-1,1]$ such that

$$k_1 + \cdots + k_n = \mathbb{1}_o$$

and such that the support of each $k_i$ is an interval of length less than $3/n$ (the $k_i$'s form a partition of the identity $\mathbb{1}_o$). By the inductive assumption $||.||$ and $|.|$ are not equivalent on $A_g \equiv \tilde{g}_{n-1} \cdot A \cdot \tilde{g}_{n-1}'$. Hence, given $r \in N$, there is an element $x_r \in A_g$ such that $|x_r| = 1$ and $||x_r|| > r$.

Since

$$x_r = \sum_{1\leq i,j\leq n} \tilde{k}_i x_r \tilde{k}_j,$$

the triangle inequality shows that for each $r \in N$ there exists $i_r$ and $j_r$, $1 \leq i_r, j_r \leq n$, such that $||\tilde{k}_{i_r} x_r \tilde{k}_{j_r}|| > r/n^2$. On the other

other hand, $|\tilde{k}_{i_r} x_r \tilde{k}_{j_r}| \leq |x_r| = 1$. Since there are only finitely many integers between 1 and n, but infinitely many $i_r$ and $j_r$, there exist $i_o$ and $j_o$ between 1 and n such that $i_o = i_r$ and $j_o = j_r$ for infinitely many $r \in N$. Hence $||\cdot||$ and $|\cdot|$ are not equivalent on $\tilde{k}_{i_o} \cdot A_g \cdot \tilde{k}_{j_o}$.

Let us set

$$g_n = k_{i_o} g_{n-1}, \ g_n' = k_{j_o} g_{n-1}'.$$

This completes the inductive construction of the sequences $\{g_n\}_{n \in N}$ and $\{g_n'\}_{n \in N}$.

Next, for each $n \in N$, let $S_n = \mathrm{supp}(g_n + g_n') \cap \sigma_A(h)$. By the construction of the nonnegative functions $g_n$ and $g_n'$, the sets $S_n$ form a decreasing sequence of nonempty, compact sets, and so the intersection $M = \bigcap_{n=1}^{\infty} S_n$, is nonempty and contains one or two points. (Since $S_n \subseteq \mathrm{supp}(g_n) \cup \mathrm{supp}(g_n')$, with each of these sets having diameter $\leq 3/n$, M has one point if $\mathrm{diam}(S_n) \to 0$, and two points otherwise.) The sets $S_n$ are nonempty because $\tilde{g}_n \neq 0$, $\tilde{g}_n' \neq 0$; indeed, for infinitely many $r \in N$ $||\tilde{g}_n a_r \tilde{g}_n'|| > r/n^2$, where $x_r = \tilde{g}_{n-1} a_r \tilde{g}_{n-1}'$ with $a_r \in A$.

Let

$$I_M = \{f \in C'(h) : f(\xi) = 0 \ \text{ if } \ \xi \in M\}.$$

We claim that if $y \in I_M$, then

$$||y|| \leq K|y|. \tag{*}$$

Indeed, let f be a continuous function on $\sigma_A(h)$ which vanishes on a neighborhood U of M. There is an integer n such that $S_n \subseteq U$ since the sequence $S_n$ is decreasing. Since f vanishes at any point where $g_n$ or $g_n'$ is strictly positive,

$$f \cdot \tilde{g}_n \cdot A \cdot \tilde{g}_n' = \tilde{g}_n \cdot A \cdot \tilde{g}_n' \cdot f$$

$$= f \cdot \tilde{g}_n' \cdot A \cdot \tilde{g}_n$$

$$= \tilde{g}_n' \cdot A \cdot \tilde{g}_n \cdot f = \{0\}.$$

Since f is normal, A is K-indecomposable, and $||\cdot||$ and $|\cdot|$ are not equivalent on $\tilde{g}_n \cdot A \cdot \tilde{g}'_n$; then $||f|| \le K|f|$ by (48.4). Since functions like $f$ are dense in the equivalent norms $||\cdot||$ and $|\cdot|$ on $I_M$, the inequality $||y|| \le K|y|$ holds for all $y$ in $I_M$.

Assume, now, that $M$ contains only one point $\xi$. Then $h = \lambda\mathbb{1} + g$, where $\lambda \in R$, $|\lambda| \le 1$ and $g \in I_M$, $|g| \le 2$. However, this is impossible, since by (*) we have

$$||h|| \le ||\mathbb{1}|| + K|g| \le ||\mathbb{1}|| + 2K,$$

contradicting the hypothesis of the lemma.

It follows therefore that $M$ contains two points, say $\xi_1$ and $\xi_2$, and then (*) establishes (1°). To prove (2°), choose a function $f \in C'(h)$ satisfying the conditions $0 \le f \le 1$, $f(\xi_1) = 1$, and $f(\xi_2) = 0$. Then, letting $g$ be defined by

$$h = h(\xi_1)f + h(\xi_2)(\mathbb{1} - f) + g,$$

we have $g \in I_M$, and since $|h(\xi_1)| \le 1$, $|h(\xi_2)| \le 1$, it follows that $|h(\xi_1)f + h(\xi_2)(\mathbb{1} - f)| \le 1$, whence $|g| \le 2$. Therefore

$$\begin{aligned} ||h|| &\le 2||f|| + ||\mathbb{1}|| + ||g|| \\ &\le 2||f|| + ||\mathbb{1}|| + 2K, \text{ by } (*) \end{aligned}$$

and so

$$||f|| \ge \frac{1}{2}(||h|| - ||\mathbb{1}|| - 2K) > K.$$

Finally, condition (3°) follows by symmetry. □

The proof of (48.5) shows that, under the hypothesis of the lemma, we can choose $k_1$ and $k_2$ in $C'(h)$ satisfying the conditions $0 \le k_1, k_2 \le 1$, $k_1(\xi_1) = k_2(\xi_2) = 1$, $k_1 \cdot k_2 = 0$, and $||\cdot||$ and $|\cdot|$ are not equivalent on the product $k_1 \cdot A \cdot k_2$. To get such functions $k_1, k_2$ we first choose $n$ so large that

$$\text{supp}(\tilde{g}_n) \cap \text{supp}(\tilde{g}'_n) = \emptyset.$$

Then we choose functions $k_1, k_2 \in C'(h)$ with $0 \le k_1, k_2 \le 1$, $k_1 \cdot k_2 = 0$ and $k_1 = 1$ on $\text{supp}(\tilde{g}_n)$, $k_2 = 1$ on $\text{supp}(\tilde{g}'_n)$. It follows that

$k_1\tilde{g}_n = \tilde{g}_n$, $\tilde{g}'_n = \tilde{g}'_n k_2$ and so $k_1 \cdot A \cdot k_2 \supseteq k_1\tilde{g}_n \cdot A \cdot \tilde{g}'_n k_2 = \tilde{g}_n \cdot A \cdot \tilde{g}'_n$. Therefore $||\cdot||$ and $|\cdot|$ not equivalent on $\tilde{g}_n \cdot A \cdot \tilde{g}'_n$ implies $||\cdot||$ and $|\cdot|$ are not equivalent on $k_1 \cdot A \cdot k_2$.

(48.6) NOTATION: With $A$, $k_1$ and $k_2$ as just described, we denote by $E$ the $||\cdot||$-closure of the set $k_1 \cdot A \cdot k_2$ (possibly smaller than the $|\cdot|$-closure). This set $E$ will remain fixed until the end of the proof of (48.1).

(48.7) LEMMA. *The set* $E$ *has the following properties:*

*(i)* $||\cdot||$ *and* $|\cdot|$ *are not equivalent on* $E$*;*

*(ii)* $||\cdot||$ *and* $|\cdot|$ *are equivalent on* $E^* \cdot E$ *and* $E \cdot E^*$*;*

*(iii)* $E \cdot E = \{0\}$*;*

*(iv) If* $u \in E$*, then* $u \cdot C(u^*u) \subseteq E$.

*Proof.* (i) and (iii) are clearly true by the construction of $E$. To prove (iv) note that if $u \in k_1 \cdot A \cdot k_2$, then $u \cdot (u^*u) \in k_1 \cdot A \cdot k_2$. Hence, if $u \in k_1 \cdot A \cdot k_2$, $u \cdot p(u^*u) \in k_1 \cdot A \cdot k_2$, where $p$ is a polynomial over $\mathbb{C}$ such that $p(0) = 0$. By continuity of multiplication and denseness of $k_1 \cdot A \cdot k_2$ in $E$ we get $u \cdot p(u^*u) \in E$ for every $u \in E$ and every such polynomial $p$. Since the expressions $p(u^*u)$ are dense in $C(u^*u)$, it follows that $u \cdot C(u^*u) \subseteq E$.

For the proof of (ii) observe that $||k_1|| > K|k_1|$ and $||k_2|| > K|k_2|$ by (48.5), 3° and 2°. Moreover, by $k_1k_2 = 0$ and definition of $E$,

$$k_1 \cdot E^* \cdot E = E^* \cdot E \cdot k_1 = \{0\}$$

and

$$k_2 \cdot E \cdot E^* = E \cdot E^* \cdot k_2 = \{0\}.$$

The assertion in (ii) now follows from (48.4), (a). □

(48.8) LEMMA. *The norms* $||\cdot||$ *and* $|\cdot|$ *are equivalent on* $u \cdot C'(u^*u)$ *whenever* $u \in A$ *and* $u^2 = 0$.

*Proof.* Suppose, to the contrary, that $||\cdot||$ and $|\cdot|$ are not equivalent on $u \cdot C'(u^*u)$. There is no loss of generality in assuming furthermore that $|u| = 1$. We represent, then, $C'(u^*u)$ as the algebra of continuous functions on $\sigma_A(u^*u) \subseteq [0,1]$.

Given $\alpha \in (0,1)$, consider the following two linear spaces:

$$I_\alpha = \{uf\colon f \in C'(u^*u) \text{ and } \operatorname{supp}(f) \subseteq [\alpha,1]\}$$

$$J_\alpha = \{uf\colon f \in C'(u^*u) \text{ and } \operatorname{supp}(f) \subseteq [0,\alpha]\}.$$

Observe that $(I_\alpha, |\cdot|)$, as a normed linear space, is isometrically isomorphic to the space

$$\hat{I}_\alpha = \{g \in C'(u^*u)\colon \operatorname{supp}(g) \subseteq [\alpha,1]\}, \text{ with norm } |\cdot|,$$

via the isomorphism $\Phi\colon uf \to (u^*u)^{1/2}f$. The mapping $\Phi$ is isometric because

$$|(u^*u)^{1/2}f|^2 = |[(u^*u)^{1/2}f]^*[(u^*u)^{1/2}f]| = |fu^*uf|$$

$$= |(uf)^*(uf)| = |uf|^2,$$

and it is surjective because the function $(u^*u)^{1/2}$ on $\sigma_A(u^*u)$ vanishes only at 0.

Since $\hat{I}_\alpha$ is $|\cdot|$-complete, $I_\alpha$ is too. Therefore, $I_\alpha$ is $||\cdot||$-closed and hence is $||\cdot||$-complete. The open mapping theorem then shows that the two norms are equivalent on $I_\alpha$.

Consider real numbers $\alpha$ and $\beta$ such that $0 < \alpha < \beta < 1$. Then each $x \in u\cdot C'(u^*u)$ can be written as $x = x_1 + x_2$, where $x_1 \in I_\alpha$, $x_2 \in J_\beta$ and $|x_1| \leq |x|$, $|x_2| \leq |x|$. Hence, if $||\cdot||$ and $|\cdot|$ were equivalent on $J_\beta$, then they would be equivalent on $u\cdot C'(u^*u)$ in contradiction to the assumption with which the proof began. Consequently, $||\cdot||$ and $|\cdot|$ are not equivalent on $J_\beta$ whenever $0 < \beta < 1$.

Now, if $0 < \beta < \alpha < 1$ and $uf \in I_\alpha$, $ug \in J_\beta$, we have $(uf)(ug)^* = 0$ and

$$(ug)^*(uf) = g^*u^*uf = u^*ug^*f = 0,$$

since $gf = 0$. On the other hand, $u^2 = 0$ implies that $up(u^*u)u = 0$ for all polynomials $p$. Hence $u\cdot C'(u^*u)\cdot u = 0$ and so $(ufu)g = (ugu)f = 0$, i.e., $(uf)(ug) = (ug)(uf) = 0$. Thus (48.4) can be applied, with $X = J_\beta$, and it shows that $||uf|| \leq 2K\cdot|uf|$ for all $uf \in I_\alpha$.

We next note that the linear space $I = \bigcup_{\alpha>0} I_\alpha$ is $|\cdot|$-dense in $u\cdot C'(u^*u)$. Indeed, simply observe that $\Phi(I)$ is $|\cdot|$-dense in $\Phi(u\cdot C'(u^*u))$ and $\Phi$ is isometric.

Let $x_n \in I$ converge in $|\cdot|$ to $x \in u\cdot C'(u^*u)$. Then $\{x_n\}$ is a $||\cdot||$-Cauchy sequence since

$$||x_n - x_m|| \leq 2K\cdot|x_n - x_m|.$$

Therefore, $\{x_n\}$ converges in $||\cdot||$ to an element of $A$ which must be $x$. We conclude that

$$||x|| = \lim_{n\to\infty}||x_n|| \leq \lim_{n\to\infty} 2K|x_n| = 2K|x|.$$

This holds for every $x \in u\cdot C'(u^*u)$. It therefore shows that $||\cdot||$ and $|\cdot|$ are equivalent in $u\cdot C'(u^*u)$, in contradiction to the assumption made at the beginning of the proof. Hence this assumption is untenable and the lemma is proved. □

Two additional technical lemmas are required before completing the proof of (48.1). Before presenting these we point out that $C(u^*u)$, as a subalgebra of $A$, coincides with $C((u^*u)^{1/2})$ so that if $v \in u\cdot C(u^*u)$, $v^*v$ can be represented as a function on $\sigma_A((u^*u)^{1/2})$.

(48.9) LEMMA. *Let* $N > 0$ *and let* $u \in E$ *satisfy* $|u| = 1$ *and* $||u|| > N$. *Then there exists* $\tilde{u} \in u\cdot C(u^*u)$ *such that* $|\tilde{u}| = 1$ *and* $||\tilde{u}|| > N/2$, *and such that* $(\tilde{u}^*\tilde{u})^{1/2}$ *as a function on* $\sigma_A((u^*u)^{1/2})$ *vanishes on a neighborhood of* 0.

*Proof*. Recall that $C(u^*u)$ consists of those functions in $C'(u^*u)$ which vanish at 0 since by (48.7), (iii), $u^2 = 0$, so $u$ is not invertible. It follows from (48.8) that there exists $M > 0$ such that $||y|| \leq M\cdot|y|$ for all $y \in u\cdot C(u^*u)$. Let $f$, $0 \leq f \leq 1$, be a continuous function on $\sigma_A((u^*u)^{1/2}) \subseteq [0,1]$ such that $\mathrm{supp}(f) \subseteq [0,\varepsilon]$ and $f$ is identically 1 on $[0,\frac{\varepsilon}{2}] \cap \sigma_A((u^*u)^{1/2})$, where $\varepsilon > 0$. Since $1 - f$ vanishes at 0, it lies in $C((u^*u)^{1/2}) = C(u^*u)$. If $\varepsilon < \min\{1,N/2M\}$, then $|u(\mathbb{1} - f)| = |(u^*u)^{1/2}(\mathbb{1} - f)| = 1$, since $1 = |u|^2 = |u^*u| = |((u^*u)^{1/2})^2| = |(u^*u)^{1/2}|^2$ which implies that $1 \in \sigma_A((u^*u)^{1/2})$. Also,

$$||u - u(\mathbb{1} - f)|| = ||uf|| \leq M\cdot|uf| \leq M\cdot\varepsilon < N/2.$$

The fact that $|uf| = |(u^*u)^{1/2}f| \leq \varepsilon$ follows since $(u^*u)^{1/2}$ is the identity function in $C(\sigma_A(u^*u)^{1/2})$ and this function is less than or equal to $\varepsilon$ on supp(f).

Therefore, $\tilde{u} = u(\mathbb{1} - f)$ is an element with the desired properties. □

(48.10) LEMMA. *Assume that* u, N *and* u *are as in* (48.9), *and let* $N \geq 4K$. *Let* $g$, $0 \leq g \leq 1$, *be a continuous function on* $\sigma_A((u^*u)^{1/2})$ *such that* $g \equiv 1$ *on a neighborhood of* 0 *and* $(\tilde{u}^*u)^{1/2}g = 0$. *Then:*

*(i) There exists* $z \in u\cdot C((u^*u)^{1/2})$ *such that* $g = \mathbb{1} - z^*z$.

*(ii) Writing* $g' = \mathbb{1} - zz^*$, *we have* $\tilde{u}g = 0$, $g'\tilde{u} = 0$, $g\tilde{u} = \tilde{u}$, *and* $\tilde{u}g' = \tilde{u}$.

*(iii) The norms* $||\cdot||$ *and* $|\cdot|$ *are equivalent on* $g'\cdot E\cdot g$.

*Proof*. (i) The function

$$f(\xi) = \left[\frac{(1 - g)(\xi)}{(u^*u)(\xi)}\right]^{1/2}$$

is continuous on $\sigma_A((u^*u)^{1/2})$ since $(u^*u)^{1/2}$ is bounded away from 0 on the support of $1 - g$. Then $z = uf$ has the required properties.

(ii) By definition of $g$, we have $0 = |(\tilde{u}^*\tilde{u})^{1/2}g|^2 = |g\tilde{u}^*\tilde{u}g| = |(\tilde{u}g)^*(\tilde{u}g)| = |\tilde{u}g|^2$, so $\tilde{u}g = 0$. Since $\tilde{u}$, $z \in E$ (see (48.7)), it follows that $z\tilde{u} = \tilde{u}z = 0$ and hence

$$g\tilde{u} = (\mathbb{1} - z^*z)\tilde{u} = \tilde{u}$$

and

$$\tilde{u}g' = \tilde{u}(\mathbb{1} - zz^*) = \tilde{u}.$$

To obtain $g'\tilde{u} = 0$, let us write $\tilde{u} = u\tilde{f}$, where $\tilde{f} \in C((u^*u)^{1/2}) = C(u^*u)$ and $z = uf$, where $f$ is as in the proof of (i). We first observe that

$$\tilde{f}(\mathbb{1} - g) = \tilde{f}.$$

To see this, note that since $(\tilde{u}^*\tilde{u})^{1/2}g = 0$, $(\tilde{u}^*\tilde{u})^{1/2} = (\tilde{f}^*\cdot(u^*u)\cdot\tilde{f})^{1/2}$ must vanish at all points $\xi$ where $g(\xi) > 0$. Now $\tilde{f}(0) = 0$ since $\tilde{f} \in C((u^*u)^{1/2}) = C(u^*u)$, and $(u^*u)(\xi) = \xi^2 > 0$ for $0 < \xi \leq 1$. Thus $\tilde{f}(\xi) = 0$ whenever $g(\xi) > 0$ and we have $\tilde{f}g = 0$ or $\tilde{f}(\mathbb{1} - g) = \tilde{f}$.

Then

$$g'\tilde{u} = (1 - uffu^*)\tilde{u} = \tilde{u} - uf^2(u^*u)\tilde{f}$$

$$= \tilde{u} - u(1 - g)\tilde{f} = \tilde{u} - u\tilde{f} = 0.$$

To prove (iii) note that since $\tilde{u} \in E$ and $E \cdot E = \{0\}$, we have, by (ii), that

$$\tilde{u} \cdot g' \cdot E \cdot g = g' \cdot E \cdot g \cdot \tilde{u} = \tilde{u} \cdot g \cdot E^* \cdot g'$$

$$= g \cdot E^* \cdot g' \cdot u = \{0\}.$$

On the other hand,

$$||\tilde{u}|| > (N/2)|\tilde{u}| \geq 2K|\tilde{u}|.$$

(iii) is now an immediate consequence of (48.4). □

COMPLETION OF THE PROOF OF (48.1)

The desired contradiction is now near at hand. We take and fix elements $u$, $g$, $g'$ and $z$ with the properties described in Lemmas (48.9) and (48.10). We have shown that the norms $||\cdot||$ and $|\cdot|$ are equivalent on the set

$$B = E^* \cdot E \cup E \cdot E^* \cup g' \cdot E \cdot g \cup C'(u^*u).$$

This implies that there exists a real constant $\gamma > 1$ such that $||y|| \leq \gamma|y|$ for all $y \in B$.

The identity

$$x = x(1 - g) + (1 - g')xg + g'xg$$

$$= xz^*z + zz^*xg + g'xg$$

gives the following estimate for $x \in E$:

$$||x|| \leq ||xz^*|| \cdot ||z|| + ||z|| \cdot ||z^*x|| \cdot ||g|| + ||g'xg||$$

$$\leq \gamma||z|| \cdot |x| + \gamma^2||z|| \cdot |x| + \gamma|x|$$

$$< \gamma^2(2||z|| + 1)|x|.$$

The second inequality holds, since $xz^* \in E \cdot E^*$, $z^*x \in E^* \cdot E$, $g \in C'(u^*u)$,

$g'xg \in g'\cdot E\cdot g$, and since, moreover, $|xz^*| \leq |x|$, $|z^*x| \leq |x|$, $|g| \leq 1$, and $|g'xg| \leq |x|$. The inequality

$$||x|| \leq \gamma^2(2||z|| + 1)\cdot|x| \quad (x \in E)$$

means that $||\cdot||$ and $|\cdot|$ are equivalent on $E$. This, of course, is not the case [48.7 (i)].

This final grand contradiction was generated by the italicized assumption made after Lemma (48.3). It follows that that assumption is untenable and, as noted there, this conclusion establishes (48.1). □

## EXERCISES

(IX.1) Show that any closed *-subalgebra of a locally C*-equivalent algebra is locally C*-equivalent.

(IX.2) Prove that the set $H_n$ in the proof of Proposition (47.5) is closed and that $H(A) = \bigcup_{n=1}^{\infty} H_n$.

(IX.3) Prove that the mapping $\phi: C(t)/(I \cap C(t)) \to B/I$ defined by $\phi(a + I \cap C(t)) = a + I$ for $a \in C(t)$ in the proof of Proposition (47.6) is a *-isomorphism.

(IX.4) Let $A$ be a Banach *-algebra. Suppose that for every $h \in H(A)$ there exists a real constant $M_h > 0$ such that $||x^*||\cdot||x|| \leq M_h||x^*x||$ for all $x \in C(h)$. Prove that $A$ is C*-equivalent.

(IX.5) Let $A$ be a Banach *-algebra with hermitian involution. Suppose that for every $h \in H(A)$ there exists a real constant $M_h > 0$ such that $||k|| \leq M_h\nu(k)$ for all hermitian $k \in C(h)$. Prove that $A$ is C*-equivalent.

# 10

# Applications of the Characterization Theorems

## §49. *Introduction.*

Numerous applications of the Gelfand-Naimark theorems appear in the literature. In this chapter we discuss a few of these. Because of space limitations we do not give full details in every case; in fact, we often indicate only the general direction the application takes. To compensate for this we have given references where the reader can find a more detailed account of the material.

## §50. *Compactifications in topology.*

Consider a topological space X. A family F of complex-valued functions on X is said to be *self-adjoint* if whenever f is in F, then the complex-conjugate $\overline{f}$ is in F. The family F is *separating* if whenever $x_1$ and $x_2$ are distinct points in X, there is a function $f \in F$ such that $f(x_1) \neq f(x_2)$. Let $C_b(X)$ denote the $C^*$-algebra of bounded continuous complex-valued functions on X under the sup-norm $||\cdot||_\infty$ and involution $f^* = \overline{f}$.

A *compactification* of the space X is a compact Hausdorff space Y together with a continuous one-to-one map $\tau$ of X onto a dense subset $\tau(X)$ of Y. It is clear that every compactification Y of X determines a closed separating self-adjoint subalgebra of $C_b(X)$ which contains the constants, namely the functions in C(Y) followed by $\tau$.

On the other hand, let A be the algebra $C_b(X)$. Then A is a commutative $C^*$-algebra with identity. By the Gelfand-Naimark theorem (7.1), A is isometrically $^*$-isomorphic to $C(\hat{A})$. We know that the structure space $\hat{A}$ of A is a compact Hausdorff space. Each point x in X determines the evaluation homomorphism $\tau(x)$ at $x$, defined by

$$\hat{f}(\tau(x)) = f(x), \quad f \in A.$$

The function $\tau$ from $X$ into $\hat{A}$ is continuous by definition of the Gelfand topology on $\hat{A}$. Since $A$ separates the points of $X$, $\tau$ is one-to-one. Furthermore, if $g \in C(\hat{A})$ and $g$ is zero on $\tau(X)$, then $g$ is the Gelfand transform of the function identically zero on $X$, so $g = 0$. This shows that $\tau(X)$ is dense in $\hat{A}$. Thus $\hat{A}$ is a compactifaction of $X$.

In general, with $A = C_b(X)$, the injection $\tau$ of $X$ into $\hat{A}$ is not a homeomorphism. It is easy to prove that $\tau$ is a homeomorphism iff the space $X$ is *completely regular*, i.e., iff for every closed subset $T$ of $X$ and $x \in X \setminus T$ there is a continuous complex-valued function $f$ in $A$ such that $f(x) = 0$ and $f|T = 1$. Indeed, since $A$ is isometrically *-isomorphic to $C(\hat{A})$, the space $X$ is completely regular iff for all closed subsets $T$ of $X$ and $x \in X \setminus T$, the point $\tau(x)$ does not belong to the closure $\overline{\tau(T)}$ of $\tau(T)$ in $\hat{A}$. This occurs iff for each closed subset $T$ of $X$, $\tau(X \setminus T)$ is the intersection of $\tau(X)$ and the open subset $\hat{A} \setminus \overline{\tau(T)}$ in $\hat{A}$, or equivalently, iff $\tau^{-1}$ is continuous.

Summarizing the discussion above we obtain the following theorem.

(50.1) THEOREM. *(Stone-Čech Compactification). Let* $X$ *be a completely regular Hausdorff space. Then there is a compact Hausdorff space* $Y$ *such that* $X$ *is homeomorphic to a dense subset of* $Y$*, and every bounded continuous complex-valued function on* $X$ *extends continuously to* $Y$.

For additional information on compactifications, see Dunford and Schwartz [2] and Larsen [1].

## §51. *Almost periodic functions and harmonic analysis.*

Let $G$ be a locally compact abelian group. We assume that the topology on $G$ is Hausdorff and that the group operation is written additively. If $f \in C_b(G)$, then $f$ is said to be *almost periodic* if the set $\{T_s(f): s \in G\}$ has compact closure in $C_b(G)$, where $T_s(f)(t) = f(t - s)$ for $t \in G$. Equivalently, $f$ is almost periodic if $\{T_s(f): s \in G\}$ is a totally bounded subset of $C_b(G)$; that is, given $\varepsilon > 0$, there exist

$s_1,\ldots,s_n$ in G such that for each $s \in G$ we have

$$||T_s(f) - T_{s_k}(f)||_\infty < \varepsilon \quad \text{for some } k = 1,2,\ldots,n.$$

It follows easily that the set AP(G) of almost periodic functions on G is a closed $^*$-subalgebra of $C_b(G)$ with identity, where the involution is the obvious one of complex conjugation. Hence AP(G) is a commutative C*-algebra with identity, and by (7.1) AP(G) is isometrically $^*$-isomorphic to $C(AP(G)^\wedge)$.

Now, in the usual manner, one sees that each point $t \in G$ defines a complex homomorphism of AP(G), namely, the complex homomorphism $\tau_t$ defined by

$$\tau_t(f) = f(t), \quad f \in AP(G).$$

Clearly, the mapping $\tau(t) = \tau_t$, $t \in G$, maps G into $AP(G)^\wedge$. It was shown in §50 that $\tau$ is continuous and one-to-one. It is not difficult to show that $\tau(G)$ is dense in $AP(G)^\wedge$. Hence $AP(G)^\wedge$ is a compactification of the locally compact abelian group G. We remark that the map $\tau\colon G \to AP(G)^\wedge$ is, in general, not a homeomorphism unless G is compact.

Actually much more can be said about $AP(G)^\wedge$. Indeed, one can extend the group structure of G to all of $AP(G)^\wedge$ in such a way that $AP(G)^\wedge$ becomes a compact abelian group. In fact, the following theorem can be proved (see Hewitt and Ross [1, p. 430]; Rudin [1, p. 30]; Larsen [1, p. 328]).

(51.1) THEOREM. *(Bohr Compactification). Let* G *be a locally compact abelian group. Then* $AP(G)^\wedge$ *is a compactification of* G, *and the group structure of* G *can be extended to* $AP(G)^\wedge$ *in such a way that* $AP(G)^\wedge$ *becomes a compact abelian group. Moreover, suppose* H *is any compact abelian group and* $\phi\colon G \to H$ *is a continuous isomorphism such that:*

*(i)* H *is a compactification of* G;

*(ii) the map* $\phi^*$, *defined by*

$$\phi^*(h)(t) = h(\phi(t)), \quad t \in G,\ h \in C(H),$$

*is an algebra isomorphism of* C(H) *onto* AP(G). *Then* H *is topologically isomorphic to* $AP(G)^\wedge$.

Utilizing the Bohr compactification of a locally compact abelian group G, one can show that AP(G) is precisely the closure in C(G) of the algebra of trigonometric polynomials on G, that is, of the algebra of finite linear combinations of the continuous characters on G. This result has far-reaching implications in abstract harmonic analysis.

Other important theorems in harmonic analysis can also be established by utilizing the Gelfand-Naimark Theorem for commutative C*-algebras. For example, each of the following is proved in Larsen [1] by utilizing (7.1):

(51.2) THEOREM. *(Plancherel theorem). Let* $G$ *be a locally compact abelian group and* $\lambda$ *a given Haar measure on* $G$. *Then there exists a Haar measure* $\mu$ *on the character group* $\hat{G}$ *and a linear subspace* $V_0$ *of* $L^2(G)$, *the square-integrable functions on* $G$ *relative to* $\lambda$, *such that:*

*(i)* $V_0 \subseteq L^1(G) \cap L^2(G)$;

*(ii)* $V_0$ *is norm dense in* $L^2(G)$;

*(iii)* $\hat{V}_0$ *is norm dense in* $L^2(\hat{G})$;

*(iv)* $||f||_2 = ||\hat{f}||_2$ *for* $f \in V_0$;

*(v) the map* $f \to \hat{f}$, $f \in V_0$, *from* $V_0$ *to* $\hat{V}_0$ *can be uniquely extended to a linear isometry of* $L^2(G)$ *onto* $L^2(\hat{G})$.

(51.3) THEOREM. *(Pontryagin duality theorem). Let* $G$ *be a locally compact abelian group with character group* $\hat{G}$. *Let* $(\hat{G})\hat{}$ *denote the character group of* $\hat{G}$. *If* $\alpha: G \to (\hat{G})\hat{}$ *is defined, for each* $t \in G$, *by*

$$\alpha(t)(\gamma) = \gamma(t), \quad \gamma \in \hat{G},$$

*then* $\alpha$ *is a topological isomorphism of* $G$ *onto* $(\hat{G})\hat{}$. *Briefly, one writes* $G \simeq \hat{\hat{G}}$.

These applications and others in harmonic analysis are truly among the most interesting. However, even the most rudimentary discussion would require more space than we have available. For more details we refer the reader to Dunford-Schwartz [2], Larsen [1], and Naimark [1].

## §52. *The spectral theorem for a bounded normal operator.*

A bounded linear operator $T$ on a complex Hilbert space $H$ is *normal* if $TT^* = T^*T$. If $H$ is finite-dimensional and $T$ is normal, the spectral theorem states that there exist complex scalars $\lambda_1,\lambda_2,\ldots,\lambda_k$, the distinct eigenvalues of $T$, and nonzero orthogonal projections $P_1,P_2,\ldots,P_k$ such that:

(1) $P_i$ is the orthogonal projection on $\text{null}(T - \lambda_i I)$, $i = 1,2,\ldots,k$, where $I$ is the identity operator.

(2) $P_i P_j = 0$ if $i \neq j$.

(3) $I = \Sigma_{j=1}^{k} P_j$.

(4) $T = \Sigma_{j=1}^{k} \lambda_j P_j$.

Furthermore, the decomposition in part (4) is unique, in the following sense. If $\lambda_1,\lambda_2,\ldots,\lambda_k$ are distinct complex numbers, and $P_1,P_2,\ldots,P_k$ are nonzero linear operators on $H$, such that (2), (3), and (4) are satisfied, then $\lambda_1,\lambda_2,\ldots,\lambda_k$ are precisely the distinct eigenvalues of $T$, and for each $j$, $P_j$ is the orthogonal projection of $H$ onto $\text{null}(T - \lambda_j I)$.

Our objective in this section is to use the Gelfand-Naimark theorem (7.1) to prove the spectral theorem for bounded normal operators on a Hilbert space which may be infinite-dimensional. As might be expected, the extension of the spectral theorem to this case requires some care and a bit of work. We begin by collecting a few facts about spectral measures. All topological spaces will be assumed Hausdorff and all Hilbert spaces to lie over the complex numbers.

Let $X$ be a locally compact space, $\mathcal{B}$ the $\sigma$-algebra of Borel sets in $X$, and $H$ a Hilbert space. A *spectral measure* (or *resolution of the identity*) on $\mathcal{B}$ is a mapping $P\colon \mathcal{B} \to B(H)$ such that:

(i) $P(\emptyset) = 0$ and $P(X) = I$;

(ii) Each $P(E)$ is a self-adjoint projection;

(iii) $P(E_1 \cap E_2) = P(E_1)P(E_2)$;

(iv) $P(E_1 \cup E_2) = P(E_1) + P(E_2)$ if $E_1 \cap E_2 = \emptyset$;

(v) For $\xi, \eta \in H$, the mapping $P_{\xi,\eta}\colon \mathcal{B} \to C$ defined by $P_{\xi,\eta}(E) = (P(E)\xi|\eta)$ is a regular complex Borel measure on $\mathcal{B}$.

Some immediate consequences of the definition are the following: since each $P(E)$ is a self-adjoint projection,

$$P_{\xi,\xi}(E) = (P(E)\xi|\xi) = ||P(E)\xi||^2, \quad \xi \in H,$$

so that $P_{\xi,\xi}$ is a nonnegative measure on $B$ with total variation $||P_{\xi,\xi}|| = P_{\xi,\xi}(X) = ||\xi||^2$. Clearly, (iii) implies that any two projections $P(E)$ commute. Also, (i), (iii) and (A.16) show that the ranges of $P(E_1)$ and $P(E_2)$ are orthogonal whenever $E_1 \cap E_2 = \emptyset$. Although P is finitely additive, it is not, in general, countably additive. Even so, we do have the following:

(52.1) PROPOSITION. *Let* X *be a locally compact space,* H *a Hilbert space, and* P *a spectral measure on* $B$.

*(i) If* $\xi \in H$, *then the map* $\mu_\xi: B \to H$ *defined by* $\mu_\xi(E) = P(E)\xi$ *is a countably additive* H-*valued measure.*

*(ii) If* $E_n \in B$, $E = \bigcup_{n=1}^{\infty} E_n$ *and* $P(E_n) = 0$ *for each* n, *then* $P(E) = 0$.

*Proof.* (i) Suppose E is a disjoint union of sets $E_n$ in $B$. Since $P(E_n)P(E_m) = 0$ when $n \neq m$, the vectors $P(E_n)\xi$ and $P(E_m)\xi$ are orthogonal by (A.16). By property (v) of spectral measures,

$$\Sigma_{n=1}^{\infty}(P(E_n)\xi|\eta) = (P(E)\xi|\eta), \qquad \eta \in H.$$

Hence, by (A.12), $\Sigma_{n=1}^{\infty} P(E_n)\xi = P(E)\xi$, where the convergence is in the norm topology of H; thus, $\mu_\xi$ is countably additive. The other properties of a measure are clear from corresponding properties of P.

(ii) Since $P(E_n) = 0$ for each n, then $P_{\xi,\xi}(E_n) = 0$ for each $\xi$ in H. Therefore, since $P_{\xi,\xi}$ is countably additive, $P_{\xi,\xi}(E) = 0$. Since $||P(E)\xi||^2 = P_{\xi,\xi}(E)$, we obtain $P(E) = 0$. □

We turn our attention now to a commutative Banach *-algebra which ultimately will provide us with the projections needed to approximate a given bounded normal operator by linear combinations of orthogonal projections. The basic construction was described in (B.2), Example (4), but we wish to look at it here in terms of a given spectral measure.

Let X be a locally compact space, H a Hilbert space, and P a spectral measure on the Borel subsets $B$ of X. Suppose f is a complex-valued Borel measurable function on X.

The *essential range* of f, denoted ess-range(f), is defined to be the smallest closed subset of the complex plane that contains f(x) for almost all $x \in X$, i.e., for all $x \in X$ except those that lie in some set $E \in B$

with $P(E) = 0$. To see that the essential range exists, recall that the usual topology of the plane has a basis consisting of a countable family $\{W_n\}$ of open disks. Letting $W$ be the union of those $W_n$ such that $P(f^{-1}(W_n)) = 0$ we see, from (52.1), (ii), that $P(f^{-1}(W)) = 0$. Furthermore, $W$ is open and is clearly the largest open subset of the plane with this property. The essential range of $f$ is now simply the complement of $W$.

If the essential range is bounded (hence compact), $f$ is said to be *essentially bounded*, and the *essential supremum* $||f||_\infty$ of $f$ is defined by

$$||f||_\infty = \sup\{|\lambda|: \lambda \in \text{ess-range}(f)\}.$$

It is straightforward to check that the algebra $D$ of all *bounded* complex-valued Borel measurable functions on $X$, under pointwise operations, involution $f^* = \overline{f}$ (complex conjugate), and supremum norm, is a commutative Banach $^*$-algebra. Also, the set

$$N = \{f \in D: ||f||_\infty = 0\}$$

is an ideal of $D$ which is *closed*, by (52.1), (ii). Therefore, $D/N$ is a Banach $^*$-algebra which, as in (B.2), will be denoted by $L^\infty(X,\mathcal{B},P)$ or simply by $L^\infty(P)$ if no confusion is possible. Clearly the quotient norm of $f + N$ in $L^\infty(P)$ coincides with $||f||_\infty$ and the spectrum of $f + N$ in $L^\infty(P)$ is the essential range of $f$. Moreover there is a function $\tilde{f} \in f + N$ such that $\tilde{f}(X)$ is the essential range of $f$. As usual, we shall identify functions in $L^\infty(P)$ which are equal almost everywhere with respect to $P$.

(52.2) THEOREM. *Let* $X$ *be a locally compact space,* $H$ *a Hilbert space, and* $P$ *a spectral measure on* $\mathcal{B}$. *Then the formula*

$$(\theta(f)\xi|\eta) = \int_X f\, dP_{\xi,\eta}, \quad \xi, \eta \in H, \tag{1}$$

*defines an isometric $^*$-isomorphism* $\theta$ *of* $L^\infty(P)$ *onto a closed commutative $^*$-subalgebra of* $B(H)$. *Further,*

$$||\theta(f)\xi||^2 = \int_X |f|^2 dP_{\xi,\xi}, \quad \xi \in H,\ f \in L^\infty(P), \tag{2}$$

*and an operator* S *in* B(H) *commutes with each* P(E) *iff* S *commutes with each* θ(f).

*Proof.* The idea of the proof is to first establish the stated results for simple functions and then, approximating arbitrary functions in $L^\infty(P)$ by simple measurable functions, extend to all of $L^\infty(P)$. To this end, let $\{E_1,\ldots,E_n\}$ be a partition of X, with $E_i \in \mathcal{B}$, and suppose s is a simple function such that $s = \alpha_i$ on $E_i$. Define an operator θ(s) on H by

$$\theta(s) = \sum_{i=1}^{n} \alpha_i P(E_i). \tag{3}$$

It is easy to verify that θ is well defined. If $\{E_1',\ldots,E_m'\}$ is another partition of X, with $E_i' \in \mathcal{B}$ and t is a simple function such that $t = \beta_i$ on $E_i'$, then

$$\theta(s)\theta(t) = \sum_{i=1}^{n} \sum_{j=1}^{m} \alpha_i\beta_j P(E_i)P(E_j')$$

$$= \sum_{i=1}^{n} \sum_{j=1}^{m} \alpha_i\beta_j P(E_i \cap E_j').$$

Since st is the simple function that equals $\alpha_i\beta_j$ on $E_i \cap E_j'$, we see that

$$\theta(s)\theta(t) = \theta(st). \tag{4}$$

A similar argument establishes that θ is linear on simple functions. Since each $P(E_i)$ is self-adjoint,

$$\theta(s)^* = \sum_{i=1}^{n} \overline{\alpha}_i P(E_i) = \theta(\overline{s}), \tag{5}$$

so that θ is a *-homomorphism on simple functions.

Given ξ, η ∈ H we have, by (3), that

$$(\theta(s)\xi|\eta) = \sum_{i=1}^{n} \alpha_i (P(E_i)\xi|\eta)$$

$$= \sum_{i=1}^{n} \alpha_i P_{\xi,\eta}(E_i) \tag{6}$$

$$= \int_X s\, dP_{\xi,\eta},$$

and from (4) and (5) that

$$\theta(s)^*\theta(s) = \theta(\bar{s})\theta(s) = \theta(\bar{s}s) = \theta(|s|^2).$$

Therefore, from (6), we obtain

$$||\theta(s)\xi||^2 = (\theta(s)^*\theta(s)\xi|\xi) = (\theta(|s|^2)\xi|\xi) = \int_X |s|^2 dP_{\xi,\xi}. \tag{7}$$

To see that $\theta$ is an isometry on simple functions note that, since $||P_{\xi,\xi}|| = ||\xi||^2$, we have from (7)

$$||\theta(s)\xi|| \leq ||s||_\infty ||\xi||. \tag{8}$$

However, if $\xi \in$ range $P(E_k)$, then

$$\theta(s)\xi = \alpha_k P(E_k)\xi = \alpha_k \xi, \tag{9}$$

because the projections $P(E_i)$ have mutually orthogonal ranges. Choosing $k$ so that $|\alpha_k| = ||s||_\infty$, we have from (8) and (9) that

$$||\theta(s)|| = ||s||_\infty. \tag{10}$$

Hence, the theorem (except for the last statement) is established for simple functions.

If $f$ is an arbitrary function in $L^\infty(P)$, then there exists a sequence $\{s_n\}$ of simple measurable functions on $X$ which converges to $f$ in the norm of $L^\infty(P)$. By (10), the sequence $\{\theta(s_n)\}$ in $B(H)$ corresponding to $\{s_n\}$ is Cauchy and hence converges to an operator which we denote by $\theta(f)$. Clearly $\theta(f)$ is well-defined, i.e., it does not depend on the sequence $\{s_n\}$, and (10) shows that $||\theta(f)|| = ||f||_\infty$ for all $f \in L^\infty(P)$.

Since each $P_{\xi,\eta}$ is a finite measure, equation (1) is a consequence of (6) with $s$ replaced by $s_n$. Similarly, (2) follows from (7). Approximating two essentially bounded measurable functions $f$ and $g$ on $X$, in the norm of $L^\infty(P)$, by simple measurable functions $s_n$ and $t_n$, we see from (4), (5) and the fact that $\theta$ is linear on simple functions, that $\theta$ is a *-isomorphism of $L^\infty(P)$ into $B(H)$. Because $\theta$ is an isometry and $L^\infty(P)$ is complete, it follows that its image $\theta(L^\infty(P))$ is a closed commutative *-subalgebra of $B(H)$.

The last statement of the theorem follows easily from the fact that if $S$ commutes with each $P(E)$, then $S$ commutes with $\theta(s)$ for each simple function s. Approximating $f \in L^{\infty}(P)$ by simple measurable functions we see that S commutes with $\theta(f)$. □

REMARK. *The formula in (1) of (52.2) is usually written in the abbreviated form*

$$\theta(f) = \int_X f\, dP, \quad f \in L^{\infty}(P).$$

We turn our attention now to the spectral theorem. Let H be a Hilbert space and $T \in B(H)$. Recall that the spectrum $\sigma(T)$ of T is the set of complex scalars such that $T - \lambda I$ is not invertible in $B(H)$. Since $B(H)$ is a $C^*$-algebra, if T belongs to a closed $^*$-subalgebra A of $B(H)$ containing the identity operator, then (8.2) shows that $\sigma(T) = \sigma_A(T)$. We shall use this fact freely.

Before proving that every bounded normal operator T on a Hilbert space induces a unique spectral measure P on the Borel subsets of the spectrum $\sigma(T)$ and that

$$T = \int_{\sigma(T)} \lambda\, dP(\lambda),$$

we will prove the following more general spectral theorem. In essence it gives a spectral measure which "reduces simultaneously" each member of an arbitrary family of commuting normal operators. It is often called the *spectral theorem for commutative* $C^*$-*algebras*.

(52.3) THEOREM. *(General spectral theorem). Let* H *be a Hilbert space and* A *a commutative* $C^*$-*subalgebra of* $B(H)$ *containing the identity operator* I. *Then:*

*(i) there exists a unique spectral measure* P *on the Borel subsets of* $\hat{A}$ *such that*

$$(T\xi|\eta) = \int_{\hat{A}} \hat{T}\, dP_{\xi,\eta}, \quad \xi, \eta \in H,\ T \in A, \tag{1}$$

*where* $\hat{A}$ *denotes the structure space of* A *and* $\hat{T}$ *is the Gelfand transform. Briefly, we write*

$$T = \int_{\hat{A}} \hat{T}\, dP.$$

*(ii)* $P(E) \neq 0$ *for each nonempty open subset* $E$ *of* $\hat{A}$.

*(iii) An operator* $S$ *in* $B(H)$ *commutes with each* $T$ *in* $A$ *iff* $S$ *commutes with each projection* $P(E)$.

*Proof.* Since $A$ is a commutative $C^*$-algebra, the Gelfand-Naimark theorem (7.1) asserts that $T \to \hat{T}$ is an isometric $*$-isomorphism of $A$ onto $C(\hat{A})$.

We establish the uniqueness of the spectral measure $P$ first. Since $\hat{T}$ ranges over all of $C(\hat{A})$ and since each of the complex Borel measures $P_{\xi,\eta}$ is regular, then equation (1) and the uniqueness assertion of the Riesz representation theorem show that each $P_{\xi,\eta}$ is uniquely determined by (1). Since $(P(E)\xi|\eta) = P_{\xi,\eta}(E)$, each projection $P(E)$ is also uniquely determined by (1).

To establish the existence of the spectral measure $P$ note that, since $||\hat{T}||_\infty = ||T||$, the functional $\phi$ on $C(\hat{A})$ defined by $\phi(\hat{T}) = (T\xi|\eta)$, for fixed $\xi, \eta \in H$, is bounded with norm $||\phi|| \leq ||\xi||\cdot||\eta||$. By the Riesz representation theorem there is a unique regular complex Borel measure $\mu_{\xi,\eta}$ on $\hat{A}$ such that

$$(T\xi|\eta) = \int_{\hat{A}} \hat{T}\, d\mu_{\xi,\eta}, \qquad \xi, \eta \in H, T \in A. \tag{2}$$

When the function $\hat{T}$ is real, $T = T^*$ so that $(T\xi|\eta) = \overline{(T\eta|\xi)}$. Therefore

$$\mu_{\xi,\eta} = \overline{\mu}_{\eta,\xi}, \qquad \xi, \eta \in H. \tag{3}$$

Since, for fixed $T$ in $A$, the functional $(T\xi|\eta)$ is sesquilinear, (2) and the uniqueness of the measures $\mu_{\xi,\eta}$ show that $\mu_{\xi,\eta}(E)$ is a sesquilinear functional on $H \times H$ for each Borel set $E$ in $\hat{A}$. Denote by $C_B(\hat{A})$ the algebra of all bounded Borel measurable functions on $\hat{A}$ with pointwise operations and supremum norm. Since $||\mu_{\xi,\eta}|| \leq ||\xi||\cdot||\eta||$, for each fixed $f \in C_B(\hat{A})$,

$$\int_{\hat{A}} f\, d\mu_{\xi,\eta}$$

is a bounded sesquilinear functional on $H \times H$. By (A.11) there exists, for each $f \in C_B(\hat{A})$, a unique operator $\psi(f)$ in $B(H)$ such that

$$(\psi(f)\xi|\eta) = \int_{\hat{A}} f\, d\mu_{\xi,\eta}, \qquad \xi, \eta \in H. \tag{4}$$

Comparing this with (2) we see that $\psi(\hat{T}) = T$ for each $T$ in $A$. Hence $\Psi$ is an extension to $C_B(\hat{A})$ of the inverse Gelfand transform $\hat{T} \to T$ which maps $C(\hat{A})$ onto $A$.

Equation (3) shows, whenever $f \in C_B(\hat{A})$ is real, that $(\psi(f)\xi|\eta) = \overline{(\psi(f)\eta|\xi)}$. Hence $\psi(f)^* = \psi(f)$ when $f$ is real.

We claim next that

$$\psi(fg) = \psi(f)\psi(g), \quad f, g \in C_B(\hat{A}) \tag{5}$$

Since $(ST)\hat{} = \hat{S}\hat{T}$ for $S, T \in A$, equation (2) gives

$$\int_{\hat{A}} \hat{S}\hat{T}\, d\mu_{\xi,\eta} = (ST\xi|\eta) = \int_{\hat{A}} \hat{S}\, d\mu_{T\xi,\eta}. \tag{6}$$

Because the set $\{\hat{S}: S \in A\}$ of Gelfand transforms coincides with $C(\hat{A})$ we can conclude that $\hat{T}\, d\mu_{\xi,\eta} = d\mu_{T\xi,\eta}$ for all $\xi, \eta \in H$ and $T \in A$. Hence, (6) remains valid if $\hat{S}$ is replaced by any $f \in C_B(\hat{A})$; thus

$$\begin{aligned} \int_{\hat{A}} f\hat{T}\, d\mu_{\xi,\eta} &= \int_{\hat{A}} f\, d\mu_{T\xi,\eta} = (\psi(f)T\xi|\eta) \\ &= (T\xi|\zeta) = \int_{\hat{A}} \hat{T}\, d\mu_{\xi,\zeta}, \end{aligned} \tag{7}$$

where $\zeta = \psi(f)^*\eta$. Applying this argument once more shows that the first and last integrals in (7) remain equal when $\hat{T}$ is replaced by any function $g \in C_B(\hat{A})$. We therefore have, for $f, g \in C_B(\hat{A})$, that

$$\begin{aligned} (\psi(fg)\xi|\eta) &= \int_{\hat{A}} fg\, d\mu_{\xi,\eta} = \int_{\hat{A}} g\, d\mu_{\xi,\zeta} \\ &= (\psi(g)\xi|\zeta) = (\psi(f)\psi(g)\xi|\eta), \end{aligned}$$

which gives (5).

We can now define our candidate for the spectral measure $P$. Suppose $E$ is a Borel subset of $\hat{A}$ and $\chi_E$ its characteristic function. Set

$$P(E) = \psi(\chi_E).$$

Clearly $P(\emptyset) = \psi(0) = 0$; and $P(\hat{A}) = I$ follows from the fact that $\psi(\hat{T}) = T$ for each $T \in A$. By (5), $P(E_1 \cap E_2) = P(E_1)P(E_2)$. When $E_1 = E_2$,

$P(E_1) = P(E_1)^2$, so each $P(E)$ is a projection. Since $\psi(f)^* = \psi(f)$ when $f$ is real, each $P(E)$ is self-adjoint. Both the finite additivity of $P$ and the relation

$$\mu_{\xi,\eta}(E) = (P(E)\xi|\eta) \tag{8}$$

follow from (4). Therefore $P$ is a spectral measure. Since (1) is immediate from (2) and (8), part (i) is proved.

(ii) Let $E$ be open in $\hat{A}$ and suppose $P(E) = 0$. If $S \in A$ and $\hat{S}$ has its support in $E$, (1) implies that $S = 0$. Since $\{\hat{S}: S \in A\} = C(\hat{A})$, it follows from Urysohn's lemma that $E = \emptyset$.

(iii) Select an operator $S \in B(H)$, vectors $\xi, \eta \in H$ and let $\zeta = S^*\eta$. Then, for any $T \in A$ and any Borel subset $E$ of $\hat{A}$ we have

$$(ST\xi|\eta) = (T\xi|\zeta) = \int_{\hat{A}} \hat{T}\, dP_{\xi,\zeta}, \tag{9}$$

$$(TS\xi|\eta) = \int_{\hat{A}} \hat{T}\, dP_{S\xi,\eta}, \tag{10}$$

$$(SP(E)\xi|\eta) = (P(E)\xi|\zeta) = P_{\xi,\zeta}(E),$$

$$(P(E)S\xi|\eta) = P_{S\xi,\eta}(E).$$

Now, if $ST = TS$ for every $T$ in $A$, the measures in (9) and (10) coincide, so that $SP(E) = P(E)S$ from the last two equations. Since the argument is reversible, (iii) is proved. □

It will be useful to have the following result, which is a consequence of the Gelfand-Naimark theorem (7.1), to prove the spectral theorem for a single operator.

(52.4) THEOREM. *Let* $T$ *be a normal operator on a Hilbert space* $H$ *and* $A$ *the* $C^*$*-subalgebra of* $B(H)$ *generated by* $T$ *and the identity operator. Then* $A$ *is commutative and the structure space* $\hat{A}$ *is homeomorphic to* $\sigma(T)$. *Moreover, the formula* $\psi(f)^\wedge = f\circ\hat{T}$ *defines an isometric* $*$*-isomorphism* $\psi$ *of* $C(\sigma(T))$ *onto* $A$ *for each* $f \in C(\sigma(T))$. *When* $f \in C(\sigma(T))$ *is the particular function* $f(\lambda) = \lambda$, *then* $\psi(f) = T$.

*Proof.* Since $T$ and $T^*$ commute, the collection of all polynomials in $T$ and $T^*$ forms a commutative $*$-subalgebra of $B(H)$ which is contained in the $C^*$-algebra generated by $T$. Since the closure of this collection is a commutative $C^*$-algebra, it coincides with $A$.

The Gelfand transform $\hat{T}$ of $T$ is a continuous function on $\hat{A}$ whose range is $\sigma(T)$ by (B.6.6). If $\phi_1, \phi_2 \in \hat{A}$ and $\hat{T}(\phi_1) = \hat{T}(\phi_2)$, then $\phi_1(T) = \phi_2(T)$. By (7.1), $\phi_1(T^*) = \phi_2(T^*)$ and it follows that $\phi_1$ and $\phi_2$ agree on all polynomials in $T$ and $T^*$. Since these are dense in $A$, $\phi_1 = \phi_2$ by (B.6.3), so $\hat{T}$ is one-to-one. Since $\hat{A}$ is a compact Hausdorff space, $\hat{T}$ is a homeomorphism of $\hat{A}$ onto $\sigma(T)$.

It follows that the mapping $f \to f \circ \hat{T}$ is an isometric $*$-isomorphism of $C(\sigma(T))$ onto $C(\hat{A})$. Therefore, each $f \circ \hat{T}$ is, by (7.1), the Gelfand transform of a unique element, say $\psi(f)$, in $A$ satisfying $||\psi(f)|| = ||f||_\infty$. When $f(\lambda) = \lambda$, then $f \circ \hat{T} = \hat{T}$; so, in this case, $\psi(f)\hat{} = \hat{T}$, from which it follows that $\psi(f) = T$. □

(52.5) THEOREM. *(Spectral theorem). Let* $H$ *be a Hilbert space and* $T$ *a bounded normal operator on* $H$. *Then there exists a unique spectral measure* $P$ *on the Borel subsets of* $\sigma(T)$ *such that*

$$T = \int_{\sigma(T)} \lambda \, dP(\lambda).$$

*Moreover, each projection* $P(E)$ *commutes with each* $S \in B(H)$ *which commutes with* $T$.

*Proof.* Let $A$ be the commutative $C^*$-subalgebra of $B(H)$ generated by $T$ and $I$. By (52.4), $\hat{A}$ is homeomorphic to $\sigma(T)$ and $\hat{T}(\lambda) = \lambda$ for all $\lambda \in \sigma(T)$. The claimed spectral measure $P$ on the Borel subsets of $\sigma(T)$ exists, by (52.3), and satisfies $T = \int_{\sigma(T)} \hat{T}(\lambda) dP(\lambda) = \int_{\sigma(T)} \lambda \, dP(\lambda)$.

To see that $P$ is uniquely determined by $T$, note that (52.2) implies

$$p(T,T^*) = \int_{\sigma(T)} p(\lambda,\overline{\lambda}) dP(\lambda),$$

where $p$ is any complex polynomial in two variables. Since these polynomials are dense in $C(\sigma(T))$ by the Stone-Weierstrass theorem (A.7), the projections $P(E)$ are uniquely determined by the above integral representation for $p(T,T^*)$, and therefore are uniquely determined by $T$, just as in the proof of uniqueness in (52.3).

Finally, if $ST = TS$, then $ST^* = T^*S$ by (A.17). Therefore $S$

commutes with each element of A. From (52.3), (iii), we see that $SP(E) = P(E)S$ for each Borel subset of $\sigma(T)$. □

Let $H$ be a Hilbert space and $T$ a normal operator in $B(H)$. The unique spectral measure $P$ in (52.5) is often called the *spectral decomposition of* $T$. It can be extended to all Borel subsets $E$ in the complex plane by setting $P(E) = 0$ whenever $E \cap \sigma(T) = \emptyset$. If $f$ belongs to the algebra $C_B(\sigma(T))$ of bounded Borel measurable functions on $\sigma(T)$, then the operator $\theta(f) = \int_{\sigma(T)} f\, dP$, given by (52.2), is denoted by $f(T)$. Hence,

$$f(T) = \int_{\sigma(T)} f(\lambda)dP(\lambda).$$

## §53. *The structure space of an abelian von Neumann algebra.*

In this section we utilize the Gelfand-Naimark theorem (7.1) to show that the structure space of an abelian von Neumann algebra is *extremally disconnected,* i.e., the closure of each open set is open. Since the space $L^\infty(\mu)$ of essentially bounded measurable functions on a finite measure space is an abelian von Neumann algebra of operators in $B(L^2(\mu))$, $L^\infty(\mu)$ has an extremally disconnected structure space. This shows how the Gelfand representation may convert an algebra of bad functions on a nice space into an algebra of nice functions on a bad space.

Let $H$ be a Hilbert space. The *weak operator topology* on $B(H)$ is the weakest topology in which every linear functional of the form $T \to (T\xi|\eta)$ is continuous. A *-subalgebra of operators in $B(H)$ which contains the identity operator and which is closed in the weak operator topology is called a *von Neumann algebra* (or a $W^*$-*algebra*).

An operator $T$ on $H$ is *positive* if $T$ is self-adjoint and $(T\xi|\xi) \geq 0$ for $\xi \in H$.

(53.1) PROPOSITION. *Let* $\{T_\alpha: \alpha \in D\}$ *be a net of positive operators in* $B(H)$ *such that* $T_\alpha \geq T_\beta$ *whenever* $\alpha \geq \beta$; *assume moreover that* $\{T_\alpha\}$ *is uniformly bounded. Then* $\{T_\alpha: \alpha \in D\}$ *converges to a positive operator* $T \in B(H)$ *in the weak operator topology. The operator* $T$ *is the least upper bound for* $\{T_\alpha: \alpha \in D\}$ *in* $B(H)$.

*Proof.* If $\xi$ is any vector in $H$, we have $(T_\alpha\xi|\xi) \geq (T_\beta\xi|\xi)$ whenever $\alpha \geq \beta$, and $(T_\alpha\xi|\xi) \leq M||\xi||^2$, where $M = \sup\{||T_\alpha||: \alpha \in D\}$. Hence, $\lim\{(T_\alpha\xi|\xi): \alpha \in D\}$ exists for every vector $\xi \in H$. If $\xi$ and $\eta$ are arbitrary vectors in $H$, it follows from the polarization identity

$$(T_\alpha\xi|\eta) = \frac{1}{4}[\sum_{n=0}^{3} i^n(T_\alpha(\xi + i^n\eta)|\xi + i^n\eta)]$$

that $\lim\{(T_\alpha\xi|\eta): \alpha \in D\}$ exists for all $\xi, \eta \in H$. Define

$$< \xi|\eta > = \lim\{(T_\alpha\xi|\eta): \alpha \in D\}.$$

Then $< , >$ is a sesquilinear functional and $|<\xi|\eta>| \leq M||\xi||\cdot||\eta||$ for all $\xi, \eta \in H$. Hence, by (A.11), there is a unique operator $T \in B(H)$ such that $<\xi|\eta> = (T\xi|\eta)$ for all $\xi, \eta$ in $H$. From the relation $(T\xi|\eta) = \lim (T_\alpha\xi|\eta): \alpha \in D\}$, which is true for all $\xi, \eta$, it follows that $T_\alpha \to T$ in the weak operator topology. If $\xi \in H$, then $(T\xi|\xi) = \lim_{\alpha\in D}(T_\alpha\xi|\xi) \geq 0$ shows that $T$ is a positive operator. Finally, let us show that $T$ is the least upper bound. If $\alpha$ is fixed in $D$, $\xi \in H$, and $\beta \geq \alpha$, then $(T_\beta\xi|\xi) \geq (T_\alpha\xi|\xi)$; hence

$$(T\xi|\xi) = \lim_{\beta\in D}(T_\beta\xi|\xi) \geq (T_\alpha\xi|\xi),$$

so that $T \geq T_\alpha$. If $S \in B(H)$ is such that $S \geq T_\alpha$ for every $\alpha \in D$, then, for arbitrary $\xi \in H$, we have $(S\xi|\xi) \geq \lim_{\alpha\in D}(T_\alpha\xi|\xi) = (T\xi|\xi)$. Thus, $S \geq T$ and the proof is complete. □

(53.2) THEOREM. *Let* $A$ *be an abelian von Neumann algebra on a Hilbert space* $H$. *Then the structure space* $\hat{A}$ *is extremely disconnected.*

*Proof.* Let $U$ be an open set in $\hat{A}$ and let

$$D = \{f \in C(\hat{A}): 0 \leq f(\phi) \leq 1 \text{ for all } \phi \in \hat{A} \text{ and } f(\phi) = 0 \text{ for } \phi \notin U\}.$$

Direct the set $D$ by $f_1 \geq f_2$ if and only if $f_1(\phi) \geq f_2(\phi)$ for all $\phi \in \hat{A}$. By the Gelfand-Naimark theorem (7.1), there exists, for each $f \in D$, a unique operator $T_f$ in $A$ for which $\hat{T}_f = f$. Then $\{T_f\}_{f\in D}$ is a net of operators in $A$ such that $0 \leq T_f \leq I$ (the identity operator) for all $f \in D$, and $T_{f_1} \geq T_{f_2}$ whenever $f_1 \geq f_2$. Hence the net $\{T_f\}_{f\in D}$ converges in the weak operator topology to a positive operator $P$, which is a least upper bound for the net $\{T_f\}_{f\in D}$. Since $A$ is weak-operator closed, $P \in A$. We assert that $\hat{P} = \chi_{\overline{U}}$. Since $P \geq T_f$ for all $f \in D$, $\hat{P}(\phi) \geq \hat{T}_f(\phi) = f(\phi)$ for all $\phi \in \hat{A}$. We consider two cases:

*Case 1.* Let $\phi_o \in U$. By Urysohn's lemma, there is a function $f \in D$ such that $f(\phi_o) = 1$. This shows that $\hat{P}(\phi_o) \geq 1$. Since $P$ is a least upper bound for the $T_f$, and $T_f \leq I$ for each $f \in D$ we have $\hat{P}(\phi_o) \leq \hat{I}(\phi_o) = 1$. It follows that $\hat{P}(\phi) = 1$ for all $\phi \in U$. Since $\hat{P}$ is continuous, $\hat{P}(\phi) = 1$ for all $\phi \in \overline{U}$.

*Case 2.* Suppose that $\phi_o \notin \overline{U}$. Using Urysohn's lemma again, there exists a continuous function $g\colon \hat{A} \to [0,1]$ such that

$$g(\phi) = \begin{cases} 1, & \text{for } \phi \in \overline{U}, \\ 0, & \text{for } \phi = \phi_o. \end{cases}$$

Let $T_g$ be the operator in $A$ such that $\hat{T}_g = g$. Since $g(\phi) \geq f(\phi)$ for all $\phi \in \hat{A}$ whenever $f \in D$, it follows that $T_g \geq T_f$ for all $f$ in $D$. We have $T_g \geq P$ because $P$ is the least upper bound of $\{T_f\}_{f\in D}$ and hence $0 = g(\phi_o) = \hat{T}_g(\phi_o) \geq \hat{P}(\phi_o)$. Since $P$ is a positive operator, $\hat{P}(\phi_o) \geq 0$; hence $\hat{P}(\phi_o) = 0$. Since the point $\phi_o$ was arbitrary, $\hat{P}(\phi) = 0$ for all $\phi \notin \overline{U}$.

We have therefore shown that $\hat{P} = \chi_{\overline{U}}$. Since $\hat{P}$ is continuous, $\overline{U}$ must be open as required. □

We remark that, in 1952, J. M. G. Fell and J. L. Kelley showed that a proof of the spectral theorem for normal operators on a Hilbert space can be based on (53.2). We refer the reader to Kadison and Ringrose [1, p. 310] for a careful discussion of the spectral theorem from this point of view.

## §54. *The C*-algebra of compact operators.*

Consider the C*-algebra $B(H)$ of all bounded linear operators on a complex Hilbert space $H$, and let $B_c(H)$ denote the C*-algebra of compact operators on $H$ (an operator is *compact* if it maps every bounded set into a relatively compact set). Then $B_c(H)$ is a closed *-ideal of $B(H)$ and hence $B(H)/B_c(H)$ is also a C*-algebra by (24.5). Applying the Gelfand-Naimark theorem (19.1) we arrive at the following result:

(54.1) THEOREM. *Let* $H$ *be a Hilbert space. The quotient algebra*

$B(H)/B_c(H)$ *of* $B(H)$ *modulo the ideal* $B_c(H)$ *of compact operators is isometrically* $*$*-isomorphic to the* $C^*$*-algebra of bounded linear operators on some Hilbert space.*

This nontrivial theorem was first established by J. Calkin in 1941 (see Naimark [1, p. 320]). Calkin directly constructed a realization of the algebra $B(H)/B_c(H)$ in the form of an algebra of operators on a Hilbert space.

§55. *On the closure of the numerical range.*

If $T$ is a bounded linear operator on a complex Hilbert space $H$, the *spatial numerical range* of $T$ is the set

$$W(T) = \{(T\xi|\xi): ||\xi|| = 1, \xi \in H\}.$$

The spatial numerical range is convex (see Halmos [1, p. 110]) and, since $W(T)$ is bounded, its closure $W(T)^-$ is compact and convex.

Roughly speaking, we use the Gelfand-Naimark theorem (19.1) in this application to give an intrinsic characterization of the closure of the numerical range of an element in an abstract $C^*$-algebra.

Let $A$ be a $C^*$-algebra with identity 1, and suppose $x \to T_x$ and $x \to S_x$ are faithful $*$-representations of $A$ on Hilbert spaces $H$ and $K$, respectively, such that $T_1$ and $S_1$ are the identity operators. It is shown in Berberian-Orland [1] that $W(T_x)^- = W(S_x)^-$ for all $x$ in $A$. Hence, we may define the *closed numerical range* of $x$, denoted $\overline{W}(x)$, to be the set $W(T_x)^-$.

Let $S(A)$ denote the set of all states on $A$; that is, the set of all linear functionals $f$ on $A$ such that $f(1) = 1$ and $f(x^*x) \geq 0$ for all $x \in A$. Then $S(A)$ is a nonempty, convex, subset of the dual of $A$ and is compact in the weak $*$-topology by (29.7). For any $x$ in $A$, write

$$V(x) = \{f(x): f \in S(A)\}.$$

Since the map $f \to f(x)$ is linear and weak $*$-continuous, $V(x)$ is a nonempty, compact, convex subset of $C$ (see §41).

(55.1) THEOREM. *If* $A$ *is a* $C^*$*-algebra with identity, then* $V(x) = \overline{W}(x)$ *for every* $x$ *in* $A$.

*Proof.* Each $f$ in $S(A)$ induces a canonical $^*$-representation of $A$ on a Hilbert space $H_f$, and the Gelfand-Naimark theorem (19.1) states that the direct sum of these representations is an isometric $^*$-representation $x \to T_x$ of $A$ as operators on the direct sum $H$ of the Hilbert spaces $H_f$. Fix $x \in A$. It is clear from the construction that $f(x) \in W(T_x)$ for each $f$ in $S(A)$; hence $V(x) \subseteq W(T_x)$. On the other hand if $\xi \in H$, $||\xi|| = 1$, then the formula $f(y) = (T_y\xi|\xi)$ defines a state of $A$, and $(T_x\xi|\xi) = f(x) \in V(x)$; varying $\xi$, we have $W(T_x) \subseteq V(x)$, and therefore $V(x) = W(T_x)$. In particular, for each $x$ in $A$, $W(T_x)$ is closed and therefore coincides with $\overline{W}(x)$. This completes the proof. □

Theorem (55.1) was proved implicitly by Bohnenblust and Karlin [1, Theorem 12] and, independently, by Lumer [1, Theorem 11]. The treatment above follows Berberian and Orland [1], where the reader will find the following interesting corollaries of (55.1), whose proofs we omit.

(55.2) COROLLARY. *Let* $A$ *be a* $C^*$*-algebra with identity,* $P(A)$ *the set of pure states of* $A$*, and, for each* $x \in A$*, let* $P(x) = \{f(x): f \in P(A)\}$. *Then* $\overline{\text{conv}}\, P(x) = \overline{W}(x)$, *for each* $x \in A$, *where* $\overline{\text{conv}}\, P(x)$ *denotes the closed convex hull of* $P(x)$.

(55.3) COROLLARY. *If* $A$ *is a commutative* $C^*$*-algebra with identity, then* $\overline{\text{conv}}\, \sigma_A(x) = \overline{W}(x)$ *for each* $x$ *in* $A$.

(55.4) COROLLARY. *Let* $A$ *be a* $C^*$*-algebra with identity, and define* $\omega(x) = \sup\{|f(x)|: f \in S(A)\}$ *for* $x \in A$. *Then* $\omega(x^n) \leq \omega(x)^n$ *for all positive integers* $n$*, and* $x$ *in* $A$.

(55.5) COROLLARY. *If* $H$ *is a Hilbert space and* $f$ *is any state on the* $C^*$*-algebra* $B(H)$ *of bounded linear operators on* $H$*, then for each* $T$ *in* $B(H)$ *there exists in* $H$ *a sequence* $\{\xi_n\}$ *of unit vectors, depending on* $T$*, such that* $f(T) = \lim_{n\to\infty}(T\xi_n|\xi_n)$.

## §56. *The Gelfand-Raikov theorem.*

The purpose of this section is to show, using the Gelfand-Naimark theorem, that a locally compact group has enough irreducible continuous unitary representations to separate the points of the group. This result is known as the Gelfand-Raikov theorem and is sometimes stated by saying that every locally compact group has a "complete system" of irreducible continuous unitary representations. We shall assume throughout that all topological groups are Hausdorff and that all Hilbert spaces are complex.

A *unitary representation* of a locally compact group $G$ is a homomorphism $U$ of $G$ into the group $U(H)$ of unitary operators on a Hilbert space $H$, i.e., $U$ is a mapping $t \to U_t$ from $G$ into $U(H)$ such that

$$U_{st} = U_s U_t \quad \text{for all } s, t \in G.$$

This condition immediately implies that $U_e = I$, where $I$ is the identity operator, and $U_{s^{-1}} = U_s^{-1} = U_s^*$. The Hilbert space $H$ is called the *space* of $U$. A unitary representation $U: G \to U(H)$ is said to be *continuous* if $t \to U_t\xi$ is a continuous map from $G$ into the Hilbert space $H$ for each $\xi \in H$. The relation

$$||U_s\xi - U_t\xi|| = ||U_{s^{-1}t}\xi - \xi||$$

shows that $U$ is continuous iff it is continuous at the identity $e$ of $G$.

The notions of *equivalence, direct sums, cyclic vector,* are defined as in the case of *-representations of *-algebras in §26. One need *not* define the essential subspace of a unitary representation since a unitary representation is automatically nondegenerate. The elementary results (26.2), (26.3), (26.10), and (26.11) are established in the same way as before.

A unitary representation $t \to U_t$ of $G$ on $H$ is said to be *irreducible* if the set $\{U_t: t \in G\}$ of operators on $H$ is topologically irreducible. Thus, $U$ is irreducible iff the only closed subspaces of $H$ invariant for $U$ are $\{0\}$ and $H$. For unitary representations we have the same criterion for irreducibility as we had previously for *-algebras (see (28.2)):

(56.1) THEOREM. *Let* $G$ *be a locally compact group and* $U$ *a unitary representation of* $G$ *on a (nontrivial) Hilbert space* $H$. *The following conditions are equivalent:*

*(a)* $U$ *is irreducible;*

*(b) Every nonzero vector in* $H$ *is a cyclic vector for* $U$;

*(c) The only bounded linear operators commuting with all* $U_t$, $t \in G$, *are of the form* $\lambda I$, $\lambda \in C$.

To prove the Gelfand-Raikov theorem we will proceed as follows:

(1°) Form the group algebra $L^1(G)$ with respect to a given left invariant Haar measure.

(2°) Show that $L^1(G)$ has a faithful $^*$-representation $f \to T_f$ on the Hilbert space $L^2(G)$, the so-called *left regular representation*.

(3°) Show that there is a one-to-one correspondence between the set of all nondegenerate $^*$-representations of $L^1(G)$ and the set of all continuous unitary representations of $G$.

(4°) Show that the correspondence between nondegenerate $^*$-representations of $L^1(G)$ and continuous unitary representations of $G$ described in (3°) preserves irreducibility.

(5°) Show that the left regular representation of $L^1(G)$ can be used to construct a C*-algebra, which by (30.7) has a complete set of irreducible representations. By composing one of these with the left regular representation we obtain an irreducible representation of $L^1(G)$ which, via the correspondence in (3°), gives an irreducible continuous unitary representation of $G$.

Let us begin by reviewing some basic facts concerning the group algebra $L^1(G)$ referred to in step (1°). Let $G$ be a locally compact group and let $C_{oo}(G)$ be the space of all complex-valued continuous functions on $G$ with compact support. Let $\mu$ be a fixed left invariant Haar measure on $G$ (we will often denote $\mu$ by $dt$). Then it is straightforward to verify that

$$||f||_p = \left(\int_G |f|^p d\mu\right)^{1/p}, \quad f \in C_{oo}(G),$$

defines a norm in the linear space $C_{oo}(G)$ for each $p$, $1 \leq p < \infty$. $L^p(G)$ is the Banach space obtained by completing $C_{oo}(G)$ with respect to this norm. The space $L^1(G)$ is a Banach algebra under convolution multiplication defined by

$$(f*g)(s) = \int_G f(st)g(t^{-1})dt = \int_G f(t)g(t^{-1}s)dt,$$

for $f, g \in L^1(G)$. The group algebra $L^1(G)$ is commutative iff $G$ is abelian.

It will be useful to recall a few basic facts involving Haar measure and integration on locally compact groups. The *modular function of* G will be denoted, as usual, by $\Delta$; it is a continuous homomorphism of G into the multiplicative group of strictly positive real numbers. Because of the left invariance of the Haar measure dt we have, for $f \in L^1(G)$, that:

$$\int_G f(t)dt = \int_G f(st)dt = \int_G f(s^{-1}t)dt.$$

Also,

$$\int_G f(t)dt = \Delta(s)\int_G f(ts)dt = \int_G \Delta(s^{-1})f(ts^{-1})dt.$$

In general, $\int_G f(t)dt \neq \int_G f(t^{-1})dt$. In fact, equality holds for all $f \in C_{oo}(G)$ iff $\Delta(t) \equiv 1$, in which case G is said to be *unimodular*. Thus, G is unimodular iff its Haar measure is both left and right invariant. Every compact group and every locally compact abelian group is unimodular.

The group algebra $L^1(G)$ becomes a Banach $^*$-algebra with respect to the (isometric) involution defined by

$$f^*(s) = \Delta(s^{-1})\overline{f(s^{-1})}, \quad s \in G.$$

If $f \in L^p(G)$, let $f_s$ and ${}_sf$ denote the *translation functions* defined on G by

$$f_s(t) = f(ts) \quad \text{and} \quad {}_sf(t) = f(st).$$

Then, from the left invariance of $\mu$,

$$||f_s||_p = \Delta(s)^{-1/p}||f||_p \quad \text{and} \quad ||{}_sf|| = ||f||_p.$$

We remark that the spaces $L^p(G)$, for $p > 1$, are, in general, not closed with respect to convolution.

A net $\{e_\lambda\}_{\lambda\in\Lambda}$ in a normed algebra A is called a *two-sided bounded approximate identity* if there exists a positive real constant K such that $||e_\lambda|| \leq K$ for all $\lambda \in \Lambda$ and

$$\lim_{\lambda\in\Lambda} e_\lambda x = x = \lim_{\lambda\in\Lambda} xe_\lambda$$

for all x in A. Although the group algebra $L^1(G)$ does not ordinarily possess an identity element (it does iff $G$ is discrete), it always contains a bounded two-sided approximate identity. This will be a consequence of the following two lemmas.

(56.2) LEMMA. *If* $f \in L^p(G)$, $1 \leq p < \infty$, *then the map* $x \to {}_xf$ *of* $G$ *into* $L^p(G)$ *is left uniformly continuous.*

*Proof.* Let $\varepsilon > 0$ and choose a continuous function $g$ on $G$ with compact support $K$ such that $||f - g||_p < \varepsilon/3$. Fix a compact symmetric neighborhood $W$ of the identity $e$ in $G$. Using the standard fact that $g$ is left uniformly continuous on $G$, Loomis [1, p. 109], there is a symmetric neighborhood $V$ of $e$ contained in $W$ such that

$$|g(s) - g(t)| < \frac{\varepsilon}{3}\cdot\mu(WK)^{-1/p} \quad \text{if} \quad st^{-1} \in V.$$

Thus,

$$||g - {}_xg||_\infty < \frac{\varepsilon}{3}\cdot\mu(WK)^{-1/p} \quad \text{if} \quad x \in V.$$

Hence $||g - {}_xg||_p < \frac{\varepsilon}{3}$ for all $x \in V$, and so

$$||f - {}_xf||_p \leq ||f - g||_p + ||g - {}_xg||_p + ||{}_xg - {}_xf||_p < \varepsilon$$

for all $x \in V$. Then, for $x = st^{-1} \in V$, we have $||{}_sf - {}_tf||_p = ||{}_xf - f||_p < \varepsilon$ as required. □

(56.3) LEMMA. *Given* $f \in L^p(G)$, $1 \leq p < \infty$, *and* $\varepsilon > 0$, *there exists a neighborhood* $V$ *of the identity in* $G$ *such that*

$$||f - u*f||_p < \varepsilon \quad \textit{and} \quad ||f - f*u||_p < \varepsilon,$$

*whenever* $u$ *is any nonnegative function in* $L^1(G)$ *such that* $u(s) = 0$ *for* $s \notin V$ *and* $\int_G u \, d\mu = 1$.

*Proof.* If $f$ and $g$ are functions on $G$ such that $f\bar{g} \in L^1(G)$, we write $(f,g) = \int_G f\bar{g} \, d\mu$. Let $g \in L^q(G)$, where $\frac{1}{p} + \frac{1}{q} = 1$. If $u$ is any nonnegative function in $L^1(G)$, then $u*f \in L^p(G)$ and hence, by Hölder's inequality, $(u*f - f)\bar{g} \in L^1(G)$, so

$$(u*f - f,g) = \int_G [(u*f)(s) - f(s)]\overline{g(s)}ds$$

$$= \int_G\int_G [u(t)f(t^{-1}s) - f(s)]\overline{g(s)}dtds.$$

Interchanging the order of integration (Fubini) and applying Hölder's inequality gives

$$|(u*f - f,g)| \leq ||g||_q \int_G ||_{t^{-1}}f - f||_p u(t)dt.$$

Then, considering the operator norm of the linear functional $g \to (u*f - f,g)$ on $L^q(G)$, we have

$$||u*f - f||_p \leq \int_G ||_{t^{-1}}f - f||_p u(t)dt.$$

By (56.2), there is a neighborhood $V$ of $e$ such that $||_{t^{-1}}f - f||_p < \varepsilon$ for all $t \in V$. Therefore if $\int_G u(s)ds = 1$ and $u(t) = 0$ for $t \notin V$, then $||u*f - f||_p < \varepsilon$.

To prove the other inequality, recall that the modular function $\Delta$ is a continuous homomorphism of $G$ into $(0,\infty)$, $\Delta(e) = 1$, and

$$\int_G u(s^{-1})\Delta(s^{-1})ds = \int_G u(s)ds$$

for any $u \in L^1(G)$. Now set $m = \int_G u(s^{-1})ds$ (we see below that $m$ is finite and nonzero), and let $g \in L^q(G)$. Then, as above,

$$|(f*u - f,g)| = \left|\int_G\int_G [f(st) - f(s)/m]u(t^{-1})\overline{g(s)}dtds\right|$$

$$\leq ||g||_q \int_G ||mf_t - f||_p [u(t^{-1})/m]dt.$$

Thus,

$$||f*u - f||_p \leq \int_G ||mf_t - f||_p [u(t^{-1})/m]dt.$$

Observe that $m \to 1$ as the neighborhoods $V$ of $e$ decrease (indeed, since $m = \int_G u(s^{-1})\Delta(s^{-1})\Delta(s)ds$, then $a = a\int_G u(s)ds = a\int_G u(s^{-1})\Delta(s^{-1})ds \leq m \leq A\int_G u(s^{-1})\Delta(s^{-1})ds = A\int_G u(s)ds = A$, where $a = \min_{s\in V} \Delta(s)$ and $A = \max_{s\in V} \Delta(s)$.) Hence, by the continuity of right translation (see exercise (X.22)), there is a neighborhood $V$ of $e$ with $||mf_t - f||_p < \varepsilon$ if $t \in V$, and then $||f*u - f||_p < \varepsilon\int_G[u(t^{-1})/m]dt = \varepsilon$. □

(56.4) PROPOSITION. *The group algebra* $L^1(G)$ *of the locally compact group* $G$ *has a two-sided approximate identity bounded by* 1.

*Proof*. The family of neighborhoods $V$ of the identity $e$ in $G$ forms a directed set under inclusion, and if $u_V$ is a nonnegative function on $G$ vanishing off $V$ and satisfying $\int_G u_V d\mu = 1$, then $u_V*f$ and $f*u_V$ converge to $f$ in the $L^p$-norm for any $f \in L^p(G)$ by (56.3). □

It should be noted that the functions $u_V$ in the proof of (56.4) can be assumed to lie in $C_{oo}(G)$ for each neighborhood $V$ of the identity $e$ in $G$.

The *left regular representation* of $L^1(G)$ on the space $L^p(G)$ is defined as follows: For each $f \in L^1(G)$, define $T_f\colon L^p(G) \to L^p(G)$ by $T_f g = f*g$ for all $g \in L^p(G)$, $1 \leq p < \infty$. Since $||f*g||_p \leq ||f||_1||g||_p$, then $||T_f|| \leq ||f||_1$ and it follows easily that the mapping $f \to T_f$ is an algebra homomorphism of $L^1(G)$ into $B(L^p(G))$. When $p = 2$, $L^2(G)$ is a Hilbert space and we have:

(56.5) PROPOSITION. *The left regular representation* $f \to T_f$ *is a faithful *-representation of* $L^1(G)$ *on the Hilbert space* $L^2(G)$.

*Proof*. Since $f \to T_f$ is a homomorphism of $L^1(G)$ into the bounded operators on $L^2(G)$, we need only show that it is faithful and that $T_{f*} = (T_f)^*$ for each $f \in L^1(G)$.

Suppose $f \in L^1(G)$ and $T_f = 0$. Then $f*g = 0$ for all $g \in L^2(G)$; we must show that $f = 0$. For each neighborhood $V$ of $e$, let $g_V$ be a nonnegative continuous function on $G$ with compact support which vanishes off of $V$ and $\int g_V(t)dt = 1$. By hypothesis $f*g = 0$ for all $g \in L^2(G)$. In particular, $f*g_V = 0$ for each neighborhood $V$ of $e$ in $G$. Since $\{g_V\}$ is an approximate identity, $||f||_1 = ||f - f*g_V||_1 \to 0$ by (56.4), and so $f = 0$. Hence $f \to T_f$ is faithful.

Let $f \in L^1(G)$. To show $T_{f^*} = (T_f)^*$, suppose that $g_1, g_2 \in L^2(G)$. Then, using properties of the modular function $\Delta$, we have

$$\begin{aligned}
(T_{f^*}g_1|g_2) &= (f^**g_1|g_2) \\
&= \int_G\int_G f^*(st)g_1(t^{-1})dt\,\overline{g_2(s)}ds \\
&= \int_G\int_G \Delta(t^{-1}s^{-1})\overline{f(t^{-1}s^{-1})}g_1(t^{-1})\overline{g_2(s)}dtds \\
&= \int_G \Delta(s^{-1})\int_G \Delta(t^{-1})\overline{f(t^{-1}s^{-1})}g_1(t^{-1})dt\,\overline{g_2(s)}ds \\
&= \int_G \Delta(s^{-1})\int_G \overline{f(ts^{-1})}g_1(t)dt\,\overline{g_2(s)}ds \\
&= \int_G\int_G \overline{f(ts)}g_1(t)dt\,\overline{g_2(s^{-1})}ds \\
&= \int_G\int_G \overline{f(ts)}g_1(t)\overline{g_2(s^{-1})}dtds.
\end{aligned}$$

On the other hand,

$$\begin{aligned}
(T_f^*g_1|g_2) &= (g_1|T_fg_2) = (g_1|f*g_2) \\
&= \int_G g_1(t)\overline{\int_G f(ts)g_2(s^{-1})ds}dt \\
&= \int_G\int_G g_1(t)\overline{f(ts)}\,\overline{g_2(s^{-1})}dsdt.
\end{aligned}$$

Hence, $(T_{f^*}g_1|g_2) = (T_f^*g_1|g_2)$ is a consequence of the Fubini theorem, and the proof is complete. □

We turn our attention next to the relation between continuous unitary representations of $G$ and nondegenerate *-representations of $L^1(G)$.

(56.6) THEOREM. *Let* $G$ *be a locally compact group. Then there exists a one-to-one correspondence between continuous unitary representations* $U$

*of* $G$ *on a Hilbert space* $H$ *and nondegenerate* $^*$*-representations* $\pi$ *of* $L^1(G)$ *on* $H$. *In one direction the correspondence is given by*

$$(\pi(f)\xi|\eta) = \int_G (U_s\xi|\eta)f(s)ds \tag{1}$$

*for all* $f \in L^1(G)$ *and* $\xi, \eta \in H$, *and in the other direction by*

$$U_s\pi(f)\xi = \pi({}_{s^{-1}}f)\xi \tag{2}$$

*for all* $s \in G$, $f \in L^1(G)$ *and* $\xi \in H$.

*Proof.* Let $U: s \to U_s$ be a continuous unitary representation of $G$ on a Hilbert space $H$. For each pair of vectors $\xi, \eta \in H$ consider, for $f \in L^1(G)$, the integral

$$\Phi_f(\xi,\eta) = \int_G (U_s\xi|\eta)f(s)ds \tag{3}$$

Since the function $s \to (U_s\xi|\eta)$ is continuous and bounded by $||\xi||\cdot||\eta||$, the function $s \to (U_s\xi|\eta)f(s)$ is integrable and so the integral in (3) exists. Clearly, $\Phi_f$ is a sesquilinear form on $H \times H$ such that

$$|\Phi_f(\xi,\eta)| \leq ||f||_1||\xi||\cdot||\eta||, \quad \xi, \eta \in H.$$

Let $\pi(f)$ be the operator on $H$ such that $\Phi_f(\xi,\eta) = (\pi(f)\xi|\eta)$ for all $\xi, \eta \in H$ (see (A.11)). Then $\pi: L^1(G) \to B(H)$ is a mapping such that

$$(\pi(f)\xi|\eta) = \int_G (U_s\xi|\eta)f(s)ds.$$

It is clear that $\pi$ is linear; further, since

$$|(\pi(f)\xi|\eta)| \leq ||f||_1||\xi||\cdot||\eta||,$$

we have $||\pi(f)|| \leq ||f||_1$. Hence $\pi$ is continuous. To show that $\pi$ is a $^*$-representation we must show that $\pi(f^*) = \pi(f)^*$ and that $\pi(f*g) = \pi(f)\pi(g)$. To show the first of these, let $f \in L^1(G)$. Then

$$\begin{aligned}(\pi(f^*)\xi|\eta) &= \int_G (U_s\xi|\eta)f^*(s)ds \\ &= \int_G (U_s\xi|\eta)\Delta(s^{-1})\overline{f(s^{-1})}ds \\ &= \int_G (\xi|U_{s^{-1}}\eta)\Delta(s^{-1})\overline{f(s^{-1})}ds \\ &= \int_G \overline{(U_{s^{-1}}\eta|\xi)}\Delta(s^{-1})\overline{f(s^{-1})}ds \\ &= \int_G \overline{(U_s\eta|\xi)}\,\overline{f(s)}ds \\ &= \overline{(\pi(f)\eta|\xi)} = (\xi|\pi(f)\eta) \\ &= (\pi(f)^*\xi|\eta),\end{aligned}$$

hence, $\pi(f^*) = \pi(f)^*$.

Now, suppose $f, g \in L^1(G)$. Then, from Fubini's theorem and left invariance of Haar measure,

$$\begin{aligned}(\pi(f*g)\xi|\eta) &= \int_G (U_s\xi|\eta)(f*g)(s)ds \\ &= \int_G (U_s\xi|\eta)\int_G f(t)g(t^{-1}s)dtds \\ &= \int_G f(t)\int_G (U_s\xi|\eta)g(t^{-1}s)dsdt \\ &= \int_G f(t)\int_G (U_{ts}\xi|\eta)g(s)dsdt \\ &= \int_G f(t)\int_G (U_s\xi|U_t^*\eta)g(s)dsdt \\ &= \int_G f(t)(\pi(g)\xi|U_t^*\eta)dt \\ &= \int_G f(t)(U_t\pi(g)\xi|\eta)dt \\ &= \int_G (U_t\pi(g)\xi|\eta)f(t)dt\end{aligned}$$

$$= (\pi(f)\pi(g)\xi|\eta);$$

hence $\pi(f*g) = \pi(f)\pi(g)$.

To see that $\pi$ is nondegenerate, suppose that $\xi$ is a nonzero vector in $H$. We wish to find $f \in L^1(G)$ such that $\pi(f)\xi \neq 0$. Since $s \to U_s$ is continuous, then $U_s\xi$ is nearly equal to $\xi$ for $s$ near the identity $e$. Therefore, if $f$ is a nonnegative function on $G$ whose support is small and $\int_G f(s)ds = 1$, then $\int_G (U_s\xi|\xi)f(s)ds$ is nearly equal to $\int_G (\xi|\xi)f(s)ds = ||\xi||^2$. In particular, $(\pi(f)\xi|\xi) \neq 0$, and consequently $\pi(f)\xi \neq 0$.

We have now shown that a continuous unitary representation of $G$ induces a nondegenerate $^*$-representation of $L^1(G)$ satisfying (1).

To establish the other direction, suppose that $\pi$ is a nondegenerate $^*$-representation of $L^1(G)$ on a Hilbert space $H$. We shall assume for now that $\pi$ is cyclic with cyclic vector $\zeta$; later this assumption will be dropped. Before proceeding further we note that, if $f \in L^1(G)$, then

$$({}_{s^{-1}}f)^* * ({}_{s^{-1}}f) = f^* * f \tag{4}$$

for all $s \in G$. Indeed, for $u \in G$,

$$\begin{aligned}
({}_{s^{-1}}f)^* * ({}_{s^{-1}}f)(u) &= \int_G ({}_{s^{-1}}f)^*(t)\, {}_{s^{-1}}f(t^{-1}u)dt \\
&= \int_G \Delta(t^{-1})\overline{f(s^{-1}t^{-1})}f(s^{-1}t^{-1}u)dt \\
&= \int_G \overline{f(s^{-1}t)}f(s^{-1}tu)dt \\
&= \int_G \overline{f(t)}f(tu)dt \\
&= \int_G \Delta(t^{-1})\overline{f(t^{-1})}f(t^{-1}u)dt \\
&= \int_G f^*(t)f(t^{-1}u)dt \\
&= (f^* * f)(u).
\end{aligned}$$

Utilizing (4) we next observe that if $\pi(f)\zeta = 0$, then $\pi({}_{s^{-1}}f)\zeta = 0$ for all $s \in G$. This is immediate from the calculation

$$(\pi({}_{s^{-1}}f)\zeta|\pi({}_{s^{-1}}f)\zeta) = (\pi(({}_{s^{-1}}f)^* * {}_{s^{-1}}f)\xi|\xi)$$

$$= (\pi(f^* * f)\zeta|\zeta)$$

$$= (\pi(f)\zeta|\pi(f)\zeta).$$

We define $U_s$, for each $s \in G$, on the dense subset $\{\pi(f)\zeta : f \in L^1(G)\}$ by the formula

$$U_s\pi(f)\zeta = \pi({}_{s^{-1}}f)\zeta.$$

This is well-defined: if $\pi(f)\zeta = \pi(g)\zeta$ for $f, g \in L^1(G)$, then $\pi(f - g)\zeta = 0$, $\pi({}_{s^{-1}}f - {}_{s^{-1}}g)\xi = 0$, and so $\pi({}_{s^{-1}}f)\zeta = \pi({}_{s^{-1}}g)\zeta$. Further, for $s \in G$, $f \in L^1(G)$, we have

$$(U_s\pi(f)\zeta|U_s\pi(f)\zeta) = (\pi({}_{s^{-1}}f)\zeta|\pi({}_{s^{-1}}f)\zeta)$$

$$= (\pi(({}_{s^{-1}}f)^* * {}_{s^{-1}}f)\zeta|\zeta)$$

$$= (\pi(f^* * f)\zeta|\zeta)$$

$$= (\pi(f)\zeta|\pi(f)\zeta),$$

so that $||U_s\xi|| = ||\xi||$ for all vectors $\xi$ of the form $\pi(f)\zeta$. Because these are dense in $H$, we can extend $U_s$ uniquely by continuity to a unitary operator, again denoted by $U_s$ on $H$. We clearly have $U_e = I$, the identity operator, and the calculation

$$U_{st}\pi(f)\zeta = \pi({}_{(st)^{-1}}f)\zeta = \pi({}_{s^{-1}}({}_{t^{-1}}f))\zeta$$

$$= U_s\pi({}_{t^{-1}}f)\zeta = U_sU_t\pi(f)\zeta$$

shows that $U_{st} = U_sU_t$ for all $s, t \in G$.

We next claim that the map $s \to U_s\xi$ from $G$ into $H$ is continuous for each $\xi$ in $H$. For $s_o, s \in G$,

$$||U_{s_o}\pi(f)\zeta - U_s\pi(f)\zeta|| = ||\pi({}_{s_o^{-1}}f - {}_{s^{-1}}f)\zeta||$$

$$\leq ||{}_{s_0^{-1}}f - {}_{s^{-1}}f||_1 ||\zeta||.$$

Therefore, given $\xi \in H$, $\varepsilon > 0$, and $s_0 \in G$, choose $f$ in $L^1(G)$ so that $||\pi(f)\zeta - \xi|| < \varepsilon/3$ and then a neighborhood $N(s_0)$ such that, for $s \in N(s_0)$, we have

$$||{}_{s_0^{-1}}f - {}_{s^{-1}}f||_1 < \varepsilon/(3||\zeta||).$$

This is possible by (56.2) and the cyclicity. We then have, for $s \in N(s_0)$ that

$$||U_{s_0}\xi - U_s\xi|| \leq 2||\xi - \pi(f)\zeta|| + ||U_{s_0}\pi(f)\zeta - U_s\pi(f)\zeta||$$

$$< 2\varepsilon/3 + \varepsilon/3 = \varepsilon$$

and so the claim is established.

We observe next that $U_s\pi(f)\xi = \pi({}_{s^{-1}}f)\xi$ for an *arbitrary* vector in $H$ and any $f \in L^1(G)$. For, if $\xi = \pi(g)\zeta$ for some $g \in L^1(G)$, then

$$U_s\pi(f)\xi = U_s\pi(f)\pi(g)\zeta = U_s\pi(f*g)\zeta$$

$$= \pi({}_{s^{-1}}(f*g))\zeta = \pi(({}_{s^{-1}}f)*g)\zeta$$

$$= \pi({}_{s^{-1}}f)\pi(g)\zeta = \pi({}_{s^{-1}}f)\xi$$

and the case for arbitrary $\xi$ follows by continuity since vectors of the form $\pi(g)\zeta$ are dense in $H$.

If $\pi$ is not a cyclic representation, we can decompose it into a direct sum, $\pi = \oplus_\alpha \pi_\alpha$, of cyclic representations $\pi_\alpha$, as in (26.10). For each $\pi_\alpha$ we form a continuous unitary representation $U_\alpha$ as described above, and then form the direct sum $U = \oplus_\alpha U_\alpha$ of the $U$. Then $U$ and $\pi$ are related by

$$U_s\pi(f)\xi = \pi({}_{s^{-1}}f)\xi, \quad \xi \in H. \tag{5}$$

For, if $\xi \in H = \oplus_\alpha H_\alpha$ and $\xi_\alpha$ denotes the projection of $\xi$ on $H_\alpha$, we know that

$$(U_\alpha)_s\pi_\alpha(f)\xi_\alpha = \pi_\alpha({}_{s^{-1}}f)\xi_\alpha,$$

from which (5) follows.

Finally, we must show that the correspondence indicated between continuous unitary representations and nondegenerate $^*$-representations is one-to-one. Suppose that $\pi$ has arisen from $U'$ and that $U$ has arisen from $\pi$ by formula (2). Then, for $f \in L^1(G)$ and vectors $\xi,\eta$,

$$(\pi(f)\xi|\eta) = \int_G (U'_s\xi|\eta)f(s)ds$$

and

$$\begin{aligned}(\pi({}_{t^{-1}}f)\xi|\eta) &= \int_G (U'_s\xi|\eta)f(t^{-1}s)ds \\ &= \int_G (U'_{t^{-1}s}\xi|U'_{t^{-1}}\eta)f(t^{-1}s)ds \\ &= (\pi(f)\xi|U'_{t^{-1}}\eta) \\ &= (U'_t\pi(f)\xi|\eta);\end{aligned}$$

but this is also $(U_t\pi(f)\xi|\eta)$ by definition, so that $U_t = U'_t$ for all $t \in G$, since $\{\pi(f)\xi\colon f \in L^1(G),\ \xi \in H\}$ is dense by (V.1).

On the other hand, suppose that $U$ has arisen from $\pi'$ and $\pi$ from $U$. The functional $h \to (\pi'(h)\xi|\eta)$ is, by (26.13) and the Riesz representation theorem, a complex-valued integral on $C_{oo}(G)$ for $\xi,\ \eta \in H$; we may write it, say, as

$$(\pi'(h)\xi|\eta) = \int_G h(s)d\mu_{\xi,\eta}(s).$$

Then

$$(\pi'(f*g)\xi|\eta) = \int_G\int_G f(t)g(t^{-1}s)dtd\mu_{\xi,\eta}(s)$$

and, interchanging the order of integration by Fubini's theorem, we obtain

$$\begin{aligned}(\pi'(f*g)\xi|\eta) &= \int_G\int_G {}_{t^{-1}}g(s)d\mu_{\xi,\eta}(s)f(t)dt \\ &= \int_G (\pi'({}_{t^{-1}}g)\xi|\eta)f(t)dt\end{aligned}$$

$$= \int_G (U_t \pi'(g)\xi|\eta) f(t)dt$$

$$= (\pi(f)\pi'(g)\xi|\eta)$$

by definition of $\pi$. But this is also equal to $(\pi'(f)\pi'(g)\xi|\eta)$ and, since the set $\{\pi'(g)\xi\colon g \in C_{oo}(G),\ \xi \in H\}$ is dense in $H$, it follows that $\pi(f) = \pi'(f)$ for all $f \in C_{oo}(G)$. By the continuity of $\pi$ and $\pi'$ (26.13) and the denseness of $C_{oo}(G)$ in $L^1(G)$, $\pi = \pi'$. Therefore the correspondence is one-to-one and the proof of (56.6) is complete. □

(56.7) PROPOSITION. *If* $U$ *and* $\pi$ *are representations which are related as in (56.6), then* $U$ *is irreducible iff* $\pi$ *is irreducible.*

*Proof*. Suppose that $U$ is a *reducible* continuous unitary representation of $G$ on a Hilbert space $H$. Then there exists an operator $P \neq \lambda I$, $\lambda \in C$, such that $PU_s = U_sP$ for all $s \in G$. For example, we can take $P$ to be the orthogonal projection on a nonzero invariant subspace. Then, for $f \in L^1(G)$ and $\xi, \eta \in H$,

$$(P\pi(f)\xi|\eta) = (\pi(f)\xi|P\eta) = \int_G (U_s\xi|P\eta)f(s)ds$$

$$= \int_G (PU_s\xi|\eta)f(s)ds$$

$$= \int_G (U_sP\xi|\eta)f(s)ds = (\pi(f)P\xi|\eta),$$

and since this holds for all $\xi, \eta \in H$, we have $P\pi(f) = \pi(f)P$ and $\pi$ is reducible by (26.3).

Conversely, suppose $\pi$ is reducible, and let $P$ commute with all $\pi(f)$. Then, for $\xi \in H$,

$$PU_s\pi(f)\xi = P\pi({}_{s^{-1}}f)\xi = \pi({}_{s^{-1}}f)P\xi$$

$$= U_s\pi(f)P\xi = U_sP\pi(f)\xi,$$

and since the vectors $\pi(f)\xi$ are dense in $H$ ($\pi$ is nondegenerate; see (Example V.1)), it follows that $PU_s = U_sP$, and so $U$ is reducible, again by (the analogue for unitary representations of) (26.3). □

Suppose that $f \to T_f$ is the (faithful, (56.5)) left regular representation of $L^1(G)$ on the Hilbert space $L^2(G)$. The set of operators

$$\{T_f + \lambda I: f \in L^1(G),\ \lambda \in C\}$$

is a $^*$-subalgebra of $B(L^2(G))$ containing the identity operator I. Let A denote its closure in the norm topology of $B(L^2(G))$. Then A is a C*-algebra.

If $\Psi$ is any $^*$-representation of A on a nonzero Hilbert space H which maps the identity operator onto itself and $\pi$ is the $^*$-representation of $L^1(G)$ on H defined by composing $\Psi$ with the left regular representation, i.e.,

$$\pi(f) = \Psi(T_f), \quad f \in L^1(G),$$

then

$$\{\pi(f): f \in L^1(G)\}' = \{\Psi(x): x \in A\}',$$

where the prime denotes the commutant (B.6.18); that is, the operators in B(H) commuting with all $\pi(f)$ coincide with the operators commuting with all $\Psi(x)$. To see this, we note that the following conditions on an operator S in B(H) are equivalent: S commutes with $\pi(f) = \Psi(T_f)$ for all f in $L^1(G)$; S commutes with $\pi(f) + \lambda I = \Psi(T_f + \lambda I)$ for all f in $L^1(G)$ and scalars $\lambda$ (because $\Psi$ maps I into I); $T_f + \lambda I \in \Psi^{-1}(\{S\}')$ for all f in $L^1(G)$ and $\lambda$; $A \subseteq \Psi^{-1}(\{S\}')$ (because of the continuity of $\Psi$ (26.13)); $\Psi(A) \subseteq \{S\}'$; S commutes with $\Psi(x)$ for all x in the C*-algebra A. In particular, it follows that the $^*$-representation $\pi$ of $L^1(G)$ is irreducible iff the $^*$-representation $\Psi$ of the C*-algebra A is irreducible.

We are now prepared to prove the Gelfand-Raikov theorem.

(56.8) THEOREM. *(Gelfand-Raikov). A locally compact group* G *always has enough irreducible continuous unitary representations to separate the points of* G*: given* $s \neq e$*, there is an irreducible continuous unitary representation* U *of* G *on a Hilbert space* H *such that* $U_s \neq I$.

*Proof.* Let s be an element in G such that $s \neq e$, and choose $f \in L^1(G)$ such that ${}_{s^{-1}}f \neq f$. For example, choose any $f \in C_{oo}(G)$ such that $f(s^{-1}) \neq f(e)$.

Let $h = {}_{s^{-1}}f - f$. Since $h \neq 0$ and the left regular representation

$f \to T_f$ of $L^1(G)$ on $L^2(G)$ is one-to-one (56.5), we have $T_h \neq 0$. Hence, by the Gelfand-Naimark completeness theorem (30.7), applied to the C*-algebra $A$ constructed above, there is an irreducible *-representation $\Psi: A \to B(H)$ on some Hilbert space $H$ such that $\Psi(I) = I$ and $\Psi(T_h) \neq 0$. Define $\pi: L^1(G) \to B(H)$ by forming the composition of $\Psi$ and $f \to T_f$:

$$\pi(f) = \Psi(T_f), \quad f \in L^1(G).$$

Since $\pi$ is an *irreducible* *-representation (by the discussion just preceding (56.8)) which is not identically zero, it is nondegenerate. By (56.7) the continuous unitary representation $U$ of $G$ which arises from $\pi$ is also irreducible. To complete the proof, we note that

$$\pi({}_{s^{-1}}f) - \pi(f) = \pi({}_{s^{-1}}f - f) = \pi(h) = \Psi(T_h) \neq 0;$$

hence $U_s\pi(f)\xi = \pi({}_{s^{-1}}f)\xi \neq \pi(f)\xi$ for some $\xi \in H$; so $U_s \neq I$ as claimed. □

## §57. *Unitary representations and positive definite functions.*

A continuous complex-valued function $\phi$ on a topological group $G$ is called *positive definite* if for any finite set of elements $s_1,\ldots,s_n$ in $G$ and any complex numbers $\lambda_1,\ldots,\lambda_n$, the inequality

$$\sum_{i=1}^{n}\sum_{j=1}^{n} \phi(s_i^{-1}s_j)\overline{\lambda}_i\lambda_j \geq 0 \qquad (1)$$

holds.

If $U$ is a continuous unitary representation of a topological group $G$ and $\xi$ is a vector in the representation space $H(U)$ of $U$, then the function

$$\phi(s) = (U_s\xi|\xi), \quad s \in G \qquad (2)$$

is positive definite. Such a function $\phi$ is called a *diagonal element* of the unitary representation $U$. The fact that $\phi$ is positive definite follows from the calculation

$$\sum_{i=1}^{n}\sum_{j=1}^{n} \phi(s_i^{-1}s_j)\overline{\lambda}_i\lambda_j = \sum_{i=1}^{n}\sum_{j=1}^{n} (U_{s_i^{-1}s_j}\xi|\xi)\overline{\lambda}_i\lambda_j$$

$$= \Big(\sum_{j=1}^{n}\lambda_jU_{s_j}\xi\Big|\sum_{i=1}^{n}\lambda_iU_{s_i}\xi\Big) = \Big|\Big|\sum_{i=1}^{n}\lambda_iU_{s_i}\xi\Big|\Big|^2 \geq 0.$$

We show in this section, by utilizing the GNS construction, that *every* positive definite function $\phi$ on G has the form (2) for some continuous unitary representation U on a Hilbert space H and a cyclic vector $\xi$ in H. We first observe some elementary properties of positive definite functions.

(57.1) PROPOSITION. *Let* $\phi$ *be a positive definite function on* G. *Then:*

*(i)* $\phi(e) \geq 0$;

*(ii)* $\phi(s^{-1}) = \overline{\phi(s)}$;

*(iii)* $|\phi(s)| \leq \phi(e)$ *for all* $s \in G$;

*(iv)* $\overline{\phi(s)}$ *is also positive definite.*

*Proof.* (i) Let A be the $n \times n$ matrix whose (i,j)-element is $\phi(s_i^{-1}s_j)$. Then the inequality (1) implies

$$(A\xi,\xi) \geq 0 \tag{3}$$

for any $\xi \in C^n$. Putting $\xi$ equal to the transpose of (1,0,...,0), we get $\phi(e) \geq 0$.

(ii) Since $(A\xi,\xi)$ is real by (3), we have $(A\xi,\xi) = (\xi,A^*\xi) = (A^*\xi,\xi)$ for all $\xi \in C^n$. From the polarization identity, we obtain $(A\xi,\eta) = (A^*\xi,\eta)$ for all $\xi,\eta$ in $C^n$. Hence $A = A^*$ and A is self-adjoint. This implies $\phi(s^{-1}) = \overline{\phi(s)}$.

(iii) Since the self-adjoint matrix

$$\begin{pmatrix} \phi(e) & \phi(s) \\ \overline{\phi(s)} & \phi(e) \end{pmatrix}$$

is positive semidefinite, its determinant, which is the product of its eigenvalues, is nonnegative. Thus $|\phi(s)| \leq \phi(e)$.

(iv) is a consequence of (1) and (ii). □

(57.2) THEOREM. *Let* G *be a topological group. If* $\phi$ *is a positive definite function on* G, *there is a continuous unitary representation* U *of* $G$ *on a Hilbert space* $H$ *and a cyclic vector* $\xi_o$ *in* $H$ *such that* $\phi(s) = (U_s\xi_o|\xi_o)$ *for all* $s \in G$.

*Proof.* Let $H_o$ denote the set of all complex-valued functions $f$ on the group $G$ such that $f(s) = 0$ for all but finitely many $s \in G$. Under pointwise addition and scalar multiplication $H_o$ is a vector space. For $f, g \in H_o$ define

$$(f|g) = \sum_{s,t\in G} \phi(t^{-1}s)f(s)\overline{g(t)}.$$

This is linear in $f$, conjugate-linear in $g$ and, since $\phi$ is positive definite,

$$(f|f) \geq 0 \quad\text{and}\quad \overline{(g|f)} = f|g).$$

Thus $(\cdot|\cdot)$ is a nonnegative semidefinite inner product on $H_o$. Let

$$K = \{f \in H_o\colon (f|f) = 0\}.$$

Then $K$ is a linear subspace of $H_o$ and we form the quotient space $H_o/K$. Define an inner product on $H_o/K$ by

$$(f + K|g + K) = (f|g),$$

and let $H$ denote the Hilbert space completion of $H_o/K$.

For each $s \in G$, define an operator $U_s$ on $H_o$ by $(U_sf)(t) = f(s^{-1}t)$. Then

$$\begin{aligned}(U_sf|U_sg) &= \sum_{t,u} \phi(u^{-1}t)f(s^{-1}t)\overline{g(s^{-1}u)}\\ &= \sum_{t,u} \phi(u^{-1}s^{-1}st)f(t)\overline{g(u)}\\ &= \sum_{t,u} \phi(u^{-1}t)f(t)\overline{g(u)}\\ &= (f|g).\end{aligned}$$

Hence $K$ is an invariant subspace of $U_s$, and therefore $U_s$ can be considered as an operator on $H_o/K$ such that

$$(U_s(f + K)|U_s(g + K)) = (f + K|g + K).$$

In particular, $U_s$ is isometric on $H_o/K$ and so it can be uniquely extended to an isometric operator, again denoted by $U_s$, on H. Clearly, the correspondence $s \to U_s$ is a representation of G on H; since $U_{s^{-1}} = U_s^{-1}$, it is a unitary representation of G.

Now, let $f_o$ be the function in $H_o$ defined by

$$f_o(s) = \begin{cases} 1, & \text{if } s = e, \\ 0, & \text{if } s \neq e, \end{cases}$$

and set $\xi_o = f_o + K$. Then $\xi_o \in H$ and, for all $s \in G$, we have

$$(U_s\xi_o|\xi_o) = (U_s(f_o + K)|f_o + K) = (U_sf_o|f_o)$$

$$= \sum_{t,u} \phi(u^{-1}t)f_o(s^{-1}t)\overline{f_o(u)}$$

$$= \phi(s).$$

It is clear that $\xi_o$ is a cyclic vector for the representation.

Finally, we must show that $s \to U_s$ is continuous, i.e., that $s \to U_s\xi$ is continuous from G into H for each $\xi \in H$. Since $\phi$ is continuous, then for *fixed* vectors $\xi = f + K$ and $\eta = g + K$ in $H_o/K$, the relation

$$(U_s\xi|\eta) = (U_sf|g) = \sum_{t,u} \phi(u^{-1}t)f(s^{-1}t)\overline{g(u)}$$

$$= \sum_{t,u} \phi(u^{-1}st)f(t)\overline{g(u)}$$

shows that $(U_s\xi|\eta)$ is a continuous function of s. Since $H_o/K$ is dense in H, given *any* $\xi, \eta \in H$, there exist sequences $\{\xi_n\}$ and $\{\eta_n\}$ in $H_o/K$ such that $\xi_n \to \xi$ and $\eta_n \to \eta$; therefore, since $||U_s|| = 1$ for each s, we have

$$|(U_s\xi|\eta) - (U_s\xi_n|\eta_n)| \leq |(U_s\xi|\eta) - (U_s\xi|\eta_n)|$$

$$+ |(U_s\xi|\eta_n) - (U_s\xi_n|\eta_n)|$$

$$\leq ||\xi||\cdot||\eta - \eta_n|| + ||\xi - \xi_n||\cdot||\eta_n||,$$

and so the function $(U_s\xi|\eta)$ is continuous for all $\xi,\eta$ in H. But then the relation

$$\|U_s\xi - U_t\xi\|^2 = \|U_{t^{-1}s}\xi - \xi\|^2$$
$$= 2\|\xi\|^2 - 2\mathrm{Re}(U_{t^{-1}s}\xi|\xi)$$

shows that $s \to U_s\xi$ is continuous for each $\xi \in H$. □

As an application of (57.2) we prove the following:

(57.3) PROPOSITION. *Any positive definite function* $\phi$ *on a topological group* G *is uniformly continuous. In fact,* $\phi$ *satisfies the inequality*

$$|\phi(s) - \phi(t)|^2 \leq 2\phi(e)\{\phi(e) - \mathrm{Re}\phi(s^{-1}t)\}$$

*for all* $s, t \in G$.

*Proof.* In view of (57.2), we can assume that $\phi$ is of the form $\phi(s) = (U_s\xi|\xi)$ for some unitary representation U of G on a Hilbert space H, and $\xi \in H$. Then we have

$$|\phi(s) - \phi(t)|^2 = |((U_s - U_t)\xi|\xi)|^2$$
$$\leq \|\xi\|^2\|U_s\xi - U_t\xi\|^2$$
$$= 2\phi(e)\{\|\xi\|^2 - \mathrm{Re}(U_s\xi|U_t\xi)\}$$
$$= 2\phi(e)\{\phi(e) - \mathrm{Re}\phi(s^{-1}t)\}. \quad \square$$

We remark that a substantial literature exists concerning positive definite functions on groups and their applications to harmonic analysis and group representations. We refer the reader to Berg, Christensen and Ressel [1], Dixmier [5, pp. 286-303], Hewitt and Ross [2, pp. 253-327], and Stewart [1] for more information.

## §58. *Completely positive mappings and Stinespring's theorem.*

In 1955 W. F. Stinespring [1] proved a remarkable theorem concerning the structure of certain linear mappings, called completely positive mappings, of a C*-algebra A into the algebra B(H) of bounded linear operators on a Hilbert space H. During the ensuing years this theorem and completely positive mappings have had profound and far-reaching consequences in the study of operator algebras, group representations, and operator theory.

The purpose of this section is to present Stinespring's theorem, whose proof is based on the fundamental GNS-construction, and observe a few of its consequences. We presume that the reader is familiar with the most elementary facts concerning the algebraic tensor product of vector spaces and linear maps.

Let A be a C*-algebra; for each positive integer n let $M_n$ denote the C*-algebra of complex $n \times n$ matrices. The algebraic tensor product $A \otimes M_n$ consists of all $n \times n$ matrices with entries from A. We therefore denote $A \otimes M_n$ by $M_n(A)$. Using the standard matrix operations in $M_n(A)$ and setting $(a_{ij})^* = (a_{ji}^*)$, it becomes a *-algebra. Further there is a unique norm on $M_n(A)$ under which it becomes a C*-algebra. The existence of this norm follows by tensoring faithful *-representations of A and $M_n$; the details of this construction are of little importance for us. The uniqueness follows from (24.2).

If A and B are C*-algebras and $\phi$ is a linear map of A into B, we can define a linear map $\phi_n\colon M_n(A) \to M_n(B)$ by applying $\phi$ to each element of $M_n(A)$. The map $\phi$ is said to be *completely positive* if each $\phi_n$ is positive, $n \geq 1$. In other words, if $I_n$ is the identity map of $M_n$ onto itself, then $\phi$ is completely positive iff $\phi \otimes I_n\colon A \otimes M_n \to B \otimes M_n$ is positive for all $n \geq 1$.

A useful characterization of the positivity of the linear maps $\phi_n$ is given by:

(58.1) PROPOSITION. *Let* $\phi\colon A \to B$ *be a linear map of* C*-*algebras. Then for any positive integer* n *the map* $\phi_n$ *is positive if and only if*

$$\sum_i \sum_j y_i^* \phi(x_i^* x_j) y_j \geq 0 \tag{1}$$

*for every* $x_1, \ldots, x_n \in A$ *and* $y_1, \ldots, y_n \in B$.

*Proof.* Assume $\phi_n\colon M_n(A) \to M_n(B)$ is positive, $x_1, \ldots, x_n \in A$, and $y_1, \ldots, y_n \in B$. Let $x \in M_n(A)$ be defined by $x_{ij} = x_i^* x_j$. It is clear that $x \geq 0$ since $x = c^*c$, where c is the matrix whose first row is $(x_1, \ldots, x_n)$ and whose other entries are 0. Now we have $(\phi(x_i^* x_j)) \geq 0$, and we denote $\phi(x_i^* x_j)$ by $a_{ij}$ and the matrix $(a_{ij})$ by a. By the positivity of a we can write $a = d^*d$ for some $d \in M_n(B)$, i.e., $a_{ij} =$

$\sum_k d_{ki}^* d_{kj}$. By (12.3) it will suffice to show that for each $k = 1,\ldots,n$

$$\sum_i \sum_j y_i^* d_{ki}^* d_{kj} y_j \geq 0.$$

However $\sum_i \sum_j y_i^* d_{ki}^* d_{kj} y_j = (\sum_i d_{ki} y_i)^* (\sum_j d_{kj} y_j)$, which is a positive element of $B$.

Conversely, assume that (1) holds for all $x_1,\ldots,x_n \in A$ and $y_1,\ldots,y_n \in B$. Let $a$ be a positive element of $M_n(A)$. As in the proof of necessity we can write $a$ as the sum of elements having the form $(a_i^* a_j)$ for some $a_1,\ldots,a_n \in A$. By the linearity of $\phi$ and (12.3) it suffices to show that $\phi(a) \geq 0$ for $a$ having this special form. Let $\pi$ be any cyclic $^*$-representation of $B$ on a Hilbert space $H$ with cyclic vector $\xi_o$. Denote by $\tilde{\pi}$ the corresponding representation of $M_n(B)$ on the Hilbert space $H^n$, i.e., if $b = (b_{ij}) \in M_n(B)$

$$\tilde{\pi}(b)(\xi_1,\ldots,\xi_n) = (\sum_j \pi(b_{1j})\xi_j,\ldots,\sum_j \pi(b_{nj})\xi_j)$$

for $\xi_1,\ldots,\xi_n \in H$. For any $\xi = (\xi_1,\ldots,\xi_n) \in H^n$ we have

$$(\tilde{\pi}(\phi(a))\xi|\xi) = \sum_k (\eta_k|\xi_k)$$

where $\eta_k = \sum_j \pi(\phi(a_{kj}))\xi_j$. Now

$$\sum_k (\eta_k|\xi_k) = \sum_j \sum_k (\pi(\phi(a_{kj}))\xi_j|\xi_k).$$

For each $k = 1,\ldots,n$ choose a sequence $\{y_{k,m}: m = 1,2,3,\ldots\}$ in $B$ such that $\lim_{m\to\infty} \pi(y_{k,m})\xi_o = \xi_k$. We then conclude

$$(\pi(\phi(a))\xi|\xi) = \lim_{m\to\infty} \sum_j \sum_k (\pi(\phi(a_{ij}))\pi(y_{j,m})\xi_o|\pi(y_{k,m})\xi_o)$$

$$= \lim_{m\to\infty} (\pi(\sum_j \sum_k y_{k,m}^* \phi(a_{ij}) y_{j,m})\xi_o|\xi_o) \geq 0,$$

since $a_{ij} = a_i^* a_j$ and a $^*$-representation preserves positivity of elements. Since $(\oplus_\alpha \pi_\alpha)^\sim = \oplus_\alpha \tilde{\pi}_\alpha$, we find by (26.10) that $\tilde{\pi}(\phi(a)) \geq 0$ for every $^*$-representation $\pi$ of $B$. By (19.1) $B$ has an isometric $^*$-representation $\pi$. For such a representation $\tilde{\pi}$ is faithful. Therefore $\phi(a) \geq 0$. □

In the next two propositions we shall see that positivity and complete positivity are equivalent if either the domain or range is commutative.

(58.2) PROPOSITION. *Let* A *and* B *be* $C^*$*-algebras. If* B *is commutative, then every positive linear map* $\phi: A \to B$ *is completely positive. In particular every positive linear functional on a* $C^*$*-algebra is completely positive.*

*Proof.* By (7.1) we may assume that $B = C_o(T)$ for some locally compact Hausdorff space T. For every $x_1,\ldots,x_n \in A$ and $y_1,\ldots,y_n \in B$ we have for $t \in T$:

$$[\sum_i \sum_j y_i^* \phi(x_i^* x_j) y_j](t) = \sum_i \sum_j \overline{y_i(t)} \phi(x_i^* x_j)(t) y_j(t)$$

$$= [\sum_i \sum_j \overline{y_i(t)} \phi(x_i^* x_j) y_j(t)](t)$$

$$= [\phi(\sum_i \sum_j \overline{y_i(t)} x_i^* x_j y_j(t)](t)$$

$$= [\phi((\sum_i y_i(t) x_i)^* (\sum_j y_j(t) x_j))](t) \geq 0.$$

By (58.1) it follows that $\phi$ is completely positive. □

(58.3) PROPOSITION. *Let* A *and* B *be* $C^*$*-algebras with* A *commutative. Then every positive linear map* $\phi: A \to B$ *is completely positive.*

*Proof.* By the Gelfand-Naimark theorems (7.1) and (19.1) we may assume that $A = C_o(T)$ for some locally compact Hausdorff space T and that B is a norm-closed subalgebra of B(H) for some Hilbert space H. By (58.1) it suffices to verify that if $\xi_1,\ldots,\xi_n$ are vectors in H, and $x_1,\ldots,x_n \in A$,

$$\sum_i \sum_j (\phi(x_i^* x_j)\xi_j | \xi_i) \geq 0.$$

For each $i, j = 1,\ldots,n$ let

$$f_{ij}(x) = (\phi(x)\xi_j | \xi_i), \quad x \in A.$$

Now, using the Riesz representation theorem, the continuous (see Exercise (X.21)) linear functionals $f_{ij}$ can be represented as

$$f_{ij}(x) = \int_T x \, d\mu_{ij}, \text{ for all } x \in A,$$

for complex, regular Borel measures $\mu_{ij}$ on T. Let $\mu = \sum_i \sum_j |\mu_{ij}|$, where $|\mu_{ij}|$ denotes the total variation of $\mu_{ij}$. By the Radon-Nikodym theorem there exist complex-valued functions $h_{ij} \in L^1(T,\mu)$ such that

$$f_{ij}(x) = \int_T xh_{ij}\, d\mu, \text{ for all } x \in A.$$

For any $\lambda_1,\ldots,\lambda_n \in C$ we have

$$\sum_i \sum_j f_{ij}(x^*x)\overline{\lambda}_i\lambda_j = \sum_i \sum_j (\phi(x^*x)\xi_j|\xi_i)\overline{\lambda}_i\lambda_j = (\phi(x^*x) \sum_j \lambda_j\xi_j|\sum_i \lambda_i\xi_i) \geq 0.$$

Since $\sum_i \sum_j f_{ij}\overline{\lambda}_i\lambda_j$ is a positive functional, $\sum_i \sum_j h_{ij}\overline{\lambda}_i\lambda_j \geq 0$ a.e. $[\mu]$. This implies that for any complex-valued simple functions $g_1,\ldots,g_n$ on T we have $\sum_i \sum_j h_{ij}\overline{g}_i g_j \geq 0$ a.e. $[\mu]$; and since measurable functions can be obtained as the pointwise limit of simple functions (see Rudin [3, p. 16]), we can conclude that $\sum_i \sum_j h_{ij}\overline{g}_i g_j \geq 0$ a.e. $[\mu]$ for any complex-valued measurable functions $g_1,\ldots,g_n$. Hence we have

$$\sum_i \sum_j (\phi(x_i^*x_j)\xi_j|\xi_i) = \sum_i \sum_j \int_T (x_i^*x_j)h_{ij}\, d\mu = \int_T \sum_i \sum_j h_{ij}\overline{x}_i x_j\, d\mu \geq 0. \quad \square$$

In general, there are positive linear maps which are not completely positive (Cf. Stinespring [1, p. 213]). The main result of this section is the following:

(58.4) THEOREM. *(Stinespring). Let* A *be a* $C^*$*-algebra with identity,* H *a Hilbert space, and* $\phi$ *a linear map of* A *into* B(H). *Then* $\phi$ *is completely positive iff* $\phi$ *has the form*

$$\phi(x) = V^*\pi(x)V, \quad x \in A,$$

*where* $\pi$ *is a* **-representation of* A *on some Hilbert space* K *and* V *is a bounded linear map of* H *into* K.

*Proof.* First, assume that $\phi$ has the form $\phi(x) = V^*\pi(x)V$, where V, $\pi$, and K are as in the theorem. To show that $\phi$ is completely positive, let $(x_{ij})$ be a positive $n \times n$ matrix over A, let $\xi_1,\ldots,\xi_n$ be vectors in H, and select elements $z_{ij} \in A$ such that $(x_{ij}) = (z_{ij})^*(z_{ij})$. Then

$$\sum_i \sum_j (\phi(x_{ij})\xi_j|\xi_i) = \sum_i \sum_j (V^*\pi(x_{ij})V\xi_j|\xi_i)$$

$$= \sum_i \sum_j (\pi(x_{ij})V\xi_j|V\xi_i)$$

$$= \sum_i \sum_j \sum_k (\pi(z_{kj})V\xi_j|\pi(z_{ki})V\xi_i)$$

$$= \sum_k ||\sum_j \pi(z_{kj})V\xi_j||^2 \geq 0,$$

which shows that the matrix $(\phi(x_{ij}))$ is a positive operator on $H \otimes C^n = H \oplus \cdots \oplus H$, and therefore $\phi_n$ is positive. Since $n$ was arbitrary, it follows that $\phi$ is completely positive.

Conversely, assume that $\phi$ is completely positive, and consider the vector space tensor product $A \otimes H$. The construction which follows parallels the GNS-construction once again. Define a bilinear form $(\cdot|\cdot)$ on $A \otimes H$ as follows: for $u = \sum_i x_i \otimes \xi_i$ and $v = \sum_j y_j \otimes \eta_j$ in $A \otimes H$, set

$$(u|v) = \sum_i \sum_j (\phi(y_j^* x_i)\xi_i|\eta_j).$$

Since $\phi$ is completely positive, we have $(u|u) \geq 0$, and hence $(\cdot|\cdot)$ is positive semidefinite. For each $x \in A$, define a linear operator $\psi(x)$ on $A \otimes H$ by

$$\psi(x)(\sum_i x_i \otimes \xi_i) = \sum_i xx_i \otimes \xi_i.$$

Then $\psi$ is easily seen to be an algebra homomorphism such that

$$(u|\psi(x)v) = (\psi(x^*)u|v)$$

for all $u, v \in A \otimes H$. Hence, for a fixed element $u \in A \otimes H$, it follows that $f(x) = (\psi(x)u|u)$ is a positive linear functional on $A$. Thus, by (22.11),

$$\begin{aligned}(\psi(x)u|\psi(x)u) &= (\psi(x^*x)u|u) = f(x^*x) \\ &\leq ||x^*x||f(e) = ||x||^2(u|u).\end{aligned} \tag{1}$$

Next, let $M = \{u \in A \otimes H: (u|u) = 0\}$. Then by the Schwarz inequality, as in the proof of (27.2), $M$ is a linear subspace of $A \otimes H$; by (1), $\psi(x)M \subseteq M$ for each $x \in A$, i.e., $M$ is invariant under each operator $\psi(x)$. Hence, the quotient space $(A \otimes H)/M$ becomes a pre-Hilbert space if we define

$$(u + M|v + M) = (u|v).$$

Let K denote the Hilbert space completion of $(A \otimes H)/M$; it follows again from (1) that there is a unique $^*$-representation $\pi$ of A on K defined by

$$\pi(x)(u + M) = \psi(x)u + M \quad \text{for} \quad x \in A,\ u \in A \otimes H.$$

Now, define a linear map V of H into K by $V\xi = e \otimes \xi + M$. Since

$$||V\xi||^2 = (\phi(e)\xi|\xi) \leq ||\phi(e)||\cdot||\xi||^2,$$

V is bounded. Finally, we have

$$(V^*\pi(x)V\xi|\eta) = (\pi(x)V\xi|V\eta) = (\psi(x)(e \otimes \xi)|e \otimes \eta)$$

$$= (x \otimes \xi|e \otimes \eta) = (\phi(x)\xi|\eta)$$

for all $\xi,\eta$ in H; hence $\phi(x) = V^*\pi(x)V$ for all $x \in A$, and the proof is complete. □

Let $\phi(x) = V^*\pi(x)V$ be as in (58.3); if $\phi(e) = I$, where I is the identity operator, then $V^*V = I$, and hence V is an isometric embedding of H in K. Therefore, H may be viewed as a subspace of K and the original equation takes the form $\phi(x) = P\pi(x)|H$, where P is the projection of K onto H. The new V is the inclusion map of H into K whose adjoint is P.

The following corollary of (58.3) was first proved by M. A. Naimark; we mention it here because it was the starting point of Stinespring's investigation. Aside from a change of terminology (measures to functionals via the Riesz representation theorem) the corollary follows easily from (58.3) and (58.2).

(58.4) COROLLARY. *Let* X *be a set,* $B$ *a* $\sigma$*-algebra of subsets of* X, H *a Hilbert space, and* $F: B \to B(H)$ *a map such that* $F(X) = I$, $F(E) \geq 0$ *for all* $E \in B$, *and* $F(\bigcup_{n=1}^{\infty} E_n) = \Sigma_{n=1}^{\infty} F(E_n)$ *(weakly). Then there exists a Hilbert space* K *which contains* H *as a subspace and a spectral measure* P *on* $B$ *such that* $F(E)Q = QP(E)Q$ *for all* $E \in B$, *where* Q *is the projection of* K *on* H.

Since every positive linear functional on a $C^*$-algebra is completely positive by (58.1), another important consequence of Stinespring's theorem is (see §§17, 27):

(58.5) COROLLARY. *(Gelfand-Naimark-Segal). Let* $A$ *be a* $C^*$*-algebra with identity* $e$*, and* $f$ *a positive functional on* $A$*. Then there exists a cyclic *-representation* $\pi$ *of* $A$ *on a Hilbert space* $H$*, unique up to unitary equivalence, and a cyclic vector* $\xi$ *in* $H$ *such that* $f(x) = (\pi(x)\xi|\xi)$ *for all* $x \in A$.

The next corollary, proved by R. V. Kadison in 1951 by different methods, is a valuable tool in the study of linear mappings of operator algebras. It is known as the "generalized Schwarz inequality."

(58.6) COROLLARY. *Let* $A$ *be a* $C^*$*-algebra with identity* $e$*,* $H$ *a Hilbert space, and* $\phi: A \to B(H)$ *a positive linear map such that* $||\phi|| \leq 1$*. Then* $\phi(x^2) \geq \phi(x)^2$ *for each self-adjoint element* $x$ *in* $A$.

*Proof.* Let $x = x^*$ in $A$, and consider the commutative $C^*$-algebra $A_o$ generated by $x$ and $e$. Restricting $\phi$ to $A_o$ we see from (58.3) that $\phi$ is completely positive. By (58.4), the map $\phi$ has the form

$$\phi(y) = V^*\pi(y)V, \quad y \in A_o,$$

where $\pi$ is a *-representation of $A_o$ on a Hilbert space $K$, and $V$ is a bounded linear map of $H$ into $K$. Since $||\phi|| \leq 1$ and $\phi(e) = V^*V$, we have $||V^*V|| \leq 1$. Hence, $||VV^*|| \leq 1$ and thus $VV^* \leq I$, where $I$ is the identity operator. Therefore,

$$\phi(x^2) = V^*\pi(x^2)V = V^*\pi(x)^*I\pi(x)V \geq V^*\pi(x)VV^*\pi(x)V = \phi(x)^2$$

as required. □

## §59. $C^*$-*algebra methods in quantum theory.*

In recent years the theory of $C^*$-algebras and von Neumann algebras has entered into the study of statistical mechanics and quantum theory. The motivation for this algebraic approach was a dissatisfaction among many physicists and mathematicians that the classical mathematical methods used in theoretical physics were too undiscriminating and too restricted in their range to solve the difficult problems encountered in these theories.

The basic principle of the algebraic approach to statistical mechanics and quantum theory is to avoid starting with a specific Hilbert space scheme and rather to emphasize that the primary objects of the theory are the fields (or observables) considered as purely algebraic quantities, together with their linear combinations, products, and limits in an appropriate topology. The representations of these objects as operators acting on a suitable Hilbert space can then be obtained in a way that depends essentially only on the states of the physical system under investigation. The principal tool needed to build the required Hilbert space and associated representation is the Gelfand-Naimark-Segal construction discussed earlier in §27.

A substantial literature has now emerged from this $C^*$-algebraic point of view and several books (see Bratteli and Robinson [1, 2], and G. Emch [1], for example) have been written with the express purpose of offering a systematic introduction to the ideas and techniques of the $C^*$-algebra approach to physical problems. The authors recommend these books to the reader who would like to pursue this subject further. They contain large bibliographies which should aid the interested reader wishing to learn more about this interesting application of operator algebras.

## EXERCISES

(X.1) Let $G$ be a locally compact abelian group. Show that the set $AP(G)$ of almost periodic functions on $G$ is a $C^*$-algebra.

(X.2) Let $P$ be a spectral measure, as defined in §52. Show that

$$|P_{\xi,\eta}(B)|^2 \leq P_{\xi,\xi}(B)P_{\eta,\eta}(B)$$

for all $\xi, \eta \in H$ and Borel sets $B$.

(X.3) If $P$ is a spectral measure, show that $P(B_1) \leq P(B_2)$ iff $B_1$ and $B_2$ are Borel sets with $P(B_2 \setminus B_1) = 0$.

(X.4) Let $T$ be a normal operator on a Hilbert space $H$, $f$ a bounded Borel function on the spectrum $\sigma(T)$, and let $S$ denote the operator $f(T)$. If $P_S$ and $P_T$ are the spectral decompositions of $S$ and $T$, respectively (see (52.5)), show that

$$P_S(B) = P_T(f^{-1}(B))$$

for each Borel subset $B$ in $\sigma(T)$.

(X.5) Let $T$ be a bounded hermitian operator on a Hilbert space $H$. Show there exists a family of orthogonal projections $\{P_t : t \in R\}$ on $H$ such that:

(a) $P_t P_u = P_u$ if $u \leq t$;

(b) every $P_t$ commutes with every operator in $B(H)$ which commutes with $T$;

(c) for every $\xi \in H$, $\lim_{\varepsilon \downarrow 0} \|P_{t-\varepsilon}\xi - P_t\xi\| = 0$;

(d) there are real numbers $a$ and $b$, $a < b$, such that $P_t = 0$ for $t \leq a$ and $P_t = I$ for $t \geq b$.

(e) for every $\xi \in H$ and every subdivision $a = t_0 < t_1 < \cdots < t_m = b$ such that $\max\{|t_j - t_{j-1}| : 1 \leq j \leq m\} < \varepsilon$, we have

$$\|T\xi - \sum_{j=1}^{m} t_{j-1}(P_{t_j} - P_{t_{j-1}})\xi\| < \varepsilon \cdot \|\xi\|;$$

(f) for $\xi, \eta \in H$, we have

$$(T\xi|\eta) = \int_a^b t\,d(P_t\xi|\eta),$$

the integral being taken as the usual Riemann-Stieltjes integral.

(X.6) Let $S$ be a bounded hermitian operator on a Hilbert space $H$. Prove, for all $\xi \in H$ and real numbers $t$, that

$$\|(e^{itS} - I)\xi\| \leq |t| \cdot \|S\xi\|.$$

(X.7) Let $G$ be a locally compact group and assume that $G$ is unimodular. Let $H = L^2(G)$, the Hilbert space of equivalence classes of measurable functions which are square-integrable with respect to Haar measure. For $f \in H$ and $s \in G$, define

$$[R(s)f](t) = f(ts), \quad t \in G,$$

and

$$[L(s)f](t) = f(s^{-1}t), \quad t \in G.$$

(a) Prove that R and L are continuous unitary representations of G on H; they are called the *right* and *left regular representation* of G, respectively.

(b) Show that R and L are irreducible iff G consists of the identity element only.

(c) Show that R and L are unitarily equivalent via the unitary operator $U: H \to H$ given by $(Uf)(s) = f(s^{-1})$, $f \in H$, $s \in G$.

(X.8) Let G be a locally compact unimodular group. Prove that a continuous complex-valued function $\phi$ on G is positive definite iff, for any bounded complex measure $\mu$ on G, we have

$$\int_G\int_G \phi(s^{-1}t)\overline{d\mu(s)}d\mu(t) \geq 0.$$

(X.9) Let $H_1$ and $H_2$ be Hilbert spaces and K the algebraic tensor product of $H_1$ and $H_2$. Then there is a unique inner product on K satisfying

$$(\xi_1 \otimes \xi_2|\eta_1 \otimes \eta_2) = (\xi_1|\eta_1)(\xi_2|\eta_2).$$

The completion of K with respect to this inner product is called the *tensor product* of the Hilbert spaces $H_1$ and $H_2$ and is denoted by $H_1 \otimes H_2$. If $T_i \in B(H_i)$ $(i = 1, 2)$ is a bounded linear operator on $H_i$, then there exists a unique bounded linear operator $T_1 \otimes T_2$ on $H_1 \otimes H_2$ satisfying

$$(T_1 \otimes T_2)(\xi_1 \otimes \xi_2) = T_1\xi_1 \otimes T_2\xi_2.$$

$T_1 \otimes T_2$ is called the *tensor product* of $T_1$ and $T_2$.

(a) If $U^i$ is a unitary representation of a topological group G on a Hilbert space $H_i$ $(i = 1, 2)$, show that the mapping $U: s \to U_s^1 \otimes U_s^2$ is a unitary representation of G on $H_1 \otimes H_2$. U is called the *tensor product* of $U^1$ and $U^2$ and is denoted $U^1 \otimes U^2$.

(b) Let $\phi_1$ and $\phi_2$ be positive definite functions on a topological group G. Show, using part (a), that the product $\phi = \phi_1 \cdot \phi_2$ is also positive definite on G.

(X.10) Give an example to show that Corollary (58.6) may fail if the hypothesis $||\phi|| \leq 1$ is omitted.

(X.11) Let A and B be unital $C^*$-algebras and $\phi$: $A \to B$ a unital positive linear mapping. ($\phi$ is *unital* if it maps the identity of A into the identity of B.) Show that $\phi(x^*x) \geq \phi(x^*)\phi(x)$ and $\phi(x^*x) \geq \phi(x)\phi(x^*)$ for every normal element $x \in A$.

(X.12) Let A and B be $C^*$-algebras. If $\phi$ is a completely positive mapping from A into B, prove that $\phi(x)^*\phi(x) \leq ||\phi||\phi(x^*x)$ for all x in A.

(X.13) A $U^*$-*algebra* is a $^*$-algebra A in which each element is a linear combination of quasi-unitary elements. If A has an identity, then a $U^*$-algebra is spanned by its unitaries. All Banach $^*$-algebras are $U^*$-algebras (cf. (22.6)). Prove that Theorem (58.3) remains valid if A is a $U^*$-algebra with identity.

(X.14) Let A be a $U^*$-algebra with identity and H a Hilbert space. If $\phi$: $A \to B(H)$ is a completely positive map which takes unitaries into unitaries, prove that $\phi$ is a $^*$-homomorphism.

(X.15) Let A be a $U^*$-algebra with identity. For $x \in A$, define

$$\gamma(x) = \sup\{||\pi(x)||: \pi \text{ a } ^*\text{-representation of } A \text{ on Hilbert space}\}.$$

Prove that:

(a) $\gamma(x)$ is finite for each $x \in A$;

(b) $\gamma$ is an algebra seminorm on A and satisfies $\gamma(x^*x) = \gamma(x)^2$ for all $x \in A$;

(c) $\gamma$ vanishes on the $^*$-radical $R^*(A)$;

(d) $A/R^*(A)$ is a normed $^*$-algebra in the norm induced by $\gamma$;

(e) the completion B of $A/R^*(A)$ in the norm $\gamma$ is a $C^*$-algebra;

(f) if H is a Hilbert space, and $\phi$: $A \to B(H)$ is a completely positive map, then $\phi$ is continuous with respect to the seminorm $\gamma$;

(g) if $\phi$: $A \to B(H)$ is a completely positive map, then there is a positive linear map $\tilde{\phi}$: $B \to B(H)$ such that $\phi = \tilde{\phi}\circ\tau$, where B is as in (e) and $\tau$ is the natural $^*$-homomorphism of A into B.

(X.16) Let $H$ be a Hilbert space. Utilize (58.4) and the results of problem (X.15) to show that every positive linear map from a commutative $U^*$-algebra with identity into $B(H)$ is completely positive.

(X.17) Let $A$ be a unital $C^*$-algebra. Let $f$ be a state on $A$ and $\pi_f$ the corresponding $^*$-representation of $A$ on a Hilbert space $H_f$. If $p$ is a state on $A$ and $\xi$ is a vector in $H_f$ such that $p(x) = (\pi_f(x)\xi|\xi)$ for all $x \in A$, prove there is a net $\{y_\alpha\}$ in $A$ such that $f(y_\alpha^* x y_\alpha) \to p(x)$ for all $x \in A$.

(X.18) Let $A$ be a unital $C^*$-algebra and let $f$ be a pure state on $A$. Choose any element $y$ in $A$ satisfying $f(y^*y) = 1$, and define a state $p$ on $A$ by $p(x) = f(y^*xy)$ for all $x \in A$. Prove that $p$ is a pure state on $A$.

(X.19) Let $A$ be a unital $C^*$-algebra and suppose that $f$ is an element of the pure state space of $A$, i.e., the weak$^*$-closure of the set of pure states on $A$. Choose any element $y \in A$ satisfying $f(y^*y) = 1$, and define a state $p$ on $A$ by $p(x) = f(y^*xy)$ for all $x \in A$. Prove that $p$ belongs to the pure state space of $A$.

(X.20) Let $A$ be a unital $C^*$-algebra and suppose that $f$ is an element of the pure state space of $A$ (see (X.19)). Consider the $^*$-representation $\pi_f$ of $A$ on a Hilbert space $H_f$ induced by $f$ and let $\xi$ be any unit vector in $H_f$. Prove that the state $p$ on $A$ defined by $p(x) = (\pi_f(x)\xi|\xi)$ for all $x \in A$ belongs to the pure state space of $A$.

(X.21) Show that every *positive* linear map between two $C^*$-algebras is continuous; that is, if $A$ and $B$ are $C^*$-algebras, $\phi: A \to B$ is linear, and $x \geq 0 \Rightarrow \phi(x) \geq 0$, then $\phi$ is continuous.

(X.22) Make the necessary modifications (using $\Delta$) of the proof of (56.2) to prove that right translation $G \to L^p(G)$ is continuous.

## Notes and Remarks

Our purpose here is to discuss several additional topics, mostly without proof, which are related in one way or another to the Gelfand-Naimark theorems and the classification of C*-algebras. Our discussion is informal and is meant primarily to inform the reader of further results which are likely to be of interest.

*Finite-dimensional* C**-algebras*. A complete description of all finite-dimensional complex C*-algebras is given by the following: If $A$ is a finite-dimensional C*-algebra, then $A$ is unital and can be decomposed into the direct sum $A = \Sigma_{k=1}^{m} \oplus A_k$, where each $A_k$ is isomorphic to the algebra of $n_k \times n_k$ complex matrices. The sequence $\{n_1, n_2, \ldots, n_m\}$ of positive integers is uniquely determined by $A$, up to permutations, and is a complete invariant for the algebraic structure of $A$ in the sense that if $B$ is another finite-dimensional C*-algebra with associated sequence $\{\overline{n}_1, \overline{n}_2, \ldots \overline{n}_{\overline{m}}\}$, then $A$ and $B$ are isomorphic iff $m = \overline{m}$ and there is a permutation $\sigma$ of $\{1,2,\ldots,m\}$ such that $\overline{n}_k = n_{\sigma(k)}$, $k = 1,2,\ldots m$. For a proof see Takesaki [1, pp. 50-51].

*The* C**-norm condition on matrix algebras.* In (16.1) it was shown that the C*-norm conditions $||x^*x|| = ||x||^2$ and $||x^*x|| = ||x^*||\cdot||x||$ on a Banach *-algebra $A$ are equivalent. The proof was nontrivial and required considerable ingenuity. We show here that if $A$ is the *-algebra of complex $n \times n$ matrices, where the involution is the usual adjoint operation, and if $A$ has a norm under which it is a unital Banach *-algebra, then the equivalence of the two C*-norm conditions can be proved quite directly. The argument is due to J. B. Deeds [1].

THEOREM. *Let* A *be the* $^*$*-algebra of complex* $n \times n$ *matrices, and suppose* $||\cdot||$ *is a norm on* A *under which* A *is a unital Banach* $^*$*-algebra. If* $||x^*x|| = ||x^*||\cdot||x||$ *for all* x *in* A, *then* $||x^*x|| = ||x||^2$ *for all* x *in* A.

*Proof.* If u is a unitary element of A, then $||u|| \geq |u|_\sigma = 1$. Similarly, $||u^*|| \geq |u^*|_\sigma = |u|_\sigma = 1$. Since $1 = ||1|| = ||u^*u|| = ||u^*||\cdot||u||$, with both $||u||$ and $||u^*||$ at least 1, it follows that $||u|| = 1$.

If $h \in A$ is hermitian, then (B.4.12) gives $|h|_\sigma = \lim_{n\to\infty} ||h^{2n}||^{1/2n} = \lim_{n\to\infty} ||h|| = ||h||$.

If $p \in A$ is positive and s is the positive square root of p (A.19), then since both are hermitian, $||p|| = |p|_\sigma = |s^2|_\sigma = |s|_\sigma^2 = ||s||^2$; hence $||p^{1/2}|| = ||p||^{1/2}$.

Now, suppose $x \in A$ is arbitrary and let $x = pu$ be its polar decomposition (A.21). Then $||x|| = ||pu|| \leq ||p||\cdot||u|| = ||p|| = ||(x^*x)^{1/2}|| = ||x^*x||^{1/2}$. Therefore, $||x||^2 \leq ||x^*x|| = ||x^*||\cdot||x||$, and so $||x|| \leq ||x^*||$. Replacing x by $x^*$, we obtain $||x^*|| = ||x||$ from which the theorem follows. □

*The Dauns-Hofmann theorems.* In an attempt to generalize the beautiful duality between locally compact Hausdorff spaces and commutative C$^*$-algebras provided by the Gelfand-Naimark theorem (7.1), one is led naturally to certain fiber spaces, all of whose fibers are C$^*$-algebras. The basic construction is as follows: Given a C$^*$-algebra A, let X denote the set of primitive ideals of A (B.5.8). For each $P \in X$, form the quotient A/P, and let B denote the disjoint union of the algebras A/P, as P runs through X:

$$B = \cup\,\{A/P\colon P \in X\}.$$

Let $\pi\colon B \to X$ be the projection defined by $\pi(x + P) = P$. For each $x \in A$, define $\hat{x}\colon X \to B$ by $\hat{x}(P) = x + P \in A/P$, and set $\hat{A} = \{\hat{x}\colon x \in A\}$. Note that $\pi\circ\hat{x}$ is the identity on X for each $x \in A$. The fibers (or stalks) of the "bundle" $\xi = (\pi,B,X)$ are the C$^*$-algebras $\pi^{-1}(P) = A/P$, $P \in X$. Let

$$\Gamma(\pi) = \{\sigma\colon X \to B \mid \pi\circ\sigma = 1_X\},$$

and note that $\hat{A} \subseteq \Gamma(\pi)$. The maps σ in Γ(π) are called cross-sections (or simply sections). Under pointwise operations $\Gamma(\pi)$ is a $^*$-algebra, and since A is semisimple, the mapping $x \to \hat{x}$ is a $^*$-isomorphism of A onto $\hat{A} \subseteq \Gamma(\pi)$.

Normally the set X of primitive ideals is given the *hull-kernel* (or *Jacobson*) topology; that is, if S is a nonempty subset of X, one defines $\overline{S} = h(k(S))$, where $k(S)$, the *kernel* of S, denotes the intersection of the ideals in S, and $h(k(S))$, the *hull* of $k(S)$, is the set of primitive ideals in X which contain $k(S)$. If S is empty, $\overline{S}$ is defined to be empty. The hull-kernel topology on X is $T_o$, and is compact if A has an identity (see Dixmier [5, pp. 69-72]).

If the hull-kernel topology on X is Hausdorff, then it is possible to put a topology on the set B so that the projection $\pi$ and each section are continuous; moreover, $\Gamma(\pi)$ becomes a $C^*$-algebra, and A is isometrically $^*$-isomorphic to $\Gamma(\pi)$ via the generalized Gelfand representation $x \to \hat{x}$ described above. This beautiful result was proved in 1961 by J. M. G. Fell [1], who also proved that *any* $C^*$-algebra A is classifiable as a $C^*$-algebra of sections of some $C^*$-bundle $\xi$ over a compact Hausdorff space $H(X)$ associated with X. Utilizing these results Fell was able to give a precise description of the group $C^*$-algebra of the $2 \times 2$ complex matrices of determinant 1. Special cases of Fell's results had been proved earlier in 1951 by I. Kaplansky [3] and some important related results were obtained independently by J. Tomiyama and M. Takesaki [1], and J. Tomiyama [1].

Motivated by these papers and also by certain problems in the representation of biregular rings by sheaves, J. Dauns and K. H. Hofmann [1] developed, during 1966-1968, a very general theory of "uniform fields" with the purpose of representing various algebraic structures (topological groups, rings, algebras, $C^*$-algebras, etc.) in the space of continuous sections. Thinking of a $C^*$-bundle loosely as a fiber space $\xi = (\pi, B, X)$, where $\pi: B \to X$ is a continuous open surjection of topological spaces, each fiber $\pi^{-1}(x)$, $x \in X$, is a $C^*$-algebra, and all algebraic operations are continuous in the respective fibers (including the involution and the norm), Dauns and Hofmann proved the following theorem:

THEOREM. *(Dauns-Hofmann). Let* A *be a* $C^*$*-algebra with identity and* X *the space of maximal ideals of the center of* A *with the hull-kernel topology. Then* A *is isometrically* $^*$*-isomorphic to the* $C^*$*-algebra of all continuous sections* $\Gamma(\pi)$ *of a* $C^*$*-bundle* $\xi = (\pi, B, X)$ *over* X. *The fiber* *above* $m \in X$ *is the quotient* $C^*$*-algebra* $A/mA$, *the isometric* $*$*-isomorphism is the Gelfand representation* $x \to \hat{x}$, *where* $\hat{x}(m) = x + mA$, *and the norm of* $\hat{x}$ *is given by*

$$||\hat{x}|| = \sup\{||\hat{x}(m)||: m \in X\}.$$

*Further, the real-valued map* $m \to ||\hat{x}(m)||$ *on* $X$ *is upper-semicontinuous for each* $x \in A$.

Although the original proof of Dauns and Hofmann was quite involved technically, a simple, self-contained, proof, using ideas of J. Varela [1], is available in the book by M. J. Dupré and R. M. Gillette [1]. Another proof, using category theory and sheaves, has been given by C. J. Mulvey [1]. The theorem is also treated in the book by G. K. Pedersen [1, pp. 106-108].

If the $C^*$-algebra $A$ has no identity, one can pass to $M(A)$, the so-called "multiplier algebra" of $A$ to obtain the corresponding result (see Dupré and Gillette [1]). When $A$ is commutative the Dauns-Hofmann theorem reduces to the classical Gelfand-Naimark theorem (7.1) so that it is, in fact, a true generalization and not simply an analogue of (7.1).

The potential value of representing a $C^*$-algebra $A$ by sections lies in the fact that one hopes to be able to transfer information from the fibers ($C^*$-algebras which are presumably better known and simpler in structure than $A$) of the representing bundle to the $C^*$-algebra of sections, and thus obtain information about the original $C^*$-algebra. A recent short paper by T. Becker [1] shows that certain pathology can enter into this process which reduces the usefulness of the Dauns-Hofmann theorem, at least in some situations.

The next theorem was also proved by Dauns and Hofmann [1].

THEOREM. *(Dauns-Hofmann). Let* $A$ *be a* $C^*$*-algebra with identity and* $\mathrm{Prim}(A)$ *the space of primitive ideals with the hull-kernel topology. If* $x \in A$ *and* $f$ *is a bounded continuous complex-valued function on* $\mathrm{Prim}(A)$, *then there exists a unique element* $fx \in A$ *such that, for all* $t \in \mathrm{Prim}(A)$,

$$(fx)(t) = f(t)x(t), \tag{1}$$

*where* $x(t)$ *denotes the canonical image of* $x$ *in the* $C^*$*-algebra* $A/t$.

The expression in (1), although written functionally, represents a product. Simplified proofs of the theorem have been published by J. Dixmier [4] and G. A. Elliott and D. Olesen [1]. Elliott [2] has also proved an abstract version of the theorem in the context of real Banach spaces. The relationship between the two Dauns-Hofmann theorems is briefly discussed in the paper by Becker [1], based on unpublished notes of K. H. Hofmann. For precise definitions and more information on sectional representations in bundles, the reader should consult K. H. Hofmann [1, 2, 3, 4], J. Dauns and K. H. Hofmann [1, 2], M. Dupré [1, 2, 3, 4, 5], M. Dupré and R. M. Gillette [1], J. M. G. Fell [1, 2, 3, 4], and the extensive bibliographies in these references.

*The Gelfand-Naimark theorems for real* $C^*$*-algebras. A real* $C^*$*-algebra* is a Banach $^*$-algebra $A$ with identity 1 over the reals such that for all $x \in A$:

(1°) $||x^*x|| = ||x||^2$;

(2°) $1 + x^*x$ is invertible

The involution is easily seen to be isometric, just as in the complex case, but unlike the situation for complex $C^*$-algebras, the invertibility of $1 + x^*x$ cannot be deduced from the other axioms. For example, if we view the complex numbers $C$ as a real Banach $^*$-algebra with absolute value norm and involution $x^* = x$ for all $x \in C$, then $C$ satisfies all of the axioms for real $C^*$-algebras except (2°), since $1 + i^*i = 0$.

Examples of real $C^*$-algebras are: any complex $C^*$-algebra (simply ignore scalar multiplication by non-real scalars); $C(X,R)$, the algebra of continuous real-valued functions on a compact Hausdorff space $X$ with pointwise operations and supremum norm; $C_\sigma(X)$, the complex-valued functions $f$ on a compact Hausdorff space $X$ with pointwise operations and supremum norm such that $f \circ \sigma = f^*$, where $\sigma: X \to X$ is a homeomorphism with $\sigma^2$ equal to the identity on $X$ and $f^*(t) = \overline{f(t)}$ for $t \in X$; $B(H)$, the bounded linear operators on a real Hilbert space $H$; and any norm-closed real $^*$-subalgebra of $B(H)$ for a real Hilbert space $H$ (to verify that $1 + x^*x$ is invertible in the subalgebra, one must pass to the "complexification" and use the corresponding fact which is true for complex $C^*$-algebras).

It turns out that these examples exhaust all real $C^*$-algebras. Indeed, R. Arens and I. Kaplansky [1] proved the following theorem which characterizes commutative real $C^*$-algebras:

THEOREM. *(Arens-Kaplansky). Let* $A$ *be a real commutative* $C^*$*-algebra with identity. Then there exists a compact Hausdorff space* $X$ *and a homeomorphism* $\sigma: X \to X$*, with* $\sigma^2$ *equal to the identity map on* $X$*, such that* $A$

*is isometrically *-isomorphic (as a real* C**-algebra) to the real* C**-algebra* $C_\sigma(X)$.

When A is an arbitrary (possibly noncommutative) real C*-algebra, L. Ingelstam [1] established the next result which corresponds to the Gelfand-Naimark theorem (19.1):

THEOREM. *(Ingelstam). Let* A *be a real* C**-algebra with identity. Then there exists a real Hilbert space* H *such that* A *is isometrically *-isomorphic as a real* C**-algebra to a norm-closed real *-subalgebra of* B(H).

A beautiful exposition of both of the above theorems is given in the small book by K. R. Goodearl [1]; it also contains a very clear and concise treatment of the Gelfand-Naimark theorems for complex C*-algebras.

For more information concerning the theorems for real C*-algebras and for related results the reader should consult S. H. Kulkarni and B. V. Limaye [1], B. Li [1, 2], T. Ono [4], and T. W. Palmer [3].

*Gelfand-Naimark theorems for Jordan algebras.* A *Jordan algebra* is an algebra A over the real or complex field with a product ab which satisfies for $a, b \in A$:

(i) $ab = ba$;

(ii) $(a^2b)a = a^2(ba)$.

If B is an associative algebra with product ab, then B can be made into a Jordan algebra by introducing the *Jordan product*

$$a \circ b = \frac{1}{2}(ab + ba).$$

Such algebras B (or algebras isomorphic to them) are called *special Jordan algebras*. Jordan algebras which are not special are called *exceptional*.

In 1934, P. Jordan, J. von Neumann, and E. Wigner classified the finite-dimensional simple Jordan algebras over the reals which were formally real, i.e., $a_1^2 + a_2^2 + \cdots + a_n^2 = 0$ implies $a_1 = a_2 = \cdots = a_n = 0$. Except for the exceptional Jordan algebra $M_3^8$, the hermitian $3 \times 3$ matrices over the Cayley numbers, these algebras were all represented as Jordan algebras of selfadjoint operators acting on a complex Hilbert space. This result was the forerunner of the various characterization theorems discussed in this section.

Before proceeding further, we describe some of the terminology which has been introduced to cover the basic Jordan algebra generalizations of C*-algebras.

First, any complex associative *-algebra A may be considered as a complex Jordan *-algebra under the Jordan product $a \circ b$. Further, the set H(A) of hermitian elements in A may be considered as a real Jordan algebra (with no involution) under the same product. When defining Jordan algebra generalizations of C*-algebras, either procedure may be applied either to abstract C*-algebras or to concrete C*-algebras of operators on a Hilbert space. In formulating the definitions in the abstract case, one wants axioms which are strong enough to give representation theorems in terms of the corresponding concrete Jordan algebras of operators. The idea, of course, is to obtain characterization theorems which parallel the Gelfand-Naimark theorems.

A *Banach Jordan algebra* is a Jordan algebra A which is also a Banach space whose norm satisfies $||a \circ b|| \leq ||a|| \cdot ||b||$ for all $a, b \in A$. A JC-*algebra* is a real Banach Jordan algebra which is a norm-closed Jordan subalgebra of the Jordan algebra of all bounded self-adjoint operators on some Hilbert space. Thus, JC-algebras are the analogues of concrete C*-algebras; their structure is quite well understood, and is close to that of C*-algebras (see D. Topping [1]). E. Størmer [1] characterized those JC-algebras which are the Jordan algebra of all hermitian elements in some C*-algebra.

In 1978, E. M. Alfsen, F. W. Shultz, and E. Størmer [1] defined a JB-*algebra* to be a real Banach Jordan algebra A with identity satisfying $||a^2|| = ||a||^2$ and $||a^2|| \leq ||a^2 + b^2||$ for all $a, b \in A$. The axiom $||a^2|| = ||a||^2$ is analogous to the usual C*-norm condition $||a^*a|| = ||a||^2$. The term JB-algebra is an analogue of the term B*-algebra. Alfsen, Shultz, and Størmer's main result asserts that the study of JB-algebras can be reduced to that of JC-algebras and the exceptional Jordan algebra $M_3^8$. More precisely, they show:

THEOREM. *(Alfsen-Shultz-Størmer). Let* A *be a* JB-*algebra. Then there is a unique norm closed Jordan ideal* J *in* A *such that* A/J *has a faithful isometric representation as a Jordan algebra of self-adjoint operators on a complex Hilbert space, while every "irreducible" representation of* A *not annihilating* J *is onto the exceptional Jordan algebra* $M_3^8$.

The proof of this result requires that many of the standard tools used in the study of abstract $C^*$-algebras be developed for JB-algebras. For example, it is shown that if $A$ is a JB-algebra and $M$ is a closed associative subalgebra containing 1, then $M$ is isometrically isomorphic to $C(X)$ for some compact Hausdorff space $X$. It follows from this that JB-algebras have a well-behaved continuous functional calculus. An enveloping JB-algebra of $A$ is constructed which is the analogue of the second dual so successfully used in $C^*$-algebra theory. JB-factors are defined (JB-algebras whose centers are the scalars) and studied, and an analogue of the GNS-representation is constructed. A comparison theory for idempotents is developed, and every Jordan ideal $J$ in a JB-algebra $A$ is shown to have an increasing approximate identity $\{e_\alpha\}$ such that $||a + J|| = \lim_\alpha ||a - e_\alpha \circ a||$ for all $a \in A$. It follows that if $J$ is a Jordan ideal in a JB-algebra $A$, then $A/J$, with its natural Jordan product and quotient norm, is a JB-algebra. These and many additional technical facts from the theory of Jordan algebras and operator algebras are utilized to establish the theorem. For details the reader is referred to the paper of Alfsen, Shultz , and Størmer.

A corollary of the theorem is that a special JB-algebra is isometrically isomorphic to a JC-algebra. In the paper of F. F. Bonsall [3] a Vidav-Palmer type theorem is obtained which is analogous to this corollary. Let $A$ be a real unital closed Jordan subalgebra of a unital Banach algebra $B$, and let $B_H$ denote the subset of elements of $B$ with real numerical range. Assume that:

(1) $A \cap iA = \{0\}$;

(2) each $a \in A$ can be written as $a = h + ik$, with $h \in B_H \cap A$ and $k \in B_H$.

Then Bonsall shows that $(A + iA) \cap B_H$ is a special JB-algebra and hence, by the Alfsen-Shultz-Størmer theorem, is isometrically isomorphic to a JC-algebra. Furthermore, if one defines a $JC^*$-*algebra* to be a complex Banach Jordan $^*$-algebra which is a closed Jordan $^*$-subalgebra of the Jordan $^*$-algebra of all bounded linear operators on some Hilbert space, then, as Bonsall shows, $A + iA$ is homeomorphically $^*$-isomorphic to a $JC^*$-algebra. The main step in the proof is showing that $A + iA$ is closed in $B$. The above mentioned corollary on special JB-algebras gives the representations as JC and $JC^*$-algebras.

Vidav-Palmer theorems for Banach Jordan algebras have also been obtained by A. Rodríguez [2] and M. A. Youngson [1, 3].

A JB*-*algebra* (or *Jordan* C*-*algebra*) is a complex Jordan *-algebra A satisfying $||\{aa^*a\}|| = ||a||^3$ for all $a \in A$, where $\{abc\}$ is the Jordan triple product, i.e.,

$$\{abc\} = abc + bca - acb.$$

M. A. Youngson [1] proved that the involution in a JB*-algebra is isometric. Every C*-algebra is a JB*-algebra with respect to its Jordan product, but there exist JB*-algebras which cannot be embedded in any C*-algebra. The self-adjoint part of a JB*-algebra is easily seen to be a JB-algebra. Conversely, J. D. M. Wright [1] showed that every JB-algebra is the self-adjoint part of a unique JB*-algebra. Hence the deep results obtained by Alfsen, Shultz, and Størmer on the structure of JB-algebras can be extended to JB*-algebras.

In another direction, R. Braun, W. Kaup, and H. Upmeier [1] have given a penetrating and incisive study of the geometric and holomorphic properties of JB*-algebras. They established a bijective correspondence between JB*-algebras and so-called symmetric tube domains in complex Banach spaces: For every unital JB*-algebra A with self-adjoint part H(A), the associated tube domain is the "generalized upper half plane"

$$D = \{a \in A: \frac{1}{2i}(a - a^*) \in \Omega\},$$

where $\Omega$ is the interior of the set of squares in the real Banach Jordan algebra H(A). They show that D is biholomorphically equivalent to $A_1$, the open unit ball of A. Hence $A_1$ is a bounded symmetric domain. Those Banach spaces that carry a JB*-algebra structure are also characterized holomorphically as follows: Let X be a complex Banach space with open unit ball $X_1$. Suppose that e is an element of X with $||e|| = 1$. Then X can be given the structure of a JB*-algebra with identity element e iff $X_1$ is holomorphically homogeneous and the holomorphic tangent cone to $X_1$ at e vanishes.

Braun, Kaup, and Upmeier [1] also give a new proof of a result of J. D. M. Wright and M. A. Youngson [2]: A surjective isometry between two unital JB*-algebras which preserves the identity is a Jordan *-isomorphism. Then the authors show, by example, that a given Banach space X may carry non-isomorphic JB*-algebra structures.

A class of noncommutative Banach Jordan algebras is studied in great detail by Braun [1]. A $C^*$-*alternative algebra* A is a unital not necessarily associative complex algebra A with conjugate-linear involution $x \to x^*$ such that the *alternative laws* $a^2b = a(ab)$, $ab^2 = (ab)b$, for all $a, b \in A$, are satisfied, and A is a complex Banach space with respect to a norm satisfying the $C^*$-condition $||a^*a|| = ||a||^2$ for all $a \in A$. A $C^*$-*Cayley algebra* is a $C^*$-alternative algebra A such that A, considered as a complex algebra, is isomorphic to a complex Cayley algebra (for the definition of Cayley algebra, see Jacobson [2, p. 17]).

$C^*$-alternative algebras arise naturally in several areas of mathematics. For example, in *geometry* $C^*$-Cayley algebras can be used to construct *Moufang planes*, i.e., projective planes which are harmonic but not Desarguesian (see Jacobson [2, p. 393]). $C^*$-alternative algebras arise in *complex analysis* as induced substructures associated with symmetric Siegel domains of the second kind in Banach spaces (see Braun, Kaup, and Upmeier [1]), and in *functional analysis* they appear in their relationship with (not necessarily associative) unital Banach $^*$-algebras.

R. Braun [1] shows that, up to isometric $^*$-isomorphism, there is precisely one $C^*$-Cayley algebra. He also shows, following Araki and Elliott [1], that $C^*$-alternative algebras are normed algebras, i.e.,

$$||ab|| \leq ||a||\cdot||b||, \quad a, b \in A.$$

If T is a compact topological space and A is a $C^*$-alternative algebra, then the Banach space $C(T,A)$ of continuous functions $f: T \to A$ with pointwise operations, supremum norm, and involution $f^*(t) = f(t)^*$, $t \in T$, is again a $C^*$-alternative algebra. Furthermore, the direct product $A \times B$ of $C^*$-alternative algebras A and B is a $C^*$-alternative algebra with respect to the involution $(a,b)^* = (b^*,a^*)$ and norm defined by $||(a,b)|| = \max\{||a||,||b||\}$. Braun [1, p. 239] established the following Gelfand-Naimark type theorem:

THEOREM. *(Gelfand-Naimark theorem). Let* A *be a unital* $C^*$-*alternative algebra. Then there exists a complex Hilbert space* H, *a compact topological space* T *and an isometric* $^*$-*homomorphism* h *of* A *into* $B(H) \times C(T,Q)$, *where* $B(H)$ *denotes the* $C^*$-*algebra of bounded linear operators on* H *and* Q *denotes the standard* $C^*$-*Cayley algebra.*

Braun's proof of this result is based on: (1) a bidualization process for $C^*$-alternative algebras; (2) construction of "factor representations"; and (3) the determination of $C^*$-alternative factors. It should be remarked that R. Paya, J. Peres, and A. Rodriguez [1] have also proved the above theorem. Their paper gives a systematic and rather complete treatment of noncommutative Jordan $C^*$-algebras.

Braun [1] closes his paper with the following characterization theorem for $C^*$-normed *nonassociative* algebras:

THEOREM. *(Characterization theorem). Let* $A$ *be a complex Banach space and an algebra with identity element* $1$ *such that* $||1|| = 1$. *Then the following conditions are equivalent:*

*(1)* $A$ *is a normed algebra with conjugate-linear vector space involution* $a \to a^*$ *such that* $1^* = 1$ *and* $||a^*a|| = ||a||^2$ *for all* $a \in A$.

*(2)* $A$ *is an algebra with conjugate-linear involution* $a \to a^*$ *such that* $ab^2 = (ab)b$, $a^2b = a(ab)$ *and* $||a^*a|| = ||a||^2$ *for all* $a, b \in A$.

*(3)* $A$ *is a normed alternative algebra such that* $A = H(A) + iH(A)$; *here* $H(A)$ *is the real Banach space of all elements* $h \in A$ *such that* $f(h)$ *is real for every complex-linear functional* $f: A \to C$ *with* $f(1) = 1 = ||f||$.

*(4) There exists an isometric *-representation*

$$h: A \to B(H) \times C(T,Q)$$

*as in the above Gelfand-Naimark theorem.*

Many additional results concerning Jordan algebra generalizations of $C^*$-algebras are known. For example, the Russo-Dye theorem, state space and order-theoretic characterizations, spectral theory, etc., have been treated from various points of view. For information on these and other topics the reader should consult the book of Hanche-Olsen and Størmer [1] and the following papers: Alfsen and Shultz [1, 2, 3]; Alfsen, Shultz, and Størmer [1]; Braun [1]; Braun, Kaup, and Upmeier [1]; Friedman and Russo [1, 2, 3]; Harris [2]; Martinez, Mojtar, and Rodriguez [1]; Rodriguez [1, 2, 3]; Shultz [1]; Størmer [1]; Topping [1]; J. D. M. Wright [1]; Wright and Youngson [1, 2]; and Youngson [1, 2, 3].

*A Gelfand-Naimark theorem for finitely generated* $C^*$*-algebras*. The paper of P. Kruszyński and S. L. Woronowicz [1] contains a noncommutative analogue, for finitely generated $C^*$-algebras, of the Gelfand-Naimark theorem.

To obtain their result the authors introduce the notions of "compact domain" and "continuous operator function." Let N be a positive integer. An N-*dimensional compact domain* is a family $D = \{D(H): H \in \mathcal{H}\}$, where $\mathcal{H}$ is the family of all separable Hilbert spaces and where, for each $H \in \mathcal{H}$, D(H) is a subset of the product $B(H)^N$. Each subset D(H) is assumed to be uniformly bounded in H, and the elements D(H) of D are related by two compatibility relations which involve representations. A *continuous operator function* on D is a family $F = \{F_H: H \in \mathcal{H}\}$, where each $F_H$ is a mapping from D(H) into B(H). For each $a \in D(H)$, $F_H(a)$ belongs to the unital $C^*$-algebra generated by the N components of a, and the mappings $F_H$ are assumed to obey a compatibility relation involving representations. The space $C(D)$ of all continuous operator functions on D is easily seen to be a unital $C^*$-algebra under pointwise operations and supremum norm. The noncommutative Gelfand-Naimark theorem obtained by Kruszyński and Woronowicz states that *every* finitely generated unital $C^*$-algebra arises in this way for an essentially unique compact domain D. It follows that a compact domain is analogous to the spectrum of a finitely generated commutative $C^*$-algebra, although it is not a direct generalization of it.

*A characterization of the $C^*$-norm condition without using the norm.* In 1965, G. R. Allan [1] characterized the norm condition $||x^*x|| = ||x||^2$ in a normed $*$-algebra without using the norm. Let A be a complex normed algebra with identity e and involution $*$. Recall that a subset B of A is *absolutely convex* if $x, y \in B$ implies $\alpha x + \beta y \in B$ for all scalars $\alpha$, $\beta$ with $|\alpha| + |\beta| \leq 1$. Let $\mathcal{B}$ denote the family of all closed and bounded subsets B of A which are absolutely convex, $B^2 \subseteq B$, and $B^* = B$. A *pre-$C^*$-algebra* is a complex normed $*$-algebra A which satisfies $||x^*x|| = ||x||^2$ for all $x \in A$. Allan proved the following:

THEOREM 1. *Let* A *be a complex normed $*$-algebra with identity* e *and continuous involution. Then* A *is a pre-$C^*$-algebra, in an equivalent norm, iff the family* $\mathcal{B}$ *has a largest element and, for each* $x \in A$, *there are sequences* $\{u_n\}$, $\{v_n\}$ *in* A *such that* $u_n(e + x^*x) \to e$, $(e + x^*x)v_n \to e$.

THEOREM 2. *Let* A *be a complex Banach $*$-algebra with identity. Then* A *is a $C^*$-algebra, in an equivalent norm, iff* $\mathcal{B}$ *has a largest element and* A *is symmetric.*

The primary value of these theorems lies in the direction they provide for suitably extending the definition of C*-algebras to the setting of non-normable topological algebras.

*Representations of topological algebras.* A *topological algebra* over the field K of real or complex numbers is a topological vector space over K which is also an algebra over K such that the ring multiplication in A is separately continuous in each factor. If A is a Fréchet space, i.e., a metrizable, complete, locally convex space, then it can be shown that multiplication is jointly continuous. In this case, if $\{U_\alpha\}$ is a neighborhood basis at 0, then for each $\alpha$ there is an index $\beta$ such that

$$U_\beta^2 \subseteq U_\alpha.$$

Correspondingly, if the topology of A is given by a family $\{p_\alpha\}$ of seminorms, then for each $\alpha$ there is a $\beta$ such that

$$p_\alpha(xy) \leq p_\beta(x)p_\beta(y).$$

Serious study of topological algebras began around 1950 to handle certain important classes of non-normable algebras which arose naturally in mathematics and physics. Although isolated results had been obtained as early as 1946, R. Arens [3] and E. A. Michael [1], in 1952, gave the first systematic treatments of the theory. Since then numerous papers have been written on the subject. We restrict our attention to a few of these which are primarily concerned with generalizations of C*-algebras.

We begin with a brief description of the work of G. R. Allan [3]. Let A be a complex, locally convex, topological *-algebra with identity e. Let $A_o$ be the class of elements in A for which the set of powers $(\varepsilon x)^n$, $n = 1,2,\ldots$, is bounded for some nonzero $\varepsilon$. If $x \in A$, the *spectrum* $\sigma_A(x)$ of x is defined to be the set of complex $\lambda$ such that $\lambda e - x$ does not have an inverse in $A_o$, together with $\infty$ if $x \in A_o$ (see Allan [2]).

Motivated by Theorems 1 and 2 of the preceding discussion, Allan lets $\mathcal{B}$ denote the family of all closed and bounded, absolutely convex, subsets B of A such that $e \in B$, $B^2 \subseteq B$ and $B^* = B$. The topological *-algebra A is called a *generalized* B*-*algebra* (GB*-*algebra*) if $\mathcal{B}$ has a largest element, A is symmetric and pseudocomplete (i.e., the linear span of each B in $\mathcal{B}$ is complete). For commutative GB*-algebras, an appropriate functional calculus and representation theory on the space of maximal ideals of $A_o$ is established. If h is a hermitian element of A, then h can be written as the difference, $h = h_+ - h_-$, where $h_+$, $h_-$ are hermitian elements which are postive in the sense of the representation theory. Further, Allan shows that each continuous linear functional f on A can be written as $f = f_+ - f_-$, where $f_+$, $f_-$ are positive linear (possibly discontinuous) functionals on A. From this it is shown that $A_o$ is dense in its GB*-algebra even when A is not commutative.

The study of GB*-algebras was continued by P. G. Dixon [1, 2]. Let H be a complex Hilbert space. An *extended* C*-*algebra* is a set A of closed operators on H such that A ∩ B(H) is a C*-algebra and such that, if a and b are in A, then the closures of a + b and ab are in A. In the first paper Dixon shows: (1) that every commutative GB*-algebra is algebraically *-isomorphic to an algebra of unbounded functions on a compact Hausdorff space, and (2) a *-algebra A can be given a locally convex GB*-topology iff it is algebraically *-isomorphic to an extended C*-algebra. In the second paper two counterexamples are presented. The first one shows that the algebra of complex Borel functions on [0,1], modulo equality almost everywhere, does not have a GB*-topology (so that not *every* function algebra arises as in (1)), and the other example concerns a GB*-algebra that is not *-isomorphic with an extended C*-algebra (so that (2) does not extend to the nonlocally convex case).

Generalizations in another direction have been given, for example, by C. Apostol [1], R. M. Brooks [1], S. M. Moore [1], and K. Schmüdgen [1]. These papers, although differing in detail and emphasis, have many points of contact with each other. To illustrate the basic approach we briefly describe the work of Schmüdgen. An LMC*-*algebra* is a complex *-algebra A endowed with a complete locally convex topology defined by a family of seminorms p such that $p(xy) \leq p(x)p(y)$ and $p(x^*x) = p(x)^2$ for all $x, y \in A$, i.e., each p is a C*-*seminorm*. All C*-algebras are, of course, LMC*-algebras. Typical of the commutative LMC*-algebras which are not C*-algebras is the algebra of all continuous functions on a completely regular noncompact space. Noncommutative examples arise naturally in the context of noncommutative integration theory and physics (the canonical example in the paper of Moore [1] is the Borchers algebra $C \oplus \Sigma_{n=1}^{\infty} S(R^{4n})$ from quantum field theory). It can be shown, in general, that every LMC*-algebra is the projective limit of a family of C*-algebras.

Most of Schmüdgen's (as well as the others') efforts are devoted to formulating and proving certain generalizations and well-known facts about C*-

algebras. However it should be made clear that, although most of the generalizations are easily anticipated, many of the proofs require novel insights and substantial reworking to recover the C*-algebra counterpart. Schmüdgen's version of the Gelfand-Naimark theorem for commutative algebras is:

THEOREM. *(Gelfand-Naimark). A commutative* LMC*-*algebra with identity is topologically and algebraically* *-*isomorphic to the algebra* C(X) *of all complex-valued continuous functions on a completely regular topological space* X, *where* C(X) *is furnished with a complete locally convex topology which, in general, is weaker than the compact-open topology.*

With this result in hand, a functional calculus is developed by Schmüdgen which yields most of the familiar results from C*-algebra theory concerning hermitian, positive, and unitary elements. A crucial role in all of this is played by a certain dense subalgebra $A_b$ of A which turns out to be a C*-algebra (Apostol's work also involves $A_b$). An element x belongs to $A_b$ iff $\sup p(x) < \infty$, where the supremum is taken over all algebra seminorms p defining the topology of A.

The works of Brooks [1] and Moore [1] contain discussions on positive functionals, pure states, cyclic *-representations, irreducible *-representations, and the usual correspondence that exists between these concepts. The GNS-construction is also shown to carry over, in modified form, for LCM*-algebras and other suitably defined topological *-algebras. For more information we refer the reader to Allan [1, 2, 3], Apostol [1], Arens [3], Brooks [1, 2], Dixon [1, 2], Husain and Malviya [1], Husain and Rigelhof [1], Husain and Warsi [1], Michael [1], S. M. Moore [1], Schmüdgen [1], Wenjen [1], Wood [1], and Żelazko [1].

*A characterization of* C*-*algebras by linear functionals.* The following generalization of the well-known Hahn-Jordan decomposition theorem was proved in 1957 by A. Grothendieck [1] and gives a nice characterization of C*-algebras in terms of their linear functionals:

THEOREM. *(Grothendieck). A unital Banach* *-*algebra* A *is a* C*-*algebra iff, for each continuous hermitian functional* f *on* A, *there exist positive functionals* g *and* h *on* A *such that* $f = g - h$ *and*

$$||f|| = ||g|| + ||h||$$

A linear functional characterization of closed two-sided ideals in $C^*$-algebras was given in 1980 by S. Takahasi [1]. In order that a $C^*$-subalgebra $B$ of a $C^*$-algebra $A$ be a two-sided ideal it is necessary and sufficient that

$$||f + g|| = ||f|| + ||g||$$

for all continuous linear functionals $f$ and $g$ on $A$ with $||f|B|| = ||f||$ and $g|B = 0$.

*A characterization of* $C^*$-*subalgebras*. Let $A$ be a $C^*$-algebra and $B$ a closed linear subspace of $A$. If $x \in A$, let $|x|$ denote the unique positive square root of $x^*x$ in $A$. If necessary, adjoin an identity $e$ to $A$. In 1978, J. A. van Casteren [1] proved that $B$ is a $C^*$-subalgebra of $A$ iff, for each $x$ in $B$, the elements $x^*$ and $|x| + e - ||x| - e|$ are in $B$. If $e$ is in $B$, then $B$ is a $C^*$-subalgebra of $A$ iff $|x|$ is in $B$ for each $x$ in $B$. R. B. Burckel and S. Saeki [1] have slightly extended van Casteren's theorem by showing that it is enough to assume, in the general setting, that $x$ in $B$ implies $|x + e| - e$ is in $B$. Their proof is elementary and elegant.

*Symmetric *-algebras*. If $A$ is a Banach *-algebra, the Shirali-Ford theorem (33.2) states that every hermitian element of $A$ has real spectrum iff $x^*x$ has positive spectrum for every $x$ in $A$. In 1979, H. Behncke [3] showed that the Shirali-Ford theorem remains true when $A$ is a unital Banach Jordan *-algebra. Recently, B. Aupetit and M. A. Youngson [1] gave a simple "subharmonic" proof of Behncke's theorem. As a corollary, they obtained the following characterization of $JB^*$-algebras:

THEOREM. *A complex unital Banach Jordan *-algebra* $A$ *is a* $JB^*$-*algebra in an equivalent norm iff there exists a real constant* $\beta \geq 1$ *such that* $||\exp(ih)|| \leq \beta$ *for all hermitian* $h$ *in* $A$.

Further information on symmetric Banach *-algebras can be found in Aupetit's book [3]. A characterization of *real* symmetric unital Banach *-algebras is given by J. Vukman [1] who shows that the inequality $|x|_\sigma^2 \leq |x^*x|_\sigma$ must be replaced by $|x|_\sigma^2 \leq |x^*x + y^*y|_\sigma$ for all $x$ and $y$.

*Characterizations of Hilbert space*. If $H$ is a real or complex Hilbert space, then the bounded linear operators $B(H)$ on $H$ form a $C^*$-algebra.

Conversely, if $E$ is a real or complex Banach space and $B(E)$ is a $C^*$-algebra for some involution $T \to T^*$, is $E$ a Hilbert space? That is, does there exist an inner product on $E$ such that the corresponding norm is equivalent to the given norm on $E$ and such that $T^*$ is the adjoint of $T$ relative to this inner product? S. Kakutani and G. W. Mackey [2] showed, in 1946, that the answer is affirmative. In fact, they showed that the assumption that $B(E)$ be a $C^*$-algebra can be weakened considerably:

THEOREM. *(Kakutani-Mackey). Let* $E$ *be a real or complex Banach space. If* $B(E)$ *admits an involution* $T \to T^*$ *such that* $T^*T = 0$ *implies* $T = 0$, *then an inner product can be introduced into* $E$ *so that it becomes a Hilbert space with norm equivalent to the given norm in* $E$ *and so that* $T^*$ *is the adjoint of* $T$ *relative to the inner product.*

A proof of this theorem for complex spaces, based on representation theory for Banach $^*$-algebras with minimal ideals, is given in the book of C. Rickart [1, p. 265]. In 1983, J. Vukman [2] proved that if $B(E)$ admits an involution $T \to T^*$ such that each unitary operator has norm one, then there exists an inner product on $E$ such that the corresponding norm is *equal* to the given norm, and $T^*$ is the adjoint of $T$ relative to the inner product. Vukman's proof is elementary in the sense that no results or methods from Banach $^*$-algebra theory are used. His arguments are also valid for real and complex Banach spaces. A consequence of his result is the following: If $E$ is a real or complex Banach space and $B(E)$ admits an involution such that each unitary element has norm one, then $B(E)$ is a $C^*$-algebra.

The Vidav-Palmer theorem (45.1) is applied in Bonsall-Duncan [3, p. 78-79] to obtain the following characterization of complex Hilbert space: Let $X$ be a complex Banach space and let $A$ be the Banach algebra $B(X)$. Then $A = H(A) + iH(A)$ iff $X$ is a Hilbert space.

*Characterizations of von Neumann algebras.* Let $H$ be a Hilbert space. A von Neumann algebra is a $^*$-subalgebra of $B(H)$ which is closed in the weak operator topology. These algebras, which are also called *rings of operators* and $W^*$-*algebras*, were introduced in 1929 by J. von Neumann. In 1946, C. Rickart [1, p. 289] initiated a program to study von Neumann algebras abstractly, without reference to the underlying Hilbert space (earlier attempts had been made by S. W. P. Steen). Given a subset $S$ of a ring $A$, the *right annihilator* of $S$ is the set of all $x$ in $A$ satisfying $sx = 0$

for all $s \in S$. Rickart studied the class of C*-algebras (now called *Rickart C*-algebras*) with the property that the right annihilator of any element is generated by a projection (i.e., a hermitian idempotent). In 1951, I. Kaplansky defined an AW*-*algebra* to be a C*-algebra A such that the left (or right) annihilator of an arbitrary subset of A is a principal ideal generated by a projection. He proceeded to show that much of the "non-spatial" theory of von Neumann algebras carries over to AW*-algebras. Even so, it turned out that AW*-algebras did not characterize von Neumann algebras and additional conditions were needed (see Pedersen [2, p. 66]) for them to be representable as weakly closed *-subalgebras of B(H).

Characterizations of general von Neumann algebras were obtained independently, in 1956, by R. V. Kadison and S. Sakai. Sakai showed (see Pedersen [2, p. 67]) that a necessary and sufficient condition for a C*-algebra to admit a *-representation as a von Neumann algebra is that it be the conjugate space of some Banach space. The necessity (the easier direction) had been obtained in 1953 by J. Dixmier. This characterization of W*-algebras forms the basis for the theory given in the book by Sakai [1].

There is an extensive literature devoted to the theory of von Neumann algebras. For more information we refer the reader to Berberian [1] (for the purely algebraic theory of W*-algebras), Dixmier [6], Kadison and Ringrose [1], Pedersen [2], Sakai [1], Stratila and Zsidó [1], and Takesaki [1].

*Characterizations of commutativity in C*-algebras.* In this final section we give several characterizations of commutativity for C*-algebras. Since commutative C*-algebras are structurally much simpler than their noncommutative counterparts, and because of the important role they play in the general theory, it is useful to have criteria which imply commutativity of a given C*-algebra. We collect some definitions which will be needed.

Let E be a real vector space with positive cone P and partial order $v \le u$ iff $u - v \in P$. The space E is said to be *lattice-ordered* if, for each $u \in E$ there exists $u^+ \in P$ such that $u^+ \ge u$ and $u^+ \le v$ whenever $v \in P$ and $u \le v$. The vector space E is said to have the *Riesz decomposition property* if $u, v, w \in P$ and $u \le v + w$ imply that there exist $v_1, w_1 \in P$ with $u = v_1 + w_1$ and $v_1 \le v$, $w_1 \le w$.

Let A be a unital C*-algebra. An element x in A is *nilpotent* if there is a positive integer n such that $x^n = 0$. A *-representation $\pi$ of A on a Hilbert space H is said to be *multiplicity-free* if the commutant $\pi(A)'$ of $\pi(A)$ is abelian. A *factor* is a von Neumann algebra whose center consists of scalar multiples of the identity operator. A $^*$-representation $\pi$ of the C*-algebra A is called a *factor representation* if the von Neumann algebra generated by $\pi(A)$ is a factor. A state on A is called a *factor state* if its associated $^*$-representation is a factor representation. The *numerical index* n(A) of A is defined by

$$n(A) = \inf\{w(x): x \in A, ||x|| = 1\}$$

where

$$w(x) = \sup\{|(x\xi|\xi)|: \xi \in H, ||\xi|| = 1\}.$$

The *normal approximate spectrum* of an element x in A, denoted $\sigma_A^n(x)$, is the set of complex scalars $\lambda$ such that the operator

$$(x - \lambda e)^*(x - \lambda e) + (x - \lambda e)(x - \lambda e)^*$$

is not bounded below (an operator T is *bounded below* if there exists a real constant $\beta > 0$ such that $T^*T \geq \beta I$).

THEOREM. *Let* A *be a unital* C*-*algebra and let* H(A) *denote the real Banach space of hermitian elements in* A. *The following are equivalent:*

*(1)* A *is commutative;*

*(2)* H(A) *is closed under products;*

*(3)* $x \in A$ *and* $x^2 = 0 \Rightarrow x = 0$;

*(4)* 0 *is the only nilpotent element in* A;

*(5) each closed left ideal of* A *is a two-sided ideal;*

*(6)* $x, y \in A$ *and* $0 \leq y \leq x \Rightarrow y^2 \leq x^2$;

*(7)* $x, y \in A$ *and* $0 \leq y \leq x \Rightarrow y^\beta \leq x^\beta$ *for some real* $\beta > 1$;

*(8) the usual order in* H(A) *is a lattice ordering;*

*(9) the usual order in* $H(A)^*$ *is a lattice ordering;*

*(10)* H(A) *has the Riesz decomposition property;*

*(11) each cyclic* $^*$-*representation of* A *is multiplicity-free;*

*(12) each factor state on* A *is pure;*

*(13) the numerical index* n(A) *of* A *is* 1;

*(14)* $\sigma_A(x) \subseteq \{\lambda \in C: \text{dist}(\lambda,\sigma_A(y)) \leq ||x - y||\}$ *for all* $x, y \in A$;

*(15)* $\sigma_A^n(x) = \sigma_A(x)$ *for all* $x \in A$.

The equivalence of (1) and (2) is nearly obvious. Parts (3), (4), and (5) are unpublished results of I. Kaplansky and are given as exercises in

Dixmier [5, p. 68] and Kadison-Ringrose [1, p. 292]. Parts (6) and (7) are due to T. Ogasawara [1]; (8) is due to S. Sherman [1]; (9) is proved in M. J. Crabb, J. Duncan, and C. M. McGregor [1]. Part (10) is due to M. Fukamiya, M. Misonou and Z. Takeda [1]. Part (11) was proved by C. F. Skau (see Williamson [2, p. 272]); (12) is due to S. Wright [1]; and (13) is proved in Crabb, Duncan, and McGregor [1]. Part (14) is due to R. Nakamoto (*Math. Japon.* 24(1979/80), 399-400), and (15) is due to Y. Kato (*Math. Japon.* 24(1979/80), 209-210).

The paper of Crabb, Duncan, and McGregor [1] is an excellent reference for the equivalence of (1) - (10). Further results on characterizations of commutativity for $C^*$-algebras are given in Duncan and Taylor [1], and Kainuma [1]. For characterizations of commutativity in general Banach algebras see Aupetit [10], Belfi and Doran [1], and Williams [1].

# Appendix A
# Functional Analysis

In this appendix we state, without proof, several results from functional analysis which are used frequently in this book. We refer the reader to Dunford and Schwartz [1], [2] for proofs and a complete discussion.

A *topological vector space* (TVS) is a vector space $E$ over $R$ or $C$ with a topology such that addition and scalar multiplication are continuous. A TVS $E$ is said to be *locally convex* if it possesses a *neighborhood basis of convex sets at zero*. A TVS in which the topology is given by a complete translation-invariant metric is called an F-*space*. A locally convex F-space is called a *Fréchet space*. A *normed space* is a vector space $E$ with a map $\xi \to ||\xi||$ such that for $\xi, \eta$ in $E$ and scalar $\lambda$: (i) $||\xi|| \geq 0$; (ii) $||\xi|| = 0$ iff $\xi = 0$; (iii) $||\lambda\xi|| = |\lambda| \cdot ||\xi||$; and (iv) $||\xi + \eta|| \leq ||\xi|| + ||\eta||$. When $E$ is complete with respect to this norm, it is called a *Banach space*. Every Banach space is a Fréchet space, but not conversely.

Let $T$ be a linear map of a normed space $E$ into a normed space $F$, where $E$ and $F$ are both real or both complex. If there is $k > 0$ such that $||T\xi|| \leq k||\xi||$ for all $\xi \in E$, then $T$ is said to be bounded. The infimum of all such values of $k$ is defined to be the *norm of* $T$, denoted $||T||$. If $F$ is the scalar field, then $T$ is called a *bounded linear functional*. The norm of a bounded linear map $T$ is easily seen to be given by

$$||T|| = \sup\{||T\xi||: \xi \in E, ||\xi|| \leq 1\}. \tag{1}$$

Further, a linear map $T: E \to F$ is bounded iff $T$ is continuous.

If $E$ is a normed space and $D$ is a dense linear subspace of $E$, then a bounded linear map $T$ defined on $D$ into a Banach space $F$ can be extended to a bounded linear map $T_o: E \to F$. Further, $||T_o|| = ||T||$, and the extension is unique.

(A.1) THEOREM. *(Hahn-Banach). Let* $E$ *be a normed space and* $M$ *a linear subspace of* $E$. *Then any bounded linear functional* $f$ *on* $M$ *can be extended to a bounded linear functional* $f_o$ *on* $E$ *such that* $||f_o|| = ||f||$.

If $E$ is a normed space and $M$ is a closed subspace of $E$, then the vector space $E/M$ is a normed space under the norm $||\xi + M|| = \inf\{||\xi + m||: m \in M\}$. Further, if $E$ is complete, then $E/M$ is complete.

(A.2) THEOREM. *(Uniform boundedness principle). Let* $E$ *be a Banach space and* $\{E_\alpha\}_{\alpha\in I}$ *a family of normed spaces, and let* $T_\alpha: E \to E_\alpha$ $(\alpha \in I)$ *be a family of bounded linear mappings. If, for each* $\xi \in E$, *the family* $\{||T_\alpha\xi||\}_{\alpha\in I}$ *is bounded, then* $\{||T_\alpha||\}_{\alpha\in I}$ *is bounded. In other words, there is a* $\beta > 0$ *such that* $||T_\alpha\xi|| \leq \beta||\xi||$ *for all* $\xi \in E$ *and* $\alpha \in I$.

If $E$ is a normed space, the set of all continuous linear functionals on $E$ is denoted by $E^*$, and is called the *dual space* or *conjugate space* of $E$. If $E^*$ is given the norm (1), then it is a Banach space.

For each $x$ in a normed space $E$, let $x'$ denote the bounded linear functional on $E^*$ defined by

$$x'(f) = f(x) \quad (f \in E^*).$$

Clearly $x'$ belongs to the dual space $E^{**}$ of $E^*$, and $||x'|| = ||x||$. Hence the map $x \to x'$ is a linear norm-preserving map of $E$ into $E^{**}$. The space $E$ is said to be *reflexive* if this map carries $E$ onto $E^{**}$.

Let $E$ be a normed space. Besides the norm topology $E$ may have other useful topologies. Indeed, for each $x \in E$, each finite subset $\{f_1,\ldots,f_n\}$ of $E^*$, and each $\varepsilon > 0$, define

$$U(x; f_1\ldots,f_n; \varepsilon) = \{y \in E: |f_k(y) - f_k(x)| < \varepsilon \text{ for } k = 1,\ldots,n\}.$$

These sets form a basis for a topology on $E$, denoted $\sigma(E,E^*)$, under which $E$ becomes a locally convex TVS. The topology $\sigma(E,E^*)$ is called the *weak topology* on E, and is the weakest topology on E for which each f in $E^*$ is continuous.

Now, consider the dual space $E^*$. For each $f \in E^*$, each finite subset $\{x_1,\dots,x_n\}$ of E, and $\varepsilon > 0$ define

$$U(f; x_1\dots,x_n; \varepsilon) = \{g \in E^*: |g(x_k) - f(x_k)| < \varepsilon \text{ for } k = 1,\dots,n\}.$$

These sets form a basis for a locally convex topology on $E^*$, denoted $\sigma(E^*,E)$. It is called the *weak *-topology* on $E^*$, and is the weakest topology on $E^*$ for which each x' in $E^{**}$ is continuous on $E^*$. When E is reflexive the topologies $\sigma(E^*,E^{**})$ and $\sigma(E^*,E)$ coincide.

(A.3) PROPOSITION. *(Banach-Alaoglu). Let* E *be a normed space. Then the closed unit ball* $E_1^* = \{f \in E^*: ||f|| \leq 1\}$ *in* $E^*$ *is compact in the weak *-topology of* $E^*$.

Let $E_1$ and $E_2$ be F-spaces and let T be a linear mapping from $E_1$ into $E_2$. T is said to be *closed* (or to have *closed graph*) if whenever $\{\xi_\alpha\}$ converges to $\xi$ and $\{T\xi_\alpha\}$ converges to $\eta$, then $T\xi = \eta$.

(A.4) THEOREM. *(Closed graph theorem). Let* $E_1$ *and* $E_2$ *be F-spaces, and let* $T: E_1 \to E_2$ *be a closed linear map. Then* T *is continuous.*

A mapping between TVS's is called *open* if the image of every open set is open. The next result is closely related to the closed graph theorem.

(A.5) THEOREM. *(Open mapping theorem). Let* $E_1$ *and* $E_2$ *be F-spaces and* $T: E_1 \to E_2$ *a continuous linear map of* $E_1$ *onto* $E_2$. *Then* T *is an open mapping.*

Let C be a convex subset of a vector space E. A convex subset B of C is said to be an *extreme subset* of C if the relations $x, y \in C$, $\alpha x + (1 - \alpha)y \in B$ and $0 < \alpha < 1$ imply that x and y are in B. If $\{x\}$ is an extreme subset of C, then x is said to be an *extreme point* of C. Hence x is an extreme point of C if $x = \alpha x_1 + (1 - \alpha)x_2$, $0 < \alpha < 1$, $x_1,x_2 \in C$ imply $x = x_1 = x_2$.

For any subset A of a vector space E the set of all elements of

E of the form $y = \Sigma_{i=1}^{n}\alpha_i a_i$, $a_i \in A$, $0 \leq \alpha_i \leq 1$ and $\Sigma_{i=1}^{n}\alpha_i = 1$ is called the *convex hull* of A.

(A.6) THEOREM. *(Krein-Milman theorem). Let* E *be a locally convex (Hausdorff) TVS and* A *a nonempty compact convex subset of* E. *Then* A *is the closed convex hull of its extreme points.*

A family F of functions on a set X is said to *separate the points* of X if for $x, y \in X$, $x \neq y$, there is an $f \in F$ such that $f(x) \neq f(y)$. For other terminology used in the next theorem see Example B.2.1.

(A.7) THEOREM. *(Stone-Weierstrass theorem). Let* X *be a locally compact Hausdorff space, and* A *a subalgebra of* $C_o(X)$, *the algebra of continuous complex-valued functions which vanish at infinity on* X. *If for every point of* X *the subalgebra* A *contains a function which does not vanish there,* A *separates the points of* X, *and is closed under complex conjugation, then* $\overline{A} = C_o(X)$, i.e., A *is uniformly dense in* $C_o(X)$.

An *inner product* on a vector space E is a function $(\xi,\eta) \to (\xi|\eta)$ from $E \times E$ into C such that:

(1) $(\xi|\eta) = \overline{(\eta|\xi)}$;

(2) $(\xi_1 + \xi_2|\eta) = (\xi_1|\eta) + (\xi_2|\eta)$;

(3) $(\lambda\xi|\eta) = \lambda(\xi|\eta)$;

(4) $(\xi|\xi) > 0$ when $\xi \neq 0$.

An *inner product space* is a vector space equipped with an inner product. A *Hilbert space* is an inner product space which is complete in the norm $||\xi|| = (\xi|\xi)^{1/2}$.

Let H be a Hilbert space. Two elements $\xi, \eta \in H$ are *orthogonal* if $(\xi|\eta) = 0$. Two subsets A and B of H are *orthogonal* if $(\xi|\eta) = 0$ for all $\xi \in A$ and $\eta \in B$. If $\xi$ and $\eta$ are orthogonal, we write $\xi \perp \eta$; similarly $A \perp B$ means that A and B are orthogonal. For a subset A of H, $A^{\perp}$ will denote the set $\{\xi \in H: \xi \perp \eta$ for all $\eta \in A\}$.

A subset A of H is said to be *orthogonal* if $(\xi|\eta) = 0$ for all $\xi, \eta \in A$ such that $\xi \neq \eta$. If also $||\xi|| = 1$ for all $\xi \in A$, then A is said to be *orthonormal*. An *orthonormal basis* of H is a maximal orthonormal subset of H.

(A.8) THEOREM. *Let* M *be a closed subspace of a Hilbert space* H. *Then* $M^{\perp}$ *is a closed subspace of* H, $M^{\perp}$ *is orthogonal to* M, $M^{\perp\perp} = M$, *and* $H = M \oplus M^{\perp}$.

If H is a Hilbert space, then H has at least one orthonormal basis, and any two orthonormal bases of H have the same cardinal number. The *dimension* of a Hilbert space is the cardinal number of any of its orthonormal bases.

(A.9) THEOREM. *(Riesz-Fréchet). If* H *is a Hilbert space and* f *is a continuous linear form on* H, *then there is a unique vector* $\eta \in H$ *such that* $f(\xi) = (\xi|\eta)$ *for all* $\xi \in H$.

(A.10) THEOREM. *Let* H *be a Hilbert space and* $T \in B(H)$, *i.e.,* T *is a bounded linear map on* H. *Then there exists a unique* $T^* \in B(H)$ *called the* <u>*adjoint*</u> *of* T *such that*

$$(T\xi|\eta) = (\xi|T^*\eta) \qquad (\xi,\ \eta \in H).$$

*Moreover, for* $S, T \in B(H)$ *and* $\lambda \in C$:

*(1)* $T^{**} = T$;

*(2)* $(S + T)^* = S^* + T^*$;

*(3)* $(\lambda T)^* = \overline{\lambda}T^*$;

*(4)* $(ST)^* = T^*S^*$;

*(5)* $||T^*T|| = ||T||^2$;

*(6)* $\text{null}(T^*) = \text{range}(T)^{\perp}$;

*(7)* $\text{null}(T^*)^{\perp} = \overline{\text{range}(T)}$.

(A.11) THEOREM. *If* H *is a Hilbert space and* $\phi$ *is a bounded sesquilinear functional, then there exists a unique* $T \in B(H)$ *such that* $\phi(\xi,\eta) = (T\xi|\eta)$ *for all* $\xi,\ \eta \in H$.

(A.12) THEOREM. *Let* H *be a Hilbert space and* $\{\xi_n\}$ *a sequence of pairwise orthogonal vectors in* H. *Then the following are equivalent:*

*i)* $\Sigma_{n=1}^{\infty}\xi_n$ *converges in the norm topology of* A;

*ii)* $\Sigma_{n=1}^{\infty}||\xi_n||^2 < \infty$;

*iii)* $\Sigma_{n=1}^{\infty}(\xi_n|\eta)$ *converges for all* $\eta \in H$.

Let H be a Hilbert space. An operator $T \in B(H)$ is *normal* if $T^*T = TT^*$; T is *self-adjoint* (or *hermitian*) if $T^* = T$. If I denotes the identity operator on H, then T is *unitary* if $T^*T = I = TT^*$. Finally,

$T$ is called a *projection* if $T^2 = T$. Obviously, self-adjoint and unitary operators are normal.

(A.13) THEOREM. *Let* $H$ *be a Hilbert space and suppose* $T \in B(H)$. *Then:*

*i)* $T$ *is normal iff* $\|T\xi\| = \|T^*\xi\|$ *for all* $\xi \in H$;

*ii) If* $T$ *is normal, then* $\text{null}(T) = \text{null}(T^*) = \text{range}(T)^\perp$

*iii) If* $T$ *is normal and* $T\xi = \lambda\xi$ *for some* $\xi \in H$ *and complex* $\lambda$, *then* $T^*\xi = \overline{\lambda}\xi$;

*iv) If* $T$ *is normal and* $\lambda$ *and* $\mu$ *are distinct eigenvalues of* $T$, *then the corresponding eigenspaces are orthogonal to each other.*

(A.14) THEOREM. *Let* $H$ *be a Hilbert space and* $U \in B(H)$. *The following are equivalent:*

*i)* $U$ *is unitary;*

*ii)* $\text{range}(U) = H$ *and* $(U\xi|U\eta) = (\xi|\eta)$ *for all* $\xi, \eta \in H$;

*iii)* $\text{range}(U) = H$ *and* $\|U\xi\| = \|\xi\|$ *for all* $\xi \in H$.

(A.15) THEOREM. *Let* $H$ *be a Hilbert space and* $P \in B(H)$ *be a projection. Then the following are equivalent:*

*i)* $P$ *is self-adjoint;*

*ii)* $P$ *is normal;*

*iii)* $\text{range}(P) = \text{null}(P)^\perp$;

*iv)* $(P\xi|\xi) = \|P\xi\|^2$ *for all* $\xi \in H$.

(A.16) THEOREM. *Let* $H$ *be a Hilbert space and* $S, T \in B(H)$. *If* $S$ *is self-adjoint, then* $ST = 0$ *iff* $\text{range}(S) \perp \text{range}(T)$.

(A.17) THEOREM. *(Fuglede-Putnam-Rosenblum). Let* $H$ *be a Hilbert space, and suppose that* $M, N, T \in B(H)$, *and that* $M$ *and* $N$ *are normal. If* $MT = TN$, *then* $M^*T = TN^*$.

*Proof.* See Rudin [2, pp. 300-301].

An operator $T$ on a Hilbert space $H$ is *positive* if $(T\xi|\xi) \geq 0$ for every $\xi \in H$. If $T$ is positive one writes $T \geq 0$.

(A.18) THEOREM. *Let* $H$ *be a Hilbert space and suppose* $T \in B(H)$. *Then* $T \geq 0$ *iff* $T^* = T$ *and* $\sigma(T) = \{\lambda: T - \lambda I$ *is singular*$\} \subseteq [0,\infty)$.

(A.19) THEOREM. *Every positive operator* T *on a Hilbert space* H *has a unique positive square root* S. *If* T *is invertible, so is* S. *Further,* S *commutes with all operators on* H *which commute with* T.

(A.20) THEOREM. *Let* H *be a Hilbert space. If* $T \in B(H)$*, then the positive square root of* $T^*T$ *is the only positive operator* P *on* H *such that* $||P\xi|| = ||T\xi||$ *for all* $\xi \in H$.

If a bounded operator T on a Hilbert space can be written in the form $T = UP$ where U is unitary and $P \geq 0$, then UP is called a *polar decomposition* of T. Since U is an isometry, (A.20) shows that the positive operator P is uniquely determined by T. It is not true that every $T \in B(H)$ has a polar decomposition in this sense.

(A.21) THEOREM. *Let* H *be a Hilbert space and* $T \in B(H)$.

*i) If* H *is finite-dimensional, then* T *has a polar decomposition* $T = UP$

*ii) If* T *is invertible, then* T *has a unique polar decomposition* $T = UP$.

*iii) If* T *is normal, then* T *has a polar decomposition* $T = UP$ *in which* U *and* P *commute with each other and with* T.

Let H be a Hilbert space. An operator $V \in B(H)$ is called a *partial* isometry if there is a closed subspace M of H such that $||V\xi|| = ||\xi||$ for $\xi \in M$ and $V(M^{\perp}) = \{0\}$. The subspace M is called the *initial domain* of V and $V(M) = V(H)$ is called the *final domain* of V.

(A.22) THEOREM. *Let* H *be a Hilbert space. Then every* $T \in B(H)$ *has a factorization* $T = VP$*, where* P *is positive and* V *is a partial isometry.*

# Appendix B
# Banach Algebras

In this appendix we provide the reader with the necessary background from the general theory of complex normed algebras to read this book. It is primarily for reference.

## B.1. *First properties*

A *normed algebra* is a normed space A over C on which there is defined a multiplication making it an associative algebra such that

$$||xy|| \leq ||x||\cdot||y||, \quad x, y \in A. \tag{1}$$

A *Banach algebra* is a normed algebra which is complete with respect to the given norm.

Relation (1) implies that multiplication is jointly continuous. If (1) is replaced by the hypothesis that multiplication is jointly continuous (or even separately continuous, the two hypotheses being equivalent by the uniform boundedness theorem), then there is $K > 0$ with $||xy|| \leq K||x||\cdot||y||$ $(x, y \in A)$. Replacing $||\cdot||$ by $||\cdot||'$, defined by $||x||' = K||x||$, we obtain a norm which satisfies (1) and is equivalent to $||\cdot||$.

A Banach algebra A is *unital* if it contains an identity element e satisfying $||e|| = 1$. It is possible for a Banach algebra to contain an identity and not be unital. In this case it can be renormed by the equivalent norm $||x||' = \sup\{||xy||: y \in A, ||y|| \leq 1\}$ so as to become unital.

Every normed algebra A can be embedded, isometrically and isomorphically, as a dense subalgebra of a Banach algebra $\tilde{A}$. The Banach algebra $\tilde{A}$ is unique up to isometric isomorphism and is called the *completion* of A.

If A is a normed algebra then the direct sum $A_e = A \oplus C$, with multiplication

$$(x,\lambda)(y,\mu) = (xy + \lambda y + \mu x, \lambda\mu) \qquad x, y \in A, \quad \lambda, \mu \in \mathbb{C},$$

and norm

$$||(x,\lambda)|| = ||x|| + |\lambda|,$$

is a unital normed algebra which is complete iff A is. The identity in $A_e$ is $e = (0,1)$, and, identifying $x$ with $(x,0)$, A is contained in $A_e$ as an ideal of codimension 1. The algebra $A_e$ is called the *unitization* of A. Elements in $A_e$ are generally written in the form $x + \lambda e$ or simply $x + \lambda$.

If A is a normed algebra and I is a closed two-sided ideal of A, then the quotient space A/I becomes a normed algebra, called the *quotient algebra*, if a product and norm are defined by:

$$(x + I)(y + I) = xy + I, \qquad x + I, y + I \in A/I$$
$$||x + I|| = \inf\{||x + u||: u \in I\}.$$

The quotient algebra is complete (resp. unital) if A is complete (resp. unital). The map $\tau: A \to A/I$ defined by $\tau(x) = x + I$ is called the *quotient mapping* or *canonical homomorphism* of A onto A/I. $\tau$ is a continuous open mapping with $||\tau|| \leq 1$.

Let A be a normed algebra. For $x$ in A define

$$\nu(x) = \lim_{n\to\infty} \sup ||x^n||^{1/n}.$$

(B.1.1) PROPOSITION. *For* x,y *in a normed algebra* A, *and* $\lambda \in C$ *the above limit superior is actually a limit and furthermore:*

*(a)* $\nu(x) = \inf\{||x^n||^{1/n}: n \in N\}$;

*(b)* $0 \leq \nu(x) \leq ||x||$;

*(c)* $\nu(\lambda x) = |\lambda|\cdot\nu(x)$;

*(d)* $\nu(xy) = \nu(yx)$ *and* $\nu(x^k) = \nu(x)^k$, $k \in N$

*(e)* *If* $xy = yx$, *then* $\nu(xy) \leq \nu(x)\nu(y)$.

*Moreover*

*(f)* $\nu(x) = ||x||$ *for every* $x \in A$ *iff* $||x^2|| = ||x||$ *for every* $x \in A$.

*Proof.* We shall prove (a), (e), and (f); the other parts will be left as simple exercises for the reader.

(a) Let $\nu = \inf\{||x^n||^{1/n}: n = 1,2,\ldots\}$; since $\nu \le ||x^n||^{1/n}$ for all $n$, we have $\nu \le \liminf_{n\to\infty} ||x^n||^{1/n}$. Let $\varepsilon > 0$, and choose a positive integer $m$ such that $||x^m||^{1/m} < \nu + \varepsilon$. For each positive integer $n$, there exists a nonnegative integer $a_n$ such that $n = a_n m + b_n$, where $0 \le b_n \le m$. Then

$$||x^n||^{1/n} = ||x^{a_n m + b_n}||^{1/n} = ||x^{a_n m} x^{b_n}||^{1/n}$$

$$\le ||x^m||^{a_n/n} ||x||^{b_n/n} < (\nu + \varepsilon)^{m a_n/n} ||x||^{b_n/n}.$$

Since $\lim_{n\to\infty} m a_n/n = 1$ and $\lim_{n\to\infty} b_n/n = 0$, we have $\limsup_{n\to\infty} ||x^n||^{1/n} \le \nu + \varepsilon$. Since $\varepsilon$ was arbitrary, $\limsup_{n\to\infty} ||x^n||^{1/n} \le \nu$. Therefore, $\nu = \lim_{n\to\infty} ||x^n||^{1/n}$.

(e) Assume $xy = yx$. Then, $(xy)^n = x^n y^n$ and hence $||(xy)^n||^{1/n} = ||x^n y^n||^{1/n} \le ||x^n||^{1/n} ||y^n||^{1/n}$. It follows that $\nu(xy) \le \nu(x)\nu(y)$.

(f) If $\nu(\cdot) = ||\cdot||$ on $A$, then $||x^2|| = \nu(x^2) = \nu(x)^2 = ||x||^2$ by (d). Conversely, assume that $||x^2|| = ||x||^2$ for all $x \in A$. Then, by iteration, we obtain $||x^{2^k}|| = ||x||^{2^k}$ for all $k = 1,2,\ldots$ . Whence $\nu(x) = \lim_{k\to\infty} ||x^{2^k}||^{1/2^k} = ||x||$. □

It can also be shown that if $xy = yx$, then $\nu(x + y) \le \nu(x) + \nu(y)$; however we will not need this. It follows that if $A$ is commutative then $\nu(\cdot)$ is a *seminorm* on $A$; that is, $\nu(\cdot)$ is a norm except possibly $\nu(x) = 0$ for nonzero $x$. Of course, whenever the norm on $A$ satisfies $||x^2|| = ||x||^2$ for all $x$, then $\nu(\cdot)$ is also a norm by part (f) of (B.1.1). It turns out that a normed algebra satisfying $||x^2|| = ||x||^2$ for all $x$ is necessarily commutative (see (B.6.16)).

## B.2. *Examples*

1) Let $X$ be a locally compact Hausdorff space and $C_o(X)$ the set of continuous complex functions on $X$ which vanish at infinity

($f$ vanishes at infinity if for each $\varepsilon > 0$ there is a compact set $K$ in $X$ such that $|f(t)| < \varepsilon$ for all $t \in X \setminus K$). With pointwise operations

$$\begin{aligned}(f + g)(t) &= f(t) + g(t), \\ (\lambda f)(t) &= \lambda f(t), \qquad\qquad f, g \in C_0(X),\ t \in X \\ (fg)(t) &= f(t)g(t)\end{aligned}$$

and norm

$$||f||_\infty = \sup\{|f(t)|: t \in X\},$$

$C_0(X)$ is a commutative Banach algebra. This algebra is unital iff $X$ is compact. When $X$ is compact we write $C(X)$ for $C_0(X)$. When $X$ consists of one point, $C_0(X)$ is the field $C$.

2) Let $A$ be the algebra of complex polynomials on $[0,1]$ with pointwise operations. If $A$ is given the sup-norm, then $A$ is a unital commutative normed algebra which is not complete. Another norm on $A$ is $||p|| = \Sigma_{k=0}^n |a_k|$, where $p(x) = \Sigma_{k=0}^n a_k x^k$. Again $A$ is a unital commutative normed algebra which is not complete.

3) Let $A(D)$ denote the continuous complex functions on $D = \{\lambda \in C: |\lambda| \leq 1\}$ which are analytic in the interior of $D$. With pointwise operations and sup-norm, $A(D)$ is a unital commutative Banach algebra which is properly contained in $C(D)$. $A(D)$ is called the *disk algebra*.

4) Let $(X,\mathcal{B},\mu)$ be a $\sigma$-finite measure space. A complex measurable function $f$ on $X$ is essentially bounded if there is a constant $M > 0$ such that $|f(t)| \leq M$ a.e.[$\mu$]. Let $L^\infty(X,\mu)$ denote the set of equivalence classes of essentially bounded measurable functions on $X$, where $f$ and $g$ are equivalent iff $f = g$ a.e.[$\mu$]. If $f \in L^\infty(X,\mu)$, the essential sup-norm of $f$ is defined by $||f|| = \inf\{M: |f| \leq M \text{ a.e.}[\mu]\}$. With pointwise operations and essential sup-norm $L^\infty(X,\mu)$ is a unital commutative Banach algebra.

5) Let $E$ be a normed space. The vector space $B(E)$ of all bounded linear operators on $E$ is a unital normed algebra under composition as multiplication and norm

$$||T|| = \sup\{||T\xi||: \xi \in E, ||\xi|| \leq 1\}, \quad T \in B(E).$$

If $\dim E > 1$, $B(E)$ is noncommutative. $B(E)$ is a Banach algebra iff $E$ is a Banach space.

6) Let $G$ be a locally compact (Hausdorff) group and $\mu$ a left Haar measure on $G$. Let $L^1(G)$ be the set of equivalence classes of complex Borel measurable functions $f$ on $G$ such that $\int_G |f| d\mu$ exists. Then $L^1(G)$ is a Banach space under pointwise linear operations and norm $||f|| = \int_G |f| d\mu$. With the convolution product

$$(f*g)(x) = \int_G f(y)g(y^{-1}x)d\mu(y), \quad f, g \in L^1(G),$$

$L^1(G)$ is a Banach algebra, called the *group algebra*, which does not in general have an identity element and which is not in general commutative. This algebra is central in harmonic analysis.

7) Let $G$ be a locally compact (Hausdorff) group, and let $M(G)$ be the Banach space of all complex Borel measures on $G$ with setwise linear operations and total variation norm. With convolution product defined on the Borel subsets $E$ of $G$ by

$$(\mu*\nu)(E) = \int_G \mu(Ex^{-1})d\nu(x), \quad \mu, \nu \in M(G),$$

$M(G)$ is a unital Banach algebra, called the *measure algebra*, which is commutative iff $G$ is. The algebra $L^1(G)$, viewed as the set of all measures on $G$ which are absolutely continuous with respect to Haar measure, is an ideal in $M(G)$.

Note that the above examples fall into three general classes: function algebras (1) - (4); operator algebras (5); and group algebras (6) - (7). The classification is according to whether multiplication is defined pointwise, by composition, or by convolution, respectively.

## B.3. *Invertible and quasi-regular elements*

Let $A$ be an algebra with identity $e$. An $x$ in $A$ is *right* (resp. *left*) *invertible* if there is $y$ in $A$ such that $xy = e$ (resp. $yx = e$); $y$ is called a *right* (resp. *left*) *inverse* of $x$. An element $x$ in $A$ is *invertible* if it is both left invertible and right invertible.

If an element $x$ in $A$ has a left inverse $y$ and a right inverse z, then $y = z$. Indeed, $z = ez = (yx)z = y(xz) = ye = y$; thus, if $x$ is invertible there is a unique element, denoted $x^{-1}$ and called the inverse of $x$, such that $xx^{-1} = e = x^{-1}x$. The set of invertible elements in A will be denoted by $A^{-1}$. If an element A is not invertible it will be called *singular*. The set of singular elements in A will be denoted by $S_A$.

Of course, the property of an element x having an inverse depends both on A and x. If x in A is contained in a subalgebra B containing e, then x may be invertible in A and singular in B.

If A does not possess an identity, one may form the unitization $A_e$; then inverses can be considered in $A_e$. However, it is often quite cumbersome to work in $A_e$, and therefore it is desirable to have a theory which applies directly to algebras without identity. Fortunately, such a theory exists.

Let A be an algebra. The mapping $(x,y) \to x + y - xy$ of $A \times A$ into A, denoted

$$x \circ y = x + y - xy,$$

is called the *circle operation* on A. An element x in A is said to be *right* (resp. *left*) *quasi-regular* if there exists y in A such that $x \circ y = 0$ (resp. $y \circ x = 0$); the element y is called a *right* (resp. *left*) *quasi-inverse* of x. An element x in A is said to be *quasi-regular* (QR) if it is both left and right quasi-regular.

Noting that the circle operation is associative with zero as identity element, it follows that if $x \in A$ has both a left and right quasi-inverse, then these elements are equal.

If x is quasi-regular in A, the unique y in A satisfying $x \circ y = 0 = y \circ x$ is called the *quasi-inverse* of x, and will be denoted by $x'$. The set $Q_A$ of quasi-regular elements in A forms a group with respect to the circle operation, quasi-inversion, and identity 0. If an element in A is not quasi-regular, it will be called *quasi-singular*.

The following elementary lemma plays an important rôle in spectral theory a little later.

(B.3.1) LEMMA. *Let* x *and* y *be elements of an algebra* A. *Then* xy *is right (resp. left) quasi-regular if and only if* yx *is right (resp. left) quasi-regular. Hence,* xy *is quasi-regular if and only if* yx *is quasi-regular.*

*Proof*. Assume $xy$ is right quasi-regular, say $(xy)\circ z = 0$ for some $z$ in $A$, and let $w = -yx + yzx$. Then the computation $(yx)\circ w = yx + w - yxw = y(z + xy - xyz)x = y((xy)\circ z)z = y\cdot 0\cdot x = 0$ shows that $yx$ is right quasi-regular. A similar argument applies to left quasi-regularity. □

An element $x$ in an algebra $A$ with identity $e$ is right (resp. left) quasi-regular iff $e - x$ is right (resp. left) invertible. This is an immediate consequence of the identity

$$e - (x\circ y) = (e - x)(e - y), \quad (x, y \in A).$$

Further, the map $\psi\colon Q_A \to A^{-1}$ defined by $\psi(x) = e - x$ is a group isomorphism of $Q_A$ onto $A^{-1}$.

When is an element in a Banach algebra in $Q_A$, and what are some of the properties of $Q_A$? We answer these questions next. Recall $\nu(x) = \lim_{n\to\infty} ||x^n||^{1/n}$.

(B.3.2) PROPOSITION. *Let* $A$ *be a Banach algebra.*

*(a) If* $x \in A$ *satisfies* $\nu(x) < 1$ *(in particular if* $||x|| < 1$), *then* $x \in Q_A$ *and the quasi-inverse* $x'$ *of* $x$ *is given by* $x' = -\Sigma_{k=1}^{\infty} x^k$. *When* $||x|| < 1$,

$$||x||/(1 + ||x||) \le ||x'|| \le ||x||/(1 - ||x||).$$

*(b) The group* $Q_A$ *is an open subset of* $A$.

*(c) The mapping* $y \to y'$ *of* $Q_A$ *onto itself is a homeomorphism.*

*Proof*. (a) If $r$ is a real number such that $\nu(x) < r < 1$, then $||x^n|| < r^n$ for large $n$; comparing with the geometric series $\sum_{k=1}^{\infty} r^k$, the series $\sum_{k=1}^{\infty} ||x^k||$ converges. Define $y_n = -\sum_{k=1}^{n} x^k$ for $n = 1,2,\ldots$ . If $n > m$, the relation

$$||y_n - y_m|| = ||\sum_{k=m+1}^{n} x^k|| \le \sum_{k=m+1}^{n} ||x^k||$$

shows that $\{y_n\}$ is a Cauchy sequence in $A$. Since $A$ is complete, $y_n \to y$ for some $y$ in $A$. It is easily checked that $xy_n = y_nx =$

$x + y_{n+1}$ for each $n$. Taking limits on $n$, and using continuity of multiplication in $A$ we obtain $xy = yx = x + y$; that is, $x + y - xy = x + y - yx = 0$. Therefore $x' = y$.

To prove the inequalities in (a) when $||x|| < 1$, observe that since $x + y - xy = 0$, then $||y|| = ||xy - x|| \leq ||x||\cdot||y|| + ||x||$ and $||x|| = ||y - xy|| \leq ||y|| + ||x||\cdot||y||$. The inequalities follow quickly from these relations.

(b) Let $x \in A$. We first show that if $y \in Q_A$ and $||x|| < (1 + ||y'||)^{-1}$, then $x + y \in Q_A$. This will follow if we prove that $x + y$ has both a left and right quasi-inverse. Since $||x - y'x|| \leq ||x||(1 + ||y'||) < 1$, (a) implies that $x - y'x$ has a quasi-inverse, say $u$. Utilizing $y \circ y' = 0$ and $u \circ (x - y'x) = 0$ we easily verify that $z = -uy' + y' + u$ is a left quasi-inverse of $x + y$. Similarly, $x - xy'$ has a quasi-inverse $v$, and the element $w = -y'v + y' + v$ is a right quasi-inverse of $x + y$. Hence, $(x + y)' = -uy' + y' + u$ is the unique quasi-inverse of $x + y$.

Now let $y \in Q_A$, and set $\varepsilon = (1 + ||y'||)^{-1}$. If $x \in B_\varepsilon(y)$, the open ball of radius $\varepsilon$ and center $y$, then $||x - y|| < (1 + ||y'||)^{-1}$, and by the preceding observation the element $x = (x - y) + y$ has a quasi-inverse, i.e., $x \in Q_A$.

(c) If $x \in A$, $y \in Q_A$ and $||x|| < (1 + ||y'||)^{-1}$ we noted in (b) that $x + y \in Q_A$. In that argument $u$ was the quasi-inverse of $x - y'x$; hence by the inequality in part (a):

$$||(x + y)' - y'|| = ||(-uy' + y' + u) - y'|| \leq ||u||(||y'|| + 1)$$

$$\leq \frac{||x - y'x||}{1 - ||x - y'x||}(||y'|| + 1)$$

$$\leq \frac{||x||(||y'|| + 1)^2}{1 - (||y'|| + 1)||x||}.$$

The continuity of the self-inverse map $y \to y'$ is an immediate consequence of this inequality. □

(B.3.3) COROLLARY. *Let* $A$ *be a unital Banach algebra.*

*(a) If* $x \in A$ *and* $\nu(x) < 1$, *then* $e - x \in A^{-1}$ *and* $(e - x)^{-1} = e + \Sigma_{k=1}^{\infty} x^k$.

*(b) The group* $A^{-1}$ *is an open subset of* A.

*(c) The map* $x \to x^{-1}$ *of* $A^{-1}$ *onto itself is a homeomorphism.*

*Proof.* For $y \in A^{-1}$ we have $(e - y)' = e - y^{-1}$; hence $y^{-1} = e - (e - y)'$ from which the corollary follows from (B.3.2). □

Let A be a unital Banach algebra. If $x \in A$ we define the *exponential function* of x by

$$\exp(x) = \sum_{n=0}^{\infty} x^n/n! \text{ where } x^0 = e.$$

The series converges absolutely for all $x \in A$. We set $\exp(A) = \{\exp(x): x \in A\}$.

(B.3.4) PROPOSITION. *Let* A *be a unital commutative Banach algebra.*

*(a) If* $x, y \in A$, *then* $\exp(x + y) = \exp(x)\exp(y)$.

*(b)* $\exp(-x) = (\exp(x))^{-1}$ *for all* $x \in A$; *hence* $\exp(A) \subseteq A^{-1}$.

*(c) If* $x \in A$ *and* $\nu(e - x) < 1$, *there exists* $y \in A$ *with* $\exp(y) = x$.

*(d)* $\exp(A)$ *is the connected component of* e *in* $A^{-1}$.

*Proof.* (a) By the binomial theorem we have

$$\exp(x + y) = \sum_{n=0}^{\infty} (x + y)^n/n! = \sum_{n=0}^{\infty} 1/n! \sum_{j+k=n} n!/(j!k!)x^j y^k$$

$$= \sum_{j=0}^{\infty} x^j/j! \sum_{k=0}^{\infty} y^k/k! = \exp(x)\exp(y).$$

The rearranging of the sums is justified by absolute convergence.

(b) By part (a) we have $\exp(x)\exp(-x) = \exp(0) = e$ for any $x \in A$; hence $\exp(-x) = \exp(x)^{-1}$.

(c) Set $y = -\Sigma_{n=1}^{\infty}(e - x)^n/n$. Since $\nu(e - x) < 1$ it follows that the series defining y is absolutely convergent. It remains to show that $\exp(y) = x$. For each $k \geq 1$, we have $y^k = \Sigma_{n=1}^{\infty} c_{nk}(e - x)^n$, where the complex coefficients $c_{nk}$ are determined by the Cauchy rule for multiplying power series; hence the $c_{nk}$ do not depend on x. Then

$$\exp(y) = \sum_{k=0}^{\infty} y^k/k! = e + \sum_{n=1}^{\infty} \sum_{k=1}^{\infty} c_{nk}/k! \; (e - x)^n,$$

the interchange in order of summation being justified by absolute convergence. Now, it is well-known from elementary analysis that $\exp(y) = x$ where $x \in C$, and therefore the coefficient of $(e - x)^n$ above is 0 when $n > 1$, and $-1$ when $n = 1$. Hence $\exp(y) = x$.

(d) We note first that for each $y \in A$, the mapping $\lambda \to \exp(\lambda y)$, $0 \leq \lambda \leq 1$, is a path in $\exp(A)$ connecting $e$ to $\exp(y)$, so $\exp(A)$ is connected. Next, if $x = \exp(z)$, and $||x - y|| < ||x^{-1}||^{-1}$, then $||e - x^{-1}y|| < 1$ and so by part (c) we have $x^{-1}y = \exp(u)$ for some $u \in A$. Hence $y = \exp(z)\exp(u) = \exp(z + u) \in \exp(A)$, and so $\exp(A)$ is open. Finally, if $x \in A^{-1}$ and $x$ is in the closure of $\exp(A)$, there exists $y \in \exp(A)$ with $||y - x|| < ||x^{-1}||^{-1}$. Then $||yx^{-1} - e|| < 1$ and so $yx^{-1} \in \exp(A)$. It follows that $x \in \exp(A)$. Summarizing, $\exp(A)$ is connected, open and closed in $A^{-1}$, and $e \in \exp(A)$. □

## B.4. *The spectrum and normed division algebras*

One of the most important and useful concepts in the theory of Banach algebras is the notion of spectrum of an element. For commutative Banach algebras with identity it leads in a natural way to the existence of maximal ideals (or equivalently multiplicative linear functionals) which are indispensable in the representation theory of such algebras.

(B.4.1) DEFINITION. *Let* $x$ *be an element of an algebra* $A$. *If* $A$ *has an identity* $e$, *the* *__spectrum of__* $x$ *is defined to be the set*

$$\sigma_A(x) = \{\lambda \in C:\ x - \lambda e \ \textit{ is singular}\}.$$

*If* $A$ *has no identity, the* *__spectrum of__* $x$ *is defined to be the set*

$$\sigma_A(x) = \{\lambda \in C:\ \lambda \neq 0 \ \textit{ and } \ x/\lambda \ \textit{ is quasi-singular}\} \cup \{0\}.$$

In case the algebra $A$ has an identity $e$ then the second definition of spectrum in (B.4.1) coincides (except for the number 0) with the first. Indeed, if $\lambda \neq 0$, then $\lambda \notin \sigma_A(x)$ if and only if $e - \lambda^{-1}x$ is invertible. Hence $\lambda \notin \sigma_A(x)$ if and only if $\lambda^{-1}x$ is quasi-regular.

If $A$ has no identity, then $0 \in \sigma_A(x)$ for all $x$ in $A$. The virtue of this can be seen in the next lemma.

(B.4.2) LEMMA. *Let* A *be an algebra without identity, and* $A_e$ *the algebra obtained from* A *by adjoining an identity. If* $x \in A$*, then the spectrum of* x *as an element in* A *coincides with the spectrum of* x *as an element in* $A_e$*; that is,* $\sigma_A(x) = \sigma_{A_e}(x)$.

*Proof.* The lemma follows easily upon noting that elements of the form $(x,0)$ in $A_e$ are singular in $A_e$. The details are left to the reader. □

Let E be a non-zero Banach space and let A be B(E). If $T \in A$, then $\sigma_A(T)$ consists of those complex numbers $\lambda$ such that the operator $T - \lambda I$ is singular, where I denotes the identity operator on E. An *eigenvalue* of T is a complex number $\lambda$ such that $T\xi = \lambda\xi$ for some nonzero vector $\xi \in E$. Every eigenvalue of T is clearly contained in $\sigma_A(T)$; however, simple examples show that when E is infinite-dimensional the opposite inclusion may fail to hold.

Consider now the Banach algebra $A = C([0,1])$ and let $f \in A$. Then it is easy to see that $\sigma_A(f) = \{f(t): t \in [0,1]\}$. More generally, if X is any compact Hausdorff space and $f \in A = C(X)$, then $\sigma_A(f) = f(X)$. If X is a locally compact, noncompact, Hausdorff space, and $f \in A = C_o(X)$, then A has no identity element, and in this case $\sigma_A(f) = f(X) \cup \{0\}$.

It is natural to inquire about the behavior of the spectrum relative to subalgebras.

(B.4.3) PROPOSITION. *Let* A *be an algebra.*

*(a) If* B *is a subalgebra of* A *and* $x \in B$*, then* $\sigma_A(x) \subseteq \sigma_B(x) \cup \{0\}$.

*(b) If* B *is a maximal commutative subalgebra of* A *and* $x \in B$*, then* $\sigma_A(x) \cup \{0\} = \sigma_B(x) \cup \{0\}$.

*Proof.* (a) Let $x \in B$. If $\lambda \in \sigma_A(x)$, then either $\lambda = 0$ or $x/\lambda$ is quasi-singular in A, hence quasi-singular in B. In either case we have $\lambda \in \sigma_B(x) \cup \{0\}$.

(b) Let $x \in B$. In view of part (a) it suffices to show that $\sigma_B(x) \subseteq \sigma_A(x) \cup \{0\}$. Suppose that $\lambda \notin \sigma_A(x) \cup \{0\}$. Then $y = x/\lambda$ is quasi-regular in A, i.e., there is $z \in A$ such that $y \circ z = 0 = z \circ y$. To complete the proof it is enough to show $z \in B$ (for then $\lambda \notin \sigma_B(x)$). However, if w is any element of B, then $y \circ w = w \circ y$. Hence, $z \circ w = (z \circ w) \circ 0 =$

$(z \circ w) \circ (y \circ z) = (z \circ y) \circ (w \circ z) = 0 \circ (w \circ z) = w \circ z$ and since $B$ is a maximal commutative subalgebra of $A$, $z \in B$. □

It is an important fact that the spectrum of an element in a normed algebra is never empty.

(B.4.4) THEOREM. *If* $x$ *is an element of a normed algebra* $A$*, then* $\sigma_A(x)$ *is nonempty.*

*Proof.* In view of (B.4.2) we may assume that $A$ has an identity $e$. Suppose, to the contrary, that $\sigma_A(x) = \emptyset$. Then $(x - \lambda e)^{-1}$ exists for all $\lambda \in C$. In particular, $x^{-1}$ exists. By the Hahn-Banach theorem there is a bounded linear functional $L$ on $A$ such that $L(x^{-1}) = 1$. Let $x(\lambda) = (x - \lambda e)^{-1}$, and define $g\colon C \to C$ by $g(\lambda) = L(x(\lambda))$. Then $g(0) = 1$. Moreover, $g$ is an entire function. Indeed, since $x(\lambda) - x(\mu) = (\lambda - \mu)x(\lambda)x(\mu)$ for $\lambda, \mu$ in $C$, it follows that

$$\lim_{\lambda\to\mu} \frac{g(\lambda) - g(\mu)}{\lambda - \mu} = \lim_{\lambda\to\mu} L(x(\lambda)x(\mu)) = L(x(\mu)^2).$$

Further $|g(\lambda)| \le ||L||\cdot||x(\lambda)||$ and since $x(\lambda) = \lambda^{-1}(\lambda^{-1}x - e)^{-1} \to 0$ as $|\lambda| \to \infty$, then $g(\lambda) \to 0$. By Liouville's theorem the bounded entire function $g$ is constant; hence $g \equiv 0$. This is a contradiction since $g(0) = 1$. □

(B.4.5) PROPOSITION. *Let* $A$ *be a Banach algebra. For each* $x \in A$ *the spectrum* $\sigma_A(x)$ *of* $x$ *is a nonempty closed and bounded (hence compact) subset of the complex plane. If* $\lambda \in \sigma_A(x)$*, then* $|\lambda| \le ||x||$.

*Proof.* By (B.4.2) we may assume that $A$ has an identity $e$. The map $f\colon C \to A$ defined by $f(\lambda) = x - \lambda e$ is continuous; moreover, by (B.3.3), the set $A^{-1}$ of invertible elements in $A$ is open. Hence $f^{-1}(A^{-1})$ is open, and its complement $\sigma_A(x)$ is closed. If $|\lambda| > ||x||$, then $||x/\lambda|| < 1$, and by (B.3.3) $(e - x/\lambda)^{-1}$ exists. Therefore, $(x - \lambda e)^{-1}$ exists and $\lambda \notin \sigma_A(x)$. □

Recall that a *division algebra* is an algebra with identity $e$ in which each nonzero element is invertible.

(B.4.6) THEOREM. *(Mazur-Gelfand). Let* $A$ *be a unital normed algebra which is also a division algebra. Then* $A = \{\lambda e: \lambda \in C\}$; *hence,* A *is isometrically isomorphic to* C. *We shall write* $A \simeq C$.

*Proof.* Let $x \in A$ be nonzero. By (B.4.4) $(x - \lambda e)^{-1}$ fails to exist for some $\lambda \in C$. Since each nonzero element in A has an inverse, it must be the case that $x - \lambda e = 0$, i.e., $x = \lambda e$. □

(B.4.7) PROPOSITION. *Let* A *be a unital Banach algebra.*

*(a) If* $||x^{-1}|| = ||x||^{-1}$ *for each* $x \in A^{-1}$, *then* $A \simeq C$

*(b) If* $||xy|| = ||x||\cdot||y||$ *for all* $x, y \in A$, *then* $A \simeq C$.

*Proof.* (a) Let $x_n \in A^{-1}$ and suppose $x_n \to x \in A \setminus \{0\}$. The hypothesis and continuity of the norm imply that $||x_n^{-1}|| = ||x_n||^{-1} \to ||x||^{-1}$. Hence, $\{x_n^{-1}\}$ is bounded, i.e., there is $K > 0$ such that $||x_n^{-1}|| < K$ for all n. Since $x_n \to x$, there is a positive integer N such that $||x_n - x|| < 1/K$ for all $n \geq N$. It follows that $||e - x_N^{-1}x|| \leq ||x_N^{-1}||\cdot||x_N - x|| < 1$. By (B.3.3), (a) $x_N^{-1}x \in A^{-1}$; hence $x = x_N(x_N^{-1}x) \in A^{-1}$.

Therefore, $A^{-1}$ is both open and closed in $A \setminus \{0\}$, and since $A \setminus \{0\}$ is connected (any two points can be joined by a path made by two line segments) it follows that $A^{-1} = A \setminus \{0\}$. An application of (B.4.6) proves (a). Part (b) is an immediate consequence of (a). □

(B.4.8) PROPOSITION. *If* x *and* y *are elements of an algebra* A, *then the sets* $\sigma_A(xy)$ *and* $\sigma_A(yx)$ *differ at most by the number* 0; *that is,* $\sigma_A(xy) \cup \{0\} = \sigma_A(yx) \cup \{0\}$.

*Proof.* Let $\lambda$ be a nonzero complex number. It follows from (B.3.1) that $\lambda \notin \sigma_A(xy)$ if and only if $xy/\lambda$ is quasi-regular if and only if $yx/\lambda$ is quasi-regular if and only if $\lambda \notin \sigma_A(yx)$. □

The next two results are concerned with invariance of the spectrum under the action of polynomials and inversion.

(B.4.9) PROPOSITION. *Let* x *be an element of an algebra* A, *and let* p *be a polynomial with complex coefficients. Then* $\sigma_A(p(x)) = p(\sigma_A(x))$.

*Proof.* If A does not have an identity we adjoin one. Let $\mu \in p(\sigma_A(x))$. Then there is a $\lambda \in \sigma_A(x)$ with $\mu = p(\lambda)$. Let $q(t) =$

$p(t) - p(\lambda)$. Then $q(\lambda) = 0$; factoring $q$ into linear factors we can write

$$q(t) = \gamma(t - \lambda)(t - \lambda_1)\cdots(t - \lambda_n)$$

for $\gamma, \lambda_1,\ldots,\lambda_n \in C$, where $\gamma \neq 0$. Then

$$q(x) = \gamma(x - \lambda e)(x - \lambda_1 e)\ldots(x - \lambda_n e),$$

and since $x - \lambda e$ is singular, $q(x)$ must be singular. Hence, $p(x) - p(\lambda)e$ is singular and $\mu = p(\lambda) \in \sigma_A(p(x))$.

On the other hand, let $\mu \in \sigma_A(p(x))$ and let $q(t) = p(t) - \mu$. Factoring $q$ we have

$$q(t) = \gamma(t - \lambda_1)(t - \lambda_2)\ldots(t - \lambda_n)$$

for $\gamma, \lambda_1,\ldots,\lambda_n \in C$, where $\gamma \neq 0$. By assumption $q(x) = p(x) - \mu e$ is singular, and since

$$q(x) = \gamma(x - \lambda_1 e)(x - \lambda_2 e)\ldots(x - \lambda_n e),$$

some factor, say $x - \lambda_k e$, must be singular. Therefore, there exists $\lambda_k \in \sigma_A(x)$ such that $q(\lambda_k) = 0$; that is, $\mu = p(\lambda_k) \in p(\sigma_A(x))$. □

(B.4.10) PROPOSITION. *Let* $x$ *be an element in an algebra* $A$.

*(a) If* $A$ *has an identity* $e$ *and* $x$ *is invertible, then* $\sigma_A(x^{-1}) = \{\lambda^{-1}: \lambda \in \sigma_A(x)\}$.

*(b) If* $A$ *has no identity and* $x$ *is quasi-regular, then* $\sigma_A(x') = \{\lambda(\lambda - 1)^{-1}: \lambda \in \sigma_A(x)\}$.

*Proof*. (a) If $x$ is invertible, then $x^{-1} - \lambda^{-1}e = -\lambda^{-1}x^{-1}(x - \lambda e)$ from which (a) follows immediately.

(b) If $x$ is quasi-regular and $\lambda \neq 0$, then (b) follows easily from the relation $\lambda^{-1}(\lambda - 1)x' = x'\circ(\lambda^{-1}x)$. The case when $\lambda = 0$ is clear since $0$ belongs to both $\sigma_A(x')$ and $\sigma_A(x)$ by definition. □

(B.4.11) PROPOSITION. *(Wielandt). If* $A$ *is a normed algebra with identity* $e$*, then there do not exist elements* $x, y$ *in* $A$ *such that* $xy - yx = e$.

*Proof*. By considering the completion we may assume that A is a Banach algebra. If $x, y \in A$ are such that $xy - yx = e$, then $xy = e + yx$ and by (B.4.9) we have $\sigma_A(xy) = 1 + \sigma_A(yx)$. By (B.4.5) and (B.4.8), $\sigma_A(xy)$ and $\sigma_A(yx)$ are nonempty (compact) subsets of C which differ at most by the number 0. This contradicts the relation $\sigma_A(xy) = 1 + \sigma_A(yx)$. □

Let A be a normed algebra. The *spectral radius* of an element x in A is the nonnegative real number defined by

$$|x|_\sigma = \sup\{|\lambda|: \lambda \in \sigma_A(x)\}.$$

We next prove a remarkable formula which expresses the spectral radius of an element in a Banach algebra in terms of the norm.

(B.4.12) THEOREM. *(Beurling-Gelfand). Let* x *be an element in a Banach algebra* A. *Then* $|x|_\sigma = \lim_{n\to\infty} ||x^n||^{1/n}$.

*Proof*. Again, we may assume that A has an identity. If $x \in A$ and $\lambda \in \sigma_A(x)$, then $|\lambda| \leq ||x||$ by (B.4.5); hence $|x|_\sigma \leq ||x||$. By (B.4.9) $\sigma_A(x^n) = \sigma_A(x)^n$ for each positive integer n; this implies that if $\lambda \in \sigma_A(x)$ then $\lambda^n \in \sigma_A(x^n)$. Hence, $|x^n|_\sigma = |x|_\sigma^n$ and so $|x|_\sigma = |x^n|_\sigma^{1/n} \leq ||x^n||^{1/n}$ for all n; thus

$$|x|_\sigma \leq \lim_{n\to\infty} ||x^n||^{1/n}. \tag{1}$$

Let us prove the opposite inequality. We may clearly suppose x is nonzero. Consider $\lambda \in C$ such that $|\lambda| < 1/||x||$; then $||\lambda x|| < 1$. By (B.3.3) $e - \lambda x \in A^{-1}$ and

$$(e - \lambda x)^{-1} = e + \sum_{k=1}^{\infty} (\lambda x)^k. \tag{2}$$

Let $E = \{\lambda \in C: 0 < |\lambda| < 1/|x|_\sigma\}$. If $\lambda \in E$, then $1/|\lambda| > |\mu|$ for all $\mu \in \sigma_A(x)$; hence $1/\lambda \notin \sigma_A(x)$. It follows that $(e - \lambda x)^{-1}$ exists for all $\lambda \in E$. If L is any bounded linear functional on A, and $f: E \to C$ is defined by $f(\lambda) = L((e - \lambda x)^{-1})$, then by the argument given in the proof of (B.4.4) f is analytic on E. The continuity of L and and (2) imply that

$$f(\lambda) = L(e) + \sum_{k=1}^{\infty} \lambda^k L(x^k) \tag{3}$$

for all complex $\lambda$ satisfying $|\lambda| < 1/||x|| \leq 1/|x|_\sigma$. Hence, the series in (3) is the Taylor expansion for $f$. Since $f$ is analytic on $E$, it follows from elementary complex analysis that $\sum_{k=1}^{\infty} \lambda^k L(x^k)$ converges absolutely on $E$. Hence

$$\lim_{k\to\infty} |L(\lambda^k x^k)| = 0 \tag{4}$$

for all $\lambda \in E$ and all $L$ in the dual space $A^*$ of $A$.

To complete the proof we shall apply the uniform boundedness principle. Let $\lambda \in E$. For $n \in N$ and $L \in A^*$, define $\psi_n(L) = L(\lambda^n x^n)$. Then, as with the usual proof concerning the canonical injection of $A$ into $A^{**}$, we have for each $n$,

$$||\psi_n|| = ||\lambda^n x^n||. \tag{5}$$

By (4), $\sup\{|\psi_n(L): n \in N\} = \sup\{|L(\lambda^n x^n)|: n \in N\} < \infty$ for each $L \in A^*$. By the uniform boundedness principle there is $K > 0$ such that $||\psi_n|| \leq K$ for all $n$. From (5) we have $||\lambda^n x^n|| \leq K$, and hence $||x^n||^{1/n} \leq K^{1/n}/|\lambda|$ for all $n$. Thus, $\lim_{n\to\infty} ||x^n||^{1/n} \leq 1/|\lambda|$ for every $\lambda \in E$. It follows that $\lim_{n\to\infty} ||x^n||^{1/n} \leq |x|_\sigma$, and this with (1) completes the proof. □

An immediate consequence of (B.4.12) is that the spectral radius of an element $x$ of a Banach algebra is independent of the containing algebra.

## B.5. *Ideals in Banach algebras*

In this section we study ideals and the Jacobson radical in Banach algebras, and show a semisimple Banach algebra has a unique norm. Recall that a linear subspace $I$ of an algebra $A$ is a *left* (resp. *right*) *ideal* if $x \in I$ and $a \in A$ imply $ax \in I$ (resp. $xa \in I$). A *two-sided ideal* (or simply an *ideal*) is a left ideal which is also a right ideal. An ideal $I$ is said to be *proper* in case $I \neq A$; it is *maximal* if it is proper and is not properly contained in any other ideal of the same kind. It is important to note that a two-sided ideal may be maximal and still contained in a larger proper left or right ideal. Also, a maximal left (resp. right) ideal may be properly contained in a larger right (resp. left) ideal. An ideal $I$ of $A$ is called *minimal* if it is different from $\{0\}$ and does not properly contain any ideal of the same kind other than $\{0\}$. In general, minimal ideals fail to exist in a Banach algebra.

For the convenience of the reader we briefly recall several methods of combining ideals. Although our definitions are given only for left ideals it will be clear that they apply equally well to right ideals and two-sided ideals.

Let $A$ be an algebra. If $I$ and $J$ are left ideals of $A$, we define the *sum* of $I$ and $J$ by

$$I + J = \{x + y: x \in I, y \in J\}.$$

More generally, if $\{I_\alpha\}_{\alpha\in\Gamma}$ is finite or infinite family of left ideals of $A$, we define the *sum* of the family by

$$\sum_{\alpha\in\Gamma} I_\alpha = \{\sum_{\alpha\in F} x_\alpha: F \text{ finite}, F \subseteq \Gamma, x_\alpha \in I_\alpha\}.$$

It follows easily that the sum is again a left ideal of $A$; in fact, the sum is equal to the left ideal generated by the union $\bigcup_{\alpha\in\Gamma} I_\alpha$ of the family.

Again, let $I$ and $J$ be left ideals in $A$. The *product* $IJ$ of $I$ and $J$ is the left ideal defined by

$$IJ = \{\sum_{k=1}^{n} x_k y_k: x_k \in I, y_k \in J\}.$$

More generally, if $I_1, I_2, \ldots, I_m$ are left ideals in $A$, then the *product* $I_1 I_2 \ldots I_m$ is the set of all finite sums of terms of the form $x_1 x_2 \ldots x_m$ with $x_j \in I_j$.

The *quotient* of a left ideal $I$ in $A$ is the two-sided ideal in $A$ defined by

$$(I : A) = \{a \in A: ax \in I \text{ for all } x \in A\} = \{a \in A: aA \subseteq I\}.$$

If $I$, $J$, and $K$ are two-sided ideals in $A$, a simple check shows that $IJ \subseteq I \cap J \subseteq I$ (or $J$) $\subseteq I + J$ and $I(J + K) = IJ + IK$.

If the algebra $A$ does not possess an identity there may be no maximal ideals. In this case the following more restrictive notion of ideal allows us to proceed. A left ideal $I$ of $A$ is called a *modular left ideal* if there exists $u \in A$ such that

$$xu + I = x + I$$

for all $x \in A$. The element $u$ is called a *right identity relative to* I. *Modular right ideal* and *left identity relative to the ideal* are defined analogously. A two-sided ideal $I$ of $A$ is said to be *modular* if the quotient algebra $A/I$ has an identity $e + I$. The element $e$ is called an *identity relative to* $I$.

If $I$ is a modular left ideal of an algebra $A$, then it is clear from the definition that there exists $u \in A$ such that $xu - x \in I$ for all $x \in A$. If $I$ is proper, the element $u$ is not in $I$; indeed, if $u \in I$, then $x = (x - xu) + xu \in I$ for all $x$ in $A$; i.e., $I = A$.

If $J$ is a proper left ideal which contains the modular ideal $I$, then $J$ is also modular; in fact, it is clear that any right identity relative to $I$ is a right identity relative to $J$. Similar statements hold for modular right ideals and modular ideals.

(B.5.1) PROPOSITION. *A two-sided ideal in an algebra* $A$ *is modular if and only if* $I$ *is both a modular left and a modular right ideal.*

*Proof.* Clearly, two-sided modularity implies left and right modularity. Conversely, suppose that $I$ has a right identity $u$ relative to $I$ and a left identity $v$ relative to $I$. Let $e = vu$, and consider the relations

$$ex - x = vux - x = (vu - v)x + (vx - x),$$

$$xe - x = xvu - x = x(vu - u) + (xu - x).$$

Since $I$ is a two-sided ideal and $u$ and $v$ are right and left identities relative to $I$, we have $ex - x \in I$ and $xe - x \in I$ for all $x \in A$. Hence, $I$ is modular. □

(B.5.2) PROPOSITION. *Let* $A$ *be a Banach algebra. Statements (a), (b) and (c) are true for left, right, or two-sided modular ideals.*

*(a) The closure of a proper modular ideal of* $A$ *is again a proper modular ideal of* $A$.

*(b) Each maximal modular ideal of* $A$ *is closed.*

*(c) Each proper modular ideal of* A *is contained in a maximal modular ideal of* A.

*(d) An element* x *in* A *has a left (resp. right) quasi-inverse if and only if* x *is not a right (resp. left) identity relative to any maximal modular left (resp. right) ideal* I *of* A.

*Proof.* The proof will be given for modular left ideals. (a) Let I be a proper modular left ideal of A. Then $\bar{I}$ is a left ideal which is modular since $I \subseteq \bar{I}$. It remains to show $\bar{I}$ is proper. Assume, to the contrary, that $\bar{I} = A$. Let u be a right identity relative to I. Then $u \in A$ and since I is dense in A we may select $x \in I$ such that $||x + u|| < 1$. By (B.3.2) x + u is quasi-regular; hence there exists $y \in A$ such that $y \circ (x + u) = 0$ or $u = -x + (yu - y) + yx$. Since u is a right identity relative to I it follows that $yu - y \in I$; hence $u \in I$ which is impossible.

(b) If I is a maximal modular left ideal of A, then $\bar{I}$ is a proper modular left ideal of A which contains I. Since I is maximal, $I = \bar{I}$.

(c) Let I be a proper modular left ideal of A with right identity u relative to I. Let T denote the set of all proper left ideals of A containing I. Every ideal J in T is modular, and u is a right identity relative to J. Order T by inclusion. Let $\{J_\alpha\}$ be a chain in T and consider the union $J = \bigcup_\alpha J_\alpha$. Since the $J_\alpha$'s form a chain, it is easily seen that J is a left ideal which contains I; moreover, $u \notin J_\alpha$ for every $\alpha$, so that J is proper. Hence, $J \in T$ and is clearly an upper bound for the chain. By Zorn's lemma there exists an M in T which is maximal in T. Therefore M is a maximal modular left ideal containing I.

(d) Let $x \in A$ and assume that x has a left quasi-inverse y. Suppose, to the contrary, that x is a right identity relative to a maximal modular left ideal I. Since $y \circ x = y + x - yx = 0$ and x is a right identity relative to I, $x = yx - y \in I$, which contradicts the fact that right identities are not in their ideals.

Conversely, suppose that x has no left quasi-inverse; let $J = \{yx - y: y \in A\}$. Then J is clearly a left ideal and x is a right identity relative to J. Moreover, $x \notin J$; for if so, then $x = yx - y$ for some $y \in A$ and $y \circ x = 0$, contrary to the hypothesis. Hence, J is proper. By (c) we may extend J to a maximal modular left ideal I. It is clear that x is a right identity relative to I since $J \subseteq I$. □

If A has an identity element, then every left, right, and two-sided ideal of A is obviously modular. Hence we have:

(B.5.3) COROLLARY. *Let* A *be a Banach algebra with identity* e. *The following statements are true for left, right, or two-sided ideals.*

*(a) A proper ideal in* A *contains no invertible element of* A.

*(b) The closure of a proper ideal in* A *is again a proper ideal in* A.

*(c) Each maximal ideal in* A *is closed.*

*(d) Each proper ideal in* A *is contained in a maximal ideal of* A.

*(e) An element* x *in* A *is contained in a maximal left (resp. right) ideal of* A *if and only if it has no left (resp. right) inverse in* A.

Some elementary properties of algebra homomorphisms are given in the next result.

(B.5.4) PROPOSITION. *Let* A *and* B *be algebras, and* $f: A \to B$ *a surjective homomorphism with kernel* K. *The following statements are true for left, right, and two-sided ideals.*

*(a) If* J *is an ideal in* B, *then* $f^{-1}(J)$ *is an ideal in* A *and* $K \subseteq f^{-1}(J)$. *Moreover,* $J = f(f^{-1}(J))$.

*(b) If* I *is an ideal in* A, *then* f(I) *is an ideal in* B. *If* $K \subseteq I$, *then* $I = f^{-1}(f(I))$.

*(c) Those ideals* I *in* A *with* $K \subseteq I$ *are bijectively paired with ideals* J *in* B *under the mutually inverse correspondences* $I \to f(I)$ *and* $J \to f^{-1}(J)$. *Moreover, these correspondences preserve inclusion and proper containment.*

*(d) If* J *is a modular ideal in* B, *then* $f^{-1}(J)$ *is a modular ideal* A.

*(e) If* I *is a modular ideal in* A *with* $K \subseteq I$, *then* f(I) *is a modular ideal in* B.

*(f) The correspondences in (c) bijectively pair modular ideals* $I$ *in* $A$ *satisfying* $K \subseteq I$ *with the modular ideals in* $B$. *These correspondences preserve inclusion and proper containment.*

*(g) The correspondences in (c) bijectively pair the maximal modular ideals* $I$ *in* $A$ *satisfying* $K \subseteq I$ *with the maximal modular ideals in* $B$.

*Proof.* The proof will be given for right ideals. (a) Let $J$ be a right ideal of $B$. It is clear that $f^{-1}(J)$ is a right ideal in $A$ and that $K \subseteq f^{-1}(J)$. Since it is always true that $f(f^{-1}(J)) \subseteq J$, it suffices to prove the opposite inclusion. Since $f$ is surjective, if $y \in J$ there is $x \in A$ with $f(x) = y$. Then $x \in f^{-1}(J)$ and $y = f(x) \in f(f^{-1}(J))$.

(b) If $I$ is a right ideal in $A$, then $f(I)$ is clearly a right ideal in $B = f(A)$. Assume that $K \subseteq I$, and let $y \in f^{-1}(f(I))$. Then there exists $x \in I$ with $f(y) = f(x)$. Hence, $y - x \in K \subseteq I$ or $y \in x + I = I$ (since $x \in I$). Thus, $f^{-1}(f(I)) \subseteq I$; since the opposite inclusion is always true, $I = f^{-1}(f(I))$.

(c) To see that $I \to f(I)$ is surjective, let $J$ be any right ideal in $B$, and set $I = f^{-1}(J)$. Then $I$ is a right ideal of $A$ and $K \subseteq I$ (indeed, if $x \in K$ then $f(x) = 0 \in J$ or $x \in f^{-1}(J) = I$). By (a), $f(I) = f(f^{-1}(J)) = J$. The fact that the map is injective and inclusion-preserving follows easily from (b).

(d) Let $J$ be a modular right ideal in $B$. Then there exists $u \in B$ such that $uy - y \in J$ for every $y \in B$. Since $f$ is surjective there exists $v \in A$ such that $f(v) = u$. If $x \in A$ is arbitrary, we have $f(vx - x) = f(v)f(x) - f(x) = uf(x) - f(x) \in J$; hence, $vx - x \in f^{-1}(J)$.

(e) Let $I$ be a modular right ideal in $A$ with $K \subseteq I$. Since $I$ is modular, there exists $u \in A$ such that $ux - x \in I$ for all $x \in A$. Setting $v = f(u)$, we have for any $y \in B$, $vy - y = f(ux - x) \in f(I)$ where $x$ is any element of $f^{-1}(y)$.

(f) is a direct consequence of parts (c), (d), and (e); and statement (g) follows from (f) and the inclusion-preserving nature of the correspondence. □

Before introducing the next class of ideals we shall digress briefly to consider the important notion of a representation of an algebra.

(B.5.5) DEFINITION. *A* *<u>representation</u>* *of an algebra* $A$ *on a vector space* $E$ *is a homomorphism* $\pi$ *of* $A$ *into the algebra* $L(E)$ *of linear* *operators on* E. *That is, for each* $x \in A$, $\pi(x)$ *is a linear operator on* E, *and for* $x, y \in A$ *and complex* $\lambda$, *we have* $\pi(x + y) = \pi(x) + \pi(y)$, $\pi(\lambda x) = \lambda\pi(x)$, *and* $\pi(xy) = \pi(x)\pi(y)$. *The vector space* E *is called the* *representation space of* $\pi$.

Some useful terminology concerning representations is given in the following definition.

(B.5.6) DEFINITION. *Let* $\pi$ *be a representation of an algebra* A *on a vector space* E.

*(a)* $\pi$ *is said to be algebraically cyclic if there exists a vector* $\xi \in E$ *such that* $E = \{\pi(x)\xi\colon x \in A\}$. *The vector* $\xi$ *is called an algebraically cyclic vector for* $\pi$.

*(b) A linear subspace* M *of* E *is said to be invariant under* $\pi$ *if* $\pi(x)(M) \subseteq M$ *for all* $x \in A$.

*(c)* $\pi$ *is said to be algebraically irreducible if* $\{0\}$ *and* E *are the only subspaces of* E *invariant under* $\pi$.

*(d) A representation* $\psi$ *of* A *on a vector space* F *is said to be algebraically equivalent to* $\pi$ *if there exists a bijective linear mapping* U *from* E *onto* F *such that* $U\pi(x) = \psi(x)U$ *for all* $x \in A$.

When discussing the above properties of representations we shall sometimes omit the term "algebraically" which appears in the various parts. Later when representations of Banach algebras are discussed it will be important to distinguish between algebraic and topological properties of representations, and then the adjectives will have to be scrupulously retained

(B.5.7) EXAMPLE. (The regular representations). Let A be an algebra. For each $a \in A$, define $\rho(a)\colon A \to A$ by

$$\rho(a)x = ax.$$

Clearly, $\rho(a) \in L(A)$ for each $a \in A$, and the mapping $\rho\colon A \to L(A)$ defined by $a \to \rho(a)$ is easily seen to be a representation of A, called the *left regular representation* of A. The kernel of $\rho$ is the set

$$\ker(\rho) = \{a \in A\colon ax = 0 \text{ for all } x \in A\}.$$

If A has an identity, then $\ker(\rho) = \{0\}$ and $\rho$ is an isomorphism. I is a left ideal of A, iff $\rho(a)(I) \subseteq I$ for all $a \in A$; thus, the left ideals in A are the subspaces of A invariant under $\rho$. Hence for any left ideal I of A we may define a linear operator $\rho(a)\hat{}$ on the quotient space A/I by

$$\rho(a)\hat{}(x + I) = ax + I.$$

Then the mapping $a \to \rho(a)\hat{}$ is a representation $\rho\hat{}$ of A in L(A/I), the algebra of linear operators on A/I. The kernel of $\rho\hat{}$ is given by

$$\ker(\rho\hat{}) = \{a \in A: \rho(a)\hat{}(x + I) = 0 \text{ for all } x \in A\} = (I:A).$$

If I is modular, then $\ker(\rho\hat{}) \subseteq I$. Indeed, let u be a right identity relative to I, and suppose that $a \in (I:A)$. Then $au \in I$ and $au - a \in I$; thus, $a = au - (au - a) \in I$.

The representation $\rho\hat{}$ is algebraically irreducible if and only if I is a maximal left ideal of A. To see this, it plainly suffices to show that there is a one-to-one correspondence between left ideals in A containing I and invariant subspaces of A/I. However, if J is a left ideal in A that contains I, let

$$J' = \{x + I \in A/I: x \in J\}.$$

Then $\rho(a)(J') \subseteq J'$ for all $a \in A$; hence $J'$ is an invariant subspace of A/I. Also, since $I \subseteq J$, then

$$J = \{x \in A: x + I \in J'\}.$$

Therefore the correspondence is one-to-one. Conversely, if J' is any subspace of A/I invariant under $\rho\hat{}$, and if J is defined as in (1), then J is a left ideal of A which contains I. Therefore, the correspondence is onto.

By considering multiplication on the right we obtain the *right regular representation* of A; in this case we work with anti-homomorphisms rather than homomorphisms, and a discussion which parallels the above can be given. □

We next introduce a class of ideals which are important in the structure theory of noncommutative Banach algebras.

(B.5.8) DEFINITION. *A two-sided ideal* I *in an algebra* A *is said to be* <u>*primitive*</u> *if there exists a maximal modular left ideal* J *in* A *such that* I = (J:A).

Several elementary properties of primitive ideals are given in the following theorem.

(B.5.9) THEOREM. *Let* A *be an algebra. Then:*

*(a) Every primitive ideal in* A *is proper.*

*(b) Every maximal modular two-sided ideal in* A *is primitive.*

*(c) Every modular two-sided ideal in* A *is contained in a primitive ideal.*

*(d) If* I *is a primitive ideal in* A, *and* $I_1$, $I_2$ *are left ideals in* A *such that* $I_1I_2 \subseteq I$, *then either* $I_1 \subseteq I$ *or* $I_2 \subseteq I$.

*(e) If* A *is commutative, an ideal in* A *is primitive if and only if it is a maximal modular ideal.*

*(f) If* A *is a Banach algebra, every primitive ideal in* A *is closed.*

*Proof.* (a) If I is a primitive ideal in A, then there is a maximal modular left ideal J in A such that I = (J:A). Since J is proper and I = (J:A) $\subseteq$ J, then I is proper.

(b) Let I be a maximal modular two-sided ideal in A. By (B.5.2), (c) there is a maximal modular left ideal J containing I. Then (J:A) is clearly a two-sided modular ideal of A which contains I; since I is a maximal, I = (J:A). Thus, I is primitive.

(c) Let I be a modular two-sided ideal in A. By (B.5.2), (c) there is a maximal modular two-sided ideal J in A containing I; the ideal J is primitive by (b).

(d) Since I is primitive there is a maximal modular left ideal J in A such that I = (J:A). Suppose that $I_2 \nsubseteq I$. Then there is an element $x \in I_2$ such that $x \notin I$; but $x \notin I$ implies $xA \nsubseteq J$ and hence $I_2A \nsubseteq J$. It follows that $I_2A + J$ is a left ideal in A properly containing J, and since J is maximal we have $A = I_2A + J$. Then

$$I_1A = I_1I_2A + I_1J \subseteq I + J \subseteq J + J \subseteq J.$$

Thus, if $y \in I_1$ then $yA \subset J$; hence $y \in (J:A) = I$ and $I_1 \subseteq I$.

(e) Assume that A is commutative. By (b) every maximal modular ideal in A is primitive. Conversely, let I be a primitive ideal in A. Then there is a maximal modular ideal J in A with I = (J:A). Since

$J$ is modular, $(J:A) \subseteq J$. Let $x \in J$; if $a \in A$, then $ax \in J$ and hence $x \in (J:A)$, i.e., $J \subseteq (J:A)$. Therefore, $I = (J:A) = J$, and $I$ is a maximal modular ideal.

(f) Let $I$ be a primitive ideal in the Banach algebra $A$ and $J$ a maximal modular left ideal in $A$ such that $I = (J:A)$. Let $\{x_n\}$ be a sequence in $I$ such that $x_n \to x$ for some $x \in A$. If $a \in A$, then $ax_n \in J$ for each $n$. Since $ax_n \to ax$ and $J$ is closed, $ax \in J$. Hence, $x \in (J:A) = I$, and $I$ is closed. □

The following theorem contains a characterization of primitive ideals in terms of irreducible representations. This characterization will be of considerable importance for several applications, and is sometimes taken as the definition of primitive ideal.

(B.5.10) THEOREM. *A two-sided ideal* $I$ *in an algebra* $A$ *is primitive if and only if* $I$ *is the kernel of a nonzero algebraically irreducible representation of* $A$ *on a vector space.*

*Proof.* Assume that $I$ is primitive, and let $J$ be a maximal modular left ideal in $A$ such that $I = (J:A)$. Then by (B.5.7), the left regular representation of $A$ on the quotient space $A/J$ is a nonzero algebraically irreducible representation with kernel $(J:A) = I$.

Conversely, assume that $\pi$ is a nonzero algebraically irreducible representation of $A$ on a vector space $E$ with kernel equal to $I$. If $\xi$ is a nonzero vector in $E$, then the set $\pi(A)\xi \equiv \{\pi(x)\xi: x \in A\}$ is a nonzero subspace of $E$ invariant under $\pi$; since $\pi$ is irreducible, $E = \pi(A)\xi$. Hence, $\pi$ is algebraically cyclic with cyclic vector $\xi$. If $J = \{a \in A: \pi(a)\xi = 0\}$, then $J$ is a modular left ideal of $A$. Indeed, $J$ is clearly a left ideal. Since $\xi$ is a cyclic vector, there exists $u \in A$ such that $\pi(u)\xi = \xi$. If $x \in A$, then $\pi(xu - x)\xi = \pi(x)\pi(u)\xi - \pi(x)\xi = 0$, and hence $xu - x \in J$; that is, $J$ is modular.

Consider the quotient space $A/J$, and define a mapping $U: A/J \to E$ by $U(x + J) = \pi(x)\xi$. Since $x_1 - x_2 \in J$ if and only if $\pi(x_1)\xi = \pi(x_2)\xi$, it follows that $U$ is well-defined and injective. Further, $U$ is linear, and since $\xi$ is a cyclic vector, $U$ is surjective. If $\psi$ denotes the left regular representation of $A$ on $A/J$ (see (B.5.7)), then for all $x \in A$ and $y + J \in A/J$ we have

$$U\psi(x)(y + J) = U(xy + J) = \pi(xy)\xi = \pi(x)\pi(y)\xi = \pi(x)U(y + J).$$

Therefore, $U\psi(x) = \pi(x)U$ for all $x \in A$ so that $\pi$ and $\psi$ are algebraically equivalent. Since $\pi$ is irreducible, then $\psi$ is irreducible, and by (B.5.7) it follows that J is a maximal modular left ideal of A.

It remains to show that $I = (J:A)$. Let $x \in (J:A)$. Then, by definition, it follows that $xy \in J$ for all $y \in A$; that is, $0 = \pi(xy)\xi = \pi(x)(\pi(y)\xi)$ for all $y \in A$. Since $\xi$ is a cyclic vector, $\pi(x) = 0$; thus, $x \in \ker(\pi) = I$, and $(J:A) \subseteq I$. On the other hand, if $x \in I = \ker(\pi)$ then, for any $y \in A$, we have $0 = \pi(x)(\pi(y)\xi) = \pi(xy)\xi = U(xy + J)$; since U is injective, $xy \in J$. Then $\psi(x)(y + J) = xy + J = J$ (the zero in $A/J$), and so $x \in \ker(\psi) = (J:A)$. Thus, $I \subseteq (J:A)$ and we have $I = (J:A)$. □

We next introduce and briefly discuss the important notion of the Jacobson radical of an algebra.

(B.5.11) DEFINITION. *The* *(Jacobson) radical* Rad(A) *of an algebra* A *is defined to be the intersection of the kernels of all algebraically irreducible representations of* A. *The algebra* A *is called* *semisimple* *if* $\text{Rad}(A) = \{0\}$, *and is called a* *radical algebra* *if* $\text{Rad}(A) = A$.

It is clear that the radical is a two-sided ideal of A. Further, if A is a radical algebra then the zero representation is the only algebraically irreducible representation of A, and conversely, if the zero representation is the only irreducible representation, then A is a radical algebra.

(B.5.12) PROPOSITION. *Let* A *be an algebra such that* $\text{Rad}(A) \neq A$. *Then:* *(a)* Rad(A) *is equal to the intersection of all primitive ideals in* A.

*(b)* Rad(A) *is equal to the intersection of all maximal modular left (resp. right) ideals in* A.

*Proof.* Statement (a) is an immediate consequence of (B.5.10) and the definition of the radical.

(b) Let T denote the set of maximal modular left ideals in A. If $J \in T$, then $(J:A) \subseteq J$. Now, for each primitive ideal I in A there is a $J \in T$ such that $I = (J:A) \subseteq J$. It now follows from part (a) that Rad(A) is contained in the intersection of the ideals in T.

To obtain the opposite inclusion, assume that $x \notin \text{Rad}(A)$. Then there exists an irreducible representation $\pi$ of A on a vector space E such that $\pi(x) \neq 0$. Select any vector $\xi \in E$ such that $\pi(x)\xi \neq 0$ and set $I = \{a \in A: \pi(a)\xi = 0\}$. Then $I \in \mathcal{T}$ (see the preceding proof) and $x \notin I$; that is, $x \notin \cap\, \mathcal{T}$. This proves (b) for left ideals. We leave the argument for right ideals to the reader. □

A *quasi-regular ideal* in an algebra is an ideal (left, right, or two-sided) I such that each element in I is quasi-regular.

(B.5.13) LEMMA. *A left (resp. right) ideal* I *in an algebra* A *each of whose elements is left (resp. right) quasi-regular is a quasi-regular ideal.*

*Proof.* Let $x \in I$ and $y \in A$ such that $y \circ x = 0$. Then $y = yx - x \in I$, and hence $y$ is left quasi-regular. Since $x$ is already a right quasi-inverse for $y$, it follows that $x \circ y = y \circ x = 0$; that is, $x$ is quasi-regular. □

(B.5.14) PROPOSITION. *Let* A *be an algebra. Then:*

*(a)* Rad(A) *is a quasi-regular ideal.*

*(b)* Rad(A) *is equal to the sum of all quasi-regular left (or right) ideals in* A.

*(c)* Rad(A) *is equal to the set of all elements* y *in* A *such that* $(\lambda + x)y$ *(or* $y(\lambda + x)$*) is quasi-regular for every scalar* $\lambda$ *and* $x \in A$.

*Proof.* (a) Let $x$ be any left quasi-singular element in A. Then $I = \{y - yx: y \in A\}$ is a proper modular left ideal in A, and hence is contained in a maximal modular left ideal J. Now, $x \notin J$. In fact, if $x \in J$, then $y = (y - yx) + yx \in J$ for all $y \in A$ which contradicts the properness of J. Since $\mathrm{Rad}(A) \subseteq J$, then $x \notin \mathrm{Rad}(A)$ and it follows that every element of Rad(A) is left quasi-regular. Then, by (B.5.13), Rad(A) is a quasi-regular ideal.

(b) Assume that $x \notin \mathrm{Rad}(A)$; then there is an algebraically irreducible representation $\pi$ of A on a vector space E such that $\pi(x) \neq 0$. Select $\xi \in E$ with $\pi(x)\xi \neq 0$. Since $\pi$ is irreducible, there exists $y \in A$ such that $\pi(yx)\xi = \xi$. Then, for any $z \in A$, $\pi(z \circ (yx))\xi = \xi$. Hence, $z \circ (yx) \neq 0$ for all $z \in A$, so that $yx$ is not quasi-regular. It follows that if I is a quasi-regular left ideal, then $x \notin I$. Therefore, Rad(A) contains every such ideal. and since Rad(A) is itself a quasi-regular left ideal (by part (a)) (b) is proved for left ideals. A parallel argument applies to right ideals.

(c) Since the element $(\lambda + x)y$ is quasi-regular if and only if $y(\lambda + x)$ is quasi-regular (see (B.3.1)), it suffices to prove the result for $(\lambda + x)y$. By (a), Rad(A) is a quasi-regular ideal and hence $(\lambda + x)y$ is quasi-regular for each scalar $\lambda$, $x \in A$ and $y \in \mathrm{Rad}(A)$. On the other hand, if $y$ is an element of $A$ such that $(\lambda + x)y$ is always quasi-regular, then the set of all elements $(\lambda + x)y$ is a quasi-regular left ideal which contains $y$. Hence, $y \in \mathrm{Rad}(A)$ by (b). □

(B.5.15) PROPOSITION. *Let* A *be an algebra. Then:*

*(a) the quotient algebra* A/Rad(A) *is semisimple.*

*(b) if* I *is a two-sided ideal in* A *such that* A/I *is semisimple, then* I *contains* Rad(A).

*Proof.* (a) If A is a radical algebra, the proposition is clear. Suppose that $\mathrm{Rad}(A) \neq A$, and consider the canonical map $\tau: A \to A/\mathrm{Rad}(A)$. If $\{J_\alpha\}$ denotes the set of maximal modular left ideals in A/Rad(A), then by (B.5.4), (g) $\{\tau^{-1}(J_\alpha)\}$ is the set of maximal modular left ideals in A. The proposition now follows from $\bigcap_\alpha J_\alpha = \tau(\tau^{-1}(\bigcap_\alpha J_\alpha)) = \tau(\bigcap_\alpha \tau^{-1}(J_\alpha)) = \tau(\mathrm{Rad}(A)) = \{0\}$ and (B.5.12) (b).

(b) Consider the canonical map $\tau: A \to A/I$. If $T = \{I_\alpha\}$ denotes the set of maximal modular left ideals in A containing I, then $I = \bigcap_\alpha I_\alpha$. Indeed, by (B.5.4), (g), we have $T = \{\tau^{-1}(J_\alpha)\}$, as $J_\alpha$ varies over all maximal modular left ideals in A/I. Since A/I is semisimple, $I = \tau^{-1}(\{0\}) = \tau^{-1}(\bigcap_\alpha J_\alpha) = \bigcap_\alpha \tau^{-1}(J_\alpha) = \bigcap I_\alpha$. Now, Rad(A), being the intersection of *all* maximal modular left ideals in A, is clearly contained in $\bigcap_\alpha I_\alpha$. □

(B.5.16) PROPOSITION. *If* A *is an algebra and* $R = \mathrm{Rad}(A)$, *then* $|x|_\sigma = |x + R|_\sigma$ *for any* x *in* A.

*Proof.* We show that $\sigma_A(x) \cup \{0\} = \sigma_{A/R}(x + R) \cup \{0\}$ for any x in A. If $\lambda \notin \sigma_A(x) \cup \{0\}$, then $x/\lambda$ is quasi-regular in A. If $(x/\lambda)\circ y = y\circ(x/\lambda) = 0$, it is clear that $y + R$ is a quasi-inverse of $x/\lambda + R$ in A/R. Hence $\lambda \notin \sigma_{A/R}(x + R) \cup \{0\}$. On the other hand if $\lambda \notin \sigma_{A/R}(x + R) \cup \{0\}$, then $x/\lambda + R$ is quasi-regular in A/R. If $y + R$ is a quasi-inverse of $x/\lambda + R$, it follows that $(x/\lambda)\circ y$ and $y\circ(x/\lambda)$ belong to R. Since R is a quasi-regular ideal, there exist u and v in A such that $((x/\lambda)\circ y)\circ u = 0 = v\circ(y\circ(x/\lambda))$. Therefore $x/\lambda$ is quasi-regular and $\lambda \notin \sigma_A(x) \cup \{0\}$. □

REMARK. While it is clearly true that $\sigma_{A/R}(x + R) \subseteq \sigma_A(x)$, the opposite inclusion may fail if A has no identity element. For example if A is the subalgebra of the complex $2 \times 2$ matrices which have zeros in the second row and $x = \begin{pmatrix} 1 & 0 \\ 0 & 0 \end{pmatrix}$, then $0 \in \sigma_A(x)$ but $0 \notin \sigma_{A/R}(x + R)$.

The next theorem concerns the Jacobson radical in Banach algebras.

(B.5.17) THEOREM. *Let* A *be a Banach algebra. Then:*

*(a)* Rad(A) *is a closed two-sided quasi-regular ideal of* A.

*(b)* A/Rad(A) *is a semisimple Banach algebra.*

*(c) If* $x \in \text{Rad}(A)$, *then* $\nu(x) = 0$.

*(d) If* A *is unital and* $\nu(ax) = 0$ *for all* $a \in A$, *then* $x \in \text{Rad}(A)$.

*Proof.* Since each primitive ideal in A is a closed two-sided ideal, (a) follows immediately from (B.5.12), (a) and (B.5.14), (a).

(b) is an immediate consequence of (B.5.15).

(c) Let $x \in \text{Rad}(A)$. If $\lambda$ is a nonzero complex number, then $x/\lambda$ is in Rad(A). Hence, by (B.5.14), (a), $x/\lambda$ is quasi-regular; thus, $\lambda \notin \text{Sp}_A(x)$. Therefore, $\text{Sp}_A(x) = \{0\}$, and by (B.4.12), $\nu(x) = \sup\{|\lambda|: \lambda \in \text{Sp}_A(x)\} = 0$.

(d) If $\nu(ax) = 0$ for all $a \in A$, then by (B.4.12), ax is quasi-regular for each a and hence, by (B.5.14), (c) $x \in \text{Rad}(A)$. □

An algebra A of linear operators on a vector space E is said to be *irreducible* if the only subspaces of E invariant under all of the operators in A are {0} and E. Irreducible algebras of operators are always semisimple as we show next.

(B.5.18) PROPOSITION. *If* A *is a nonzero irreducible algebra of linear operators on a vector space* E, *then* A *is semisimple.*

*Proof.* For each nonzero $x \in E$, the set $Ax = \{Tx: T \in A\}$ is a subspace of E which is invariant under all of the operators in A. Hence, either $Ax = \{0\}$ or $Ax = E$. The set $F = \{x \in E: Ax = \{0\}\}$ is a subspace of E which is invariant under all of the operators in A; therefore, either $F = \{0\}$ or $F = E$. But the second case would mean that $A = 0$, contrary to assumption. Consequently, $F = \{0\}$ and $Ax = E$ for every $x \in E$, $x \neq 0$. This reasoning also applies to an arbitrary two-sided ideal $I \neq \{0\}$ in A, because for such an ideal I in A the sets Ix and

$\{x \in E: Ix = \{0\}\}$ are subspaces of $E$ which are invariant under all operators in $A$. Therefore, $Ix = E$ for any such ideal $I$ and arbitrary nonzero $x$ in $E$.

Suppose now that $I$ denotes the radical of $A$ and that $T \in I$. If $T \neq 0$, then there exists $x \in E$ such that $Tx \neq 0$, and this means that $ATx = E$. Hence there exists an operator $S \in A$ for which $STx = x$. But $ST$ has a left quasi-inverse since $T \in I$. Suppose that $R$ is this quasi-inverse. Then $R + ST - RST = 0$, and therefore $Rx + STx - RSTx = 0$. However this implies that $x = STx = RSTx - Rx = 0$, which contradicts the assumption $Tx \neq 0$. Consequently $I = \{0\}$ and $A$ is semisimple. □

(B.5.19) PROPOSITION. *Let* $(A, ||\cdot||)$ *be an irreducible Banach algebra of linear operators on a vector space* $E$. *Let* $y$ *be a fixed nonzero element of* $E$ *and, for each* $x \in E$, *define*

$$|x| = \inf\{||T||: T \in A, Ty = x\}.$$

*Then* $(E, |\cdot|)$ *is a Banach space and each operator* $T$ *in* $A$ *is bounded relative to* $|\cdot|$ *with bound* $|T| \leq ||T||$.

*Proof.* Let $J = \{T \in A: Ty = 0\}$. Then, as in the proof of (B.5.10), $J$ is a modular left ideal of $A$. Also, if $U$ is any element of $A$ not in $J$, then $Uy \neq 0$. Therefore, for any $T \in A$, there exists $S \in A$ such that $SUy = Ty$. Since $T = SU + (T - SU)$ and $T - SU \in J$, it follows that $J$ is a maximal left ideal of $A$. Hence $J$ is closed and $A/J$ is a Banach space. It is easily verified that the mapping $\phi: A/J \to E$ defined by $\phi(T + J) = Ty$ is a linear isomorphism of $A/J$ onto $E$, and that the norm

$$|x| = \inf\{||T||: T \in A, Ty = x\}$$

on $E$ is just the quotient norm on $A/J$ transferred by $\phi$ to $E$. Consequently $(E, |\cdot|)$ is a Banach space. Moreover,

$$\begin{aligned} |Tx| &= \inf\{||S||: S \in A, Sy = Tx\} \\ &\leq \inf\{||TS||: S \in A, Sy = x\} \leq ||T||\cdot|x| \end{aligned}$$

and therefore $|T| \leq ||T||$. □

We turn our attention next to an important class of semisimple algebras.

(B.5.20) DEFINITION. *An algebra* A *is said to be* <u>*primitive*</u> *if the zero ideal* $\{0\}$ *is a primitive ideal; that is,* A *is primitive if there exists a maximal modular left ideal* J *in* A *such that* $(J:A) = \{0\}$.

It follows from (B.5.10) that an algebra A is primitive if and only if it admits a faithful irreducible representation on a vector space. More specifically, the algebra A is primitive if and only if there is a maximal modular left ideal J in A such that A is isomorphic to an irreducible algebra of operators on the quotient space $A/J$ (see (B.5.7)). Thus, the study of primitive algebras reduces to a study of irreducible algebras of linear operators on a vector space and, in the case of Banach algebras (see (B.5.19)), to the study of irreducible algebras of bounded linear operators on a Banach space, where the algebra is also a Banach algebra under a norm which is not less than the operator bound. In particular, (B.5.18) shows that every primitive algebra is semisimple. Finally, if A is any algebra and I is a primitive ideal in A, then $A/I$ is clearly a primitive algebra.

(B.5.21) DEFINITION. *Let* $\{A_\alpha\}_{\alpha\in\Gamma}$ *be a family of Banach algebras. Let* Q *denote the set of all functions* f *defined on* $\Gamma$ *with* $f(\alpha) \in A_\alpha$ *for each* $\alpha$, *and such that* $|f|$ *is finite, where* $|f| = \sup\{||f(\alpha)||: \alpha \in \Gamma\}$, $||f(\alpha)||$ *being the norm in* $A_\alpha$. *With the natural pointwise operations and* $|f|$ *as norm,* Q *is a Banach algebra, called the* <u>*normed direct sum*</u> *of the algebras* $A_\alpha$. *Any subalgebra* R *of* Q *such that for each* $\alpha \in \Gamma$ *the elements* $f(\alpha)$ *exhaust* $A_\alpha$ *as* f *ranges over* R *is called a* <u>*normed subdirect sum*</u> *of the algebras* $A_\alpha$.

A normed subdirect sum is clearly a normed algebra; however, it is not necessarily a Banach algebra relative to the norm $|f|$.

It is well known that any semisimple algebra is isomorphic with a subdirect sum of primitive algebras. The corresponding structure theorem for semisimple Banach algebras is given next.

(B.5.22) THEOREM. *If* A *is a semisimple Banach algebra, then* A *is continuously isomorphic with a normed subdirect sum of primitive Banach algebras.*

*Proof.* Let Prim(A) denote the family of primitive ideals of A. For each $J \in$ Prim(A), A/J is a primitive Banach algebra. For $x \in A$ and $J \in$ Prim(A), define $\hat{x}(J) = x + J \in A/J$. Since A is semisimple, then $\hat{x}(J) = 0$ for every $J \in$ Prim(A) is equivalent to $x = 0$. Further, since $||\hat{x}(J)|| \leq ||x||$, we have $|\hat{x}| \leq ||x||$, where $|\hat{x}|$ is the norm in the normed direct sum Q of the family {A/J: $J \in$ Prim(A)}. Hence the mapping $x \to \hat{x}$ defines a continuous isomorphism of A into Q. Clearly, the image of A in Q under this isomorphism is a normed subdirect sum of the algebras A/J. □

(B.5.23) DEFINITION. *An algebra* A *of linear operators on a vector space* E *is said to be* k-*fold transitive on* E *if, for arbitrary linearly independent vectors* $x_1,\dots,x_k$ *and arbitrary* $y_1,\dots,y_k$ *in* E, *there exists* $T \in A$ *such that* $Tx_i = y_i (i = 1,\dots k)$. *If* A *is* k-*fold transitive for every* k, *then* A *is said to be strictly dense on* E.

It is obvious that an algebra is irreducible if and only if it is 1-fold transitive. The following important theorem shows that if an algebra is 2-fold transitive, then it is already strictly dense.

(B.5.24) THEOREM. *(Jacobson). Let* A *be an algebra of linear operators on a vector space* E. *If* A *is* 2-*fold transitive, then* A *is strictly dense on* E.

*Proof.* The proof is by induction. Suppose that A is k-fold transitive and let $x_1,\dots,x_{k+1}$ be any linearly independent set of $k + 1$ vectors in E. To prove $(k + 1)$-fold transitivity it suffices to show that for each i, with $1 \leq i \leq k + 1$, there exists $T_i \in A$ such that $T_ix_i \neq 0$ while $T_ix_j = 0$ if $i \neq j$ (indeed, if $y_1,\dots y_{k+1}$ are arbitrary vectors in E, we may select $S_i \in A$ such that $S_iT_ix_i = y_i$ and, if we set $T = \Sigma_{j=1}^{k+1} S_jT_j$, then $Tx_i = y_i$ for $i = 1,\dots,k + 1$).

To establish the assertion we may plainly restrict our attention to the case $i = k + 1$. Select, by k-fold transitivity, an element $S \in A$ such that $Sx_i = 0$ for $i = 1,\dots,k - 1$ and $Sx_{k+1} \neq 0$. If $Sx_k = 0$, then S is the desired element of A. Further, if $Sx_k$ and $Sx_{k+1}$ are linearly independent, then by 2-fold transitivity we may select $T \in A$ such that $TSx_k = 0$ while $TSx_{k+1} \neq 0$. Then TS is the desired element of A. Hence suppose that $Sx_{k+1} = \lambda Sx_k$ for some scalar $\lambda$. Since the vectors $x_1,\dots,x_{k-1},x_{k+1} - \lambda x_k$ are linearly independent, there exists $U \in A$ such that $Ux_i = 0$ for $i = 1,\ldots,k-1$ and $U(x_{k+1} - \lambda x_k) \neq 0$. As before, if $Ux_k = 0$, $U$ has the desired property and, if $Ux_k$, $Ux_{k+1}$ are linearly independent, choose $V \in A$ such that $VUx_k = 0$ and $VUx_{k+1} \neq 0$. Then $VU$ has the desired property. So, suppose that $Ux_{k+1} = \mu Ux_k$ for some $\mu$. Then $\lambda \neq \mu$. Finally, select $W \in A$ such that $WUx_k = Sx_k$ and set $T = S - WU$. Then $Tx_i = 0$ for $i = 1,\ldots,k$ and $Tx_{k+1} = (\lambda - \mu)Sx_k \neq 0$. Consequently, $A$ is $(k + 1)$-fold transitive. □

If $E$ is a vector space, an *endomorphism* of the additive group of $E$ is a mapping $D$ of $E$ into itself such that $D(x + y) = Dx + Dy$ for all $x, y \in E$.

(B.5.25) LEMMA. *(Schur's lemma). Let* $A$ *be an irreducible algebra of linear operators on a vector space and let* $T$ *be the set of endomorphisms of the additive group of* $E$ *which commute with each element of* $A$. *Then* $T$ *is a division algebra.*

*Proof.* Let $D \in T$. We first show that $D$ is a linear operator on $E$. It suffices to show that $D(\lambda x) = \lambda Dx$ for $x \in E$ and complex $\lambda$. Let $y$ be a nonzero vector in $E$; since $A$ is irreducible there exists $T \in A$ such that $Ty = x$. Then $(\lambda T)y = \lambda x$ and $Dx = DTy = TDy$. Hence, $D(\lambda x) = D((\lambda T)y) = (\lambda T)Dy = \lambda(TDy) = \lambda Dx$.

Suppose now that $D \in T$ and that $D \neq 0$. Then $D(E)$ is a nonzero linear subspace of $E$. If $T \in A$ is arbitrary, we assert that $T(D(E)) \subseteq D(E)$. Indeed, if $y \in D(E)$ then $y = Dx$ for some $x \in E$; hence $Ty = TDx = DTx \in D(E)$. Since $A$ is irreducible, $D(E) = E$. Now let $N = \ker(D)$. Then $N$ is a linear subspace. If $T \in A$ and $x \in N$, then $0 = TDx = D(Tx)$, so that $Tx \in N$. Thus $N$ is invariant under $A$ so $N = \{0\}$ or $N = E$. But $D(E) \neq \{0\}$ so $N = \{0\}$. Hence $D$ is a bijective linear mapping of $E$ onto $E$ whose inverse is clearly also in $T$. □

(B.5.26) PROPOSITION. *Let* $A$ *be an irreducible Banach algebra of linear operators on a vector space* $E$ *and let* $T$ *be as in* (B.5.25). *Then* $T$ *is isomorphic to the complex numbers.*

*Proof.* By (B.5.25), $T$ is a division algebra and each of its elements is a linear operator on $E$. Also, by (B.5.19) a norm $|\cdot|$ can be defined on $E$ making it a Banach space, and such that each $T \in A$ is bounded with bound $|T|$ satisfying $|T| \leq ||T||$. Indeed, if $y$ is a nonzero element of $E$, set $|x| = \inf\{||S||: S \in A, Sy = x\}$ for $x \in E$. Now, for $D \in \mathcal{T}$, set $z = Dy$ and, for arbitrary $x \in E$, choose any $S \in A$ such that $Sy = x$. Then

$$|Dx| = |DSy| = |SDy|$$
$$= |Sz| \leq |S|\cdot|z| \leq |z|\cdot||S||.$$

Taking the greatest lower bound over all $S \in A$ such that $Sy = x$, we obtain $|Dx| \leq |z|\cdot|x|$. Therefore $\mathcal{T}$ consists of bounded linear operators and hence is a normed division algebra. The proposition now follows from (B.4.6). □

We proved in (B.5.24) that a 2-fold transitive algebra $A$ of operators on a vector space is strictly dense. It is easy to see that (B.5.24) may fail if we weaken the hypothesis to 1-fold transitivity. On the other hand, if $A$ is a (complex) Banach algebra we prove next that 1-fold transitivity implies strict density.

(B.5.27) THEOREM. *(Rickart-Yood). Let* $A$ *be an irreducible Banach algebra of linear operators on a vector space* $E$. *Then* $A$ *is strictly dense on* $E$.

*Proof.* It suffices, by (B.5.24), to prove that $A$ is 2-fold transitive, and for this it is enough to show that for any pair $u, v$ of linearly independent vectors there exists $T \in A$ such that $Tu = 0$ and $Tv \neq 0$. Assume, to the contrary, that $Tu = 0$ implies $Tv = 0$. For any $x \in E$, select $S \in A$ such that $Su = x$ and define $Dx = Sv$. Since $S_1u = S_2u$ implies $S_1v = S_2v$, it follows that $Dx$ is independent of the choice of $S$. Further, for any $T \in A$, $TDx = TSv$, where $Su = x$. Since $(TS)u = Tx$ we have $D(Tx) = TSv$; hence $TD = DT$. Now, if $S_1u = x$ and $S_2u = y$, then $(S_1 + S_2)u = x + y$ and so $D(x + y) = (S_1 + S_2)v = S_1v + S_2v = Dx + Dy$. Hence $D \in \mathcal{T}$ and, by (B.5.26), $D = \alpha I$ for some complex number $\alpha$, where $I$ is the identity operator on $E$. It follows that $Tv = \alpha Tu$ or $T(v - \alpha u) = 0$ for all $T \in A$. But then $v - \alpha u = 0$, which contradicts the linear independence of the pair $u, v$. Hence there is an element $T \in A$ such that $Tu = 0$ and $Tv \neq 0$. □

Our purpose in the remainder of this section is to prove that every semisimple Banach algebra has a unique norm topology. Let $||\cdot||_1$ and

$||\cdot||_2$ be two norms with respect to which a given algebra A is a normed algebra. Recall that these norms are *equivalent* if there exist positive numbers $\alpha$, $\beta$ such that $||x||_1 \leq \alpha||x||_2 \leq \beta||x||_1$ for all $x \in A$.

(B.5.28) PROPOSITION. *Let* $||\cdot||_1$ *and* $||\cdot||_2$ *be two norms with respect to which an algebra* A *is a Banach algebra. Then* $||\cdot||_1$ *and* $||\cdot||_2$ *are equivalent if and only if* $||x_n||_1 \to 0$ *and* $||a - x_n||_2 \to 0$ *imply that* $a = 0$.

*Proof*. Assume there are positive numbers $\alpha$ and $\beta$ such that $||x||_1 \leq \alpha||x||_2 \leq \beta||x||_1$ for each $x \in A$. If $||x_n||_1 \to 0$ and $||a - x_n||_2 \to 0$, then $||a||_1 \leq ||a - x_n||_1 + ||x_n||_1 \leq \alpha||a - x_n||_2 + ||x_n||_1 \to 0$ and hence $a = 0$.

Conversely, assume that $||x_n||_1 \to 0$ and $||a - x_n||_2 \to 0$ imply $a = 0$. Let $A_1$ (resp. $A_2$) denote the Banach algebra A with norm $||\cdot||_1$ (resp. $||\cdot||_2$). Let T be the identity mapping of $A_1$ onto $A_2$, and let $\{y_n\} \subseteq A_1$, $y \in A_1$, and $z \in A_2$ satisfy $||y_n - y||_1 \to 0$, $||z - Ty_n||_2 \to 0$. We assert that $z = Ty$. Setting $x_n = y_n - y$ and $a = z - y$, we obtain $||x_n||_1 \to 0$, $||a - x_n||_2 \to 0$. Therefore $a = 0$ and $z = y = Ty$. This simply means that the graph of T is closed; hence by the closed graph theorem there is a positive number $\alpha$ such that $||x||_2 = ||Tx||_2 \leq \alpha||x||_1$ for each $x \in A$. By symmetry there exists a positive number $\beta$ such that $||x||_1 \leq \beta||x||_2$ for each $x \in A$. We conclude that the norms $||\cdot||_1$ and $||\cdot||_2$ are equivalent. □

The preceding proposition (which is true for arbitrary Banach spaces) suggests the following definition: Let $||\cdot||_1$ and $||\cdot||_2$ be two norms with respect to which an algebra A is a normed algebra. For each $x \in A$, let

$$s(x) = \inf_{a \in A}\{||a||_1 + ||x - a||_2\}.$$

Then s is called the *separating function* for the two norms, and if $s(x) = 0$, then x is called a *separating element*. The set of all separating elements of A will be denoted by $S_A$.

(B.5.29) LEMMA. *Let* $||\cdot||_1$ *and* $||\cdot||_2$ *be two norms with respect to which an algebra* A *is a normed algebra. Then for each* $x, y \in A$ *and complex* $\lambda$:

*(a)* $s(x + y) \le s(x) + s(y)$.

*(b)* $s(\lambda x) = |\lambda| s(x)$.

*(c)* $s(x) \le \min\{||x||_1, ||x||_2\}$.

*(d)* $s(xy) \le s(x)\{||y||_1 + ||y||_2\}$ *and* $s(xy) \le s(y)\{||x||_1 + ||x||_2\}$.

*(e)* $S_A$ *is a closed (in both norms) two-sided ideal of* A, *called the* <u>*separating ideal*</u>.

*(f) If* A *is complete relative to both norms, then these two norms are equivalent if and only if* $S_A = \{0\}$.

*Proof.* (a) Let $\varepsilon > 0$. Then there exists $a \in A$ such that $||a||_1 + ||x - a||_2 < s(x) + \varepsilon/2$; also there exists $b \in A$ such that $||b||_1 + ||y - b||_2 < s(y) + \varepsilon/2$. It follows that $s(x + y) \le ||a + b||_1 + ||(x + y) - (a + b)||_2 \le \{||a||_1 + ||x - a||_2\} + \{||b||_1 + ||y - b||_2\} < s(x) + s(y) + \varepsilon$. Since $\varepsilon$ was arbitrary, $s(x + y) \le s(x) + s(y)$.

The proof of (b) is obvious. To prove (c), we note that $s(x) \le ||a||_1 + ||x - a||_2$ for each $a \in A$; in particular, for $a = x$ and $a = 0$. Replacing a first by x and then by 0 we obtain $s(x) \le ||x||_1$ and $s(x) \le ||x||_2$. This says that $s(x) \le \min\{||x||_1, ||x||_2\}$.

(d) Observe that for each $a \in A$,

$$\begin{aligned} s(xy) &\le ||ay||_1 + ||xy - ay||_2 \le ||a||_1||y||_1 + ||x - a||_2||y||_2 \\ &\le ||a||_1||y||_1 + ||x - a||_2||y||_2 + ||x - a||_2||y||_1 + ||a||_1||y||_2 \\ &= \{||a||_1 + ||x - a||_2\}\{||y||_1 + ||y||_2\}. \end{aligned}$$

Therefore, taking the greatest lower bound over all $a \in A$, $s(xy) \le s(x)\{||y||_1 + ||y||_2\}$. The second part of (d) is proved in a similar way.

(e) By (a) and (b) the set $S_A$ is closed under addition and scalar multiplication. Part (d) implies that $S_A$ is closed under multiplication by elements of A. Hence $S_A$ is a two-sided ideal. By (c), $S_A$ is topologically closed with respect to each of the two norms.

Finally, part (f) is a direct consequence of (B.5.28). □

(B.5.30) LEMMA. *Let* A *be an algebra which is a Banach algebra under two norms* $||\cdot||_1$, $||\cdot||_2$ *and let* I *be a two-sided ideal which is closed relative to both norms. Then in order for the induced quotient norms on*

$A/I$ *to be equivalent it is necessary that* $S_A \subseteq I$ *and sufficient that* $S_A = I$.

*Proof.* Let $\tau$ denote the natural homomorphism of $A$ onto $A/I$. Let $y \in S_A$ and choose $\{x_n\} \subseteq A$ such that $||y - x_n||_1 \to 0$ and $||x_n||_2 \to 0$. It then follows from the definition of the quotient norm that $||\tau(y) - \tau(x_n)||_1 \to 0$ and $||\tau(x_n)||_2 \to 0$. Now, if the norms $||\tau(x)||_1$, $||\tau(x)||_2$ are equivalent, this implies, by (B.5.28), that $\tau(y) = 0$. Hence $y \in I$ and so $S_A \subseteq I$. This proves the necessity.

On the other hand, assume that $S_A = I$ and let $\tau(z)$ be any element in the separating ideal for the induced norms on $A/I$. Then there exists $\{\tau(x_n)\} \subseteq A/I$ such that $||\tau(z) - \tau(x_n)||_1 \to 0$ and $||\tau(x_n)||_2 \to 0$. Select $z \in \tau(z)$, $x_n \in \tau(x_n)$ and $z_n \in S_A$ such that $||x_n||_2 \to 0$ and $||z - x_n - z_n||_1 \to 0$. Since $z_n \in S_A$, there exists $y_n \in A$ such that $||z_n - y_n||_1 < 1/n$ and $||y_n||_2 < 1/n$. Define $w_n = x_n + y_n$. Then $||z - w_n||_1 \leq ||z - z_n - x_n||_1 + ||z_n - y_n||_1$ and $||w_n||_2 \leq ||x_n||_2 + ||y_n||_2$. Therefore $||z - w_n||_1 \to 0$ and $||w_n||_2 \to 0$; that is, $z \in S_A = I$. Hence $\tau(z) = 0$ and it follows that the separating ideal for the induced norms on $A/I$ is $\{0\}$. Now apply (B.5.29), (f). □

The next proposition shows that the uniqueness of the norm problem for semisimple Banach algebras can be reduced to the primitive case.

(B.5.31) PROPOSITION. *(C. Rickart). If every primitive Banach algebra has a unique norm topology, then every semisimple Banach algebra has a unique norm topology.*

*Proof.* Let $A$ be a semisimple Banach algebra and let $\mathrm{Prim}(A)$ denote the set of all primitive ideals in $A$; that is, all ideals of the form $(J:A)$ where $J$ is a maximal modular left ideal of $A$. Since $J$ is modular, $(J:A)$ is contained in $J$ (see (B.5.7)) and is closed with respect to any norm under which $A$ is a Banach algebra. It follows that the ideals in $\mathrm{Prim}(A)$ intersect in the zero element and that $A/P$ is a primitive Banach algebra for every $P \in \mathrm{Prim}(A)$. Now assume that $A$ is a Banach algebra under two norms with separating ideal $S_A$. If the Banach algebra $A/P$ has unique norm, then it follows from (B.5.30) that $S_A \subseteq P$. If this is true for every $P \in \mathrm{Prim}(A)$, then $S_A \subseteq \cap \mathrm{Prim}(A) = \{0\}$ and the two norms are equivalent by (B.5.29), (f). □

Let E, F and G be vector spaces. Recall that a mapping $\phi: E \times F \to G$ is *bilinear* if it is linear in each of its two variables. The following well-known lemma is an easy consequence of the uniform boundedness theorem and its proof will be left to the reader.

(B.5.32) LEMMA. *Let* E *and* F *be Banach spaces,* G *a normed space, and* $\phi$ *a bilinear mapping of* $E \times F$ *into* G *such that the linear mappings* $\phi(x,\cdot)$ *and* $\phi(\cdot,y)$ *are continuous for all* $x \in E$ *and all* $y \in F$. *Then* $\phi$ *is continuous on* $E \times F$.

(B.5.33) LEMMA. *Let* Z *be a Banach space and* X, Y *closed subspaces of* Z *such that* $X + Y = Z$. *Then there exists* $\alpha > 0$ *such that if* $z \in Z$ *then there is an* $x \in X$ *with:*

*(i)* $||x|| \leq \alpha||z||$;

*(ii)* $z - x \in Y$.

*Proof.* The mapping $(x,y) \to x + y$ from $X \times Y$ onto Z is continuous and hence is open by the open mapping theorem. Therefore there is $\delta > 0$ such that if $w \in Z$ with $||w|| \leq \delta$ then there exist $x_1 \in X$, $y_1 \in Y$ with $||x_1|| \leq 1$, $||y_1|| \leq 1$ and $x_1 + y_1 = w$. The lemma then follows if we set $\alpha = \delta^{-1}$, $w = z||z||^{-1}\delta$, and $x = (\alpha||z||)x_1$. □

The uniqueness of the norm of a semisimple Banach algebra will be a consequence of (B.5.32) and the following theorem.

(B.5.34) THEOREM. *(B. E. Johnson). Let* $(A,||\cdot||)$ *be an irreducible Banach algebra of operators on a normed space* $(E,||\cdot||')$. *Suppose that the mappings* $x \to Sx$ *of* E *into* E *are continuous for each* $S \in A$. *Then there exists a constant* M *such that* $||Sx||' \leq M||S||\cdot||x||'$ *for all* $S \in A$ *and all* $x \in E$.

*Proof.* If $E = \{0\}$ or if $A = \{0\}$, the theorem is obviously true; so we may assume that $E \neq \{0\}$ and that $A \neq \{0\}$. Let x be an arbitrary element of E. If the mapping $\psi: A \to E$ defined by $\psi(S) = Sx$ is continuous, then, for any $T \in A$, the map $\psi_T(S) = STx$ is also continuous since it is the composition of the continuous mappings $S \to ST$ and $\psi$.

Since $A$ is irreducible, if $x \neq 0$ we can, by a suitable choice of $T$, make $Tx$ any particular vector in $E$; hence if $\psi$ is continuous for one nonzero $x \in E$, it is continuous for all $x \in E$. We shall show that $\psi$ is continuous by assuming (to the contrary) that it is continuous only for $x = 0$ and deducing a contradiction.

The normed space $E$ is clearly infinite-dimensional for otherwise $A$ would be a finite-dimensional algebra and every linear mapping (in particular $\psi$) would be continuous. Since $E$ is infinite-dimensional we may choose a linearly independent sequence $x_1, x_2, \ldots$ from $E$ with $||x_i||' = 1$.

We now show that for each $\beta > 0$, $\varepsilon > 0$, and for every positive integer $m$ there is an $S \in A$ such that:

(1°) $||S|| < \varepsilon$.

(2°) $Sx_1 = Sx_2 = \cdots = Sx_{m-1} = 0$.

(3°) $||Sx_m||' > \beta$.

Set $J_i = \{T \in A: Tx_i = 0\}$; then, as in the proof of (B.5.19), $J_i$ is a maximal modular left ideal of $A$ and $I = (J_1 \cap J_2 \cap \ldots \cap J_{m-1}) + J_m$ is a left ideal of $A$ containing $J_m$. Since $x_1, \ldots, x_m$ are linearly independent over $C$ we can find, by the density theorem (B.5.27), an element $T \in A$ such that $Tx_1 = Tx_2 = \cdots = Tx_{m-1} = 0$ and $Tx_m = x_m \neq 0$. It is clear that $T \in I$, $T \notin J_m$ so that $I$ contains $J_m$ properly and, by maximality of $J_m$, $I = A$. Now, consider the number $\alpha$ obtained by applying (B.5.33) with $Z = A$, $X = J_1 \cap J_2 \cap \ldots \cap J_{m-1}$, and $Y = J_m$. By the discontinuity of the map $S \to Sx_m$ we can find an element $S_o \in A$ satisfying (1°) with $\varepsilon$ replaced by $\varepsilon/\alpha$, and (3°). Then by (B.5.33) there exists $S \in J_1 \cap J_2 \cap \ldots \cap J_{m-1}$ (so that (2°) holds for $S$), such that $S_o - S \in J_m$ (i.e., $S_o x_m = Sx_m$) and $||S|| \leq \alpha ||S_o|| < \varepsilon$.

Now choose, by induction, a sequence $S_1, S_2, \ldots$ in $A$ such that:

(i) $||S_n|| < 2^{-n}$.

(ii) $S_n x_1 = \cdots = S_n x_{n-1} = 0$.

(iii) $||S_n x_n||' \geq n + ||S_1 x_n + \cdots + S_{n-1} x_n||'$.

Set $T_i = \Sigma_{n>i} S_n$ for each $i \geq 0$. Since $S_n \in J_i$ for $n > i$ and $J_i$ is closed in $A$, we see that $T_i \in J_i$; that is, $T_i x_i = 0$, and $T_o = S_1 + \cdots + S_i + T_i$. Therefore it follows from (iii) that

$$\begin{aligned} ||T_o x_i||' &= ||S_1 x_i + \cdots + S_i x_i + T_i x_i||' \\ &\geq ||S_i x_i||' - ||S_1 x_i + \cdots + S_{i-1} x_i||' \\ &\geq i. \end{aligned}$$

Since $||x_i||' = 1$ this contradicts the hypothesis that $x \to T_o x$ is a bounded linear operator on $E$.

We have now shown that for each fixed $x \in E$ the map $(S,x) \to Sx$ of $A$ into $E$ is continuous. Since we also have, by hypothesis, that $(S,x) \to Sx$ is a continuous map of $E$ into $E$ for each fixed $S \in A$, (B.5.32) implies that $(S,x) \to Sx$ is continuous on $A \times E$ and the proof is complete. □

(B.5.35) THEOREM. *(Uniqueness of norms). Let* $A$ *be a semisimple algebra. Let* $||\cdot||_1$ *and* $||\cdot||_2$ *be norms on* $A$ *such that* $(A,||\cdot||_1)$ *and* $(A,||\cdot||_2)$ *are Banach algebras. Then the norms* $||\cdot||_1$ *and* $||\cdot||_2$ *are equivalent.*

*Proof.* By (B.5.31) it suffices to prove the theorem for primitive $A$. Hence we may assume that $A$ is faithfully represented as an irreducible algebra of operators on a vector space $E$. Since $A$ is a Banach algebra under the norms $||\cdot||_1$ and $||\cdot||_2$, let $|\cdot|_1$ and $|\cdot|_2$ be the respective Banach space norms on $E$ given by (B.5.19). Then each operator $T$ in $A$ is bounded relative to $|\cdot|_1$ and $|\cdot|_2$, and the operator bound satisfies $|T|_1 \leq ||T||_1$ and $|T|_2 \leq ||T||_2$. Now, suppose that $||S_n||_1 \to 0$ and $||T - S_n||_2 \to 0$ for $S_n$, $T \in A$. Then for each $x \in E$ we have $|Tx - S_n x|_2 \to 0$. However by (B.5.34) we see that $||S_n||_1 \to 0$ implies $|S_n x|_2 \to 0$. Since $|Tx|_2 \leq |Tx - S_n x|_2 + |S_n x|_2$ it follows that $Tx = 0$ for each $x \in E$, that is, $T = 0$. The theorem now follows from (B.5.28). □

REMARK. In 1982, B. Aupetit gave a very short "subharmonic proof" of an extension of (B.5.35) which avoids irreducible representations (see *J. Functional Analysis*, 47(1982), 1-6). More specifically, utilizing E. Vesentini's theorem (see Aupetit [3, p. 9]) and a Liouville-type theorem for bounded subharmonic functions in conjunction with the closed graph theorem, Aupetit proved the following result from which B. E. Johnson's "uniqueness of the norm" theorem follows as an immediate corollary.

(B.5.36) THEOREM. *(Aupetit). Let* A *and* B *be Banach algebras with* B *semisimple. If* T *is a linear mapping from* A *onto* B *such that* $|Tx|_\sigma \leq |x|_\sigma$ *for every* x *in* A, *then* T *is continuous.*

It is important to point out that the techniques used by Aupetit to prove this theorem also carry over to so-called "Banach Jordan algebras", thus showing that, when semisimple, they too have a unique complete norm topology. Since the approach using irreducible representations is known to fail in this situation, Aupetit's techniques are of considerable interest. It appears likely from results which have already been obtained that such techniques will be useful in the solution of several existing problems in Banach algebra theory. For more details concerning the use of subharmonic functions in Banach algebras and harmonic analysis the reader should consult Aupetit [3].

## B.6. *The Gelfand theory of commutative Banach algebras*

We turn our attention in this section to the beautiful Gelfand structure theory of commutative Banach algebras. Among other things, it will be shown that every semisimple commutative Banach algebra can be realized as an algebra of continuous complex-valued functions on a locally compact Hausdorff space.

Unless we state otherwise, A will denote a fixed commutative Banach algebra which may or may not have an identity element.

A *multiplicative linear functional* (MLF) (*complex homomorphism* or *character*) on A is a nonzero linear functional $\phi$ on A satisfying $\phi(xy) = \phi(x)\phi(y)$ for all $x, y \in A$; that is, $\phi$ is an algebra homomorphism of A onto the complex numbers C. The set of all MLF's on A will be denoted by $\hat{A}$.

(B.6.1) PROPOSITION. *If* $\phi \in \hat{A}$, *then* $K = \ker(\phi)$ *is a maximal modular ideal of* A, *and* A/K *is a field. Moreover, if* A *has an identity* e, *then* $\phi(e) = 1$.

*Proof*. It is easily checked that K is a proper ideal of A. Since $\phi: A \to C$ is surjective, A/K is isomorphic to C; in particular, A/K is a field. Since A/K has an identity, K is modular, and having co-dimension 1, is certainly maximal. The last statement of the proposition is obvious. □

The next result is concerned with the relation between MLF's on A and those on $A_e$, the algebra obtained from A by adjoining an identity.

(B.6.2) PROPOSITION. *Let* $A_e$ *be the commutative Banach algebra obtained from* A *by adjoining an identity, and identity* A *with* $\{(x,0): x \in A\}$. *Then every* $\phi \in \hat{A}$ *can be extended uniquely to an* MLF $\phi_e$ *on* $A_e$. *Moreover, when restricted to* A, *every* $\phi \in \hat{A}_e$ *is an* MLF *on* A *except for the functional* $\phi_\infty((x,\lambda)) = \lambda$.

*Proof*. If $\phi \in \hat{A}$, define $\phi_e: A_e \to C$ by $\phi_e((x,\lambda)) = \phi(x) + \lambda$ for $(x,\lambda) \in A_e$. Then $\phi_e \in \hat{A}_e$ and $\phi = \phi_e|_A$, hence $\phi_e$ is an MLF on $A_e$ which extends $\phi$. The uniqueness of $\phi_e$ is immediate from the fact that any element in $\hat{A}_e$ which extends $\phi$ must equal $\phi$ on A and take the value 1 at the identity $(0,1) \in A_e$. The remainder of the proposition is proved similarly. □

(B.6.3) PROPOSITION. *Each* $\phi \in \hat{A}$ *is continuous; in fact,* $||\phi|| \leq 1$. *If* A *is unital, then* $||\phi|| = 1$.

*Proof*. Assume that $|\phi(x)| > ||x||$ for some $x \in A$, and let $\lambda = \phi(x)$. Since $||x/\lambda|| < 1$, $x/\lambda$ has a quasi-inverse y by (B.3.2). Hence, $x/\lambda + y - xy/\lambda = 0$. Applying $\phi$ to this relation gives $1 + \phi(y) - \phi(y) = 0$, a contradiction. Thus, we have $|\phi(x)| \leq ||x||$ for all $x \in A$, that is, $||\phi|| \leq 1$. If A has an identity e, and $||e|| = 1$, then $||\phi|| \geq 1$ since $\phi(e) = 1$. □

There is a one-to-one correspondence between elements in $\hat{A}$ and maximal modular ideals in A.

(B.6.4) PROPOSITION. *Every maximal modular ideal* I *in* A *is the kernel of a unique* MLF $\phi$ *on* A. *Conversely, the kernel of a* MLF *on* A *is a maximal modular ideal in* A.

*Proof*. The last statement of the proposition was proved in (B.6.1).

Let $I$ be a maximal modular ideal of $A$. By (B.5.2), (b), $I$ is closed and therefore $A/I$ is a Banach algebra. Since $I$ is modular, $A/I$ has an identity $e + I$. Since $I$ is maximal, $A/I$ has no nontrivial ideals. We assert that $A/I$ is a field. Indeed, if $y + I$ is a nonzero element in $A/I$, then $M = \{xy + I: x \in A\}$ is a nonzero ideal in $A/I$; hence $M = A/I$. Since $e + I \in A/I$, there exists $x \in A$ such that $xy + I = e + I$; that is, $x + I$ is the multiplicative inverse of $y + I$. By (B.4.6) $A/I \simeq C$. The quotient map $\tau: A \to A/I \simeq C$, which has kernel $I$, is an MLF on $A$; set $\phi = \tau$.

It remains to show that $\phi$ is unique. Since $e$ is an identity relative to $I$, $0 = \phi(ex - x) = \phi(e)\phi(x) - \phi(x)$ for all $x \in A$. Since $\phi(x) \neq 0$ for some $x \in A$, then $\phi(e) = 1$. If $\phi'$ is another element in $\hat{A}$ with kernel $I$, then $\phi'(e) = 1$ also. Hence, $\phi$ and $\phi'$ coincide on $I$ and at $e$. However, every $x \in A$ can be written in the form $x = y + \lambda e$ for some $y \in I$, $\lambda \in C$; in fact, simply let $\lambda = \phi(x)$ and $y = x - \lambda e$. Since $\phi$ and $\phi'$ are linear they are equal. □

The relation between MLF's on $A$ and points in the spectra of elements in $A$ is given in the following.

(B.6.5) PROPOSITION. *Let* $x \in A$. *If* $A$ *has an identity* $e$, *then* $\lambda \in \sigma_A(x)$ *iff there is* $\phi \in \hat{A}$ *such that* $\phi(x) = \lambda$. *If* $A$ *has no identity and* $\lambda \neq 0$, *then* $\lambda \in \sigma_A(x)$ *iff there is* $\phi \in \hat{A}$ *such that* $\phi(x) = \lambda$.

*Proof.* Assume that $A$ has an identity $e$, and that $\lambda \in C$. If $\lambda \in \sigma_A(x)$, then $x - \lambda e$ is singular; it follows that $I = \{(x - \lambda e)a: a \in A\}$ is a proper ideal containing $x - \lambda e$. By (B.5.3), (d) there is a maximal ideal $J$ in $A$ such that $I \subset J$. By (B.6.4), $J$ is the kernel of a unique $\phi \in \hat{A}$. Since $\phi(x - \lambda e) = 0$, we have $\phi(x) = \lambda\phi(e) = \lambda$.

Conversely, if $x - \lambda e$ is invertible, then for every $\phi \in \hat{A}$ we have $\phi(x - \lambda e) \neq 0$, since otherwise $1 = \phi(e) = \phi((x - \lambda e)(x - \lambda e)^{-1}) = 0$. Hence, $\phi(x) \neq \lambda$. This proves the proposition when $A$ has an identity. The last statement is an immediate consequence of the first together with (B.4.2) and (B.6.2). □

For each $x$ in $A$ define a complex-valued function $\hat{x}: \hat{A} \to C$ by $\hat{x}(\phi) = \phi(x)$. The function $\hat{x}$ is called the *Gelfand transform* (or

*Fourier transform*) of the element $x$. It is the abstract analogue of the usual Fourier transform in harmonic analysis. Several elementary properties of the Gelfand transform are given next.

(B.6.6) PROPOSITION. *The Gelfand transform has the following properties for* $x, y \in A$ *and* $\lambda \in C$:

*(a)* $(x + y)\hat{} = \hat{x} + \hat{y}$;

*(b)* $(\lambda x)\hat{} = \lambda\hat{x}$;

*(c)* $(xy)\hat{} = \hat{x}\hat{y}$;

*(d)* $|\hat{x}(\phi)| \leq ||x||$ *for all* $\phi \in \hat{A}$;

*(e)* $\sigma_A(x) = \hat{x}(\hat{A}) \cup \{0\}$, *if* $A$ *has no identity;*

*(f)* *If* $A$ *has an identity,* $\sigma_A(x) = \hat{x}(\hat{A})$;

*(g)* $x$ *has a quasi-inverse iff* $\hat{x}(\phi) \neq 1$ *for all* $\phi \in \hat{A}$;

*(h)* *If* $A$ *has an identity,* $x \in A^{-1}$ *iff* $\hat{x}(\phi) \neq 0$ *for all* $\phi \in \hat{A}$.

*Proof*. Properties (a), (b), and (c) are immediate from the definition of the Gelfand transform.

(d) Since $||\phi|| \leq 1$ for all $\phi \in \hat{A}$ (B.6.3), we have $|\hat{x}(\phi)| = |\phi(x)| \leq ||\phi||\cdot||x|| \leq ||x||$.

(e) and (f) are direct consequences of (B.6.5).

(g) By (e), $\hat{x}(\phi) \neq 1$ for all $\phi \in \hat{A}$ iff $1 \notin \sigma_A(x)$ iff $x \in Q_A$.

(h) By (f), $\hat{x}(\phi) \neq 0$ for all $\phi \in \hat{A}$ iff $0 \notin \sigma_A(x)$ iff $x \in A^{-1}$. □

We now introduce a topology on the set $\hat{A}$ in such a way that each of the functions $\hat{x}: \hat{A} \to C$ becomes continuous.

The *Gelfand topology* on $\hat{A}$ is defined to be the weakest topology on $\hat{A}$ under which all of the functions $\hat{x}$ are continuous. A typical basic neighborhood of $\phi_0 \in \hat{A}$ is obtained by selecting $\varepsilon > 0$ and a finite subset $F$ of $A$, and letting

$$U_{F,\varepsilon}(\phi_0) = \{\phi \in \hat{A}: |\hat{x}(\phi) - \hat{x}(\phi_0)| < \varepsilon \text{ for } x \in F\}.$$

Equivalently, the Gelfand topology is the relative topology which $\hat{A}$ inherits as a subset of the dual space $A^*$ with the weak*-topology. The *structure space* of $A$ is the set $\hat{A}$ with the Gelfand topology.

In view of (B.6.4) and (B.6.5) the structure space $\hat{A}$ is often called the *maximal ideal space* of $A$ or the *spectrum* of $A$, respectively. The terms carrier space and character space are also used by some authors.

(B.6.7) PROPOSITION. *The structure space* $\hat{A}$ *of* $A$ *is a locally compact Hausdorff space. If* $A$ *has an identity element,* $\hat{A}$ *is compact. If* $A$ *does not possess an identity, each of the functions* $\hat{x}$ *on* $\hat{A}$ *vanishes at infinity.*

*Proof.* Since the weak*-topology is Hausdorff, it is clear that $\hat{A}$ is Hausdorff. By the Alaoglu-Banach theorem the closed unit ball $B = \{f \in A^*: ||f|| \leq 1\}$ is weak*-compact. By (B.6.3), $\hat{A} \subseteq B$. Assume that $\phi$ is an element of the weak*-closure $cl(\hat{A})$ of $\hat{A}$. We assert that $\phi(xy) = \phi(x)\phi(y)$ for all $x, y \in A$. If $x = 0$ or $y = 0$ this is clear. Assume that $x \neq 0$, $y \neq 0$; since $\phi \in cl(\hat{A})$, given $\varepsilon > 0$ there is $\psi \in \hat{A}$ such that

$$|\phi(x) - \psi(x)| < \varepsilon/3||y||, \ |\phi(y) - \psi(y)| < \varepsilon/3||x||, \ |\phi(xy) - \psi(xy)| < \varepsilon/3.$$

Since $\psi(xy) = \psi(x)\psi(y)$ we have

$$|\phi(xy) - \phi(x)\phi(y)| \leq |\phi(xy) - \psi(xy)| + |\psi(xy) - \psi(x)\psi(y)| +$$

$$|\psi(x)|\cdot|\psi(y) - \phi(y)| + |\phi(y)|\cdot|\psi(x) - \phi(x)| < \varepsilon.$$

Since $\varepsilon$ was arbitrary, we have $\phi(xy) = \phi(x)\phi(y)$. It follows that either $\phi \in \hat{A}$ or $\phi \equiv 0$. Hence, we consider two cases.

Case 1. If $0 \notin cl(\hat{A})$, then $\hat{A}$ is a weak*-closed subset of $B$ and hence is a weak*-compact Hausdorff space.

Case 2. If $0 \in cl(\hat{A})$, then $cl(\hat{A}) = \hat{A} \cup \{0\}$; since $\hat{A}$ is obtained by removing one element from the compact Hausdorff space $cl(\hat{A})$, $\hat{A}$ is locally compact and Hausdorff.

If $A$ has an identity $e$, then $\phi(e) = 1$ for all $\phi \in \hat{A}$; hence Case 2 cannot occur. Finally, if $\hat{A}$ is not compact, each $\hat{x}$ vanishes at infinity, since the point at infinity in $\hat{A}$ is the zero homomorphism. □

(B.6.8) COROLLARY. *Let* $A_e$ *be the commutative Banach algebra obtained from* $A$ *by adjoining an identity, and define* $\phi_\infty \in \hat{A}_e$ *by* $\phi_\infty((x,\lambda)) = \lambda$. *Then* $\hat{A}_e = \hat{A} \cup \{\phi_\infty\}$, $\hat{A}$ *is a locally compact Hausdorff space, and* $\hat{A}_e$ *is the one-point compactification of* $\hat{A}$. *Further, the mapping* $\phi \to \phi_e$ (B.6.2) *is a homeomorphism of* $\hat{A}$ *onto* $\hat{A}_e \setminus \{\phi_\infty\}$. *Finally, if* $\hat{A}$ *is compact,* $\phi_\infty$ *is an isolated point of* $\hat{A}_e$.

*Proof.* The corollary follows easily from the proposition and (B.6.2). The details are left to the reader. □

(B.6.9) COROLLARY. *If* A *and* B *are commutative Banach algebras which are algebraically isomorphic, then their respective structure spaces are homeomorphic.*

*Proof.* If $f: B \to A$ is an algebra isomorphism, then the map $F: \hat{A} \to \hat{B}$ defined by $F(\phi) = \phi \circ f$ is easily checked to be a homeomorphism. □

(B.6.10) THEOREM. *(Gelfand). The mapping* $x \to \hat{x}$, *called the* __*Gelfand representation*__, *is a homomorphism of* A *into* $C_0(\hat{A})$. *Moreover, if* $||\cdot||_\infty$ *denotes the sup-norm on* $C_0(\hat{A})$, *then* $||\hat{x}||_\infty \leq ||x||$, *and hence* $x \to \hat{x}$ *is continuous. If* $x \in A$, *then* $||\hat{x}||_\infty = |x|_\sigma = \nu(x)$.

*Proof.* The first part of the theorem is a consequence of (B.6.6) and (B.6.7). The last statement follows from (B.6.6), (e), (B.4.12) and the observation $||\hat{x}||_\infty = \sup\{|\phi(x)|: \phi \in \hat{A}\} = \sup\{|\lambda|: \lambda \in \sigma_A(x)\} = |x|_\sigma$. □

The reader should keep in mind that for some algebras it is possible for $||\hat{x}||_\infty = 0$ even for nonzero x.

An element x in an algebra A is called *nilpotent* if $x^n = 0$ for some positive integer n. An element x in a Banach algebra A is said to be a *generalized nilpotent element* if $\nu(x) = 0$.

In general, the Jacobson radical is a subset of the set of generalized nilpotent elements (see (B.5.17)). We see next that in the presence of commutativity the two sets coincide.

(B.6.11) PROPOSITION. *Consider the radical* Rad(A) *of the commutative Banach algebra* A. *Then:*

*(a)* Rad(A) *is the kernel of the Gelfand representation* $x \to \hat{x}$. *Hence,* $x \to \hat{x}$ *is injective if and only if* A *is semisimple.*

*(b)* Rad(A) *is equal to the set of all generalized nilpotent elements of* A.

*Proof.* (a) Let $\Psi: A \to C_0(\hat{A})$ denote the map $x \to \hat{x}$. By (B.6.4) the maximal modular ideals of A are precisely the subsets of A of the form

$\{x \in A: \hat{x}(\phi) = 0\} = \{x \in A: \phi(x) = 0\}$. Hence, $\text{Rad}(A) = \bigcap_{\phi \in \hat{A}} \{x \in A: \hat{x}(\phi) = 0\}$ $= \{x \in A: \hat{x} = 0\} = \ker(\Psi)$.

(b) We have $x \in \text{Rad}(A)$ if and only if $\phi(x) = 0$ for all $\phi \in \hat{A}$ if and only if $||\hat{x}||_\infty = 0$ if and only if $\nu(x) = 0$ by (B.6.10). □

An immediate consequence of (B.6.10) and (B.6.11) is that every commutative semisimple Banach algebra is isomorphic to an algebra of continuous complex-valued functions vanishing at infinity on a suitable locally compact Hausdorff space (with sup norm). In general, this representation is neither surjective nor norm-preserving. By imposing additional structure on the Banach algebra these deficiences will be removed.

(B.6.12) PROPOSITION. *Consider the commutative Banach algebra* $C_o(\hat{A})$ *and let* B *denote the subalgebra* $\{\hat{x}: x \in A\}$. *Then the following statemet-s are equivalent.*

*(a) The Gelfand representation* $x \to \hat{x}$ *is a homeomorphic isomorphism of* A *onto* B.

*(b) There exists a positive constant* $\beta$ *such that* $||x^2|| \geq \beta||x||^2$ *for all* $x \in A$.

*(c) The algebra* A *is semisimple, and* B *is closed in* $C_o(\hat{A})$.

*Proof.* (a) implies (b): If $x \to \hat{x}$ is a homeomorphic isomorphism, then it has a continuous inverse; hence, there is a positive constant k such that $||x|| \leq k||\hat{x}||_\infty$. Therefore, $||x||^2 \leq k^2||\hat{x}||_\infty^2 = k^2||\hat{x}^2||_\infty \leq k^2||x^2||$, and setting $\beta = 1/k^2$ we have $||x^2|| \geq \beta||x||^2$ for all $x \in A$.

(b) implies (c): Assume there exists $\beta > 0$ such that $||x^2|| \geq \beta||x||^2$ for all $x \in A$. Then $||x^4|| \geq \beta||x^2||^2 \geq \beta^3||x||^4$, and by induction, $||x^{2^n}|| \geq \beta^{2^n-1}||x||^{2^n}$. Hence, $||x^{2^n}||^{1/2^n} \geq \beta^{1-2^{-n}}||x||$, and by (B.6.10) we see that $||\hat{x}||_\infty = \lim_{n\to\infty} ||x^{2^n}||^{1/2^n} \geq \beta||x||$. We now have $\beta||x|| \leq ||\hat{x}||_\infty \leq ||x||$ from which the statements in (c) are immediate.

(c) implies (a): If A is semisimple and B is closed in $C_o(\hat{A})$, then $x \to \hat{x}$ is a bijective continuous mapping of A onto the Banach space B. It follows from the open mapping theorem that $x \to \hat{x}$ has a continuous inverse, hence is a homeomorphism. □

If for each $x \in A$ there exists an element $y \in A$ such that $\hat{y}(\phi) = \overline{\hat{x}(\phi)}$ for all $\phi \in \hat{A}$, then A is said to be *self-adjoint*.

(B.6.13) THEOREM. *If* A *is self-adjoint, then the Gelfand representation* $x \to \hat{x}$ *is an isometric isomorphism of* A *onto* $C_o(\hat{A})$ *if and only if* $||x^2|| = ||x||^2$ *for all* $x \in A$.

*Proof.* If $x \to \hat{x}$ is an isometric isomorphism, then $||x|| = ||\hat{x}||_\infty$ and hence $||x^2|| = ||\hat{x}^2||_\infty = ||\hat{x}||_\infty^2 = ||x||^2$ for all $x \in A$.

Conversely, assume that $||x^2|| = ||x||^2$ for all $x \in A$. By (B.6.10) $x \to \hat{x}$ is a homomorphism of A onto a subalgebra B in $C_o(\hat{A})$. By (B.1.1) and (B.6.10) we have $||\hat{x}||_\infty = ||x||$ for each $x \in A$; hence $x \to \hat{x}$ is an isometry (in particular, it is injective). It follows that B is a closed subalgebra of $C_o(\hat{A})$. It remains to show that $x \to \hat{x}$ is surjective, i.e., $B = C_o(\hat{A})$. Let $\phi$ and $\psi$ be distinct elements in $\hat{A}$; then $\phi(x) \neq \psi(x)$ for some $x \in A$. Hence, $\hat{x}(\phi) \neq \hat{x}(\psi)$ and B separates the points of $\hat{A}$. Since A is self-adjoint, B contains the complex conjugate of each of its functions. Since elements in $\hat{A}$ are nonzero, it is clear that for each $\phi \in \hat{A}$ there exists $\hat{x} \in B$ with $\hat{x}(\phi) \neq 0$. By the Stone-Weierstrass theorem, $B = C_o(\hat{A})$. □

The next theorem may be viewed as a substantial generalization of (B.6.3). It implies, in particular, that a semisimple commutative Banach algebra has a unique complete norm topology; hence, it provides a very simple proof of (B.5.35) when A is commutative.

(B.6.14) THEOREM. *Let* A *and* B *be commutative Banach algebras with* B *semisimple. If* $f:A \to B$ *is a homomorphism, then* f *is continuous.*

*Proof.* To show that f is continuous we shall apply the closed graph theorem. Let $x_n \in A$ and assume that $x_n \to x$ in A, and that $f(x_n) \to y$ in B. If $\phi \in \hat{B}$, then it is clear that $\phi \circ f \in \hat{A}$; hence by (B.6.3) $\phi \circ f$ is continuous. Therefore, $(\phi \circ f)(x_n) \to (\phi \circ f)(x)$. On the other hand, since $f(x_n) \to y$ in B we have $(\phi \circ f)(x_n) \to \phi(y)$. It follows that $\phi(f(x)) = \phi(y)$, or $f(x) - y \in \ker(\phi)$. Since $\phi \in \hat{B}$ was arbitrary, $f(x) - y \in \mathrm{Rad}(B)$, and since B is semisimple, $f(x) = y$. Hence, the graph of f is closed; by the closed graph theorem f is continuous. □

(B.6.15) COROLLARY. *Let* A *be a semisimple commutative algebra. Then there is at most one complete norm (within equivalence) on* A *under which* A *is a Banach algebra.*

*Proof.* If $(A,||\cdot||_1)$ and $(A,||\cdot||_2)$ are Banach algebras, then (B.6.14) implies that the identity map of each of these algebras onto the other is continuous. It follows that $||\cdot||_1$ and $||\cdot||_2$ are equivalent. □

In the following theorem and its corollary we discuss conditions on the norm of a given Banach algebra which imply that the algebra is commutative.

(B.6.16) THEOREM. *(Le Page, Hirschfeld-Żelazko). If* $A$ *is a normed algebra over* $\mathbb{C}$ *satisfying* $||x||^2 \leq \alpha||x^2||$ *for all* $x$ *in* $A$ *and some constant* $\alpha$, *then* $A$ *is commutative.*

*Proof.* The inequality $||x||^2 \leq \alpha||x^2||$ extends to the completion of $A$ so that we may assume $A$ is a Banach algebra. Iterating the given estimate, $||x||^{2^n} \leq \alpha^{2^n-1}||x^{2^n}||$ or $||x|| \leq \alpha^{1-1/2^n}||x^{2^n}||^{1/2^n} \to \alpha\nu(x)$ as $n \to \infty$.

If $A$ has no identity, adjoin one to obtain $A_e$, which contains $A$ isometrically as a closed two-sided ideal. (We do not claim that $||x|| \leq \alpha\nu(x)$ holds on all of $A_e$.) Let $x$ and $y$ be arbitrary elements of $A$, and set $z(\lambda) = \exp(\lambda x)y\exp(-\lambda x)$, $\lambda \in \mathbb{C}$. If $L$ is any continuous linear functional on $A$, define $f(\lambda) = L(z(\lambda))$. Then $f$ is an entire function of $\lambda$. Since $\exp(-\lambda x) = \exp(\lambda x)^{-1}$ (in $A_e$ if $A$ has no identity element) and $z(\lambda) \in A$,

$$||z(\lambda)|| \leq \alpha\nu(\exp(\lambda x)\ y\ \exp(-\lambda x)) = \alpha\nu(y) \text{ by (B.4.8).}$$

Therefore $f$ is bounded and entire. By Liouville's Theorem, $f$ is constant, so $f(\lambda) = L(y) = L(z(0))$. Differentiating $f$ with respect to $\lambda$ by the product rule and setting $\lambda = 0$, we have

$$0 = f'(0) = L(\exp(\lambda x)(xy - yx)\exp(-\lambda x))|_{\lambda=0} = L(xy - yx).$$

By the Hahn-Banach Theorem, $xy - yx = 0$ since $L$ was arbitrary. □

As a corollary we obtain a relation between the spectral radius and the norm which implies commutativity. This is the form which the principal theorem of Hirschfeld-Żelazko takes.

(B.6.17) COROLLARY. *(Hirschfeld-Żelazko). If* $A$ *is a complex normed algebra satisfying* $||x|| \leq \alpha\nu(x)$ *for all* $x$ *in* $A$ *and some constant* $\alpha$, *then* $A$ *is commutative.*

*Proof.* Squaring the relation $||x|| \leq \alpha\nu(x)$, we have $||x||^2 \leq \alpha^2\nu(x)^2 = \alpha^2\nu(x^2) \leq \alpha^2||x^2||$; so the hypothesis of Theorem 5.1 holds with the constant $\alpha^2$. □

If S is a nonempty subset of a Banach algebra A, the *commutant* (or *centralizer*) of S is defined as the set

$$S' = \{x \in A: xs = sx \text{ for all } s \in S\}.$$

We say S *commutes* if any two elements of S commute with each other. The set (S')' is called the *bicommutant* of S in A, denoted briefly by S".

The following result, whose proof is left to the reader, contains elementary properties of commutants.

(B.6.18) PROPOSITION. *Let* S, T *be nonempty subsets of a Banach algebra* A. *Then:*

*(i)* $S \subseteq S''$;

*(ii)* $S \subseteq T$ *implies* $T' \subseteq S'$;

*(iii)* $S''' = S'$;

*(iv)* S' *is a closed subalgebra of* A;

*(v)* S *commutes iff* $S \subseteq S'$;

*(vi)* S *commutes iff* S" *commutes.*

(B.6.19) REMARK. Let H be a Hilbert space and consider the Banach algebra B(H) of bounded linear operators on H. If A is a *-subalgebra of B(H) and $\xi \in H$, then $\xi$ is called a *separating vector* for A if the conditions $x \in A$, $x\xi = 0$ imply that $x = 0$. We say that $\xi$ is a *cyclic vector* for A if the linear subspace $A\xi = \{x\xi: x \in A\}$ is dense in H. If $\xi$ is a cyclic vector for A, then $\xi$ is a separating vector for the commutant A'. To see this, suppose $\xi$ is cyclic for A, $x' \in A'$, and $x'\xi = 0$. Then $0 = xx'\xi = x'x\xi$ for all $x \in A$. Since $A\xi = \{x\xi: x \in A\}$ is dense in H, it follows that $x' = 0$. Hence $\xi$ is separating for A'. □

As a simple application of Gelfand theory, we observe that the spectral radius is "subadditive" and "submultiplicative" on commuting elements of a Banach algebra.

(B.6.20) PROPOSITION. *Let* x *and* y *be elements in a Banach algebra such that* $xy = yx$. *Then:*

*(i)* $|x + y|_\sigma \le |x|_\sigma + |y|_\sigma$;

*(ii)* $|xy|_\sigma \le |x|_\sigma |y|_\sigma$.

*Proof.* We may assume that A has an identity. Consider the commuting set $S = \{x, y\}$, and let $B = S''$. By (B.6.18), B is a closed commutative subalgebra of A; thus, for $z \in B$, $\sigma_B(\hat{z}) = \hat{z}(\hat{B})$ by (B.6.6), (f). Since $(x + y)^\wedge = \hat{x} + \hat{y}$ and $(xy)^\wedge = \hat{x}\hat{y}$, we have

$$\sigma_B(x + y) \subseteq \sigma_B(x) + \sigma_B(y) \quad \text{and} \quad \sigma_B(xy) \subseteq \sigma_B(x)\sigma_B(y).$$

Both (i) and (ii) will now follow immediately from (B.4.12) if we can show that $\sigma_B(z) = \sigma_A(z)$ for any $z \in B$. To prove this, suppose that $z \in B$ and that z is invertible in A. We must show that $z^{-1} \in B$. However, since $z \in B = S''$, then $zt = tz$ for every $t \in S'$ and hence $t = z^{-1}tz$ or $tz^{-1} = z^{-1}t$. This says that $z^{-1} \in S'' = B$. □

## B.7. *Factorization in Banach algebras*

The possibility of decomposing a given algebraic expression into a product of two (or more) factors enjoying special properties has proved very useful in algebra and analysis. In this section we prove a factorization theorem which states that if A is a Banach algebra with bounded left approximate identity, then for $c \in A$ there are elements a, b in A such that $c = ab$, b belongs to the closed left ideal generated by c, and b is arbitrarily close to c. Actually a slight, but very useful, generalization of this result to the setting of Banach A-modules will be treated here.

Let A be a Banach algebra and X a Banach space. Then X is called a *left Banach* A*-module* if there is a bilinear product $(a,x) \to ax$ from $A \times X$ into X such that

(i) $(ab)x = a(bx)$,

(ii) $||ax|| \le K||a||\cdot||x||$

for all $a \in A$, $x \in X$ and some $K \ge 1$.

Right Banach A-modules are defined similarly. A *submodule* of a left Banach A-module $X$ is a subspace $E$ of $X$ such that $ay \in E$ for all $a \in A$, $y \in E$.

The Banach algebra $A$ has a *bounded left approximate identity* (b.l.a.i.) if there is a net $\{e_\alpha\}$ in $A$ and a real constant $M$ such that $||e_\alpha|| \leq M$ for all $\alpha$, and $e_\alpha a \to a$ for all $a \in A$. Equivalently, $A$ has a b.l.a.i. if there is a real constant $M$ such that for each finite subset $\{a_1,\ldots,a_n\}$ of $A$ and each $\varepsilon > 0$ there exists $e \in A$ with $||ea_i - a_i|| < \varepsilon$, $i = 1,\ldots,n$, and $||e|| \leq M$. A bounded right approximate identity is defined similarly.

(B.7.1) THEOREM. *(Cohen-Hewitt). Let* $A$ *be a Banach algebra with b.l.a.i., and let* $X$ *be a left Banach A-module. Then* $AX = \{ax\colon a \in A, x \in X\}$ *is a closed submodule of* $X$.

*Proof*. First note that $X$ is a left Banach $A_e$-module if we define $ex = x$ for $x \in X$. The closed linear span $X_0$ of the set $\{ax\colon a \in A, x \in X\}$ is $\{x \in X\colon \lim_\alpha e_\alpha x = x\}$. Given $x \in X_0$, following the original argument of Cohen, we shall define a sequence $\{e_n\}$ of elements in $A$ with $||e_n|| \leq M$ so that, if $\beta = M^{-1}$, then

$$\begin{aligned} b_n &= [(1+\beta)e - \beta e_n]^{-1}\cdots[(1+\beta)e - \beta e_1]^{-1} \\ &= (1+\beta)^{-n}e + a_n \quad (a_n \in A) \end{aligned} \tag{1}$$

converges in $A_e$ to a limit $b \in A$ and $y_n = b_n^{-1}x$ converges in $X_0 \subseteq X$ to an element $y$.

Observe that since $||e_n|| \leq M$ then $e - \beta(1+\beta)^{-1}e_n$ is invertible, so that $(1+\beta)e - \beta e_n$ is also. By the series expansion for inverses (see B.3.3) it follows that

$$[(1+\beta)e - \beta e_n]^{-1} = (1+\beta)^{-1}e + a$$

for some $a \in A$. Hence the relation (1) defines the elements $b_n$ and $a_n$ in the spaces $A_e$ and $A$, respectively. Then we have $x = \lim_{n\to\infty} b_n y_n = by$.

We turn now to the problem of defining the sequence $\{e_n\}$. This is done as follows: Set $a_0 = 0$, $b_0 = e$ and choose $\{e_n\}_{n=1}^{\infty}$ inductively so that for all $n \geq 0$

$$||e_{n+1}|| \leq M,$$

$$||e_{n+1}a_n - a_n|| < M(1 + \beta)^{-n-1},$$

$$||x - e_{n+1}x|| < ||b_n^{-1}||^{-1}M(1 + \beta)^{-n-1}.$$

Since $b_n - a_n \to 0$, to prove that $\{b_n\}$ is convergent, it suffices to show that $\{a_n\}$ is Cauchy. Now

$$a_{n+1} = b_{n+1} - (1 + \beta)^{-n-1}e$$

$$= [(1 + \beta)e - \beta e_{n+1}]^{-1}[(1 + \beta)^{-n}e + a_n] - (1 + \beta)^{-n-1}e,$$

so that

$$||a_{n+1} - a_n|| =$$

$$||[(1 + \beta)e - \beta e_{n+1}]^{-1}[\beta(1 + \beta)^{-n-1}e_{n+1} + \beta e_{n+1}a_n - \beta a_n]||.$$

Expanding $[e - (1 + \beta)e_{n+1}]^{-1}$ in a power series as in (B.3.3) and using $||e_{n+1}|| \leq M$, we obtain $[(1 + \beta)e - \beta e_{n+1}]^{-1}|| \leq M$. Since

$$||\beta(1 + \beta)^{-n-1}e_{n+1}|| \leq (1 + \beta)^{-n-1}$$

and

$$\beta||e_{n+1}a_n - a_n|| \leq (1 + \beta)^{-n-1},$$

it follows that $||a_{n+1} - a_n|| \leq 2M(1 + \beta)^{-n-1}$. Hence for $m > n$

$$||a_n - a_m|| \leq ||a_n - a_{n+1}|| + ||a_{n+1} - a_{n+2}|| + \cdots + ||a_{m-1} - a_m||$$

$$\leq 2M(1 + \beta)^{-n-1}(1 + (1 + \beta)^{-1} + \cdots + (1 + \beta)^{n-m+1})$$

$$\leq 2M(M + 1)(1 + \beta)^{-n-1} \to 0 \quad \text{as} \quad n \to \infty,$$

and so $\{a_n\}$ is Cauchy in $A$.

Furthermore, $y_{n+1} = b_n^{-1}[(1 + \beta)e - \beta e_{n+1}]x$, so that

$$\begin{aligned}||y_{n+1} - y_n|| &= ||b_n^{-1}(\beta x - \beta e_{n+1}x)|| \\ &\le \beta||b_n^{-1}||\cdot||x - e_{n+1}x|| \\ &\le (1+\beta)^{-n-1}\end{aligned}$$

and hence, as with the $a_n$'s, the sequence $\{y_n\}$ is Cauchy, as required. □

(B.7.2) COROLLARY. *Let* $S$ *be a compact subset of* $X_o$. *Then there exists* $a$ *in* $A$ *and a compact subset* $T$ *of* $X_o$ *such that* $S = aT$.

*Proof*. Let $B$ denote the Banach space of continuous functions from $S$ into $X_o$ under the natural linear operations and norm $||f|| = \sup\{||f(s)||: s \in S\}$. Then $B$ becomes a left Banach A-module with multiplication defined by $(a\cdot f)(s) = af(s)$, for $a \in A$, $s \in S$, $f \in B$. If $\{e_\alpha\}$ is a b.l.a.i. in $A$, then $e_\alpha x \to x$ for each $x \in X_o$ and, since $\{e_\alpha\}$ is bounded, $e_\alpha x \to x$ uniformly on compact subsets of $X_o$. If $f \in B$, then $f(S)$ is compact; so $e_\alpha f(s) \to f(s)$ uniformly on $S$. Hence, $B_o = B$. Let $i \in B$ be the identity, i.e., $i(s) = s$ for all $s \in S$. Then by (B.7.1), there exists $a \in A$ and $g \in B$ such that $i = ag$ and, in particular, $S = ag(S)$. Setting $T = g(S)$, the corollary follows. □

(B.7.3) REMARK. If $\{x_n\}$ is a sequence in $X_o$ which converges to 0, then Corollary (B.7.2) can be applied with $S = \{0,x_1,x_2,\ldots\}$. Let $g$ be as in the proof of (B.7.2) and set $y_n = g(x_n)$. Then $ay_n = x_n$, $ag(0) = 0$, and $y_n \to g(0)$. Set $z_n = y_n - g(0)$. Then $az_n = x_n$ and $z_n \to 0$. This factoring of null sequences has several useful applications. □

## B.8. *The holomorphic functional calculus*

If $x$ is a fixed element in a Banach algebra $A$ with identity $e$, then $A$ clearly contains all polynomials in $x$. More generally, $A$ contains all "entire functions in $x$"; that is, if $f(\lambda) = \Sigma_{n=0}^{\infty} a_n\lambda^n$ represents an entire function of a complex variable $\lambda$, then $A$ contains $f(x)$. Indeed, the series $\Sigma_{n=0}^{\infty} a_n x^n$ is majorized by the convergent series $\Sigma_{n=0}^{\infty} |a_n|\cdot||x||^n$. Therefore, to each entire function $f$ there corresponds an "abstract" entire function $f(x) = \Sigma_{n=0}^{\infty} a_n x^n$ in $A$. Further, if $A$ is commutative, for any $\phi \in \hat{A}$

$$f(x)^\wedge(\phi) = \phi(f(x)) = \Sigma_{n=0}^{\infty} a_n \phi(x)^n = f(\phi(x)) = (f \circ \hat{x})(\phi)$$

that is, $f(x)^\wedge = f \circ \hat{x}$. It follows that the subalgebra $\{\hat{x}: x \in A\}$ of $C(\hat{A})$ is closed under the application of entire functions.

Our objective in this section is to show that, given a Banach algebra $A$ with identity and an $x \in A$, the above ideas extend to make holomorphic functions act in $A$.

If $U$ is an open set containing the spectrum of $x$ then, to each complex-valued holomorphic function $f$ defined on $U$, we shall associate an element $f(x)$ in $A$ in such a way that the mapping $f \to f(x)$ has many nice properties. The details are given in the following theorem.

(B.8.1) THEOREM. *Let* $A$ *be a Banach algebra with identity* $e$, $x \in A$, *and* $U$ *an open subset of the complex plane which contains* $\sigma_A(x)$. *Let* $H(U)$ *denote the algebra of all complex-valued functions (under pointwise operations) which are defined and holomorphic in* $U$. *Then there exists a mapping* $f \to f(x)$ *of* $H(U)$ *into* $A$ *such that:*

*(i) If* $f(\lambda) = \mu$ *for some* $\mu \in C$, *then* $f(x) = \mu e$;

*(ii) If* $f(\lambda) = \lambda$, *then* $f(x) = x$;

*(iii)* $f \to f(x)$ *is a homomorphism of* $H(U)$ *into* $A$;

*(iv) If* $f \in H(U)$ *is represented by the power series* $\Sigma_{n=0}^{\infty} c_n \lambda^n$ *throughout the open set* $U$ *containing* $\sigma_A(x)$, *then* $f(x) = \Sigma_{n=0}^{\infty} c_n x^n$.

*(v) If* $f_n \in H(U)$ *and* $f_n \to f$ *uniformly on compact subsets of* $U$, *then* $f \in H(U)$ *and* $||f_n(x) - f(x)|| \to 0$ *as* $n \to \infty$;

*(vi)* $\sigma_A(f(x)) = f(\sigma_A(x))$ *for all* $f \in H(U)$.

*Proof.* By a theorem of complex analysis (see Rudin [3, p. 287]) there is a cycle $\Gamma$ (a formal sum $\gamma_1 + \gamma_2 + \cdots + \gamma_m$ of closed, rectifiable paths) in $U \setminus \sigma_A(x)$ such that

$$\frac{1}{2\pi i}\int_\Gamma (\zeta - \lambda)^{-1} d\zeta = \sum_{j=1}^{m} \int_{\gamma_j} (\zeta - \lambda)^{-1} d\zeta = \begin{cases} 1 & \text{if } \lambda \in \sigma_A(x) \\ 0 & \text{if } \lambda \notin U \end{cases}$$

In this situation we say that $\Gamma$ *surrounds* $\sigma_A(x)$ in $U$, and by Cauchy's Theorem it follows that for every $f \in H(U)$,

$$f(\lambda) = \frac{1}{2\pi i}\int_\Gamma (\zeta - \lambda)^{-1} f(\zeta)\, d\zeta, \quad \lambda \in \sigma_A(x).$$

For $f \in H(U)$ we now *define* $f(x)$ by setting

$$f(x) = \frac{1}{2\pi i}\int_\Gamma (\zeta e - x)^{-1} f(\zeta)d\zeta \tag{1}$$

It must be shown that $f(x)$ is a well-defined element of A. If $\zeta \in \Gamma$, then $\zeta \notin \sigma_A(x)$ and hence $(\zeta e - x)^{-1}$ exists in A. Moreover, since $\Gamma$ is a compact subset of the plane, $(\zeta e - x)^{-1}$ is a uniformly continuous function of $\zeta$ on $\Gamma$ with values in A.

For each partition $\Delta = \{\Delta_k\}_{k=1}^m$, $\Delta_k = \{\zeta_{k1}, \zeta_{k2}, \ldots \zeta_{kn_k}\}$ of $\Gamma$ define $y_\Delta$ by

$$y_\Delta = \sum_{k=1}^m \frac{1}{2\pi i} \sum_{j=1}^{n_k} (\zeta_{kj} e - x)^{-1} f(\zeta_{kj})(\zeta_{kj} - \zeta_{kj-1}). \tag{2}$$

Clearly $y_\Delta$ belongs to A. A routine argument shows that $\{y_\Delta\}$ is a Cauchy net in A, and hence has a limit y in A. To see that the integral in (1) defining $f(x)$ exists, and is equal to y, simply repeat the usual proof for the Riemann-Stieltjes integral of a continuous function, replacing the absolute value by the norm $||\cdot||$ in A. No other modifications are needed. To complete the definition it must be shown that $f(x)$ does not depend on the particular cycle $\Gamma$ provided that the cycle surrounds $\sigma_A(x)$ in U in the sense stated above. Let $L \in A^*$; since L is linear and continuous

$$L(f(x)) = \frac{1}{2\pi i}\int_\Gamma L((\zeta e - x)^{-1}) f(\zeta)d\zeta.$$

As in the proof of (B.4.4), $L((\zeta e - x)^{-1})$ is a holomorphic function of $\zeta$ if $\zeta \notin \sigma_A(x)$. Therefore, the number $L(f(x))$ is independent of $\Gamma$ by the Cauchy integral theorem. Since L was an arbitrary element of $A^*$, the Hahn-Banach theorem (A.1) implies that $f(x)$ is independent of $\Gamma$. We are now prepared to establish properties (i) - (vi).

(i) and (ii). We begin by showing that if f is a constant function or the identity function, the integral definition of $f(x)$ yields the expected element of A. Since these functions are entire, we may take $\Gamma$ to be any positively oriented circle containing $\sigma_A(x)$ in its interior. Specifically, choose $r > ||x||$ and let $\Gamma$ be centered at 0 and of radius r. If $|\zeta| = r$, we have

$$(\zeta e - x)^{-1} = \zeta^{-1}e + \zeta^{-2}x + \zeta^{-3}x^2 + \cdots .$$

Therefore, if $f(\lambda) = \mu$,

$$f(x) = \frac{1}{2\pi i}\int_\Gamma (\zeta^{-1}e + \zeta^{-2}x + \zeta^{-3}x^2 + \cdots)\mu d\zeta = \frac{1}{2\pi i}\int_\Gamma \zeta^{-1}e\mu d\zeta = \mu e;$$

and if $f(\lambda) = \lambda$,

$$f(x) = \frac{1}{2\pi i}\int_\Gamma (\zeta^{-1}e + \zeta^{-2}x + \zeta^{-3}x^2 + \cdots)\zeta d\zeta = \frac{1}{2\pi i}\int_\Gamma \zeta^{-1}x d\zeta = x.$$

(iii) Let $f, g \in H(U)$ and $\Gamma$ a cycle surrounding $\sigma_A(x)$ in $U$. Since, for any scalar $\beta$,

$$(\beta f + g)(x) = \frac{1}{2\pi i}\int_\Gamma (\zeta e - x)^{-1}(\beta f + g)(\zeta)d\zeta = \beta f(x) + g(x),$$

the mapping $f \to f(x)$ is linear. For a proof that $(fg)(x) = f(x)g(x)$ we refer the reader to Rudin [2, p. 244].

(iv) We first observe by (i) and (iii) that for every positive integer $n$,

$$x^n = \frac{1}{2\pi i}\int_\Gamma (\zeta e - x)^{-1}\zeta^n d\zeta \tag{3}$$

We may assume that $f$ is defined on the disk of convergence of the series. Let $\Gamma$ be a circle with center at 0 containing $\sigma_A(x)$ in its interior and contained in the open set $U$ on which $f$ is holomorphic and represented by $\Sigma_{n=0}^\infty c_n\lambda^n$. Since this series converges uniformly on $\Gamma$, we have, by (3), that

$$f(x) = \frac{1}{2\pi i}\int_\Gamma (\zeta e - x)^{-1}f(\zeta)d\zeta = \sum_{n=0}^\infty c_n\left(\frac{1}{2\pi i}\int_\Gamma (\zeta e - x)^{-1}\zeta^n d\zeta\right) = \sum_{n=0}^\infty c_n x^n.$$

(v) Select an open set $V$ with boundary $\Gamma$, a cycle surrounding $\sigma_A(x)$ in $U$, such that

$$\sigma_A(x) \subseteq V \subseteq V \cup \Gamma \subseteq U.$$

Since $\{f_n\}$ converges uniformly to $f$ on compact subsets of $U$, $f$ is holomorphic on $U$. Hence, $f \in H(U)$ and $\{f_n\}$ converges uniformly on $\Gamma$.

Therefore we have

$$2\pi||f_n(x) - f(x)|| = ||\int_\Gamma [f_n(\zeta) - f(\zeta)](\zeta e - x)^{-1}d\zeta||$$

$$\leq \beta|\Gamma|\cdot||f_n - f||_\Gamma \to 0,$$

as $n \to \infty$, where $\beta = \sup\{||(\zeta e - x)^{-1}||: \zeta \in \Gamma\}$, $|\Gamma|$ denotes the length of $\Gamma$ and $||f_n - f||_\Gamma = \sup\{|f_n(\zeta) - f(\zeta)|: \zeta \in \Gamma\}$.

(vi) Let $\lambda \in \sigma_A(x)$. Then $A(x - \lambda e)$ is a proper left ideal in A (a similar discussion can be given for $(x - \lambda e)A$ and right ideals). Since, for a suitable cycle $\Gamma$,

$$f(x) - f(\lambda)e = \frac{1}{2\pi i}\int_\Gamma [(\zeta e - x)^{-1} - (\zeta - \lambda)^{-1}e]f(\zeta)d\zeta$$

$$= \left(\frac{1}{2\pi i}\int_\Gamma [(\zeta e - x)(\zeta - \lambda)]^{-1}f(\zeta)d\zeta\right)(x - \lambda e)$$

$$\in A(x - \lambda e),$$

it follows that $f(\lambda) \in \sigma_A(f(x))$. Hence $f(\sigma_A(x)) \subseteq \sigma_A(f(x))$.

To establish the opposite inclusion, suppose $\mu \notin f(\sigma_A(x))$ and set $h(\zeta) = (f(\zeta) - \mu)^{-1}$. Then h is holomorphic on some open neighborhood of $\sigma_A(x)$. Since $h\cdot(f - \mu)$ is 1 on an open neighborhood of $\sigma_A(x)$, it follows from (iii) that h(x) is a two-sided inverse to $f(x) - \mu e$ in A. Therefore $\mu \notin \sigma_A(f(x))$, so that $\sigma_A(f(x)) \subseteq f(\sigma_A(x))$. □

REMARK. If the Banach algebra A is *commutative* in (B.8.1), then the "spectral mapping property" given in part (vi) translates via the Gelfand representation theory, into $f(x)\hat{} = f\circ\hat{x}$ for all $f \in H(U)$.

We next observe that the holomorphic functional calculus is unique.

(B.8.2) PROPOSITION. *(Uniqueness). Let* A *be a Banach algebra with identity* e, $x \in A$, *and* U *an open subset containing* $\sigma_A(x)$. *Suppose that* $\Phi$ *is a mapping of* H(U) *into* A *with the properties:*

*1°)* $\Phi(f) = e$ *if* $f(\lambda) = 1$;

*2°)* $\Phi(f) = x$ *if* $f(\lambda) = \lambda$;

*3°)* $\Phi$ *is a homomorphism;*

*4°)* $\Phi$ *is continuous with respect to uniform convergence on compact subsets of* U.

*Then* $\Phi(f) = f(x)$ *for all* $f \in H(U)$.

*Proof.* From 1°), 2°) and 3°), $\Phi(p) = p(x)$ for all polynomials $p$. It then follows from 1°) and 3°) that $\Phi(1/q) = q(x)^{-1}$ if $q$ is a polynomial with no zeros in $U$. Hence $\Phi(r) = r(x)$ for all rational functions $r \in H(U)$. By 4°) and the Runge Theorem (see Rudin [3], p. 288) from complex analysis which states that each $f \in H(U)$ is the limit of a sequence of rational functions with poles outside of $U$, the convergence being uniform on each compact subset of $U$, we obtain $\Phi(f) = f(x)$ for every $f \in H(U)$. □

(B.8.3) THEOREM. *(Composition). Let* $A$ *be a Banach algebra with identity* $e$*,* $y \in A$*, and* $V$ *an open set containing* $\sigma_A(y)$*. Let* $g \in H(V)$ *and choose an open set* $U$ *such that* $g(V) \subseteq U$*. If* $f \in H(U)$*, then* $(f \circ g)(y) = f(g(y))$.

*Proof.* For $g$, $U$ and $V$ as described, define a mapping $\Phi\colon H(U) \to A$ by $\Phi(f) = (f \circ g)(y)$. From (B.8.1), (vi), we have $\sigma_A(g(y)) = g(\sigma_A(y)) \subseteq U$. Also, because of properties of the functional calculus, $\Phi$ satisfies 1°) - 4°) of (B.8.2), with $g(y)$ in place of $x$. It follows from (B.8.2) that $\Phi(f) = f(g(y))$ for all $f \in H(U)$. □

We complete this section with the following result which extends the holomorphic functional calculus, for commutative Banach algebras, to the case where the algebra may not possess an identity. The reader should recall the notation and results of (B.6.8).

(B.8.4) THEOREM. *Let* $x$ *be an element of a commutative Banach algebra* $A$*, and* $f$ *a complex-valued function which is holomorphic in an open set containing the closure of the range of* $\hat{x}$*. If either*

*(a)* $f(0) = 0$*, or*

*(b)* $\hat{A}$ *is compact and* $\hat{x}(\phi) \neq 0$ *for all* $\phi \in \hat{A}$*,*

*then there is an element* $f(x)$ *in* $A$ *such that* $f(x)^\wedge = f \circ \hat{x}$.

*Proof.* If $\hat{A}$ is compact and $\hat{x}(\phi) \neq 0$ for all $\phi \in \hat{A}$, then $\hat{x}(A)$ is a compact subset of the plane which does not contain the origin. Select disjoint open sets $U$ and $V$ such that $\hat{x}(\hat{A}) \subseteq U$ and $0 \in V$. Now, define a complex-valued function $g$ on $U \cup V$ by

$$g(\lambda) = \begin{cases} f(\lambda), & \text{if } \lambda \in U, \\ 0, & \text{if } \lambda \in V. \end{cases}$$

Then $g$ is holomorphic, its domain contains the closure of the range of $\hat{x}$, and $g(0) = 0$. Hence, we may restrict our attention to the hypothesis of part (a).

Thus, assume that $f(0) = 0$. Since $\sigma_{A_e}(x) = \hat{x}(\hat{A}) \cup \{0\}$, it follows from (B.8.1) and the remark after its proof that there exists an element $f(x)$ in $A_e$ such that $f(x)\hat{\ }(\phi) = f(\hat{x}(\phi))$ for all $\phi \in \hat{A}_e$. Since $x \in A$, we have $f(x)\hat{\ }(\phi_\infty) = f(\hat{x}(\phi_\infty)) = f(0) = 0$, and so $f(x) \in A$. □

## BANACH ALGEBRA EXERCISES

### *GENERALITIES*

(B.1) Verify the inequality $||xy|| \leq ||x||\cdot||y||$ for Examples (1) - (5). Prove that the spaces in Examples 1, 3, 4, and 5 are complete.

(B.2) Prove that if $A$ is a normed algebra and $\{x_n\}$, $\{y_n\}$ are Cauchy sequences in $A$, then $\{x_n y_n\}$ is also Cauchy.

(B.3) Let $A$ be a Banach algebra and suppose that $\{x_n\}$ is a sequence in $A$. Prove that if the series $\Sigma_{n=1}^{\infty}||x_n||$ converges, then the series $\Sigma_{n=1}^{\infty}x_n^2$ converges in $A$.

(B.4) Let $A$ be a normed algebra, and $n$ a fixed positive integer. Prove that the mapping $x \to x^n$ of $A$ to itself is continuous.

(B.5) Give an example of a Banach algebra $A$ with identity and an element $x \in A$ such that $||x^n|| < ||x||^n$ for some positive integer $n$.

(B.6) Let $A$ be a normed algebra and $S$ a nonempty subset of $A$. Prove that the closed subalgebra generated by $S$ is equal to the closure of the subalgebra generated by $S$.

(B.7) Find the Banach algebra completions of the normed algebras described in Example 4.

(B.8) Let $A$ be a normed algebra. Prove that:

(a) the closure of a subalgebra of $A$ is a subalgebra of $A$.

(b) the closure of a commutative subalgebra of $A$ is commutative.

(c) a maximal commutative subalgebra of $A$ is closed.

(d) the center $Z(A) = \{x \in A: xy = yx$ for all $y \in A\}$ of $A$ is a closed subalgebra of $A$.

(B.9) Let $A$ and $B$ be normed algebras, and $f: A \to B$ a continuous homomorphism. Prove that $f$ can be extended uniquely to a continuous homomorphism from the completion of $A$ into the completion of $B$.

(B.10) Let $A$ be a normed algebra without identity, and consider the algebra $A_e$ obtained from $A$ by adjoining an identity $e$.

(a) Define two additional norms on $A_e$ (each distinct from the one given in Section (B.1)) which make $A_e$ into a normed algebra with $||e|| = 1$.

(b) Find a norm on $A_e$ which does not satisfy $||xy|| \leq ||x||\cdot||y||$.

(c) Find a norm on $A_e$ under which $A_e$ is a normed algebra but $||e|| \neq 1$.

(d) Prove that $A_e$, under the norm in Section (B.1), is complete if and only if $A$ is complete. Also, show that $I = \{(x,0): x \in A\}$ is a closed two-sided maximal ideal in $A_e$.

(B.11) Let $A$ be the algebra of all complex matrices of the form $\begin{pmatrix} \alpha & \beta \\ 0 & 0 \end{pmatrix}$ with the usual linear operations and matrix product. Find the algebra $A_e$ obtained from $A$ by adjoining an identity $e$.

(B.12) Let $A$ be a normed algebra without identity. If $\tilde{A}$ denotes the completion of $A$, prove that $(\tilde{A})_e = (A_e)^{\sim}$.

(B.13) Let $A$ be a normed algebra and $I$ a closed ideal in $A$ such that both $I$ and $A/I$ are complete. Prove that $A$ is a Banach algebra.

(B.14) Prove parts (b), (c), and (d) of Proposition (B.1.1).

(B.15) Let $A$ be a normed algebra satisfying $||x^2|| = ||x||^2$ for all $x \in A$. Prove by induction that $||x^{2^k}|| = ||x||^{2^k}$ for all $k = 1,2,\ldots$ and $x \in A$.

(B.16) Give an example of a Banach algebra $A$ which shows that the function $f$ defined on the positive integers by $f(n) = ||x^n||^{1/n}$, for a fixed element $x$ in $A$, can be monotone increasing.

(B.17) Find an element $x$ in a Banach algebra $A$ such that $x^n \neq 0$ for all positive integers $n$, but $\nu(x) = 0$.

(B.18) Let $D_n([0,1])$ denote the set of all n-times continuously differentiable complex-valued functions on $[0,1]$. Furnish $D_n([0,1])$ with pointwise operations and norm

$$||f||_n = \sum_{k=0}^{n} \frac{1}{k!} \sup\{|f^{(k)}(t)|: t \in [0,1]\}.$$

Prove that $D_n([0,1])$ is a commutative Banach algebra with identity. Show by an example that one cannot remove the term $1/k!$ from the definition of $||f||_n$.

(B.19) Prove that every finite-dimensional algebra becomes a Banach algebra under a suitable norm.

(B.20) Let $A$ be a Banach algebra such that every homomorphism of a Banach algebra into $A$ is continuous. Prove that $A$ contains no nonzero element $y$ with $y^2 = 0$.

(B.21) Show that if $X$ is a locally compact, non-compact, Hausdorff space, then the Banach algebra $C_o(X)$ (see Example 1) does not possess an identity.

(B.22) Let $X$ be a compact metric space. Prove that the Banach algebra $C(X)$ is separable, connected, but, in general, is not locally compact. Can you find necessary and sufficient conditions on $X$ for $C(X)$ to be locally compact?

(B.23) Let $X$ be a closed subset of the complex plane. A continuous complex-valued function $f$ on $X$ is said to have *compact support* if there exists a compact subset $K$ of $X$ such that $f(x) = 0$ whenever $x \notin K$. Prove that the set $C_{oo}(X)$ of continuous complex-valued functions on $X$ with compact support is a normed algebra under pointwise operations and sup-norm. Prove that $C_{oo}(X)$ is a Banach algebra if and only if $X$ is compact.

(B.24) Let $E$ be a normed space. Prove that the normed algebra $B(E)$ of bounded linear operators on $E$ is a Banach algebra if and only if $E$ is a Banach space.

(B.25) Let $E$ be a Banach space and consider the Banach algebra $B(E)$. If $\{T_n\}$ is a sequence of operators in $B(E)$ such that $\lim_{n\to\infty} |f(T_n x)|^{1/n} = 0$ for all $x \in E$ and $f \in E^*$, then $\lim_{n\to\infty} ||T_n||^{1/n} = 0$.

(B.26) Let $A$ be a normed algebra with identity $e$. Let $x \in A$ and set $\alpha_n = ||x^{2^n} - e||$. Prove that if $\limsup \alpha_n < 1$, then for some $n$, $x^{2^n} = e$. If in addition every $\alpha_k < 2$, show that $x = e$.

*INVERTIBLE AND QUASI-REGULAR ELEMENTS*

(B.27) Let A be an algebra with identity e. If x is a nilpotent element of A (i.e., $x^n = 0$ for some positive integer n) show that the element e + x is invertible in A.

(B.28) Let A be an algebra with identity e.

(a) If x, y $\in$ A and at least two elements of the set {x,y,xy,yx} are invertible in A, prove that all of them are invertible.

(b) If A is finite dimensional and xy is invertible, prove that both x and y are invertible.

(c) Show that (b) is false if the assumption of finite dimensionality is dropped.

(B.29) Let A be a normed algebra with identity. Show that $\nu(x^{-1}yx) = \nu(y)$ for every y $\in$ A and every invertible x $\in$ A.

(B.30) Give an example of a normed algebra A with identity e and a singular element x $\in$ A satisfying $||x - e|| < 1$.

(B.31) Give an example of a Banach algebra A and a sequence $\{x_n\}$ of invertible elements of A which converges to a singular element x $\in$ A.

(B.32) Does there exist a Banach algebra whose group of invertible elements is closed? What if the word "closed" is replaced by "dense"? What about noncommutative algebras?

(B.33) Let A be a Banach algebra and r > 0. Let $\{a_n\}$ be a sequence in A such that the sequence $\{r^n||a_n||\}$ is bounded. Prove that the power series $\Sigma_{n=1}^{\infty} a_n x^n$ converges for all x $\in$ A satisfying $||x|| < r$. If $0 < s < r$, prove that $\Sigma_{n=1}^{\infty} a_n x^n$ converges uniformly for all x $\in$ A such that $||x|| < s$.

(B.34) Let A be a Banach algebra with identity e. Prove that if $\exp(x + y) = \exp(x)\exp(y)$ for all x, y $\in$ A, then A is commutative.

(B.35) An element x in an algebra A is said to be idempotent if $x^2 = x$ and nilpotent if $x^n = 0$ for some positive integer n. Prove that:

(a) A nonzero idempotent element cannot be nilpotent.

(b) Every nonzero nilpotent element is a divisor of zero in A.

(c) 0 is the only idempotent element in A which is quasi-regular.

(d) Every nilpotent element in A is quasi-regular.

(B.36) Let A be an algebra. Prove that:

(a) If each element of A is idempotent, then A is commutative.

(b) If A has no nonzero nilpotent elements, then all idempotents are in the center of A.

(c) If $x^2 + x$ belongs to the center of A for all $x \in A$, then A is commutative.

(B.37) Let A be an algebra and let x, y be quasi-regular elements of A with quasi-inverses $x'$, $y'$. Prove that $x + y$ is quasi-regular if and only if $x'y'$ is quasi-regular.

(B.38) Let A be an algebra. Prove each of the following:

(a) $x \circ y - y \circ x = yx - xy$.

(b) $(x + y) \circ (z + w) = x \circ z + x \circ w + y \circ w - (x + y + z + w)$

(c) $x \circ (\Sigma_{k=1}^{n} y_k) = (\Sigma_{k=1}^{n} x \circ y_k) - (n - 1)x$.

(d) $x \circ y + (-x) y = 2y$.

(e) If x has y as a quasi-inverse, then $xy = yx$.

(f) If x has y as a right quasi-inverse and x commutes with y, then y is a quasi-inverse of x.

(B.39) Prove that if x is an element in an algebra A such that $x^n$ is quasi-regular for some positive integer n, then x is quasi-regular. Show that the converse is false.

(B.40) Prove that if x and y are elements of an algebra A such that $x = xy$ and y is quasi-regular, then $x = 0$.

(B.41) Let I be a two-sided ideal in an algebra A. Prove that if a coset $x + I$ in $A/I$ contains an element y which is invertible in A (resp. quasi-regular in A), then $x + I$ is invertible in $A/I$ (resp. quasi-regular in $A/I$).

(B.42) Prove that the set $Q_A$ of quasi-regular elements in an algebra forms a group with respect to the circle operation, quasi-inversion, and identity 0.

(B.43) Prove that if every element of a commutative algebra A is quasi-regular, with exactly one exception, then A is a field.

(B.44) Give an example of a quasi-regular element x in a Banach algebra A such that $||x|| \geq 1$.

*THE SPECTRUM*

(B.45) Prove Lemma (B.4.2). Give an example of a Banach algebra A such that $\sigma_A(x)$ consists of exactly one point for each $x \in A$.

(B.46) Let A be an algebra with identity e. If $x \in A$ is an idempotent element distinct from 0 and e, show that $\sigma_A(x) = \{0,1\}$. Examine $\sigma_A(x)$ if x is an element in A satisfying $x^2 = e$.

(B.47) Give an example of a Banach algebra A with identity e, and a subalgebra B of A containing e such that the inclusion $\sigma_A(x) \subseteq \sigma_B(x)$ is proper for some $x \in B$.

(B.48) Let $A = M_n(C)$ denote the (noncommutative) Banach algebra of all $n \times n$ complex matrices $x = (x_{ij})$ with norm $||x|| = \max\{\Sigma_{j=1}^n |x_{ij}|: 1 \leq i \leq n\}$.

(a) Let $x \in A$. Prove that $\sigma_A(x)$ consists of at most n elements.

(b) Prove that $M_n(C)$ is not a division algebra if $n > 1$.

(c) Consider the element

$$x = \begin{pmatrix} a & b \\ c & d \end{pmatrix} \in M_2(C).$$

Prove that $\sigma_A(x)$ is a one-point set if and only if $(a + d)^2 = 4\det(x)$.

(B.49) Let $A = M_n(R)$ denote the Banach algebra of real $n \times n$ matrices, where the norm is defined as in Exercise (B.48). Consider the element $x = \begin{pmatrix} 0 & -1 \\ 1 & 0 \end{pmatrix} \in M_2(R)$. Prove that $\sigma_A(x) = \{\lambda \in R: x - \lambda e \text{ is singular}\} = \emptyset$.

(B.50) Exercise (B.49) shows that Theorem (B.4.4) fails for real Banach algebras. Find another element x in $M_2(R)$ such that $\sigma_A(x) = \emptyset$.

(B.51) Give an example of an algebra A with identity over the complex numbers and an element $x \in A$ such that $\sigma_A(x) = \emptyset$. Hence, Theorem (B.4.4) may fail for non-normed algebras.

(B.52) Give an example of a Banach algebra A with identity and elements $x, y \in A$ such that $\sigma_A(xy) \neq \sigma_A(yx)$.

(B.53) Let A be a Banach algebra and let T be an isomorphism of A onto itself (considered only as an algebra). Prove that $\nu(x) = \nu(Tx)$ for all $x \in A$.

(B.54) Show that every closed bounded subset of the complex plane is the spectrum of some element in a suitable Banach algebra.

(B.55) Let A be a Banach algebra with identity e. Give a simple direct proof that any element $x \in A$ such that $\nu(x) = 0$ is singular.

(B.56) If A is an algebra and I is a two-sided ideal, prove, for each $x \in A$, that $\sigma_A(x + I) \subseteq \bigcap_{a \in x+I} \sigma_A(a)$.

(B.57) Let A be a Banach algebra with identity e, and let x be an invertible element of A such that $||x|| = ||x^{-1}|| = 1$. Prove that $\sigma_A(x)$ is contained in the unit circle $\{\lambda \in C: |\lambda| = 1\}$.

(B.58) Consider the Banach algebra $A = C^2$ with norm $||(\lambda,\mu)|| = |\lambda| + |\mu|$ and pointwise operations. Interpret the formula $||x||_\sigma = \nu(x)$ in this special case and give a direct proof.

(B.59) Let A be a Banach algebra with identity e. If x is a boundary point of the set of singular elements of A, is 0 a boundary point of the spectrum of x?

(B.60) Let A be a Banach algebra with identity e. Let T be any component of the group G of invertible elements in A. Prove that the union of the spectra of all elements of T is equal to the set of all nonzero complex numbers.

(B.61) Let A be an algebra with identity e. Prove the "resolvent equation" $x(\lambda) - x(\mu) = (\lambda - \mu)x(\lambda)x(\mu)$ for all $\lambda, \mu \notin \sigma_A(x)$, where $x \in A$ and $x(\lambda) = (x - \lambda e)^{-1}$.

(B.62) Let A be an algebra over a field F such that A has no nonzero nilpotent elements of order 2. If A is isomorphic to F as a vector space, prove that A is isomorphic to F as an algebra. In particular, A has an identity element.

(B.63) Let A and B be Banach algebras with identities e and e' respectively. If $T: A \to B$ is a linear map such that $T(xy + yx) = T(x)T(y) + T(y)T(x)$ for all $x, y \in A$ and $Te = e'$, prove that:

(a) $\sigma_B(Tx) \subseteq \sigma_A(x)$ for all $x \in A$.

(b) $\nu(Tx) \leq \nu(x)$ for all $x \in A$.

(B.64) An element z of a normed algebra A is said to be a *topological divisor of zero* if there is a sequence $\{z_n\}$ of norm 1 in A such that $zz_n \to 0$ or $z_nz \to 0$. (a) Prove that every topological divisor of zero is singular. (b) Prove that every boundary point of the set of singular elements is a topological divisor of zero.

(B.65) Denote by $\partial(E)$ the topological boundary of a subset $E$ of the complex plane. Let $B$ be a closed subalgebra of a Banach algebra $A$. (a) Prove that for every $x \in B$, $\partial(\sigma_B(x)) \subset \partial(\sigma_A(x))$. (b) Prove that if $\sigma_B(x)$ contains no interior points, $\sigma_A(x) = \sigma_B(x)$.

## *IDEALS AND THE RADICAL*

(B.66) Let $A$ be an algebra, and $\{I_\alpha\}_{\alpha\in\Gamma}$ a family of left ideals of $A$. Prove that the sum $\sum_{\alpha\in\Gamma} I_\alpha$ is the left ideal of $A$ generated by the union $\bigcup_\alpha I_\alpha$.

(B.67) Let $I$, $J$, and $K$ be ideals in an algebra $A$. Prove that:

(a) $I(J + K) = IJ + IK$ and $(I + J)K = IK + JK$.

(b) If $I \supseteq J$, then $I \cap (J + K) = J + (I \cap K)$.

(B.68) Let $A$ and $B$ be algebras and $f: A \to B$ a homomorphism of $A$ onto $B$. Prove for all ideals $I$, $J$ in $A$ that:

(a) $f(I + J) = f(I) + f(J)$.

(b) $f(IJ) = f(I)f(J)$.

(c) $f(I \cap J) \subseteq f(I) \cap f(J)$.

(d) If $\ker(f) \subseteq I \cup J$, then $f(I \cap J) = f(I) \cap f(J)$.

(e) $f((I:A)) \subseteq (f(I):B)$, and equality holds if $\ker(f) \subseteq I$.

(f) $f(\mathrm{Rad}(A)) \subseteq \mathrm{Rad}(B)$.

(g) If $\ker(f) \subseteq \mathrm{Rad}(A)$, then $\mathrm{Rad}(A) = f^{-1}(\mathrm{Rad}(B))$.

(B.69) Let $A$ be an algebra and $I$ a modular left (resp. right) ideal of $A$. If $J$ is a two-sided modular ideal of $A$, prove that $I \cap J$ is a modular left (resp. right) ideal of $A$.

(B.70) If $I$ is a modular right ideal of an algebra $A$ and $e$ is a left identity relative to $I$, prove that $e^n$ is a left identity relative to $I$ for all positive integers $n$. Similarly for modular left ideals and right identities.

(B.71) Let $A$ be an algebra. Prove that:

(a) If $f$ is a homomorphism of $A$ into an algebra $B$ and $K = \ker(f)$, then $A/K$ is isomorphic to $f(A)$, i.e., $A/K \simeq f(A)$.

(b) If $f$ is a homomorphism of $A$ onto an algebra $B$ and $I$ is any ideal in $A$ such that $\ker(f) \subseteq I$, then $A/I \simeq f(A)/f(I)$.

(c) If $I$ and $J$ are ideals in $A$ with $J \subseteq I$, then $A/I \simeq (A/J)/(I/J)$.

(d) If $I$ and $J$ are ideals in $A$, then $(I + J)/J \simeq I/(I \cap J)$.

(e) If $I$, $J$ are ideals in $A$ and $\tau: A \to A/J$ is the canonical homomorphism, then $\tau^{-1}(\tau(I)) = I + J$.

(B.72) Assume that $A$ is an algebra with identity $e$ and that $A$ has exactly one maximal ideal. Prove that the only idempotent elements in $A$ are 0 and $e$.

(B.73) Let $e$ be an idempotent in an algebra $A$. Prove that $\mathrm{Rad}(eAe) = e\mathrm{Rad}(A)e$.

(B.74) Let $A$ be an algebra.

(a) Prove that 0 is the only idempotent element in $\mathrm{Rad}(A)$.

(b) Prove that the center of $A$ is equal to the intersection of all maximal commutative subsets of $A$.

(B.75) Let $I$ be an ideal in an algebra $A$, and $f$ a homomorphism of $I$ onto an algebra $B$ with identity $e$. If $I$ is contained in the center of $A$, prove that there exists a unique homomorphism $g: A \to B$ such that $g|I = f$.

(B.76) Prove part (b) of Proposition (B.5.12) for right ideals.

(B.77) Let $A$ be an algebra with identity 1. If $e$ is an idempotent of $A$ and $I$ is a two-sided ideal of $A$, prove that $eIe = I \cap (eAe)$.

(B.78) Let $A$ be an algebra and $A_e$ the algebra obtained from $A$ by adjoining an identity $e$. Let $I$ be an arbitrary maximal right ideal of $A_e$ distinct from $A$. Prove that the set $A \cap I$ is a maximal modular right ideal of $A$, and the mapping $I \to A \cap I$ is a bijective correspondence between the family of all maximal right ideals of $A_e$ distinct from $A$ and the family of all maximal modular right ideals of $A$.

(B.79) Let $A$ be an algebra with the property that to every ordered pair of elements $x$, $y$ there are elements $u$ and $v$ such that $xy = ux = yv$. Prove that every ideal in $A$ is two-sided.

(B.80) Let $A$ be an algebra and $I$ a two-sided ideal of $A$. Prove that the following statements are equivalent:

(a) $I$ is a maximal modular right ideal.

(b) $I$ is a maximal modular left ideal.

(c) $A/I$ is a division algebra.

(B.81) Let $A$ be an algebra. The *strong radical* of $A$, denoted $R(A)$, is the intersection of all maximal modular two-sided ideals of $A$. The *strict radical* of $A$, denoted $S(A)$, is the intersection of all those two-sided ideals of $A$ which are maximal modular right ideals of $A$. The algebra $A$ is called *strongly semisimple* (resp. *strictly semisimple*) if $R(A)$ (resp. $S(A)$) is the zero ideal.

(a) Prove that $\mathrm{Rad}(A) \subseteq R(A) \subseteq S(A)$.

(b) Show that if each maximal modular right ideal is two-sided, then all three radicals coincide.

(c) Give an example of an algebra which is strongly semisimple but not strictly semisimple.

(d) Prove that any subalgebra of a strictly semisimple Banach algebra is strictly semisimple.

(e) If $A$ is a Banach algebra and $N = \{x \in A: \nu(x) = 0\}$, prove that $N \subseteq S(A)$.

(f) Give an example of a noncommutative algebra in which every maximal modular right ideal is a two-sided ideal.

(B.82) Let $A$ be a commutative algebra. An ideal $I$ of $A$ is said to be *prime* if $xy \in I$ implies that either $x \in I$ or $y \in I$. Prove that:

(a) If $A^2 \neq \{0\}$, and $M$ is a maximal ideal of $A$, then $M$ is not a prime ideal if and only if $A^2 \subseteq M$.

(b) If $A$ has an identity, then every maximal ideal is prime.

(c) If $A$ contains a nonprime maximal ideal, then $A^2$ is contained in the intersection of all such ideals.

(d) If $A$ is a commutative Banach algebra such that $A^2 \neq \{0\}$, then $A$ contains a nonprime maximal ideal if and only if $A^2 \neq A$. Further, each nonprime maximal ideal is a maximal linear subspace of $A$ which contains $A^2$.

(e) If $A$ is a commutative algebra without identity and $M$ is a maximal ideal in $A$, then $M$ is a modular ideal if and only if it is a prime ideal.

(B.83) Let $A$ be an algebra with identity $e$, in which every maximal left (or right) ideal is principal, being generated by an idempotent. Prove that the Jacobson radical $\mathrm{Rad}(A)$ is zero.

(B.84) Let $A$ be a Banach algebra and let $I$ be a modular two-sided ideal of $A$. If $e$ is a relative identity for $I$, prove that $||e - x|| \geq 1$ for all $x \in I$.

(B.85) Prove that the radical of a Banach algebra $A$ consists of all $x$ in $A$ such that $\nu(xy) = 0$ for all $y$ in $A$.

(B.86) Prove, without using Zorn's lemma, that in a separable commutative Banach algebra with identity $e$, every proper ideal $I$ of $A$ is contained in a maximal ideal.

(B.87) Let $A$ be a commutative normed algebra, and $B$ a proper closed subalgebra of $A$. Prove that there is a unique closed ideal $I$ of $A$ such that:

(a) $I \subseteq B$; and

(b) $I$ is not contained in any larger ideal which is in $B$.

(B.88) A Banach algebra $A$ with Jacobson radical $R$ is said to be *decomposable* if there exists a subalgebra $B$ of $A$ with the property that $A = B + R$ and $B \cap R = \{0\}$ (i.e., $A = B \oplus R$). If there exists a closed subalgebra $B$ of $A$ such that $A = B \oplus R$, then $A$ is called *strongly decomposable*. Prove that $A$ is decomposable (resp. strongly decomposable) if and only if $A_e$ is.

(B.89) Let $A$ be a Banach algebra under two norms $||\cdot||_1$, $||\cdot||_2$ and let $s$ be the separating function for these norms. Prove that if $x$ is an element in the center of $A$, then $\nu(x) \leq s(x)$.

(B.90) Let $A$ be a normed algebra under each of the norms $||\cdot||_1$, $||\cdot||_2$ and let $e$ be an idempotent in $A$. If $S_e$ is the separating ideal for the two norms restricted to the subalgebra $eAe$, prove that $S_e = eSe$.

(B.91) Prove that the center of a primitive Banach algebra is either $\{0\}$ or is isometrically isomorphic to the complex numbers.

(B.92) Let $A$, $B$, and $C$ be algebras. Let $f$ and $g$ be homomorphisms of $A$ onto $B$, and $A$ into $C$, respectively, such that $\ker(f) \subseteq \ker(g)$. Prove that there is a unique homomorphism $h$ of $B$ into $C$ such that $g = h \circ f$.

(B.93) Let $A, B_1,\ldots,B_n$, and $C$ be algebras, with the $B_i$ simple. Let $f_i$ be a homomorphism of $A$ onto $B_i$ and let $g$ be a homomorphism of $A$ into $C$; let $K_i = \ker(f_i)$, and $K = \ker(g)$. Prove that the condition $\cap K_i \subseteq K$ is necessary and sufficient for the existence of homomorphisms $h_i$ of $B_i$ into $C$ such that $g = \sum_{i=1}^{n} h_i \circ f_i$.

(B.94) Let $A$ be an algebra. For $a \in A$ let $I_a = \{y \in A: y(ax - xa) = 0$ for all $x \in A\}$. Prove that $I_a$ is a two-sided ideal of $A$.

(B.95) Prove that any isomorphism of a semi-simple Banach algebra onto itself is continuous.

(B.96) Prove Lemma (B.5.32).

(B.97) Let $A$ be a Banach algebra and $I$ a closed two-sided ideal of $A$. Prove that:

(a) $\sigma_{A/I}(x + I) \subseteq \sigma_A(x)$ for all $x \in A$.

(b) $\nu_{A/I}(x + I) \leq \nu_A(x)$ for all $x \in A$.

(B.98) A *Boolean ring* is a ring $A$ such that $a^2 = a$ for all $a$ in $A$. Prove that:

(a) $a = -a$, for all $a \in A$;

(b) $A$ is commutative;

(c) if $f$ is an injective map of the Boolean ring $A$ into a ring $B$ such that $f(ab) = f(a)f(b)$, then $f(a + b) = f(a) + f(b)$.

## *COMMUTATIVE BANACH ALGEBRAS*

(B.99) Let $A$ be a commutative Banach algebra with identity. Prove that an element $x \in A$ has an inverse in $A$ if and only if $\hat{x}$ has an inverse in $C(\hat{A})$, and that $(x^{-1})\hat{} = 1/\hat{x}$.

(B.100) Show by an example that completeness cannot be dropped in the statement of Proposition (B.6.3).

(B.101) Give an example of a commutative Banach algebra $A$ with identity such that:

(a) the Gelfand representation is not injective.

(b) the Gelfand representation is not surjective.

(B.102) Prove that if $A$ is a commutative Banach algebra, then the spectral radius $|\cdot|_\sigma$ is a continuous function on A.

(B.103) Prove that a continuous multiplicative linear functional $f$ on a normed algebra $A$ has norm $||f|| \leq 1$.

(B.104) Give a complete proof of Corollary (B.6.8).

(B.105) Let $A$ be a commutative Banach algebra and let $B$ be a subset of $A$ whose linear span is dense in $A$. Prove that the weak topology on $\hat{A}$ generated by the functions $\hat{x}$ with $x$ in $B$ is equal to the Gelfand topology on $\hat{A}$.

(B.106) Give an example of a nontrivial commutative Banach algebra which is not semisimple and which is not a radical algebra.

(B.107) Let $A$ be a commutative semisimple Banach algebra with identity $e$ and norm $||\cdot||$, and let $B$ be a subalgebra containing $e$ which is a Banach algebra with norm $||\cdot||_1$. Prove that there exists $M > 0$ such that $||x|| \leq M||x||_1$ for all $x \in B$.

(B.108) Let $A$ be a commutative Banach algebra with identity $e$. Prove that if $A$ is separable, then the structure space $\hat{A}$ is metrizable.

(B.109) Give an example of a Banach algebra with identity which has uncountably many idempotents.

(B.110) Prove that if $x$ and $y$ are elements in a Banach algebra $A$ such that $xy = yx$, then $\sigma_A(xy) \subseteq \sigma_A(x)\sigma_A(y)$.

(B.111) Let $A$ be a commutative Banach algebra with identity and let $x_1,\ldots,x_n$ be elements in $A$. Prove that either there exists $\phi \in \hat{A}$ such that $\phi(x_i) = 0$ $(1 \leq i \leq n)$, or else there exist elements $y_1,\ldots,y_n$ in $A$ such that $\Sigma_{i=1}^n x_i y_i = 1$.

(B.112) Let $A$ be a commutative Banach algebra with identity, and $U$ an open set in the complex plane $C$. Prove that the set $T = \{x \in A: \sigma_A(x) \subset U\}$ is open in $A$.

(B.113) Let $A$ be a commutative Banach algebra with identity $e$. Prove that if $A$ contains a proper idempotent element, then the structure space $\hat{A}$ is not connected (the converse is also true; it is a nontrivial theorem of Shilov).

(B.114) For $p$, $1 \leq p < \infty$, the Banach space $\ell^p$ of sequences $x = \{x_i\}$

satisfying $||x|| = (\Sigma_{i=1}^{\infty}|x_i|^p)^{1/p} < \infty$, together with pointwise multiplication, is a commutative Banach algebra without identity. Identify the maximal modular ideals of $\ell^p$.

(B.115) Consider the Banach algebra of Exercise (B.114) with $p = 1$, and let A be the algebra obtained by adjoining an identity. Show that A is a semisimple commutative Banach algebra with identity, but the subalgebra $B = \{\hat{x}: x \in A\}$ is not closed in $C(\hat{A})$.

(B.116) Let A and B be Banach algebras with identity each satisfying $||x^2|| = ||x||^2$ for all x. Prove that if A and B are algebraically isomorphic, then A and B are isometric as Banach spaces.

(B.117) Let $B_1$ be a commutative Banach algebra with identity e and S a nonempty subset of $B_1$. Let $A_1$ be the smallest closed subalgebra of $B_1$ which contains S, and $A_2$ the smallest closed subalgebra which contains S and the identity e. Prove that either $A_1 = A_2$ or $A_1$ is a maximal ideal in $A_2$.

(B.118) Let A be a commutative Banach algebra, B a closed subalgebra and I a maximal modular ideal of A not containing B. Prove that $B \cap I$ is a maximal modular ideal in B.

(B.119) Let A be a commutative Banach algebra and B a closed subalgebra. Prove that $\text{Rad}(B) = B \cap \text{Rad}(A)$. What if A is non-commutative?

(B.120) Let A be a commutative normed algebra and B its completion. If an element x in A has a quasi-inverse modulo every closed maximal modular ideal in A, prove that x has a quasi-inverse in B.

(B.121) Give an example to show in the preceding problem that the element x need not have a quasi-inverse in A.

(B.122) A normed algebra A is called *topologically semisimple* if the intersection of its closed maximal modular ideals is zero.

(a) If A is a commutative semisimple Banach algebra, prove that every subalgebra B of A is a topologically semisimple normed algebra.

(b) If A is a commutative normed algebra whose completion is semisimple, prove that A is topologically semisimple.

(B.123) Let A be a commutative Banach algebra with identity e. Prove that a linear functional $\phi$ on A, with $\phi(e) = 1$, is multiplicative if and only if $\phi(x) \neq 0$ for all invertible elements x in A.

(B.124) Let F be a set of generators for a commutative Banach algebra A with identity e. Let $\phi \in A^*$, $\phi(e) = 1$, and assume that $\phi(\exp(x)) \neq 0$ for all x in the linear span of F. Prove that $\phi$ is multiplicative.

(B.125) If A is a normed algebra in which 0 is the only quasi-nilpotent element and $\nu(\cdot)$ is subadditive and submultiplicative, prove that A is commutative.

(B.126) If A is a Banach algebra and $\nu(\cdot)$ is subadditive and submultiplicative, prove $\nu(xy - yx) = 0$ for all x, y in A.

(B.127) If A is a Banach algebra and $\nu(\cdot)$ is subadditive and submultiplicative, prove that A/Rad(A) is commutative.

(B.128) In a noncommutative normed algebra, prove $\inf_{x\neq 0} \nu(x/||x||) = 0$ and $\inf_{x\neq 0} ||x^2||/||x||^2 = 0$.

(B.129) Prove every noncommutative finite-dimensional normed algebra contains a nonzero nilpotent element.

(B.130) Let A be a Banach algebra with identity e which satisfies $||xy|| \leq \alpha||yx||$ for all $x, y \in A$ and some $\alpha > 0$. Prove that A is commutative.

(B.131) Let A be a Banach algebra with identity e. If $a \in A$ satisfies $||(a + \lambda e)x|| \leq ||x(a + \lambda e)||$ for all $x \in A$, $\lambda \in C$, prove that a is in the center of A.

(B.132) Let A be a Banach algebra with identity e. If each singular element z in A belongs to a closed proper two-sided ideal in A, prove that xy - yx is in the radical for all $x, y \in A$.

(B.133) Let A be a commutative Banach algebra with identity e which is finitely generated. Prove that A has exactly one maximal ideal if and only if there exist elements $x_1,\ldots,x_n$ in A satisfying the following two conditions:

(a) $\nu(x_i) = 0$, $i = 1,\ldots,n$.

(b) the set $\{e,x_1,\ldots,x_n\}$ generates A.

(B.134) Let A be a commutative normed algebra with identity e, and B its completion. Let M and N be maximal ideals in A and B respectively. If the bar denotes closure in B, prove that:

(a) $\overline{N \cap A} = N$.

(b) $\overline{M}$ is a maximal ideal in B.

(B.135) (Arens' multiplication). Let A be a Banach algebra. For $f \in A^*$, $a \in A$ define $f_a \in A^*$ by $f_a(x) = f(ax)$. For $f \in A^*$, $F \in A^{**}$, define $f_F \in A^*$ by $f_F(x) = F(f_x)$. Show that:

(a) $f_{\hat{a}} = f_a$ (here $\hat{a} \in A^{**}$ the natural embedding).

(b) $||f_a|| \leq ||f||\cdot||a||$.

(c) $||f_F|| \leq ||f||\cdot||F||$.

For $F, G \in A^{**}$, define FG by $(FG)(f) = F(f_G)$. Show that:

(d) $A^{**}$ is a Banach algebra.

(e) the natural embedding $A \rightarrow A^{**}$ is an isomorphism into.

(f) If A is commutative, $B = \{\hat{x}: x \in A\}$ is in the center of $A^{**}$, i.e., $ab = ba$ for $a \in A^{**}$, $b \in B$.

## *APPROXIMATE IDENTITIES*

(B.136) Prove the equivalence of the two definitions for bounded left approximate identities stated just preceding Theorem (B.7.1).

(B.137) A bounded left approximate identity $\{e_\lambda\}_{\lambda\in\Lambda}$ in a normed algebra A is *sequential* if $\Lambda$ is the set of positive integers with the usual order. If A is separable and has a bounded left approximate identity, prove that it has a sequential left approximate identity bounded by the same constant.

(B.138) Give an example of an incomplete, commutative, normed algebra which has no bounded approximate identity.

(B.139) Let A be a normed algebra with a right identity. Prove that if A has left approximate identity (bounded or not), then A has a two-sided identity.

(B.140) Let X be a locally compact, noncompact, Hausdorff space. Construct a bounded approximate identity for the Banach algebra $C_o(X)$.

(B.141) Let A denote the Banach algebra of all complex-valued functions f on $[0,1]$ with continuous first derivatives satisfying $f(0) = 0$, where the norm of an element $f$ is defined by

$$||f|| = \sup_{0\leq t\leq 1} |f(t)| + \sup_{0\leq t\leq 1} |f'(t)|.$$

Show that $A$ does not admit an approximate identity (bounded or otherwise).

(B.142) Let $A$ be a normed algebra and $I$ a closed ideal in $A$. Prove that if $I$ and $A/I$ have bounded left approximate identities, then $A$ has a bounded left approximate identity.

(B.143) Let $A$ be a normed algebra and $I$ a closed two-sided ideal in $A$. Prove that if both $I$ and $A/I$ have bounded left approximate identities, then $A$ does too.

(B.144) Let $A$ be a normed algebra with bounded left approximate identity $\{e_\alpha\}$, $V$ a left normed $A$-module such that $e_\alpha v \to v$ for all $v \in V$, and $X$ a normed space. Let $\psi$ be a continuous bilinear mapping of $A \times V$ into $X$. Prove that $\psi(a,v) = \lim_\alpha \psi(e_\alpha, av)$ for all $a \in A$, $v \in V$ iff there exists a real constant $\beta$ such that

$$||\psi(a,v)|| \leq \beta||av||, \quad a \in A, \ v \in V.$$

(B.145) Let $A$ be a normed space which is a normed algebra with bounded left approximate identity $\{e_\alpha\}$ for each of two multiplications $*$ and $\circ$. Prove that these multiplications coincide iff there exists a real constant $\beta$ such that $||a\circ b|| \leq \beta||a*b||$ for all $a, b \in A$.

(B.146) Prove that a normed algebra $A$ with bounded left approximate identity is commutative iff there exists a real constant $\beta$ such that $||ba|| \leq \beta||ab||$ for $a, b \in A$.

(B.147) Let $f$ be a linear functional on a normed algebra $A$ with bounded left approximate identity. If there exists a real constant $\beta$ such that $|f(a)| \leq \beta\nu(a)$ for $a \in A$, prove that $f(ab) = f(ba)$ for $a, b \in A$.

(B.148) Let $A$ be a normed algebra with bounded left approximate identity $\{e_\alpha\}$. Prove that an element $a \in A$ is in the center of $A$ iff there exists a real constant $\beta$ so that $||xay|| \leq \beta||xy||$ for $x, y \in A$.

# Bibliography


AARNES, J. F. and KADISON, R. V.

[1] Pure states and approximate identities. Proc. Amer. Math. Soc. 21(1969), 749-752. MR 39 #1980 Zbl. 177, 413.

AKEMANN, C. A.

[1] A Gelfand representation theory for C*-algebras. Pacific J. Math. 39(1971), 1-11. MR 48 #6950 Zbl. 203, 445.

AKEMANN, C. A. and RUSSO, B.

[1] Geometry of the unit sphere of a C*-algebra and its dual. Pacific J. Math. 32(1970), 575-585. MR 41 #5980 Zbl. 194, 442.

ALFSEN, E. M. and EFFROS, E. G.

[1] Structure in real Banach spaces. I. Ann. of Math. (2), 96(1972), 98-128. MR 50 #5432 Zbl. 248.46019.

[2] Structure in real Banach spaces. II. Ann. of Math. (2), 96(1972), 129-173. MR 50 #5432 Zbl. 248.46019.

ALFSEN, E. and SHULTZ, F. W.

[1] Noncommutative spectral theory for affine function spaces on convex sets. Mem. Amer. Math. Soc. no. 172, Amer. Math. Soc., Providence, R. I., 1976.

[2] State spaces of Jordan algebras. Acta Math. 140(1978), 155-190. MR 57 #12630 Zbl. 397.46066.

[3] On noncommutative spectral theory and Jordan algebras. Proc. London Math. Soc. (3), 38(1979), 497-516.

ALFSEN, E., SHULTZ, F. W., and STØRMER, E.

[1] A Gelfand-Neumark theorem for Jordan algebras. Advances in Math. 28(1978), 11-56. MR 58 #2292 Zbl. 397.46065.

ALLAN, G. R.

[1] A note on B*-algebras. Proc. Cambridge Philos. Soc. 61(1965), 29-32. MR 30 #2353 Zbl. 135, 357.

[2] A spectral theory for locally convex algebras. Proc. London Math. Soc. (3), 15(1965), 399-421. MR 31 #619 Zbl. 138, 382.

[3] On a class of locally convex algebras. Proc. London Math. Soc. (3), 17(1967), 91-114. MR 34 #4937 Zbl. 147, 335.

AMBROSE, W.
[1] Structure theorems for a special class of Banach algebras. Trans. Amer. Math. Soc. 57(1945), 364-386. MR 7, 126 Zbl. 60, 269.

APOSTOL, C.
[1] b*-algebras and their representation. J. London Math. Soc. (2), 3(1971), 30-38. MR 44 #2040 Zbl. 207, 444.

ARAKI, H. and ELLIOTT, G.
[1] On the definition of C*-algebras. Publ. Res. Inst. Math. Sci. Kyoto Univ. 9(1973), 93-112. MR 50 #8085 Zbl. 272.46033.

ARENS, R.
[1] On a theorem of Gelfand and Neumark. Proc. Nat. Acad. Sci. U.S.A. 32(1946), 237-239. MR 8, 279 Zbl. 60, 271.
[2] Representation of *-algebras. Duke Math. J. 14(1947), 269-282. MR 9, 44 Zbl. 29, 402.
[3] A generalization of normed rings. Pacific J. Math. 2(1952), 455-471. MR 14, 482 Zbl. 47, 358.

ARENS, R. and KAPLANSKY, I.
[1] Topological representation of algebras. Trans. Amer. Math. Soc. 63(1948), 457-481. MR 10, 7 Zbl. 32, 7.

ARVESON, W.
[1] An invitation to C*-algebras. Graduate Texts in Mathematics #39. Springer-Verlag, Berlin-Heidelberg-New York, 1976. MR 58 #23621 Zbl. 344.46123.

ASIMOW, L. A. and ELLIS, A. J.
[1] On hermitian functionals on unital Banach algebras. Bull. London Math. Soc. 4(1972), 333-336. MR 48 #2763 Zbl. 267.46037.

AUBERT, K. E.
[1] A representation theorem for function algebras with application to almost periodic functions. Math. Scand. 7(1959), 202-210. MR 22 #8314 Zbl. 92, 320.

AUPETIT, B.
[1] Almost commutative Banach algebras. Notices Amer. Math. Soc. 18 (1971), 191.
[2] Continuité du spectre dans les algèbres de Banach avec involution. Pacific J. Math. 56(1975), 321-324. MR 51 #11117 Zbl. 306.46072.
[3] Caractérisation spectrale des algèbres de Banach commutatives. Pacific J. Math. 63(1976), 23-35. MR 54 #3409 Zbl. 309.46045.
[4] Sur les conjectures de Hirschfeld et Żelazko dans les algebres de Banach. Bull. Soc. Math. France 104(1976), 185-193. MR 54 #8290 Zbl. 347.46052.
[5] Caractérisation spectrale des algèbres de Banach de dimension finie. J. of Functional Analysis 26(1977), 232-250. MR 56 #12887 Zbl. 359.46035.
[6] Continuité uniforme du spectre dans les algèbres de Banach avec involution. C. R. Acad. Sci. Paris, Sér. A-B, 284(1977), A1125-A1127. MR 55 #8806 Zbl. 361.46043.

[7] Continuité et uniforme continuite du spectre dans les algèbres de Banach. Studia Math. 61(1977), 99-114. MR 56 #12883.

[8] Le théorème de Russo-Dye pour les algèbres de Banach involutives. C. R. Acad. Sci. Paris, Sér. A-B, 284(1977), A151-A153. (English Summary). MR 55 #13243 Zbl. 341.46041.

[9] Nouvelles méthodes analytiques dans la théorie des algèbres de Banach. Ann. Sci. Math. Québec 2(1978), 169-196. MR 80d: 46093 Zbl. 396.46046.

[10] Propriétes spectrales des algèbres de Banach. Lecture Notes in Mathematics, Vol. 735, 192 pp. Springer-Verlag, Berlin-Heidelberg-New York, 1979. MR 81i: 46055 Zbl. 409.46054.

[11] The uniqueness of the complete norm topology in Banach algebras and Banach Jordan algebras. J. Functional Analysis 47(1982), 1-6. Zbl. 488.46043.

[12] Analytic multivalued functions in Banach algebras and uniform algebras. Advances in Math. 44(1982), 18-60. MR 84b: 46059 Zbl. 486.46041.

AUPETIT, B. and YOUNGSON, M. A.

[1] On symmetry of Banach Jordan algebras. Proc. Amer. Math. Soc. 91(1984), 364-366. Zbl. 526.46048.

BAILEY, D. W.

[1] On symmetry in certain group algebras. Pacific J. Math. 24(1968), 413-419. MR 39 #6085 Zbl. 172, 184.

BAKER, C. W.

[1] A closed graph theorem for Banach bundles. Rocky Mountain J. of Math. 12(1982), 537-543. MR 84h: 46010.

[2] The Pedersen ideal and the representation of C*-algebras. Rocky Mountain J. of Math. 13(1983), 699-707.

BARNES, B. A.

[1] Locally B*-equivalent algebras. Trans. Amer. Math. Soc. 167(1972), 435-442. MR 45 #5763 Zbl. 245.46082.

[2] Strictly irreducible *-representations of Banach *-algebras. Trans. Amer. Math. Soc. 170(1972), 459-469. MR 46 #2441 Zbl. 275.46048.

[3] Locally B*-equivalent algebras. II. Trans. Amer. Math. Soc. 176(1973), 297-303. MR 47 #9296 Zbl. 271.46050.

[4] Representations of B*-algebras on Banach spaces. Pacific J. Math. 50(1974), 7-18. MR 49 #7788 Zbl. 285.46053.

[5] When is a representation of a Banach *-algebra Naimark-related to a *-representation? Pacific J. Math. 72(1977), 5-25. MR 56 #16385 Zbl. 385.46034.

BECKER, T.

[1] A few remarks on the Dauns-Hofmann theorems for C*-algebras. Arch. Math. (Basel) 43(1984), 265-269.

BEHNCKE, H.

[1] A remark on C*-algebras. Comm. Math. Phys. 12(1969), 142-144. MR 39 #4685 Zbl. 172, 411.

[2] A note on the Gel'fand-Naimark conjecture. Comm. Pure Appl. Math. 23(1970), 189-200. MR 41 #2404 Zbl. 188, 446.

[3] Hermitian Jordan Banach algebras. J. London Math. Soc. (2) 20(1979), 327-333. MR 81f: 46065 Zbl. 405.46039.

[4] Finite-dimensional representations of JB algebras. Proc. Amer. Math. Soc. 88(1983), 426-428. MR 84e: 46079.

BEHNCKE, H. and BÖS, W.
[1] JB algebras with an exceptional ideal. Math. Scand. 42(1978), 306-312. MR 81h: 46062b.

BEHNCKE, H. and CUNTZ, J.
[1] Local completeness of operator algebras. Proc. Amer. Math. Soc. 62 (1976), 95-100. MR 55 #1077 Zbl. 347.46064.

BELFI, V. A. and DORAN, R. S.
[1] Norm and spectral characterizations in Banach algebras. L'Enseignment Math. 26(1980), 103-130. MR 81j: 46071 Zbl. 441.46039.

BÉLLISSARD, J. and IOCHUM, B.
[1] Homogeneous self-dual cones versus Jordan algebras. The theory revisited. Ann. Inst. Fourier (Grenoble) 28(1978), 27-67. MR 80b: 46082 Zbl. 365.46040.

BERBERIAN, S. K.
[1] Baer *-rings. Springer-Verlag, New York-Berlin, 1972. MR 55 #2983 Zbl. 242.16008.
[2] Lectures in functional analysis and operator theory. Graduate Texts in Mathematics #15. Springer-Verlag, Berlin-Heidelberg-New York, 1974. MR 54 #5775 Zbl. 296.46002.

BERBERIAN, S. K. and ORLAND, G.
[1] On the closure of the numerical range of an operator. Proc. Amer. Math. Soc. 18(1967), 499-503. MR 35 #3459 Zbl. 173, 421.

BERG, C., CHRISTENSEN, J. P. R., and RESSEL, P.
[1] Harmonic analysis on semigroups: Theory of positive definite and related functions. Graduate Texts in Mathematics #100. Springer-Verlag, Berlin-Heidelberg-New York, 1984.

BERGLUND, M. C. F.
[1] Ideal C*-algebras. Duke Math. J. 40(1973), 241-257. MR 48 #881 Zbl. 265.46055.

BERKSON, E.
[1] Some characterizations of C*-algebras. Illinois J. Math. 10(1966), 1-8. MR 32 #2922 Zbl. 132, 108.
[2] Prehermitian elements and B*-algebras. Math. Ann. 195(1972), 192-198. MR 45 #907 Zbl. 215, 485.

BOHNENBLUST, H. F. and KARLIN, S.
[1] Geometrical properties of the unit sphere of Banach algebras. Ann. of Math. (2) 62(1955), 217-229. MR 17, 177 Zbl. 67, 350.

BONIC, R. A.
[1] Symmetry in group algebras of discrete groups. Pacific J. Math. 11 (1961), 73-94. MR 22 #11281 Zbl. 172, 183.

BONSALL, F. F.
[1] The numerical range of an element of a normed algebra. Glasgow Math. J. 10(1969), 68-72. MR 39 #4675 Zbl. 175, 138.
[2] A survey of Banach algebra theory. Bull. London Math. Soc. 2(1970), 257-274. MR 43 #917 Zbl. 207, 442.

[3] Jordan algebras spanned by Hermitian elements in a Banach algebra. Math. Proc. Cambridge Philos. Soc. 81(1977), 3-13. MR 54 #13577 Zbl. 343.46032.

BONSALL, F. F. and DUNCAN, J.

[1] Dual representations of Banach algebras. Acta Math. 117(1967), 79-102. MR 34 #4929 Zbl. 148, 376.

[2] Dually irreducible representations of Banach algebras. Quart. J. Math. Oxford (2), 19(1968), 97-111. MR 38 #535 Zbl. 153, 446.

[3] Numerical ranges of operators on normed spaces and elements of normed algebras. Cambridge Univ. Press, 1971. MR 44 #5779 Zbl. 207, 448.

[4] Numerical ranges. II. Cambridge Univ. Press. New York-London. 1973. MR 56 #1063 Zbl. 262.47001.

[5] Complete normed algebras. Springer-Verlag, Berlin-Heidelberg-New York, 1973. MR 54 #11013 Zbl. 271.46039.

BOYADZIEV, H. N.

[1] Commutativity in Banach algebras and elements with real spectra. C. R. Acad. Bulgare Sci. 29(1976), 1401-1403. MR 55 #8802 Zbl. 352.46030.

[2] Characterization of B*-equivalent Banach algebras by means of their positive cones. Serdica 3(1977), 20-24. MR 58 #2312 Zbl. 365.46046.

[3] Partially ordered B*-equivalent Banach algebras. Serdica 4(1978), 12-18. MR 80i: 46047 Zbl. 433.46050.

[4] Order characterization of some Banach Jordan algebras. C. R. Acad. Bulgare Sci. 32(1979), 1019-1022. MR 81i: 46068.

[5] A note on a paper of S. Watanabe. Acta Sci. Math. 46(1983), 303-304.

BRATTELI, O. and ROBINSON, D. W.

[1] Operator algebras and quantum statistical mechanics. I. Springer-Verlag. Berlin-Heidelberg-New York. 1979. MR 81a: 46070 Zbl. 421.46048.

[2] Operator algebras and quantum statistical mechanics. II. Springer-Verlag. Berlin-Heidelberg-New York. 1981. MR 82k: 82013 Zbl. 463.46052.

BRAUN, R. B.

[1] A Gel'fand-Neumark theorem for C*-alternative algebras. Math. Z. 185(1984), 225-242. MR 85i: 46096 Zbl. 514.46047.

[2] Structure and representations of noncommutative C*-Jordan algebras. Manuscripta Math. 41(1983), 139-171. MR 84e: 46060 Zbl. 512.46055.

BRAUN, R. B., KAUP, W., and UPMEIER, H.

[1] A holomorphic characterization of Jordan C*-algebras. Math. Z. 161 (1978), 277-290. MR 58 #12398 Zbl. 385.32002.

BRIDGES, D. S.

[1] Constructive functional analysis. Research notes in mathematics #28. Pitman. London. 1979. MR 82k: 03094 Zbl. 401.03027.

BROOKS, R. M.

[1] On locally m-convex *-algebras. Pacific J. Math. 23(1967), 5-23. MR 35 #7126 Zbl. 156, 142.

[2] On commutative locally m-convex algebras. Duke Math. J. 35(1968), 257-267. MR 37 #759 Zbl. 177, 175.

BROWDER, A.

[1] States, numerical range, etc. Proc. of the Brown Informal Analysis Seminar, 39 pp. Summer, 1969. (multilith notes).

BROWN, L. G., DOUGLAS, R. G., and FILLMORE, P. A.

[1] Extensions of C*-algebras and K-homology. Ann. of Math. 105(1977), 265-324. MR 56 #16399 Zbl. 376.46036.

BUNCE, J. W.

[1] The ordered vector space structure of JC-algebras. Proc. London Math. Soc. (3) 22(1971), 359-368. MR 44 #4530 Zbl. 212, 155.

[2] A note on two-sided ideals in C*-algebras. Proc. Amer. Math. Soc. 28(1971), 635-636. MR 43 #2520 Zbl. 211, 439.

BUNCE, L. J.

[1] Type I JB-algebras. Quart. J. Math. Oxford Ser. (2) 34(1983), 7-19. MR 84b: 46056 Zbl. 539.46045.

[2] The theory and structure of dual JB algebras. Math Z. 180(1982), 525-534. Zbl. 518.46053.

[3] On compact action in JB-algebras. Proc. Edinburgh Math. 26(1983), 353-360.

BURCKEL, R. B.

[1] A simpler proof of the commutative Glickfeld-Berkson theorem. J. London Math. Soc. (2), 2(1970), 403-404. MR 42 #2303 Zbl. 196, 152.

[2] Characterizations of C(X) among its subalgebras. Marcel Dekker, New York 1972. MR 56 #1068 Zbl. 252.46047.

BURCKEL, R. B. and SAEKI, S.

[1] Closure and holomorphic properties in B*-algebras via functions which operate. J. Math. Anal. Appl. 98(1984), 211-219. Zbl. 532.46032.

BUSBY, R. C.

[1] On structure spaces and extensions of C*-algebras. J. Functional Analysis 1(1967), 370-377. MR 37 #771 Zbl. 165, 154.

[2] Double centralizers and extensions of C*-algebras. Trans. Amer. Math. Soc. 132(1968), 79-99. MR 37 #770 Zbl. 165, 155.

VAN CASTEREN, J. A.

[1] A characterization of C*-algebras. Proc. Amer. Math. Soc. 72(1978), 54-56. MR 81h: 46069 Zbl. 348.46048.

CHING, W.

[1] The structure of standard C*-algebras and their representations. Pacific J. Math. 67(1976), 131-153. MR 55 #1078 Zbl. 339.46040.

CHOI, M. D.

[1] A Schwarz inequality for positive linear maps on C*-algebras. Illinois J. Math. 18(1974), 565-574. MR 50 #8089 Zbl. 293.46043.

CHU, C. H.

[1] On convexity theory and C*-algebras. Proc. London Math. Soc. (3) 31(1975), 257-288. MR 56 #12895.

CIVIN, P. and YOOD, B.

[1] Involutions on Banach algebras. Pacific J. Math. 9(1959), 415-436. MR 21 #4365 Zbl. 86, 96.

CORDES, H. O.
[1] On a class of C*-algebras. Math. Ann. 170(1967), 283-313. MR 35 #749 Zbl. 154, 151.
[2] On a generalized Fredholm theory. J. Reine Angew. Math. 227(1967), 121-149. MR 39 #1983 Zbl. 177, 410.
[3] An algebra of singular integral operators with two symbol homomorphisms. Bull. Amer. Math. Soc. 75(1969), 37-42. MR 38 #2637 Zbl. 197, 121.
[4] Banach algebras and partial differential operators. Lecture Notes, Lund, 1971.
[5] On compactness of commutators of multiplications and convolutions, and boundedness of pseudo-differential operators. J. Functional Analysis 18(1975), 115-131. MR 51 #13770 Zbl. 306.47024.
[6] Elliptic pseudo-differential operators - and abstract theory. Lecture Notes in Math. Vol. 756. Springer-Verlag. Berlin-Heidelberg-New York, 1979. MR 81j: 47041 Zbl. 417.35004.

CORDES, H. O. and HERMAN, E.
[1] Gel'fand theory of pseudo-differential operators. Amer. J. Math. 90 (1968), 681-717. MR 56 #12991 Zbl. 169, 471.

COWLING, M.
[1] The Fourier-Stieltjes algebra of a semisimple group. Colloq. Math. 41(1979), 89-94. MR 81e: 43005 Zbl. 425.43012.

CRABB, M. J., DUNCAN, J., and McGREGOR, C. M.
[1] Characterizations of commutativity for C*-algebras. Glasgow Math. J. 15(1974), 172-175. MR 50 #14252 Zbl. 301.46047.

CUNTZ, J.
[1] Locally C*-equivalent algebras. J. Functional Analysis 23(1976), 95-106. MR 56 #6398 Zbl. 343.46038.

DALES, H. G.
[1] Exponentiation in Banach *-algebras. Proc. Edinburgh Math. Soc. (2), 20(1976), 163-165. MR 54 #5840 Zbl. 331.46046.

DAUNS, J.
[1] The primitive ideal space of a C*-algebra. Canadian J. Math. 26 (1974), 42-49. MR 49 #1131 Zbl. 279.46030.

DAUNS, J. and HOFMANN, K. H.
[1] Representation of rings by sections. Mem. Amer. Math. Soc. no. 83, American Mathematical Soc., Providence, R. I., 1968. MR 40 #752 Zbl. 174, 57.
[2] Spectral theory or algebras and adjunction of identity. Math. Ann. 179(1969), 175-202. MR 40 #734 Zbl. 169, 360.

DEEDS, J. B.
[1] Uniqueness of symmetric norms for matrix rings. Linear Algebra and Appl. 5(1972), 383-391. MR 46 #5383 Zbl. 243.15023.

DELAROCHE, C.
[1] Sur les centres des C*-algebres. C. R. Acad. Sci. Paris. Sér. A-B, 265(1967), A465-A466. MR 36 #6950 Zbl. 161, 110.

[2] Sur les centres des C*-algèbres. Bull. Sci. Math. (2), 91(1967), 105-112. MR 38 #3729 Zbl. 179, 181.
[3] Sur les centres des C*-algèbres. II. Bull. Sci. Math. (2), 92(1968), 111-128. MR 38 #2612 Zbl. 179, 181.
[4] Extensions des C*-algèbres. Bull. Soc. France Mém. No. 29. Supplément au Bull. Soc. France, Tome 100. Paris, 1972. 142 pp. MR 58 #30290 Zbl. 248.46047.

DIVINSKY, N. J.
[1] Rings and Radicals, University of Toronto Press, Toronto, 1965. MR 33 #5654 Zbl. 138, 263.

DIXMIER, J.
[1] Champs continus d'espaces hilbertiens et de C*-algèbres. II. J. Math. Pure Appl. (9), 42(1963), 1-20. MR 27 #603 Zbl. 127, 333.
[2] Traces sur les C*-algèbres. Ann. Inst. Fourier (Grenoble) 13(1963), 219-262. MR 26 #6807 Zbl. 118, 113.
[3] Traces sur les C*-algèbres. II. Bull. Sci. Math. (2), 88(1964), 39-57. MR 31 #6132 Zbl. 132, 108.
[4] Ideal center of a C*-algebra. Duke Math. J. 35(1968), 375-382. MR 37 #5703 Zbl. 179, 180.
[5] C*-algebras. North-Holland Mathematical Library, Vol. 15. North-Holland Publishing Co., Amsterdam 1977. MR 56 #16388 Zbl. 372. 46058.
[6] Von Neumann algebras. North-Holland Mathematical Library, Vol. 27. North-Holland Publishing Co., Amsterdam 1981. MR 50 #5482 Zbl. 473.46040.

DIXMIER, J. and DOUADY, A.
[1] Champs continus d'espaces hilbertiens et de C*-algèbres. Bull. Soc. Math. France 91(1963), 227-284. MR 29 #485 Zbl. 127, 331.

DIXON, P. G.
[1] Generalized B*-algebras. Proc. London Math. Soc. (3), 21(1970), 693-715. MR 43 #3811 Zbl. 205, 426.
[2] Generalized B*-algebras. II. J. London Math. Soc. (2), 5(1972), 159-165. MR 46 #4214 Zbl. 233.46072.
[3] An embedding theorem for commutative $B_0$-algebras. Studia Math. 41 (1972), 163-168. MR 47 #2369 Zbl. 233.46061.
[4] A symmetric normed *-algebra whose completion is not symmetric. (Preprint).
[5] A characterisation of closed subalgebras of B(H). Proc. Edinburgh Math. Soc. (2), 20(1976-77), 215-217. MR 55 #11070 Zbl. 347.46074.

DORAN, R. S.
[1] Construction of uniform CT-bundles, Notices Amer. Math. Soc. 15 (1968), 551.
[2] Representations of C*-algebras by uniform CT-bundles, Thesis, Univ. of Washington, 1968.
[3] A generalization of a theorem of Civin and Yood on Banach *-algebras. Bull. London Math. Soc. 4(1972), 25-26. MR 46 #2442 Zbl. 237.46063.
[4] An application of idempotents in the classification of complex algebras. Elem. Math. 35(1980), 16-17. MR 81b: 46065 Zbl. 427.46030.

DORAN, R. S. and TILLER, W.
[1] Extensions of pure positive functionals on Banach *-algebras. Proc. Amer. Math. Soc. 82(1981), 583-586. MR 82f: 46062 Zbl. 474.46046.
[2] Continuity of the involution in a Banach *-algebra. Tamkang J. Math. 13(1982), 87-90. MR 84d: 46086 Zbl. 492.46045.

DORAN, R. S. and WICHMANN, J.
[1] The Gelfand-Naimark theorems for C*-algebras. L'Enseignement Math. 23(1977), 153-180. MR 58 #12395 Zbl. 367.46052.
[2] Approximate identities and factorization in Banach modules. Lecture Notes in Mathematics Vol. 768. Springer-Verlag. Berlin-Heidelberg-New York, 1979. MR 83e: 46044 Zbl. 418.46039.

DOUGLAS, R. G.
[1] Banach algebra techniques in operator theory. Academic Press. New York and London 1972. MR 50 #14335 Zbl. 247.47001.

DUNCAN, J.
[1] The continuity of the involution on Banach *-algebras. J. London Math. Soc. 41(1966), 701-706. MR 34 #3351 Zbl. 144, 170.

DUNCAN, J. and TAYLOR, P. J.
[1] Norm inequalities for C*-algebras. Proc. Roy. Soc. Edinburgh, Sect. A, 75(1975/76), 119-129. MR 56 #12896 Zbl. 331.46050.

DUNFORD, N. and SCHWARTZ, J. T.
[1] Linear operators. I. J. Wiley, New York, 1958. MR 22 #8302 Zbl. 84, 104.
[2] Linear operators. II. J. Wiley, New York, 1963. MR 32 #6181 Zbl. 128, 348.
[3] Linear operators. III. J. Wiley, New York, 1972. MR 54 #1009 Zbl. 243.47001.

DUPRÉ, M. J.
[1] Hilbert bundles with infinite dimensional fibres. Recent advances in the representation theory of rings and C*-algebras by continuous sections. Sem. Tulane Univ., New Orleans, La. 1973, 165-176. Mem. Amer. Math. Soc. no. 148, Amer. Math. Soc., Providence, R. I. 1974. MR 50 #10829 Zbl. 272.55024.
[2] Classifying Hilbert bundles. I. J. Functional Analysis 15(1974), 244-278. MR 49 #11266 Zbl. 275.46050.
[3] Classifying Hilbert bundles. II. J. Functional Analysis. 22(1976), 295-322. MR 54 #3435 Zbl. 327.46064.
[4] Duality for C*-algebras. Proc. Conf. on mathematical foundations of quantum theory, Loyola Univ., New Orleans, La. (A. R. Marlow, Ed.), 329-338. Academic Press, New York, 1978. MR 80a: 46034.
[5] The classification and structure of C*-algebra bundles. Mem. Amer. Math. Soc. no. 222, Amer. Math. Soc., Providence, R. I., 1979. MR 83c: 46069 Zbl. 416.46053.

DUPRÉ, M. J. and GILLETTE, R. M.
[1] Banach bundles, Banach modules, and automorphisms of C*-algebras. Research Notes in Mathematics #92, Pitman, London, 1983. Zbl. 536.46048.

EDWARDS, C. M.
[1] On Jordan W*-algebras. Bull. Sci. Math. (2) 104(1980), 393-403. MR 82g: 46097 Zbl. 456.46057.

[2] Multipliers of JB-algebras. Math. Ann. 249(1980), 265-272.
MR 81j: 46073 Zbl. 428.46036.

EFFROS, E. G.
[1] On two numerical range formulae (unpublished note).

ELLIOTT, G. A.
[1] A weakening of the axioms for a $C^*$-algebra. Math. Ann. 189(1970), 257-260. MR 43 #2521 Zbl. 194, 156.
[2] An abstract Dauns-Hofmann-Kaplansky multiplier theorem. Canadian J. Math. 27(1975), 827-836. MR 53 #6334 Zbl. 294.46048.

ELLIOTT, G. A. and OLESEN, D.
[1] A simple proof of the Dauns-Hofmann theorem. Math. Scand. 34(1974). 231-234. MR 50 #8091 Zbl. 289.46042.

EMCH, G. G.
[1] Algebraic methods in statistical mechanics and quantum field theory. Wiley-Interscience, 1972. Zbl. 235.46085.

FEICHTINGER, H. G. and RINDLER, H.
[1] Symmetrie der Wienerschen Algebra und Gruppenstruktur. Österreich. Akad. Wiss. math.-naturw. Kl., Anzeiger (1976), 89-91. Zbl. 352.43002.

FELDMAN, J.
[1] Seminar notes. University of California, Berkeley, Cal., 1962 (dittoed notes).

FELL, J. M. G.
[1] The structure of algebras of operator fields. Acta Math. 106(1961), 233-280. MR 29 #1547 Zbl. 101, 93.
[2] An extension of Mackey's method to Banach *-algebraic bundles. Mem. Amer. Math. Soc. no. 90, Amer. Math. Soc., Providence, R. I., 1969. MR 41 #4255 Zbl. 194, 443.
[3] An extension of Mackey's method to algebraic bundles over finite groups. Amer. J. Math. 91(1969), 203-238. MR 40 #735 Zbl. 191, 25.
[4] Induced representations and Banach *-algebraic bundles. Lecture Notes in Mathematics Vol. 582. Springer-Verlag. Berlin-Heidelberg-New York, 1977. MR 56 #15825 Zbl. 372.22001.

FELL, J. M. G. and DORAN, R. S.
[1] Representations of groups, algebras, and Banach algebraic bundles. I. II. Academic Press (to appear).

FORD, J. W. M.
[1] A square root lemma for Banach *-algebras. J. London Math. Soc. 42 (1967), 521-522. MR 35 #5950 Zbl. 145, 387.

FOUNTAIN, J. B., RAMSAY, R. W., and WILLIAMSON, J. H.
[1] Functions of measures on compact groups. Proc. Roy. Irish Acad. Sect. A 76(1976), 235-251. MR 57 #7034 Zbl. 309.43010.

FRIEDMAN, Y. and RUSSO, B.
[1] Contractive projections on $C_0(K)$. Trans. Amer. Math. Soc. 273(1982), 57-74. MR 83i: 46062.

[2] Contractive projections on C*-algebras. Symp. Pure Math. 38, Part 2, Kingston Ont. 1980, 615-618(1982). MR 84g: 46080 Zbl. 525.46029.
[3] Function representation of commutative operator triple systems. J. London Math. Soc. (2) 27(1983), 513-524. MR 84h: 46095 Zbl. 543.46046.
[4] Contractive projections on operator triple systems. Math. Scand. 52 (1983), 279-311. MR 84m: 46090.
[5] Solution of the contractive projection problem. J. Functional Analysis 60(1985), 56-79.

FUKAMIYA, M.
[1] On B*-algebras. Proc. Japan Acad. 27(1951), 321-327. MR 13, 756 Zbl. 44, 327.
[2] On a theorem of Gelfand and Neumark and the B*-algebra. Kumamoto J. Sci. Ser. A. 1(1952), 17-22. MR 14, 884 MR 15, 1139 (errata) Zbl. 49, 86.

FUKAMIYA, M., MISONOU, M. and TAKEDA, Z.
[1] On order and commutativity of B*-algebras. Tohoku Math. J. (2) 6(1954), 89-93. MR 16, 376 Zbl. 57, 97.

GANGOLLI, R.
[1] On the symmetry of $L_1$-algebras of locally compact motion groups and the Wiener Tauberian theorem. J. Functional Analysis 25(1977), 244-252. MR 57 #6284 Zbl. 347.43005.

GARDNER, L. T.
[1] An elementary proof of the Russo-Dye theorem. Proc. Amer. Math. Soc. 90(1984), 171. Zbl. 528.46041.

GELFAND, I. M.
[1] Normierte Ringe, Mat. Sbornik 9(1941), 3-24. (Russian summary). MR 3, 51 Zbl. 24, 320.

GELFAND, I. M. and NAIMARK, M. A.
[1] On the embedding of normed rings into the ring of operators in Hilbert space. Mat. Sbornik 12(1943), 197-213 (Russian summary). MR 5, 147 Zbl. 60, 270.
[2] Normed rings with involutions and their representations. Izvestiya Akademii Nauk SSSR Ser. Math. 12(1948), 445-480 (Russian). MR 10, 199 Zbl. 31, 34.

GIL DE LAMADRID, J.
[1] Extending positive definite forms. Proc. Amer. Math. Soc. 91(1984), 593-594. MR 85g: 46068.

GLASER, W.
[1] Symmetrie von verallgemeinerten $L^1$-algebren, Arch. Math. (Basel), 20(1969), 656-660. MR 41 #7448 Zbl. 189, 448.

GLICKFELD, B. W.
[1] A metric characterization of C(X) and its generalization to C*-algebras. Illinois J. Math. 10(1966), 547-556. MR 34 #1865 Zbl. 154, 389.

GLIMM, J. G. and KADISON, R. V.
[1] Unitary operators in C*-algebras. Pacific J. Math. 10(1960), 547-556. MR 22 #5906 Zbl. 152, 330.

GOODEARL, K. R.
[1] Notes on real and complex C*-algebras. Birkhäuser, Boston, 1982. MR 85d: 46079 Zbl. 495.46039.

GRASSELLI, J.
[1] Selbstadjungierte Elemente der Banach-Algebra ohne Einheit. Publ. Dept. Math. (Ljubljana) 1(1964), 5-21. MR 32 #1570.

GREEN, W. A.
[1] Ambrose modules. Recent advances in the representation theory of rings and C*-algebras by continuous sections. Sem. Tulane Univ., New Orleans, La. 1973, 109-133, Mem. Amer. Math. Soc., no. 148, Amer. Math. Soc., Providence, R. I. 1974. MR 50 #8092 Zbl. 272.46039.

GROTHENDIECK, A.
[1] Un résultant sur le dual d'une C*-algèbre. J. Math. Pures Appl. 36 (1957), 97-108. MR 19, 665.

GUDDER, S. P. and MICHEL, J. R.
[1] Representations of Baer *-semigroups. Proc. Amer. Math. Soc. 81 (1981), 157-163. MR 82j: 20127 Zbl. 465.47031.

GUDDER, S. P. and SCRUGGS, W.
[1] Unbounded representations of *-algebras. Pacific J. Math. 70(1977), 369-382. MR 58 #2345 Zbl. 374.46045.

HALMOS, P. R.
[1] A Hilbert space problem book. Graduate Texts in Mathematics #19. Springer-Verlag. Berlin-Heidelberg-New York, 1974. MR 34 #8178 Zbl. 144, 387.

HALPERN, H.
[1] A generalized dual for a C*-algebra. Trans. Amer. Math. Soc. 153 (1971), 139-156. MR 42 #5058 Zbl. 209, 152.

HANCHE-OLSEN, H. and STØRMER, E.
[1] Jordan operator algebras. Monographs and Studies in Mathematics, 21. Pitman, Boston-London, 1984.

HARRIS, L. A.
[1] Banach algebras with involution and Möbius transformations. J. Functional Analysis 11(1972), 1-16. MR 50 #5480 Zbl. 239.46058.
[2] A generalization of C*-algebras. Proc. London Math. Soc., (3), 42 (1981), 331-361. MR 82e: 46089 Zbl. 476.46054.

HERSTEIN, I. N.
[1] Group rings as *-algebras. Publ. Math. Debrecen 1(1950), 201-204. MR 12, 475.
[2] Topics in algebra. Blaisdell, New York, 1964. MR 50 #9456.
[3] Noncommutative rings. (Carus Monographs #15) Mathematical Association of America, 1968. MR 37 #2790.
[4] On the multiplicative group of a Banach algebra. Symp. Math. 8(1970), 227-232. MR 50 #8079 Zbl. 239, 277.

HEWITT, E. and ROSS, K. A.
[1] Abstract harmonic analysis. I. Springer-Verlag. Berlin 1963. MR 28 #158 Zbl. 115, 106.
[2] Abstract harmonic analysis. II, Springer-Verlag. Berlin 1970. MR 41 #7378 Zbl. 213, 401.

HILLE, E. and PHILLIPS, R. S.
[1] Functional analysis and semi-groups. Amer. Math. Soc. Colloq. Publ. 31., Providence, R. I. 1957. MR 19, 664 Zbl. 78, 100.

HOFFMAN, K.
[1] Fundamentals of Banach algebras. Instituto de Mathemática da Universidade do Parana, Curitiba, Brasil, 1962. MR 28 #1504 Zbl. 115, 105.

HOFMANN, K. H.
[1] Gelfand-Naimark theorems for non-commutative topological rings. General topology and its relations to modern analysis and algebra II. Proc. Second Sympos. General Topology Appl. Prague 1967, pp. 184-189. Zbl. 204, 54
[2] The duality of compact semigroups and C*-bigebras. Lecture Notes in Math. Vol. 129. Springer-Verlag, Berlin-Heidelberg-New York 1970. MR 55 #5786 Zbl. 211, 433.
[3] Representations of algebras by continuous sections. Bull. Amer. Math. Soc. 78(1972), 291-373. MR 50 #415 Zbl. 237.16018.
[4] Some bibliographical remarks on "Representations of algebras by continuous sections" (Bull. AMS 78(1972), 291-373). Recent advances in the representation theory of rings and C*-algebras by continuous sections. Sem. Tulane Univ., New Orleans, La. 1973, 177-182. Mem. Amer. Math. Soc., no. 148, Amer. Math. Soc., Providence, R. I. 1974. MR 50 #416 Zbl. 291.16018.
[5] Banach bundles, Darmstadt Notes, 1974.
[6] Banach bundles and sheaves in the category of Banach spaces, preprint.

HULANICKI, A.
[1] On the spectral radius of hermitian elements in group algebras. Pacific J. Math. 18(1966), 277-287. MR 33 #6426 Zbl. 172, 184.
[2] On the spectral radius in group algebras. Studia Math. 34(1970), 209-214. MR 41 #5984 Zbl. 189, 447.
[3] On symmetry of group algebras of discrete nilpotent groups. Studia Math. 35(1970), 207-219. MR 43 #3814 Zbl. 205, 131.
[4] On the spectrum of convolution operators on groups with polynomial growth. Invent. Math. 17(1972), 135-142. MR 48 #2304 Zbl. 264.43007.

HULANICKI, A., JENKINS, J. W., LEPTIN, H., and PYTLIK, T.
[1] Remarks on Wiener's Tauberian theorems for groups with polynomial growth. Colloq. Math. 35(1976), 293-304. MR 53 #13469 Zbl. 338.43005.

HUSAIN, T. and MALVIYA, B. D.
[1] Representations of locally m-convex *-algebras. Math. Japon. 17 (1972), 39-47. MR 48 #9406 Zbl. 245.46067.

HUSAIN, T. and RIGELHOF, R.
[1] Representations of MQ*-algebras. Math. Ann., 180(1969), 297-306. MR 40 #745 Zbl. 157, 445.

HUSAIN, T. and WARSI, S. A.
[1] Representations of BP*-algebras. Math. Japon. 21(1976), 237-247. MR 55 #3796 Zbl. 344.46104.

INGELSTAM, L.
[1] Real Banach algebras. Ark. Math. 5(1964), 239-270. MR 30 #2358 Zbl. 149, 97.
[2] Symmetry in real Banach algebras. Math. Scand. 18(1966), 53-68. MR 34 #6555 Zbl. 143, 157.

INOUE, A.
[1] Locally C*-algebras. Mem. Fac. Sci. Kyushu Univ. Ser. A 25(1971), 197-235. MR 46 #4219 Zbl. 227.46060.
[2] Unbounded representations of symmetric *-algebras. J. Math. Soc. Japan 29(1977), 219-232. MR 55 #11055 Zbl. 345.46044.

JACOBSON, N.
[1] Structure of rings. Second ed. Amer. Math. Soc. Colloq. Publ. 37, Amer. Math. Soc., Providence, R. I., 1964. MR 36 #5158.
[2] Structure and representations of Jordan algebras. Amer. Math. Soc. Colloq. Publ. 39, Amer. Math. Soc., Providence, R. I., 1968. MR 40 #4330.

JENKINS, J. W.
[1] On the spectral radius of elements in a group algebra. Illinois J. Math. 15(1971), 551-554. MR 44 #4538 Zbl. 222.43002.
[2] On the characterization of Abelian W*-algebras. Proc. Amer. Math. Soc. 35(1972), 436-438. MR 46 #6055 Zbl. 266.46047.
[3] A characterization of growth in locally compact groups. Bull. Amer. Math. Soc. 79(1973), 103-106. MR 47 #5172 Zbl. 262.22004.
[4] Nonsymmetric group algebras. Studia Math. 45(1973), 295-307. MR 49 #5855 Zbl. 222.22008.

JOHNSON, B. E.
[1] The uniqueness of the (complete) norm topology. Bull. Amer. Math. Soc. 73(1967), 537-539. MR 35 #2142 Zbl. 172, 410.
[2] AW*-algebras are QW*-algebras. Pacific J. Math. 23(1967), 97-99. MR 35 #7135 Zbl. 152, 328.

JORDAN, P., Von NEUMANN, J., and WIGNER, E.
[1] On an algebraic generalization of the quantum mechanical formalism. Ann. of Math. 35(1934), 29-64.

KADISON, R. V.
[1] A representation theory for commutative topological algebras. Mem. Amer. Math. Soc. no. 7, Amer. Math. Soc., Providence, R. I., 1951. MR 13, 360 Zbl. 42, 348.
[2] Irreducible operator algebras. Proc. Nat. Acad. Sci. U.S.A. 43(1957), 273-276. MR 19, 47 Zbl. 78, 115.
[3] Lectures on operator algebras. Cargèze Lectures in theoretical physics, F. Lurgat, Ed. Gordon and Breach, New York, 1967. MR 37 #5014.

KADISON, R. V. and RINGROSE, J. R.
[1] Fundamentals of the theory of operator algebras. Vol. 1. Elementary

theory. Pure and Applied Mathematics, 100. Academic Press, Inc. New York-London, 1983. MR 85j: 46099 Zbl. 518.46046.

KAINUMA, D.
[1] A property that breaks down Choda's scholium on commutativity for C*-algebras. Math. Japon. 25(1980), 641-642. MR 82b: 46080. Zbl. 451.46041.

KAKUTANI, S. and MACKEY, G. W.
[1] Ring and lattice characterization of complex Hilbert space. Bull. Amer. Math. Soc. 52(1946), 727-733. MR 8, 31 Zbl. 60, 263.

KAPLANSKY, I.
[1] Normed algebras. Duke Math. J. 16(1949), 399-418. MR 11, 115 Zbl. 33, 187.
[2] Topological representation of algebras. II. Trans. Amer. Math. Soc. 68(1950), 62-75. MR 11, 317 Zbl. 35, 303.
[3] The structure of certain operator algebras. Trans. Amer. Math. Soc. 70(1951), 219-255. MR 13, 48 Zbl. 42, 349.
[4] Symmetry of Banach algebras. Proc. Amer. Math. Soc. 3(1952), 396-399. MR 14, 58 Zbl. 47, 110.
[5] Functional Analysis. Some aspects of analysis and probability. 2-34, Surveys in Applied Mathematics Vol. 4, J. Wiley, New York, 1958. MR 21 #286 Zbl. 87, 311.
[6] Algebraic and analytic aspects of operator algebras. Regional Conference Series in Math., No. 1. Amer. Math. Soc., Providence, R. I. 1970. MR 47 #845 Zbl. 217, 449.
[7] Rings of Operators, W. A. Benjamin, Inc. New York-Amsterdam, 1968. MR 39 #6092 Zbl. 174, 185.

KASPAROV, G. G.
[1] Hilbert C*-modules: theorems of Stinespring and Voiculescu. J. of Operator Theory 4(1980), 133-150. MR 82b: 46074 Zbl. 456.46059.

KATZNELSON, Y.
[1] Algèbres caractérisées par les fonctions qui opèrent sur elles. C. R. Acad. Sci. Paris 247(1958), 903-905. MR 20 #5436 Zbl. 82, 327.
[2] Sur les algèbres dont les éléments non négatifs admettent des racines carrées. Ann. Sci. École Norm. Sup. (3), 77(1960), 167-174. MR 22 #12403 Zbl. 99, 102.

KAUP, W. and UPMEIER, H.
[1] Jordan algebras and symmetric Siegel domains in Banach spaces. Math. Z. 157(1977), 179-200. MR 58 #11532 Zbl. 357.32018.

KELLEY, J. L. and VAUGHT, R. L.
[1] The positive cone in Banach algebras. Trans. Amer. Math. Soc. 74 (1953), 44-55. MR 14, 883 Zbl. 50, 110.

KITCHEN, J. W. and ROBBINS, D. A.
[1] Tensor products of Banach bundles. Pacific J. Math. 94(1981), 151-169. Zbl. 426.46056.
[2] Gelfand representation of Banach modules. Dissert. Math. 203(1982). MR 85g: 46060 Zbl. 544.46041.

[3] Sectional representations of Banach modules. Pacific J. Math. 109 (1983), 135-156. MR 85a: 46026 Zbl. 477.46044.

KÖNIG, H.
[1] A functional calculus for Hermitian elements of complex Banach algebras. Arch. Math. (Basel) 28(1977), 422-430. MR 57 #3856 Zbl. 364.46039.

KOVACS, I. and MOCANU, GH.
[1] Some remarks on square roots in a Banach algebra. Ann. Fac. Sci. Univ. Nat. Zäire (Kinshasa) Sect. Math.-Phys. 2(1976), 227-232. MR 55 #8795 Zbl. 334.46052.

KRAUSS, F. and LAWSON, T. C.
[1] Examples of homogeneous C*-algebras. Recent advances in the representation theory of rings and C*-algebras by continuous sections. Sem. Tulane Univ., New Orleans, La. 1973, 153-164. Mem. Amer. Math. Soc. no. 148, Amer. Math. Soc., Providence, R. I. 1974. MR 50 #8093 Zbl. 272.46042.

KREIN, M.
[1] On positive additive functionals in linear normed spaces. Comm. Inst. Sci. Math. Méc. Univ. Kharkoff [Zapiski Inst. Mat. Mech.] (4), 14 (1937), 227-237 (Russian with French summary). Zbl. 22, 232.

KRUSZYŃSKI, P. and WORONOWICZ, S. L.
[1] A non-commutative Gelfand-Naimark theorem. J. Operator Theory 8 (1982), 361-389. MR 84b: 46068 Zbl. 499.46036.

KUBO, F.
[1] On theorems of Phelps, Russo and Dye. Math. Japon. 20(1975), special issue, 69-71. MR 52 #15028 Zbl. 318.46072.

KUGLER, W.
[1] On the symmetry of generalized $L^1$-algebras. Math. Z. 168(1979), 241-262. MR 82f: 46074 Zbl. 394.43004.

KULKARNI, S. H. and LIMAYE, B. V.
[1] Gelfand-Naimark theorems for real Banach *-algebras. Math. Japon. 25(1980), 545-558. MR 82c: 46071 Zbl. 457.46043.

LANCE, E. C.
[1] Quadratic forms on Banach spaces. Proc. London Math. Soc. (3), 25(1972), 341-357. MR 46 #7856 Zbl. 241.46055.

LANDESMAN, E. M. and RUSSO, B.
[1] The second dual of a C*-ternary ring. Canad. Math. Bull. 26(2) (1983) 241-246. Zbl. 526.46056.

LARSEN, R.
[1] Banach algebras. Marcel Dekker, Inc., New York, 1973. MR 58 #7010 Zbl. 264.46042.

LAZAR, A. J. and TAYLOR, D. C.
[1] A Dauns-Hofmann theorem for $\Gamma(K)$. Recent advances in the representation theory of rings and C*-algebras by continuous sections.

Sem. Tulane Univ., New Orleans, La. 1973, 135-144. Mem. Amer. Math. Soc., Providence, R. I. 1974. MR 55 #11060 Zbl. 272.46040.

LEE, R. Y.

[1] On the C*-algebras of operator fields. Indiana Univ. Math. J. 25 (1976), 303-314. MR 53 #14150 Zbl. 306.46073.

[2] Full algebras of operator fields trivial except at one point. Indiana Univ. Math. J. 26(1977), 351-372. MR 55 #3812 Zbl. 326.46029.

LENARD, A.

[1] Function algebras without non-trivial positive linear forms. Notices Amer. Math. Soc. 23(1976), 224.

LEPTIN, H.

[1] Verallgemeinerte $L^1$-Algebren, Math. Ann. 159(1965), 51-76. MR 39 #1909 Zbl. 141, 317.

[2] Verallgemeinerte $L^1$-Algebren und projektive Darstellungen lokal Kompakter Gruppen I. Invent. Math. 3(1967), 257-281. MR 37 #5328 Zbl. 179, 183.

[3] Verallgemeinerte $L^1$-Algebren und projecktive Darstellungen lokal Kompakter Gruppen II. Invent. Math. 4(1967), 68-86. MR 37 #5328 Zbl. 179, 183.

[4] Darstellungen verallgemeinerter $L^1$-Algebren. Invent. Math. 5(1968), 192-215. MR 38 #5022 Zbl. 175, 446.

[5] On group algebras of nilpotent Lie groups. Studia Math. 47(1973), 37-49. MR 48 #9262 Zbl. 258.22009.

[6] On symmetry of some Banach algebras. Pacific J. Math. 53(1974), 203-206. MR 51 #6432 Zbl. 295.46085.

[7] Symmetrie in Banachschen Algebren. Arch. Math. (Basel) 27(1976), 394-400. MR 54 #5841 Zbl. 327.46060.

[8] Ideal theory in group algebras of locally compact groups. Invent. Math. 31(1976), 259-278. MR 53 #3189 Zbl. 328.22012.

[9] Lokal Kompakte Gruppen mit symmetrischen Algebren. Symposia Math. 22 (Convegno sull' Analisi Armonica e Spazi di Funzioni su Gruppi Localmente Compatti, INDAM, Rome, 1976), pp. 267-281, Academic Press, London, 1977. MR 58 #6058 Zbl. 394.43003.

LEPTIN, H. and POGUNTKE, D.

[1] Symmetry and non-symmetry for locally compact groups. J. Functional Anal. 33(1979), 119-134. MR 81e: 43010 Zbl. 414.43004.

LI, B.

[1] Real C*-algebras. Acta Math. Sinica. 18(1975), 216-218. (Chinese). MR 58 #30294 Zbl. 369.46056.

[2] Real operator algebras. Sci. Sinica. 22(1979), 733-746. MR 80i: 46050.

LIUKKONEN, J. R. and MISLOVE, M. W.

[1] Symmetry in Fourier-Stieltjes algebras. Math. Ann. 217(1975), 97-112. MR 54 #8163 Zbl. 295.43005.

LOOMIS, L. H.

[1] An introduction to abstract harmonic analysis. D. Van Nostrand, New York, 1953. MR 14, 883 Zbl. 52, 117.

LUDWIG, J.
[1] A class of symmetric and a class of Wiener group algebras. J. Functional Analysis 31(1979), 187-194. MR 81a: 43007 Zbl. 402.22003.

LUMER, G.
[1] Semi-inner-product spaces. Trans. Amer. Math. Soc. 100(1961), 29-43. MR 24 #A2860 Zbl. 102, 327.

MAGYAR, Z.
[1] A sharpening of the Berkson-Glickfeld theorem. Proc. Edinburgh Math. Soc. 26(1983), 275-278. MR 85c: 46057 Zbl. 539.46038.
[2] Conditions for hermiticity and for existence of an equivalent C*-norm. Acta Sci. Math. 46(1983), 305-310. Zbl. 535.46032.

MAGYAR, Z. and SEBESTYÉN, Z.
[1] On the definition of C*-algebras. II. Canad. J. Math. 37(1985), to appear.

MARTINEZ, J., MOJTAR, A., and RODRIGUEZ, A.
[1] On a nonassociative Vidav-Palmer theorem. Quart. J. Math. Oxford (2), 32(1981), 435-442. MR 83a: 46059 Zbl. 446.46043.

MAXWELL, G.
[1] Representation of algebras with involution. Canadian J. Math. 24(1972) 592-597. MR 45 #8688 Zbl. 215-484.

MAZUR, S.
[1] Sur les anneaux linéaires. C. R. Acad. Sci. Paris, 207(1938), 1025-1027. Zbl. 20, 201.

McCHAREN, E. A.
[1] A characterization of dual B*-algebras. Proc. Amer. Math. Soc. 37 (1973), 84. MR 46 #6048 Zbl. 252.46090.

MICHAEL, E. A.
[1] Locally multiplicatively-convex topological algebras. Mem. Amer. Math. Soc. no. 11, Amer. Math. Soc., Providence, R. I., 1952. MR 14, 482 Zbl. 47, 355.

MINGO, J. A.
[1] Jordan subalgebras of Banach algebras. J. London Math. Soc. (2) 21 (1980), 162-166. MR 81h: 46092 Zbl. 417.46053.

MISONOU, Y.
[1] On a weakly central operator algebra. Tôhoku Math. J. 4(1952), 194-202. MR 14, 566.

MOCANU, GH.
[1] A representation theorem for the space of Hermitian elements in a Banach algebra. An. Univ. Bucuresti Mat. 26(1977), 57-63. MR 57 #10430 Zbl. 389.46036.

MOORE, R. T.
[1] Banach algebras of operators on locally convex spaces. Bull. Amer. Math. Soc. 75(1969), 68-73. MR 38 #5018 Zbl. 189, 133.

[2] Adjoints, numerical ranges, and spectra of operators on locally convex spaces. Bull. Amer. Math. Soc. 75(1969), 85-90. MR 39 #805 Zbl. 189, 133.
[3] Generation of equicontinuous semigroups by hermitian and sectorial operators. I. Bull. Amer. Math. Soc. 77(1971), 224-229. MR 43 #3846a Zbl. 212, 161.
[4] Hermitian functionals on B-algebras and duality characterizations of C*-algebras. Trans. Amer. Math. Soc. 162(1971), 253-265. MR 44 #803 Zbl. 213, 140.

MOORE, S. M.
[1] Locally convex *-algebras. Revista Colombiana Mat. 10(1976), 99-120. MR 58 #17860 Zbl. 347.46061.

MOSAK, R. D.
[1] Banach algebras. The Univ. of Chicago Press. Chicago, 1975. MR 54 #3406 Zbl. 331.46040.

MULVEY, C. J.
[1] A non-commutative Gel'fand-Naimark theorem. Math. Proc. Camb. Philos. Soc. 88(1980), 425-428. MR 82m: 46062 Zbl. 507.46058.

MUNEO, C.
[1] A characterization of multiplicative linear functionals on complex *-algebras. Sci. Rep. Hirosaki Univ. 22(1975), 49-52. MR 54 #941.

NAIMARK, M. A.
[1] Normed algebras. Wolters-Noordhoff Publishing Co., Groningen, 1972. MR 55 #11042 Zbl. 255.46025.

NAKAMURA, M.
[1] A remark on a theorem of Gelfand and Neumark. Tôhoku Math. J., (2), 2(1950), 182-187. MR 12, 719 Zbl. 41, 236.

NAMSRAJ, N. and BATOR, U.
[1] On ideals and quotients of Hermitian algebras. Comment. Math. Univ. Carolinae 18(1977), 87-91. MR 55 #13245 Zbl. 342.46045.

NELSON, E.
[1] Topics in dynamics I: Flows. Math. Notes, Princeton University Press, Princeton, N. J., 1969. MR 43 #8091 Zbl. 197, 107.

OGASAWARA, T.
[1] A theorem on operator algebras. J. Sci. Hiroshima Univ. Ser. A 18 (1955), 307-309. MR 17, 514 Zbl. 64, 367.

ONO, T.
[1] Note on a B*-algebra. J. Math. Soc. Japan. 11(1959), 146-158. MR 22 #5905 Zbl. 97, 107.
[2] Note on a B*-algebra. II. Bull. Nagoya Inst. Tech. 21(1969), 93-95. (Japanese with English summary). MR 47 #2379.
[3] Note on a B*-algebra. III. Bull. Nagoya Inst. Tech. 22(1970), 119-122. (Japanese with English summary). MR 50 #2924.
[4] A real analogue of the Gelfand-Neumark theorem. Proc. Amer. Math. Soc. 25(1970), 159-160. MR 41 #2407 Zbl. 189, 445.

PAGE, W.
[1] Characterization of B*- and A*-algebras. Rev. Roum. Math. Pures Appl. 18(1973), 1241-1244. MR 48 #6949 Zbl. 273.46038.

PALMER, T. W.
[1] Characterizations of C*-algebras. Bull. Amer. Math. Soc. 74(1968), 538-540. MR 36 #5709 Zbl. 159, 185.
[2] Characterizations of C*-algebras. II. Trans. Amer. Math. Soc. 148 (1970), 577-588. MR 41 #7447 Zbl. 198, 180.
[3] Real C*-algebras. Pacific J. Math. 35(1970), 195-204. MR 42 #5055 Zbl. 208, 382.
[4] The Gelfand-Naimark pseudo-norm on Banach *-algebras. J. London Math. Soc. (2), 3(1971), 59-66. MR 43 #7932 Zbl. 207, 443.
[5] *-representations of U*-algebras. Indiana Univ. Math. J., 20(1971), 929-933. MR 53 #14146 Zbl. 198, 183.
[6] Hermitian Banach *-algebras. Bull. Amer. Math. Soc. 78(1972), 522-524. MR 45 #7481 Zbl. 255.46045.
[7] The reducing ideal is a radical. Pacific J. Math. 43(1972), 207-219. MR 47 #5607 Zbl. 248.46045.

PAYA, R., PERES, J., and RODRIGUEZ, A.
[1] Noncommutative Jordan C*-algebras. Manuscripta Math. 37(1982), 87-125. Zbl. 483.46049.
[2] Type I factor representations of non-commutative JB*-algebras. Proc. London Math. Soc. (3) 48(1984), 428-444.
[3] Type I representations of non-commutative JB*-algebras. Proc. London Math. Soc., III (to appear).

PEDERSEN, G. K.
[1] Applications of weak * semicontinuity in C*-algebra theory. Duke Math. J. 39(1972), 431-450. MR 47 #4012 Zbl. 244.46073.
[2] C*-algebras and their automorphism groups. Academic Press, London-New York 1979. MR 81e: 46037 Zbl. 416.46043.

POGUNTKE, D.
[1] Nichtsymmetrische sechsdimensionale Liesche Gruppen. J. Reine Angew. Math. 306(1979), 154-176. MR 80i: 43006 Zbl. 395.22011.

POPA, S.
[1] On the Russo-Dye theorem. Mich. Math. J. 28(1981), 311-315. MR 82j: 46074 Zbl. 499.46037.

POWER, S. C.
[1] Commutator ideals and pseudo-differential C*-algebras. Quart. J. Math. Oxford (2), 31(1980), 467-489. MR 82c: 47033 Zbl. 422.46050.

PROSSER, R. T.
[1] On the ideal structure of operator algebras. Mem. Amer. Math. Soc. no. 45, Amer. Math. Soc., Providence, R. I., 1963. MR 27 #1846 Zbl. 125, 67.

PTÁK, V.
[1] On the spectral radius in Banach algebras with involution. Bull. London Math. Soc. 2(1970), 327-334. MR 43 #932 Zbl. 209, 444.
[2] Banach algebras with involution. Manuscripta Math. 6(1972), 245-290. MR 45 #5764 Zbl. 229.46054.

[3] Hermitian algebras. Bull. Acad. Polon. Sci. Sér. Sci. Math. Astronom. Phys. 20(1972), 995-998. (Russian summary). MR 47 #7445 Zbl. 244.46071.

PUTTER, P. S. and YOOD, B.
[1] Banach Jordan *-algebras. Proc. London Math. Soc., III. Ser. 41(1980), 21-44. MR 81i: 46066 Zbl. 428.46041.

REID, G. A.
[1] A generalization of W*-algebras. Pacific J. Math. 15(1965), 1019-1026. MR 33 #4701 Zbl. 136, 113.

RICKART, C. E.
[1] General theory of Banach algebras. D. van Nostrand, Princeton, N. J. 1960 (reprinted with corrections. Robert E. Krieger, 1974). MR 22 #5903 Zbl. 95, 97.

RIEDEL, N.
[1] Metric completions of dimension groups and the Gelfand-Naimark-Segal construction. Quart. J. Math. Oxford Ser. (2) 33(1982), 365-377. MR 83j: 46071 Zbl. 504.46039.

ROBERTSON, A. G.
[1] A note on the unit ball in C*-algebras. Bull. London Math. Soc. 6 (1974), 333-335. MR 50 #8095 Zbl. 291.46042.
[2] On the density of the invertible group in C*-algebras. Proc. Edinburgh Math. Soc. (2) 20(1976), 153-157. MR 54 #5845.

RODRIGUEZ, A.
[1] A structure theorem for Jordan isomorphisms of C*-algebras. (Spanish). Revists mat: Hisp.-Amer., (4) 37(1977), 114-128. MR 58 #30296 Zbl. 374.46049.
[2] A Vidav-Palmer theorem for Jordan C*-algebras and related topics. J. London Math. Soc., II. Ser. 22(1980), 318-332. MR 81j: 46096 Zbl. 483.46050.
[3] Non-associative normed algebras spanned by hermitian elements. Proc. London Math. Soc. (3), 47(1983), 193-224. MR 85c: 46058
[4] The uniqueness of the complete norm topology in complete normed non-associative algebras. J. Functional Analysis 60(1985), 1-15.

RUDIN, W.
[1] Fourier analysis on groups. J. Wiley, New York 1962. MR 27 #2808 Zbl. 107, 96.
[2] Functional analysis. McGraw-Hill. New York 1973. MR 51 #1315 Zbl. 253.46001.
[3] Real and complex analysis. Second edition, McGraw-Hill, New York, 1974. MR 49 #8783 Zbl. 333.28001.

RUSSO, B. and DYE, H. A.
[1] A note on unitary operators in C*-algebras. Duke Math. J. 33(1966), 413-416. MR 33 #1750 Zbl. 171, 115.

SAKAI, S.
[1] C*-algebras and W*-algebras. Springer-Verlag. Berlin-Heidelberg-New York, 1971. MR 56 #1082 Zbl. 219, 292.

SAWÓN, Z.

[1] Some remarks about the three spaces problem in Banach algebras. Czech. Math. J. 28(103) (1978), 56-58. MR 57 #7174 Zbl. 418.46037.

SCHATZ, J. A.

[1] Representation of Banach algebras with an involution. Canadian J. Math. 9(1957), 435-442. MR 19, 870 Zbl. 78, 291.

[2] Review of Fukamiya [2], Math. Rev. 14(1953), 884.

SCHMÜDGEN, K.

[1] Über LMC*-Algebren. Math. Nachr. 68(1975), 167-182. MR 52 #15030 Zbl. 315.46042.

SEBESTYÉN, Z.

[1] A weakening of the definition of C*-algebras. Acta Sci. Math. (Szeged), 35(1973), 17-20. MR 48 #12070 Zbl. 272.46034.

[2] On the definition of C*-algebras. Publ. Math. Debrecen 21(1974), 207-217. MR 51 #11118 Zbl. 303.46053.

[3] Remarks on the paper of H. Araki and G. A. Elliott "On the definition of C*-algebras." Ann. Univ. Sci. Budapest Eötvös Sect. Math. 17(1974), 35-39. MR 52 #3975 Zbl. 328.46067.

[4] Some local characterization of boundedness and of C*-equivalent algebras. Ann. Univ. Budapest Eötvös. Sect. Math. 18(1975), 197-207. MR 54 #11070 Zbl. 348.46037.

[5] On a problem of Araki and Elliott. Ann. Univ. Sci. Budapest. Eötvös Sect. Math. 18(1975), 209-221. MR 54 #13578 Zbl. 355.46042.

[6] On extendibility of *-representation from *-ideals. Acta Sci. Math. (Szeged) 40(1978), 169-174. MR 58 #2314 Zbl. 423.46047.

[7] Every C*-seminorm is automatically submultiplicative. Period. Math. Hungar. 10(1979), 1-8. MR 80c: 46065 Zbl. 413.46046.

SEGAL, I. E.

[1] Postulates for general quantum mechanics. Ann. of Math. (2), 48 (1947), 930-948. MR 9, 241 Zbl. 34, 66.

[2] Irreducible representations of operator algebras. Bull. Amer. Math. Soc. 53(1947), 73-88. MR 8, 520 Zbl. 31, 360.

SHERMAN, S.

[1] Order in operator algebras. Amer. J. Math. 73(1951), 227-232. MR 13, 47 Zbl. 42, 350.

SHIRALI, S.

[1] Symmetry in complex involutory Banach algebras. Duke Math. J. 34 (1967), 741-745. MR 36 #1988 Zbl. 183, 142.

[2] Symmetry of Banach *-algebras without identity. J. London Math. Soc. (2) 3(1971), 143-144. MR 43 #2516 Zbl. 236.46063.

SHIRALI, S. and FORD, J. W. M.

[1] Symmetry in complex involutory Banach algebras. II. Duke Math. J. 37(1970), 275-280. MR 41 #5977 Zbl. 183, 142.

SHULTZ, F. W.

[1] On normed Jordan algebras which are Banach dual spaces. J. Functional Analysis 31(1979), 360-376. MR 80h: 46096 Zbl. 421.46043.

SIMMONS, G.

[1] Introduction to topology and modern analysis. McGraw-Hill, New York, 1963. MR 26 #4145 Zbl. 105, 306.

SIMS, B.

[1] A characterization of Banach-star-algebras by numerical range. Bull. Austral. Math. Soc. 4(1971), 193-200. MR 43 #2518 Zbl. 207, 443.

SINCLAIR, A. M.

[1] The states of a Banach algebra generate the dual. Proc. Edinburgh Math. Soc. (2) 17(1971), 341-344. MR 47 #828 Zbl. 233.46062.

SPAIN, P. G.

[1] On commutative V*-algebras. Proc. Edinburgh Math. Soc. (2), 17 (1970/71), 173-180. MR 44 #2045 Zbl. 215, 206.

[2] V*-algebras with weakly compact unit spheres. J. London Math. Soc. (2) 4(1971), 62-64. MR 45 #914 Zbl. 217, 449.

[3] On commutative V*-algebras. II. Glasgow Math. J., 13(1972), 129-134. MR 47 #7461 Zbl. 245.46099.

STACY, P. J.

[1] Type $I_2$ JBW-algebras. Quart. J. Math. Oxford Ser. (2) 33(1982), 115-127. MR 84b: 46057.

STEEN, L. A.

[1] Highlights in the history of spectral theory. Amer. Math. Monthly 80(1973), 359-381. MR 47 #5643 Zbl. 264.46001.

STERBOVÁ, D.

[1] On the spectral radius in locally multiplicatively convex topological algebras. Sb. Praci Prirodoved. Fak. Univ. Palackeho v Olomouci Mat. 16(1977), 155-160. MR 58 #30287 Zbl. 417.46052.

[2] On the spectral properties of elements of the type exp(ih) in hermitian Banach algebras. Sb. Praci Prirodoved. Fak. Univ. Palackeho v Olomouci Mat. 17(1978), 109-114. MR 81a: 46069 Zbl. 433.46041.

[3] On Banach *-algebras without unit. Sb. Praci Prirodoved. Fak. Univ. Palackeho v Olomouci Mat. 18(1979), 21-28. MR 81j: 46088. Zbl. 447.46048.

STEWART, J.

[1] Positive definite functions and generalizations, an historical survey. Rocky Mountain J. Math. 6(1976), 409-434. MR 55 #3679 Zbl. 337.42017.

STINESPRING, W. F.

[1] Positive functions on a C*-algebra. Proc. Amer. Math. Soc. 6(1955), 211-216. MR 16, 1033 Zbl. 64, 367.

STONE, M. H.

[1] Applications of the theory of Boolean rings to general topology. Trans. Amer. Math. Soc. 41(1937), 375-481. Zbl. 17, 135.

STØRMER, E.

[1] On the Jordan structure of C*-algebras. Trans. Amer. Math. Soc. 120 (1965), 438-447. MR 32 #2930 Zbl. 136, 114.

[2] Jordan algebras of type I. Acta Math. 115(1966), 165-184. MR 35 #754.

[3] Irreducible Jordan algebras of self-adjoint operators. Trans. Amer. Math. Soc. 130(1968), 153-166. MR 36 #700.

STRATILĂ, S. and ZSIDÓ, L.
[1] Lectures on von Neumann algebras. Abacus Press, Turnbridge Wells, Kent, England 1979.

SUZUKI, N.
[1] Representation of certain Banach *-algebras. Proc. Japan Acad. 45 (1969), 696-699. MR 41 #9000 Zbl. 197, 397.
[2] Every C-symmetric Banach *-algebra is symmetric. Proc. Japan Acad. 46(1970), 98-102. MR 43 #2519 Zbl. 205, 137.

SZAFRANIEC, F. H.
[1] Subnormals in C*-algebras. Proc. Amer. Math. Soc. 84(1982), 533-534. MR 83a: 46068 Zbl. 506.47019.

TAKAHASI, S.
[1] A simple proof of the Stone-Weierstrass theorem for CCR-algebras with Hausdorff spectrum. Math. Scand. 38(1976), 304-306. MR 54 #13581 Zbl. 334.46066.
[2] Remarks on Stone-Weierstrass problem for C*-algebras. Bull. Fac. Sci. Ibaraki Univ. Ser. A. No. 9(1977), 61-63. MR 56 #6406 Zbl. 355.46043.
[3] Continuous linear functional on closed two-sided ideals of C*-algebras. Bull. Fac. Sci. Ibaraki Univ. Ser. A 12(1980), 1-3. MR 81h: 46078 Zbl. 434.46037.

TAKEMOTO, H.
[1] On a characterization of AW*-modules and a representation of Gelfand type of non-commutative operator algebras. Michigan J. Math. 20 (1973), 115-127. MR 47 #7457 Zbl. 277.46059.

TAKESAKI, M.
[1] Theory of operator algebras I. Springer-Verlag. Berlin-Heidelberg-New York, 1979. MR 81e: 46038 Zbl. 436.46043.

TOMITA, M.
[1] Representations of operator algebras. Math. J. Okayama Univ. 3(1954), 147-173. MR 15, 968 Zbl. 55, 105.

TOMIYAMA, J.
[1] Topological representation of C*-algebras. Tôhoku Math. J. (2), 14 (1962), 187-204. MR 26 #619.

TOMIYAMA, J. and TAKESAKI, M.
[1] Applications of fibre bundles to the certain class of C*-algebras. Tôhoku Math. J. (2), 13(1961), 498-522. MR 25 #2465 Zbl. 113, 97.

TOPPING, D. M.
[1] Jordan Algebras of self-adjoint operators. Mem. Amer. Math. Soc. no. 53, Amer. Math. Soc., Providence, R. I. 1965. MR 32 #8198 Zbl. 137, 102.
[2] An isomorphism invariant for spin factors. J. Math. Mech. 15(1966), 1055-1064. MR 33 #6430 Zbl. 154, 388.

TORRANCE, E.
[1] Maximal C*-subalgebras of a Banach algebra. Proc. Amer. Math. Soc. 25(1970), 622-624. MR 41 #4265 Zbl. 198, 180.

VARELA, J.
[1] Duality of C*-algebras. Recent advances in the representation theory of rings and C*-algebras by continuous sections. Sem. Tulane Univ., New Orleans, La. 1973, 97-108. Mem. Amer. Math. Soc. no. 148, Amer. Math. Soc., Providence, R. I. 1974. MR 50 #5490 Zbl. 272.46038.
[2] Sectional representation of Banach modules. Math. Z. 139(1974), 55-61. MR 50 #5473 Zbl. 275.46041.

VIDAV, I.
[1] Eine metrische Kennzeichnung der selbstadjungierten Operatoren. Math. Z. 66(1956), 121-128. MR 18, 912 Zbl. 71, 115.
[2] Sur un système d'axiomes caractérisant les algèbres C*. Glasnik Mat.-Fiz. Astronom. Drustvo Mat. Fiz. Hrvatski Ser. II 16(1961), 189-193. (Serbo-Croatian summary). MR 25 #3386 Zbl. 108, 116.

VOWDEN, B. J.
[1] On the Gelfand-Neumark theorem. J. London Math. Soc. 42(1967), 725-731. MR 36 #702 Zbl. 152, 329.

VUKMAN, J.
[1] Real symmetric Banach *-algebras. Glas. Mat. Ser. III 16(36)(1981), 91-103. MR 83a: 46060 Zbl. 478.46048.
[2] Involution on L(X). Glas. Mat. Ser. III 17(37)(1982), 65-72. MR 84f: 46028 Zbl. 501.46022.
[3] A characterization of real and complex Hilbert space. Glas. Mat. Ser. III 18(38)(1983), 103-106. Zbl. 517.46017.
[4] A characterization of real and complex Hilbert spaces among all normed spaces. Bull. Austral. Math. Soc. 27(1983), 339-345. MR 84i: 46033 Zbl. 542.46011.

WENJEN, C.
[1] On semi-normed *-algebras. Pacific J. Math. 8(1958), 177-186. MR 20 #2626 Zbl. 80, 327.
[2] A remark on a problem of M. A. Naimark. Proc. Japan Acad. 44(1968), 651-655. MR 38 #5011 Zbl. 172, 177.

WICHMANN, J.
[1] Hermitian *-algebras which are not symmetric. J. London Math. Soc. (2), 8(1974), 109-112. MR 50 #8088 Zbl. 287.46070.
[2] On commutative B*-equivalent algebras, Notices Amer. Math. Soc. 22 (1975), A-178.
[3] On the symmetry of matrix algebras. Proc. Amer. Math. Soc. 54(1976), 237-240. MR 52 #8947 Zbl. 324.46062.
[4] The symmetric radical of an algebra with involution. Arch. Math. (Basel), 30(1978), 83-88. MR 58 #2313 Zbl. 399.16007.

WILLIAMS, J. P.
[1] On commutativity and the numerical range in Banach algebras. J. Functional Analysis 10(1972), 326-329. MR 50 #14229 Zbl. 248.46041.

WILLIAMSON, J. H.
[1] A theorem on algebras of measures on topological groups. Proc. Edinburgh Math. Soc. 11(1958-1959), 195-206. MR 22 #2851 Zbl. 101, 94.
[2] Algebras in analysis. J. H. Williamson, Editor, Conference Proceedings, Academic Press, New York 1975. MR 52 #14888.

WOOD, A.
[1] Numerical range and generalized B*-algebras. Proc. London Math. Soc. (3) 34(1977), 245-268. MR 55 #11044 Zbl. 342.46037.

WRIGHT, J. D. MAITLAND.
[1] Jordan C*-algebras. Michigan Math. J. 24(1977), 291-302. MR 58 #7108 Zbl. 384.46040.

WRIGHT, J. D. MAITLAND and YOUNGSON, M. A.
[1] A Russo-Dye theorem of Jordan C*-algebras. Functional Analysis: Survey and recent results, Proc. Conf., Paderborn 1976, (1977), 279-282. MR 58 #7102 Zbl. 372.46060.
[2] On isometries of Jordan algebras. J. London Math. Soc. (2), 17(1978), 339-344. MR 58 #2294 Zbl. 384.46041.

WRIGHT, S.
[1] On factor states. Rocky Mountain J. Math. 12(1982), 569-579. MR 83j: 46075 Zbl. 534.46043.

YLINEN, K.
[1] Vector space isomorphisms of C*-algebras. Studia Math. 46(1973), 31-34. MR 50 #14255 Zbl. 251.46064.
[2] A note on compact elements in C*-algebras. Proc. Amer. Math. Soc. 35 (1972), 305-306. MR 45 #5775 Zbl. 257.46085.

YOOD, B.
[1] Seminar on Banach algebras. University of California Notes, Berkeley, 1956-1957.
[2] Faithful *-representations of normed algebras. Pacific J. Math. 10 (1960), 345-363. MR 22 #1826 Zbl. 94, 96.
[3] Faithful *-representations of normed algebras. II. Pacific J. Math. 14(1964), 1475-1487. MR 30 #2362 Zbl. 205, 426.
[4] On axioms for B*-algebras. Bull. Amer. Math. Soc. 76(1970), 80-82. MR 40 #6273 Zbl. 188, 446.

YOUNGSON, M. A.
[1] A Vidav theorem for Banach Jordan algebras. Math. Proc. Cambridge Philos. Soc. 84(1978), 263-272. MR 58 #12397 Zbl. 392.46038.
[2] Equivalent norms on Banach Jordan algebras. Math. Proc. Cambridge Philos. Soc. 86(1979), 261-269. MR 80i: 46049 Zbl. 409.46056.
[3] Hermitian operators on Banach Jordan algebras. Proc. Edinburgh Math. Soc. (2), 22(1979), 169-180. MR 80j: 46090 Zbl. 414.46034.
[4] Nonunital Banach Jordan algebras and C*-triple systems. Proc. Edinburgh Math. Soc. (2) 24(1981), 19-29. MR 82j: 46071 Zbl. 451.46033.

ŻELAZKO, W.
[1] Banach algebras. Elsevier, Amsterdam, 1973. MR 56 #6389 Zbl. 211, 437.

ZETTL, H. H.
[1] A characterization of ternary rings of operators. Advances in Math. 48(1983), 117-143. MR 84h: 46093 Zbl. 517.46049.

# Selected Hints and References

(I.1) For nonzero $x \in A$, $||x||^2 = ||x^*x|| \leq ||x^*||\cdot||x||$; divide by $||x||$ to obtain $||x|| \leq ||x^*||$ and then replace $x$ by $x^*$ to obtain $||x^*|| \leq ||x||$.

(I.2) Same argument as (I.1).

(I.3) (a) Since $||xy|| \leq ||x||\cdot||y||$ it follows that $\frac{||xy||}{||y||} \leq ||x||$ for all $x$ and nonzero $y$. Since $||x^*x|| = ||x||^2$ we obtain the result by taking $y = x^*$.

(b) Every element of the form $x^*x$ is hermitian; since $e^*e = e^*$ is hermitian, $e^* = e$ [also, see (21.1)].

(c) See (24.1), proof of part (a).

(I.4) Cf. K. R. Goodearl [1, p. 5].

(I.5) Cf. K. R. Goodearl [1, p. 6].

(I.6) Compute.

(II.1) Cf. R. S. Doran and J. Wichmann [2, p. 67].

(II.2) Cf. S. Sakai [1, p. 6].

(II.3) Cf. I. Kaplansky [Ann. of Math. 53(1951), p. 236].

(II.4) Cf. R. Larsen [1, p. 86].

(II.5) Cf. Exercise (II.4).

(II.6) Cf. O. Bratteli and D. W. Robinson [1, p. 63].

(II.7) Cf. R. Larsen [1, pp. 77-78].

(II.8) Utilize the Arens' argument in the proof of (7.1); see (22.17). For a novel proof, see R. Phelps [Proc. Amer. Math. Soc. 16(1965), 381-382]. A generalization to noncommutative $C^*$-algebras is given in O. Bratteli and D. W. Robinson [1, p. 211].

(II.9) Use theorem (7.1).

(II.10) Cf. W. Żelazko [1, p. 130].

(II.11) Cf. H. W. Ellis and D. G. Kuchner [Canad. Math. Bull. 3(1960), 173-184].

(III.1) Cf. O. Bratteli and D. W. Robinson [1, p. 29 and p. 37].

(III.2) Cf. (12.9), (b).

(III.3) Cf. (12.8).

(III.4) Use (12.5).

(III.5) Functional calculus.

(III.6) Use an approximate identity in the ideal I.

(III.7) Let

$$x = \begin{pmatrix} 2 & -1 \\ -1 & 1 \end{pmatrix} \text{ and } y = \begin{pmatrix} 1 & 0 \\ 0 & 0 \end{pmatrix}.$$

(III.8) For a related result, see (28.4).

(III.9) Consider the compact subsets of X ordered by inclusion. For each compact $K \subseteq X$, choose a function $f_K \in C_o(X)$, equal to 1 on K and with norm 1.

(III.10) Cf. R. S. Doran and J. Wichmann [2, p. 230 and p. 275].

(III.11) Set $p = (x^*x)^{1/2}$ and $u = xp^{-1}$. The continuity of $x \to u$ and $x \to p$ follow from the fact that the square root is continuous (Cf. O. Bratteli and D. W. Robinson [1, p. 34]).

(III.12) Cf. W. Żelazko [1, p. 132] or use (19.1).

(III.13) Note that $\phi$ is positive iff $||\phi|| = 1 = \phi(e)$; i.e., (22.18). For a complete proof see S. K. Berberian [2, p. 260].

(III.14) See §27.

(III.15) Cf. G. K. Pedersen [2, p. 15].

(IV.7) For any element $f_1(b) \in B_1$, define $g: B_1 \to B_2$ by $g(f_1(b)) = f_2(b)$.

(IV.9) Utilize the map $f(b) = b + I$, where $b \in B$.

(IV.10) Part (b): Let $\phi_{ij}(a)$ denote the matrix in $M_n(A)$ having $a$ as its ij-th entry and zeros elsewhere. For any ideal $J$ in $M_n(A)$, let $I$ be the set of elements in $A$ which appear as entries in the matrices in $J$. Given any $a \in I$, say $a$ is the rs-th entry of a matrix $N$ in $J$, it follows that $\phi_{ij}(a) = \phi_{ir}(1)N\phi_{sj}(1) \in J$.

(IV.14) Cf. P. Civin and B. Yood [1].

(IV.15) Cf. I. Kaplansky [Ann. of Math. 53(1951), p. 248].

(IV.16) Cf. B. A. Barnes [Canad. J. Math. 21(1969), p. 88].

(IV.17) Cf. B. A. Barnes [Canad. J. Math. 21(1969), p. 88-89].

(IV.18) Cf. R. Haas [Pi Mu Epsilon J. 5(1971), 195-198].

(IV.19) Cf. R. Haas [Pi Mu Epsilon J. 5(1971), 195-198].

(IV.21) Cf. S. Shirali [Pacific J. Math. 27(1968), p. 402].

(IV.23) Look at the $n \times n$ matrices over the example constructed in §21.

(IV.25) Let $B$ be a Banach algebra without identity and with a bounded two-sided approximate identity. Let $A = B \times B$ be the *-Banach algebra constructed in (22.1). Then $A$ is an algebra, without identity, which admits no nonzero positive functionals.

(IV.28) Cf. R. S. Doran and W. Tiller [2].

(IV.29) Cf. R. S. Doran and W. Tiller [2].

(IV.30) Cf. J. Duncan [1].

(IV.31) Cf. P. Civin and B. Yood [1, pp. 420-421].

(IV.32) Cf. D. W. Bailey [1].

(IV.34) Let $\phi \in \hat{B}$ and note that $|\phi(f(xy))|^2 \leq \phi(f(xx^*))\phi(f(y^*y))$. The result then follows easily from $|\phi(f(xy)|^2 = \phi(|f(xy)|^2)$.

(IV.37) Cf. S. K. Berberian [2, Cayley transform].

(IV.38) This follows from (22.16).

(IV.40) Let $z$ be $(x - y)^*$.

(IV.41) Cf. C. E. Rickart [Ann. of Math. 47(1946), p. 536].

(IV.42) Let $I$ be a left ideal. If $x \in I \setminus \{0\}$, then $x^*x \in I$ and $||(x^*x)(x^*x)|| = ||(x^*x)^*(x^*x)|| = ||x^*x||^2 = ||x||^4 > 0$.

(IV.43) Cf. J. A. Erdos [Illinois J. Math. 15(1971), 682-693].

(IV.44) Compute.

(IV.45) Cf. W. Żelazko [1, p. 15].

(IV.46) The construction in (13.1) shows that an approximate identity in $A$ exists which extends one in $B$. Now use (22.17) and Hahn-Banach. For details see G. K. Pedersen [2, p. 43].

(IV.47) Use (24.11) to reduce to the commutative case. Then use functional calculus and $x$ to construct $y$.

(V.1) Cf. R. D. Mosak [1, p. 74].

(V.3) Cf. (IV.25) and its hint.

(V.4) Apply (B.5.26) to closure of $\pi(A)$ in $B(H)$, noting that $\pi(A)' = C \cdot I \supseteq \pi(A)$ or cf. R. D. Mosak [1, p. 70].

(V.5) Cf. R. D. Mosak [1, p. 75].

(V.6) Cf. Theorem 2, p. 365 of R. S. Bucy and G. Maltese, Math. Ann. 162 (1966), 364-367.

(V.7) Cf. J. Dieudonné, *Treatise on analysis*, Vol. II, Academic Press, New York, 1970, p. 355.

(V.8) Cf. O. Bratteli and D. W. Robinson [1, p. 44].

(V.9) Use continuity of $\pi$ (26.13) and first consider $\xi$ in the dense (V.1) set $\pi(A)H$ or cf. R. S. Doran and W. Tiller [1].

(V.10) Cf. J. Dixmier [5, p. 38].

(V.11) Cf. J. Dixmier [5, p. 40].

(V.12) Cf. J. Dixmier [5, pp. 47-48].

(V.14) Cf. R. S. Doran and J. Wichmann [2, p. 72].

(V.16) Use (22.18) and Hahn-Banach.

(V.18) Cf. J. Williamson [2, p. 292].

(V.21) Cf. J. Gil de Lamadrid [1].

(VI.1) It is enough to show that every MFL on $A$ is hermitian. Let $\phi \in \hat{A}$. Then $\phi(I) \equiv 0$ and so the map $\psi: A/I \to C$ defined by $\psi(x + I) = \phi(x)$ is a well-defined MLF on $A/I$. Since $A/I$ is symmetric, $\phi(x^*) = \psi(x^* + I) = \overline{\psi(x + I)} = \overline{\phi(x)}$ for all $x \in A$.

(VI.2) Cf. P. Civin and B. Yood [1, p. 423].

(VI.3) Cf. P. Civin and B. Yood [1, p. 420].

(VI.4) Cf. R. V. Kadison [1, p. 27].

(VI.5) Assume, to the contrary, that $e$ belongs to the closure of $I + I^*$. Choose $\{x_n\}$ and $\{y_n\}$ in $I$ such that $x_n + y_n^* \to e$. Then $z_n = \frac{1}{2}(x_n + y_n) \in I$ and $z_n + z_n^* \to e$ (since $e^* = e$). Pick index $m$ such that $||e - (z_m + z_m^*)|| < 1$, and write $z_m = h_1 + ih_2$, $h_1, h_2 \in H(A)$. Set $h = 2h_1$ and choose $k \in H(A)$ such that $k^{-1}$ exists, $kh = hk$, and $h = k^2$ (Ford's lemma). Let $q = 2h_2$. Since $k, q \in H(A)$, then $k^{-1}qk^{-1} \in H(A)$. Hence, the element $ie - k^{-1}qk^{-1}$ is invertible, and so $[k(e + ik^{-1}qk^{-1})k]^{-1}$ exists. However, $[k(e + ik^{-1}qk^{-1})k] = k^2 + iq = 2h_1 + i(2h_2) = 2(h_1 + ih_2) = 2z_m$. Therefore, $z_m^{-1}$ exists and $e = z_m^{-1}z_m \in I$; this contradicts the assumption that $I$ is proper. □

(VI.7) Note that $||x|| = |x|_\sigma \leq |x^*x|_\sigma^{1/2} = 0$ using (33.1), (a), or cf. I. Kaplansky [1, p. 430].

(VI.10) In (22.1) take $0 \neq \phi \in B^*$ and set $f(x,y) = \phi(x) - \phi(y)$.

(VI.11)- (VI.14) Cf. D. Sterbová [2, pp. 112-113]; for (VI.14) and (VI.13) see (33.3).

(VI.15) Cf. MR 39 #4685.

(VII.1) For an elementary proof of (c) see Bernau and Smithies [Proc. Camb. Philos. Soc. 59(1963), 727-729]; for part (h) let $H = C^2$, $T = \begin{pmatrix} 1 & 1 \\ 0 & 0 \end{pmatrix}$. Also cf. P. Halmos [1, problems 172-176].

(VII.2) Let $\eta = T\xi$, where $\xi \in R(T)^\perp$. Then for any positive number $t$, $T(\xi + t\eta) = (1 + t)\eta$, so that

$$t(1 + t)||\eta||^2 = ((1 + t)\eta|\xi + t\eta) = (T(\xi + t\eta)|\xi + t\eta)$$

$$\leq ||\xi + t\eta||^2 = ||\xi||^2 + t^2||\eta||^2.$$

Hence, $t||\eta||^2 \leq ||\xi||^2$, so that $\eta = 0$. Thus $T = 0$ on $R(T)^{\perp}$ and $T = I$ on $R(T)$, so $T$ is a projection. □

(VII.3) Since $T\xi = \xi$, we have $(\xi|T^*\xi) = (T\xi|\xi) = ||\xi||^2$. Hence,

$$\begin{aligned} ||\xi - T^*\xi||^2 &= ||\xi||^2 - (\xi|T^*\xi) - (T^*\xi|\xi) + ||T^*\xi||^2 \\ &= ||\xi||^2 - 2\mathrm{Re}(\xi|T^*\xi) + ||T^*\xi||^2 \\ &= ||\xi||^2 - 2||\xi||^2 + ||T^*\xi||^2 \\ &= ||T^*\xi||^2 - ||\xi||^2 \leq 0. \end{aligned}$$

So, $\xi = T^*\xi$. □

(VII.4) Cf. R. V. Kadison [Pacific J. Math. 26(1968), p. 168]. The result follows from

$$\begin{aligned} ||(T^* - S^*)\xi||^2 &= ||T\xi||^2 - ||S\xi||^2 + (\xi|(S - T)S^*\xi) \\ &\quad + ((S - T)S^*\xi|\xi) \\ &\leq ||(S - T)\xi||(||S\xi|| + ||T\xi||) \\ &\quad + 2||(S - T)S^*\xi||\cdot||\xi||. \end{aligned}$$

□

(VII.5) For normal $x \in A$, $||x^*||\cdot||x|| = ||x^*x|| = |x^*x|_\sigma \leq |x^*|_\sigma|x|_\sigma = |x|_\sigma^2 \leq ||x||^2$. Thus $||x^*|| \leq ||x||$ and then $||x|| \leq ||x^*||$. □

(VII.6) Compute.

(VII.7) - (VII.10) Cf. H. Araki and G. A. Elliott [1].

(VIII.4) $\sigma_A(x) = \{0,1\}$ and $V(x) = \mathrm{conv}\{0,1,\frac{1}{2}(1 + i), \frac{1}{2}(1 - i)\}$.

(VIII.5) Cf. F. F. Bonsall and J. Duncan [3, pp. 70-71].

(VIII.6) Cf. F. F. Bonsall and J. Duncan [3, p. 76].

(X.6) First observe that $|e^{it} - 1| \leq |t|$ holds for all real $t$. Let $\{E_\lambda\colon \lambda \in R\}$ be the spectral measure for the operator $S$; then

$$\begin{aligned} ||(e^{itS} - I)\xi||^2 &= (\{e^{itS} - I\}^*\{e^{itS} - I\}\xi|\xi) \\ &= \int_{-\infty}^{\infty}\{e^{-it\lambda} - 1\}\{e^{it\lambda} - 1\}d(E_\lambda\xi|\xi) \\ &\leq \int_{-\infty}^{\infty} t^2\lambda^2 d(E_\lambda\xi|\xi) = t^2(S^2\xi|\xi) = t^2||S\xi||^2. \end{aligned}$$

Taking positive square roots of this inequality completes the proof. □

(X.8) Suppose the double integral inequality holds. Let $s_1,\ldots,s_n$ be a finite number of elements in $G$ and $\lambda_1,\ldots,\lambda_n$ be complex numbers.

Then $\mu = \Sigma_{i=1}^{n} \lambda_i \delta_{s_i}$, where $\delta_s$ is the Dirac measure at $s$, is a bounded complex measure on $G$. Hence

$$\sum_{i,j=1}^{n} \phi(s_i^{-1} s_j)\overline{\lambda}_i \lambda_j = \int_G \int_G \phi(s^{-1}t)\overline{d\mu(s)}d\mu(t) \geq 0.$$

Conversely, by (57.2), $\phi(s) = (U_s\xi|\xi)$ for a continuous unitary representation $U$ on a Hilbert space $H$. Since $\mu$ is a bounded complex measure on $G$, we can define a bounded linear operator $U_\mu$ on $H$ so that

$$U_\mu = \int_G U_s d\mu(s).$$

Then $\int_G \int_G \phi(s^{-1}t)\overline{d\mu(s)}d\mu(t) = \int_G \int_G (U_t\xi|U_s\xi)\overline{d\mu(s)}d\mu(t) = (U_\mu\xi|U_\mu\xi) = ||U_\mu\xi||^2 \geq 0$. □

(X.9) (b): By (57.2), $\phi_i(s) = (U_s^i\xi_i|\xi_i)$ for a unitary representation $U^i$ $(i = 1, 2)$. Let $U = U^1 \otimes U^2$ and $\xi = \xi_1 \otimes \xi_2$. Then

$$\begin{aligned}\phi(s) = \phi_1(s)\phi_2(s) &= (U_s^1\xi_1|\xi_1)(U_s^2\xi_2|\xi_2)\\ &= ((U^1 \otimes U^2)_s(\xi_1 \otimes \xi_2)|\xi_1 \otimes \xi_2)\\ &= (U_s\xi|\xi)\end{aligned}$$

is positive definite. □

(X.10) Let $A$ be the $C^*$-algebra of $2 \times 2$ complex matrices and $B(H)$ the complex numbers. Consider the positive linear map $\phi: A \to B(H)$ defined by $\phi(x) = \text{trace}(x)$. Then $||\phi|| = \phi(e) = 2$, where $e$ is the identity matrix. Setting $x = e$ in (58.6) gives $\phi(x^2) = \phi(e) = 2$, while $\phi(x)^2 = 4$. □

(X.11) Let $H$ be the underlying Hilbert space of $B$ and let $x$ be a fixed normal operator in $A$. Then $C^*(x)$, the $C^*$-algebra generated by $x$, is commutative. By Stinespring's theorem (58.3), $\phi$ restricted to $C^*(x)$ admits a decomposition $\phi(y) = V^*\pi(y)V$ for all $y \in C^*(x)$, where $\pi$ is a $*$-representation of $C^*(x)$ on a Hilbert space $K$, and $V$ is an isometry from $H$ into $K$. Hence

$$\phi(x^*)\phi(x) = V^*\pi(x^*)VV^*\pi(x)V \leq V^*\pi(x^*)\pi(x)V = V^*\pi(x^*x)V = \phi(x^*x).$$

Replace $x$ by $x^*$ to obtain the inequality $\phi(x)\phi(x^*) \leq \phi(xx^*) = \phi(x^*x)$. □

(X.13) - (X.16) Cf. W. L. Paschke [Proc. Amer. Math. Soc. 34(1972), 412-416].

(X.17) The subspace $\pi_f(A)\xi_f$ is dense in $H_f$, where $\xi_f$ is a unit cyclic vector in $H_f$. Choose a net $\{y_\alpha\}$ in $A$ such that $||\pi_f(y_\alpha)\xi_f|| = 1$ and $||\pi_f(y_\alpha)\xi_f - \xi|| \to 0$. Setting $\xi_\alpha = \pi_f(y_\alpha)\xi_f$ and using $f(x) = (\pi_f(x)\xi_f|\xi_f)$, gives $f(y_\alpha^*xy_\alpha) = (\pi_f(y_\alpha)^*\pi_f(x)\pi_f(y_\alpha)\xi_f|\xi_f) = (\pi_f(x)\xi_\alpha|\xi_\alpha)$. Then

$$|f(y_\alpha^*xy_\alpha) - p(x)| = |(\pi_f(x)\xi_\alpha|\xi_\alpha) - (\pi_f(x)\xi|\xi)|$$
$$\leq |(\pi_f(x)\xi_\alpha|\xi_\alpha - \xi)| + |(\pi_f(x)(\xi_\alpha - \xi)|\xi)|$$
$$\leq ||x||\cdot||\xi_\alpha||\cdot||\xi_\alpha - \xi|| + ||x||\cdot||\xi_\alpha - \xi||\cdot||\xi||$$
$$= ||x||(1 + ||\xi||)||\xi_\alpha - \xi|| \to 0. \quad \square$$

(X.18) Since $f$ is a state on $A$ there is a $^*$-representation $\pi_f$ of $A$, unique up to unitary equivalence, on a Hilbert space $H_f$ and a unit vector $\xi_f \in H_f$ such that $\pi_f(A)\xi_f$ is dense in $H_f$, and $f(x) = (\pi_f(x)\xi_f|\xi_f)$. Let $\xi = \pi_f(y)\xi_f$. Then, for each $x \in A$, $p(x) = f(y^*xy) = (\pi_f(y^*xy)\xi_f|\xi_f) = (\pi_f(x)\xi|\xi)$. Setting $x = e$ and using $f(y^*y) = 1$ gives $1 = p(e) = (\pi_f(e)\xi|\xi) = ||\xi||^2$ or $||\xi|| = 1$. Since $f$ is pure, $\pi_f$ is irreducible and so $\pi_f(A)\xi$ is dense in $H_f$. This, together with $p(x) = (\pi_f(x)\xi|\xi)$ for all $x \in A$, implies (by uniqueness) that $\pi_f$ and $\pi_p$ are unitarily equivalent, and hence $\pi_p$ is irreducible. Thus $p$ is a pure state. $\square$

(X.19) Since $f$ belongs to the pure state space of $A$, there is a net $\{f_\alpha\}$ of pure states on $A$ such that $f_\alpha \to f$ in the weak$^*$-topology. Since $f_\alpha(y^*y) \to f(y^*y) = 1$, assume that $f_\alpha(y^*y) > 0$ for all $\alpha$. Set $\gamma_\alpha = (f_\alpha(y^*y))^{-1/2}$ and $y_\alpha = \gamma_\alpha y$. Then $p_\alpha(x) = f_\alpha(y_\alpha^*xy_\alpha)$ for $x \in A$. Since $f_\alpha(y_\alpha^*y_\alpha) = f_\alpha(\gamma_\alpha y^*\gamma_\alpha y) = \gamma_\alpha^2 f_\alpha(y^*y) = (f_\alpha(y^*y))^{-1}f_\alpha(y^*y) = 1$, each $p_\alpha$ is a pure state on $A$ by (X.18). If $x \in A$, then $p_\alpha(x) = f_\alpha(y_\alpha^*xy_\alpha) = f_\alpha(y^*xy)/f_\alpha(y^*y) \to f(y^*xy) = p(x)$. Thus, $p_\alpha$ is a net of pure states on $A$ such that $p_\alpha \to p$ in the weak$^*$-topology; hence $p$ belongs to the pure state space of $A$. $\square$

(X.20) The subspace $\pi_f(A)\xi_f$ is dense in $H_f$, where $\xi_f$ is the associated unit cyclic vector in $H_f$. Hence, there is a net $\{y_\alpha\}$ in $A$ such that $||\pi_f(y_\alpha)\xi_f|| = 1$ and $||\pi_f(y_\alpha)\xi_f - \xi|| \to 0$. Then $f(y_\alpha^*y_\alpha) = (\pi_f(y_\alpha^*y_\alpha)\xi_f|\xi_f) = ||\pi_f(y_\alpha)\xi_f||^2 = 1$, and, by (X.19), the states $p_\alpha$ defined by $p_\alpha(x) = f(y_\alpha^*xy_\alpha)$, for $x \in A$, are in the weak$^*$-closure of the pure states on A. For each $x \in A$,

$$p_\alpha(x) = (\pi_f(x)\pi_f(y_\alpha)\xi_f | \pi_f(y_\alpha)\xi_f) \to (\pi_f(x)\xi|\xi) = p(x);$$

thus p is in the pure state space of A. □

(X.21) Use (12.8).

(B.11) $A_e$ is the algebra of all complex matrices of the form $\begin{pmatrix} \alpha & \beta \\ 0 & \gamma \end{pmatrix}$.

(B.16) Cf. T. Kato, *Perturbation theory for linear operators*, vol. 132, Grundlehren der Math., Springer-Verlag, New York, 1966, p. 28.

(B.17) Exercise (II.7).

(B.20) Assume $y \in A$ with $y \neq 0$, $y^2 = 0$. Let B be an infinite-dimensional Banach space and set $xz = 0$ for $x, z \in B$; choose a discontinuous linear functional $\phi$ on the Banach algebra B. Define a map $f: B \to A$ by $f(x) = \phi(x)y$. Show that f is a (discontinuous) homomorphism.

(B.22) Cf. W. J. Pervin, *Foundations of General Topology*, Academic Press, New York, 1964, pp. 115-116.

(B.24) Cf. S. Goldberg, *Unbounded linear operators*, McGraw-Hill, New York, 1966, pp. 12, 21.

(B.25) Let $\varepsilon > 0$, $x \in E$ and $f \in E^*$. Then there is a constant $M(x,f) > 0$ such that $|f(T_n x)| \leq M(x,f)\varepsilon^n$ for all $n = 1,2,\ldots$ Set $S_n = T_n/\varepsilon^n$ and note that $|f(S_n x)| \leq M(x,f)$ for all n. The Banach-Steinhaus theorem gives $M > 0$ such that $||S_n|| \leq M$ for all n, i.e., $||T_n|| \leq M\varepsilon^n$. Thus $\overline{\lim_{n\to\infty}} ||T_n||^{1/n} \leq \varepsilon$; since $\varepsilon$ was arbitrary the result follows. □

(B.26) Cf. P. R. Chernoff [Proc. Amer. Math. Soc. 23(1969), 386-387].

(B.28) Cf. S. Kurepa [Math. Mag. 41(1968), pp. 70-71].

(B.29) Since $\nu(xy) = \nu(yx)$ for all $x, y \in A$, then $\nu(y) = \nu(x^{-1}xy) = \nu(x^{-1}yx)$. □

(B.30) Cf. R. Fuster and A. Marquina, Amer. Math. Monthly 91(1984), 49-51.

(B.34) Let $E = \{\exp(x): x \in A\}$. The hypothesis implies that E is an abelian group. The set $S = \{z \in A: ||e - z|| < 1\}$ is contained in E (if $z \in S$, then $z = \exp(y)$ with $y = -\Sigma_{n=1}^{\infty}(e - z)^n/n$ ). For nonzero $x, y \in A$, $e + x/2||x||$ and $e + y/2||y||$ are in S, hence in $E$ and therefore commute; then $xy = yx$. If $x = 0$ or $y = 0$ the result is clear. □

(B.37) The formulas $x' \circ (x + y) \circ y' = x'y'$ and $x + y = x \circ (x'y') \circ y$ give the result.

(B.51) Look at the rational functions.

(B.53) Use (B.4.12).

(B.55) If $x$ were invertible, then $e = x^n (x^{-1})^n$ for all positive integers $n$, so that $||e|| \leq ||x^n|| \cdot ||x^{-1}||^n$. Then $||x^n||^{1/n} \geq ||e||^{1/n} / ||x^{-1}||$ and as $n \to \infty$, $0 \geq 1/||x^{-1}||$, a contradiction. □

(B.59) No. Look, for example, at the function $f(\lambda) = \lambda$ in the algebra of bounded complex functions on the unit disk.

(B.60) For $x \in T$, let $V = \{\lambda x: \lambda \in C, \lambda \neq 0\}$. Then $V$ is an open, connected set of invertible elements. Since $x \in V$, then $V \subseteq T$. All elements in $T$ are invertible so 0 does not belong to the spectrum of any element of $T$. Let $\zeta \neq 0$ be given, and let $\mu \in \sigma_A(x)$. Then $\zeta\mu^{-1}x \in V \subseteq T$ and $\zeta \in \sigma_A(\zeta\mu^{-1}x)$. □

(B.61) $x(\lambda) = x(\lambda)(x - \mu e)x(\mu) = x(\lambda)(x - \lambda e + (\lambda - \mu)e)x(\mu) = (e + (\lambda - \mu)x(\lambda))x(\mu) = x(\mu) + (\lambda - \mu)x(\lambda)x(\mu)$. □

(B.62) Cf. R. S. Doran [4].

(B.64) (b): Since the set of invertible elements is the complement of the sets of singular elements, these sets have the same boundary. If $x$ is in the boundary of $A^{-1}$, there is a sequence $\{y_n\} \subseteq A^{-1}$ converging to $x$. The sequence $\{y_n^{-1}\}$ must be unbounded and it can be assumed that $||y_n^{-1}|| \to \infty$. Let $z_n = y_n^{-1}/||y_n^{-1}||$.

(B.65) (a) Assume that $A$ has an identity $e$ which lies in $B$. If $\lambda \in \partial(\sigma_B(x))$, $x - \lambda e$ lies in the boundary of the set of singular elements. Thus by (B.64) $x - \lambda e$ is a topological divisor of zero ). in $B$, hence in $A$. Then $\lambda \in \sigma_A(x)$. Since $\sigma_A(x) \subseteq \sigma_B(x)$, $C \setminus \sigma_B(x) \subseteq C \setminus \sigma_A(x)$; and so $\lambda \in C \setminus \sigma_A(x)$, i.e., $\lambda \in \partial(\sigma_A(x))$.
(b) $\sigma_B(x) = \partial(\sigma_B(x))$. Then $\sigma_B(x) \subseteq \partial(\sigma_A(x)) \subseteq \sigma_A(x) \subseteq \sigma_B(x)$.

(B.75) Choose $z \in I$ such that $f(z) = e$. Let $g: A \to B$ be defined by $g(x) = f(xz)$. Verify that $g$ is the required unique extension of $f$.

(B.79) Cf. E. Hille and R. S. Phillips, *Functional analysis and semi-groups*, vol. 31, AMS Colloq. Pub., Providence, R. I., 1957, p. 703.

(B.80) Cf. E. Luchins [Pacific J. Math. 9(1959), p. 551].

(B.81) Cf. E. Luchins [Pacific J. Math. 9(1959), 551-554].

(B.82) Cf. P. Porcelli, *Linear spaces of analytic functions*, Rand McNally, Chicago, 1966, pp. 87-88.

(B.83) Cf. R. J. Loy [J. Australian Math. Soc., 9(1969), 275-286].

(B.85) Cf. F. F. Bonsall and J. Duncan [5, p. 126].

(B.86) Cf. J. Dieudonné, *Treatise on analysis*, vol. II, Academic Press, New York, 1970, p. 298.

(B.88) Cf. W. G. Badé and P. C. Curtis [Amer. J. Math. 82(1960), p. 853].

(B.89) Cf. C. E. Rickart [1, p. 71].

(B.90) Cf. C. E. Rickart [1, p. 71].

(B.91) Cf. C. E. Rickart [1, p. 61].

(B.92) Cf. R. C. Buck [Proc. Amer. Math. Soc. 2(1950), 135-137].

(B.93) Cf. R. C. Buck [Proc. Amer. Math. Soc. 2(1950), 135-137].

(B.97) Cf. proof of (B.5.16).

(B.98) Cf. C. E. Rickart [Bull. Amer. Math. Soc. 54(1948), 758-764].

(B.100) See (22.14).

(B.109) Cf. E. Hille and R. S. Phillips, *Functional analysis and semigroups*, vol. 31, Amer. Math. Soc. Colloq. Pub., Amer. Math. Soc., Providence, R. I., 1957, p. 176.

(B.110) See (B.6.20).

(B.113) Cf. F. F. Bonsall and J. Duncan [5, p. 37].

(B.116) Cf. M. Nagasawa [Kōdai Math. Sem. Reports 11(1959), p. 182].

(B.117) Cf. B. Yood [Duke Math. J. 16(1949), p. 158].

(B.120) If $x$ does not have a quasi-inverse in $B$, there is (in B) a maximal modular ideal $I$ such that $\hat{x}(\phi) = -1$ (where $\ker \phi = I$). Then $I \cap A$ is a closed maximal modular ideal in $A$, and $x$ does not have a quasi-inverse modulo $I \cap A$. $\square$

(B.122) Cf. K. Hoffman [1, p. 41].

(B.123) Cf. A. Browder, *Introduction to function algebras*, W. A. Benjamin, New York, 1970, p. 53.

(B.124) Cf. A. Browder reference in (B.123), p. 54.

(B.125) - (B.132): Cf. V. A. Belfi and R. S. Doran [1, pp. 114-116].

(B.133) Cf. Y. Domar [Math. Scand. 14(1968), p. 198].

(B.134) Cf. R. J. Loy [J. Australian Math. Soc. 9(1969), p. 277].

(B.135) Cf. R. Arens [Proc. Amer. Math. Soc. 2(1951), 839-848]; P. Civin and B. Yood [Pacific J. Math. 11(1961), 847-870].

(B.136) - (B.148): Cf. R. S. Doran and J. Wichmann [2].

# Symbol Index

# Example Index

# Author Index

# Subject Index

Milton Keynes UK
Ingram Content Group UK Ltd.
UKHW020011071024
449327UK00031B/2741

9 780367 451639